EU Environmental Policy Handbook

A Critical Analysis of EU Environmental Legislation

Making it accessible to environmentalists and decision makers

Editor: Stefan Scheuer

EUROPEAN ENVIRONMENTAL BUREAU (EEB)

Foreword

I welcome the publication of this handbook, which I hope will become an indispensable tool and work of reference for everyone interested in EU environmental policy. This includes not only those who are involved professionally but also citizens who are interested in learning what the EU has done to protect the environment and to improve the quality of life in Europe, and what can be achieved in future.

Well structured, concise and forward-looking, the handbook describes the history and current status of EU environmental law, but also looks to the future by analysing the strengths and weaknesses of the actions taken so far.

EU environmental law has been decisive in improving the state of the environment in Europe. On average, 80% of what becomes national environmental legislation in EU Member States originates at EU level. I hope this handbook will be read in particular by those who are responsible for putting EU environmental law into practice. Understanding how laws have been developed and what they set out to achieve is important to improve implementation.

All stakeholders interested in the further development of our policies will find this handbook extremely useful. It will make them see where we come from and where we are today – and why we are pushing for a comprehensive and holistic approach to environmental policymaking.

The European Environmental Bureau is a key stakeholder in the process of formulating, monitoring and evaluating EU environmental policies. It defends the interests of more than 140 citizens' organisations across Europe in order to protect, preserve and improve our environment and protect our health and our resource base for future prosperity.

With this handbook, the EEB has shown once again, its value as an advocate of the environment in the EU.

Stavros Dimas
Member of the European Commission, Responsible for Environmental Policies

Contents

Foreword **3**

I. Introduction **7**

II. Executive Summary **11**

III. EU Environmental Policies: A short history of the policy strategies **17**

IV. Setting the objectives, targets and timetables: Sectoral legislation **31**

    IV.1 Introduction 32

    IV.2 Nature 36

        IV.2.1 Introduction 36

        IV.2.2 Birds Directive and Habitats Directive 37

        IV.2.3 Natura 2000 40

        IV.2.4 Implementation deficits and link with other environmental legislation 41

        IV.2.5 Way forward and the role of ECOs (Environmental Citizens' Organisations) 43

    IV.3 Air 46

        IV.3.1 Introduction 47

        IV.3.2 The overarching EU environmental and health objectives for clean air 49

        IV.3.3 The air quality Directives: the Framework Directive on ambient air quality assessment and the four Daughter Directives 50

        IV.3.4 National emission ceilings for certain atmospheric pollutants45 60

        IV.3.5 Road vehicles 65

        IV.3.6 Large Combustion Plants 70

    IV.4 Waste 77

        IV.4.1 Introduction - Waste facts 79

        IV.4.2 Legal and policy context of EU waste policy 81

        IV.4.3 Voluntary Instruments 105

IV.4.4 Standardisation (see also chapter V.4.5) 106

IV.4.5 Taxes and charges 107

IV.4.6 Waste Policy today and tomorrow - New approach in the 2005 Strategy?......... 108

IV.4.7 Implementation and decision-making 111

IV.4.8 Links and crossovers with other EU legislation 113

IV.4.9 Tips for ECO action (Environmental Citizens' Organisation).................... 116

IV.4.10 The Court of Justice on important waste cases 119

IV.5 Water **125**

IV.5.1 Introduction and Summary 126

IV.5.2 State of the aquatic environment in Europe 128

IV.5.3 From 1975 to the WFD in 2000 (see also chapter IV.5.7) 129

IV.5.4 Implementation deficits and way forward 130

IV.5.5 Environmental objectives under the Water Framework Directive WFD......... 131

IV.5.6 Tools and measures under the WFD 142

IV.5.7 The WFD - a new umbrella for environmental laws 146

IV.5.8 Assessment of the WFD 151

IV.5.9 Outlook and ECO action (Environmental Citizens' Organisation) 152

V. Horizontal legislation – providing tools to achieve
environmental objectives **157**

V.1 Introduction **158**

V.2 Public Involvement in Environmental Decision-Making **160**

V.2.1 Introduction 160

V.2.2 The Aarhus "Public Participation" Convention 161

V.2.3 Public Information on Pollution - Pollutant Release and Transfer Registers
(PRTRs) 167

V.3 Environmental Measures on Production, Planning and Management **172**

V.3.1 Introduction 174

V.3.2 IPPC-Directive 175

V.3.3 Environmental Liability - first step towards "Making the polluters pay" 197

V.3.4 Greenhouse Gas Emissions Trading Directive 208

6

V.3.5 Environmental Impact Assessment...228

V.3.6 Environmental Management and Audit Scheme.......................................247

V.4 Product Policy...**258**

V.4.1 Introduction 261

V.4.2 The EU Ecolabel – the 'Flower'...263

V.4.3 Integrated Product Policy...276

V.4.4 The Ecodesign framework for energy using products (EuP) Directive...........282

V.4.5 New Approach and Standardisation..289

V.4.6 Chemicals and Pesticides Regulations...306

V.4.7 Motor fuels and environmentally relevant legislation......................323

**VI. Environmental Legislation – information on structure,
implementation and enforcement**..**333**

VII. Authors..**337**

Key words...**341**

I.

Introduction

*By **Stefan Scheuer***[1]

---

[1] EU Policy Director, European Environmental Bureau

Protection of the environment is a well-established policy in the European Union. Following closely on from the development of Europe's Internal Market and starting in 1967 with the Directive for harmonised classification and the labelling of dangerous chemicals, environmental protection objectives and principles were finally given their own chapter in 1987, in the Treaty establishing the European Union. Today the vast majority of national environmental policies and laws have their origins in EU law.

Despite this development, environmental protection largely happens at national level. Most EU environmental laws setting minimum standards then leave it up to Member States on how to achieve them. Increasingly, however, EU environmental laws are based on internal market rules, which seriously restrict Member States from going beyond a prescribed level of protection. Together with the general domination of internal market rules in the EU, national environmental policy initiatives are quite restricted. Thus governments wanting to introduce new environmental protection often seek to influence and promote the development of EU policies. They remain the main initiators of EU environmental policies and laws in order to achieve their desired level of protection via the EU. "Brussels" is thus a trading and negotiation place for policies rather than the "real" initiator.

National environmental politics change quite frequently, with changes in government, but the EU has provided continuity, with a steady output of and increasing volume of environmental laws. Policy approaches nevertheless have changed in response to specific environmental challenges, conflicts with national sensitivities, in specific economy developments and policies, and failure to adapt to administrative challenges and to achieve set targets.

In response to the growing influence and ambition of EU environmental policies and their own failure to deliver, Member States have, on several occasions, initiated attempts to renationalise environmental competencies and to introduce greater flexibility into EU laws - with varying results. Many important new laws, setting environmental targets and deadlines, were passed in the 1990s, but have since become rather complex and descriptive, due to an increasing emphasis on process rather than on setting measurable targets.

With the 2004 enlargement of the EU involving 10 new Member States, and a general slow-down in economic growth coupled with high unemployment, European environmental policies have come under increasing attack from many sides. The Sustainable Development Strategy launched in 2001 in Gothenburg by heads of states and governments recently became marginalised under the Economic Growth Strategy established in Lisbon in 2000. The post-enlargement Commission - which took office at the end of 2004 - has been reluctant to make proposals for new environmental laws, emphasising the need to focus on increasing the competitiveness of Europe's economy and simplifying existing EU legislation. This one-sided and imbalanced agenda has already faced criticism from heads of states and governments, who instead placed the economic dimension together with the social and environmental dimensions, in the context of sustainable development, and highlighted the important contribution of environment policy to growth and jobs (*EU Presidency 2005*).

While this EU orientation discourse goes on, the Commission itself notes "*the worsening of unsustainable trends, notably, the growing pressure on natural resources, biodiversity and the climate...*" and that environmental policy integration "*has been limited so far*" (*CEC 2005a and 2005b*). This is a diplomatic way of saying that Europe is not meeting one of its core objectives and economic and fiscal policies continue to pave the way for the relentless transfer of environmental damage and health risk costs from the individual business to the general public.

It is clear that with currently 25 Member States, any EU decision making process takes more time and the outcome might be less predictable. Any attempt to simplify existing rules might result in even more complex requirements or deregulation following the lowest common denominator. In light of the growing environmental problems there is no time to become complacent and leave environmental issues on the back-burner. A lot has to be done - and can be done. Firstly, the market must be made to work for the environment. Today's main economic drivers are labour costs, while costs of natural resource use is marginal. A substantial shift of taxes on labour to natural resource use is necessary. Environmentally harmful subsidies must be abolished. Such measures have the potential to achieve an absolute decoupling of environmental damage from economic growth and at the same time provide higher employment. This can be encouraged by EU institutions, but the real action has to happen at national level. Secondly, the EU must close the biggest gaps in environmental protection, like stock-taking and controlling of chemical risks, setting soil protection standards and putting in place a marine protection strategy – all of which are already in progress and could be delivered in forthcoming years. Finally, existing EU law must be fully applied and respected.

Many important environmental laws from the last 10-15 years are starting to 'bite': deadlines are approaching, many objectives have still not been achieved and measures have been put in place too late or not at all. Laws are often not properly understood, badly transposed into national law and the administration ill-equipped to deliver. National authorities and courts are often unable to interpret EU law and fail to provide citizens and their organisations - and thus the environment for which they are a voice - with their proper rights. Nothing could be more devastating for Europe's acceptability among, and accountability towards, its citizens than failing to apply and enforce EU environmental law. Any environmental law which is not respected by a majority of Member States over some time risks becoming obsolete and is a wasted opportunity.

Quite rightly, the Commission President has identified the correct application of EU laws on the ground as a top priority for meeting his aims of achieving long-term prosperity, sustaining solidarity and social justice, and achieving tangible practical benefits in relation to the quality of life of EU citizens (CEC 2005c).

This book presents an overview and assessment of Europe's environmental policies and legislation, aiming to make them more accessible for environmental citizens' organisations (ECOs) and regional initiatives, environmental and other affected authorities, and decision-makers at EU and national levels.

It focuses on environmental legislation and policies over the last 15 years and sets a framework for appreciating and improving the understanding of the manifold interrelations and linkages between the many different laws. Its aim is to encourage the integration of different environmental policies and the widening of narrow sectoral views. It helps with the identification and use of synergies of the different policies, in order to increase their effectiveness with a view to achieving better application and enforcement of existing laws.

Therefore, a conceptual framework has been chosen to allow a logical access to, and linkages with, the different laws. Thirty years of environmental policy history is presented, following the progression and development of the 6 European Environmental Action Programmes. Then the main environmental objectives for nature, air, waste and water are explained and assessed. This is followed by EU laws providing instruments and tools to achieve environmental objectives. These horizontal laws are further subdivided into laws setting rules for industrial production, administrative planning and management and laws setting rules for products. All chapters offer an explanation of the main elements of the legislation, provide a critical assessment and conclude with suggestions for action.

I hope that this book will encourage better application and use of the many and important environmental laws which exist in today's EU.

The selection of legislation covered and political priorities closely follow the EEB's work agenda and 30 years of experience of working on environmental policies. Some laws are of course better designed or more ambitious than others in protecting the environment, but overall they present a unique opportunity to make further progress. It all depends with what energy, accuracy and intelligence they are transposed and enforced on the ground. Only when the positive elements of these laws are used and defended do they have a chance of surviving.

The authors of the different chapters are EEB staff, experts from among EEB's members, as well as some external experts, whom I want to thank for their great work. I would also like to thank Caroline Bretelle for all her help in editing and proofreading the text.

Brussels September 2005

Stefan Scheuer
EU Policy Director

REFERENCES:

CEC (2005a)2004 Environmental Policy Review, Commission Communication (2005) 17

CEC (2005b)The 2005 review of the EU sustainable development strategy: initial stocktaking and future orientations, Commission Communication (2005) 37

CEC (2005c) Strategic Objectives 2005-2009? Commission Communication (2005) 12

EU Presidency (2005) EU Summit Brussels, Luxembourg EU Presidency Conclusions, 22 and 23 March 2005

II.

Executive Summary

*By **Stefan Scheuer*** [2]

[2] EU Policy Director, European Environmental Bureau

This book provides a short outline of EU environmental policy history, analyses and presents some 60 pieces of EU environment and nature protection legislation, most of which have been adopted in the last 15 years, establishing links between the different pieces of legislation. Against the current decline in environmental policy and the trend towards its subordination to jobs and economic growth, the book encourages an informed and intelligent use of existing legislative networks and calls for resistance to attempts to water it down.

THE ENVIRONMENTAL CHALLENGE REMAINS

Europe's environmental policy started in 1973, following the 1972 UN Conference on Environment, addressing the public and scientific concerns about the "limits of growth". Today - 30 years later and with an impressive body of environmental legislation in place - similar or identical concerns remain: environmental progress through technology has been outweighed by growing consumption and natural resource use. Well known environmental problems of increasing natural resource use, biodiversity loss, destruction of natural habitats and long-term pollution of most environmental media, persist.

There have been ups and downs in Europe's environmental policy making over time, but so far the resulting legislation has not lead to any real structural change which would significantly impact economic and fiscal policies. Attempts to integrate environmental policies into other policies and to correct wrong market mechanisms were pursued in the 1990s but failed to deliver.

This book builds on the thesis that existing and legally binding environmental targets and deadlines, a lot of which were set in the 1990s, have great potential and once properly enforced and supported by existing instruments, can lead to real change. Following the thesis that environmental protection derives from clear objectives, targets and deadlines in order to develop the necessary power to achieve structural policy changes, the main part of the book analyses legislation setting objectives, targets and timetables for different environment media (sectoral legislation). The second part analyses horizontal legislation, needed to provide the set of instruments required to achieve the environmental objectives.

GENERAL OBJECTIVES, SPECIFIC TARGETS AND DEADLINES AND PRINCIPLES

The EU provides a hierarchy of objectives and targets starting with the Treaty's objective of a *high level of protection and improvement of the quality of the environment.* Following on top of this hierarchy is the objective *to halt biodiversity decline by 2010,* which is further specified by the EU's nature and environmental media protection legislation.

The most concrete and enforceable way of specifying those general objectives is through **Environmental Quality Standards** and **Emission Limit Values**. Both have a different theoretical and historical background and both have their strengths and weaknesses. Emission Limit Values can drive technological progress in order to minimise emissions but fall short of achieving structural change, dealing with diffuse pollution and promoting use of natural resource saving alternatives to provide societal services. Environmental Quality Standards on the other hand should present the "ecological truth" and thus are a constant driver to move towards sustainable development, but often suffer from lack of data or scientific uncertainty. A combination of the strengths of both approaches–minimise emissions to achieve or go beyond set environmental quality standards – seems to be the most sensible option to implement a precautionary policy as required to address long-term and complex threats to the ecosystem.

13

The establishment of **Natura 2000**, a European network of nature conservation areas, as a "safety net" for Europe's biodiversity, has a high public profile, but apart from the already missed 2000 target to designate sites, suffers from lack of enforceable targets with deadlines and contradictory EU infrastructure and agriculture funding policies. Better integration with existing air quality and aquatic ecosystem targets would provide a great potential to strengthen nature protection. Since 2005 payment of farming subsidies is conditional to compliance with EU nature conservation obligations. Better use of existing environmental assessment rules for projects and plans as well as complaint and court procedures will be another way to halt detrimental developments. Finally great uncertainty exists about the financing of Natura 2000 for 2007-13. For many countries (especially Southern and Eastern) sufficient EU co-financing is a pre-requisite for success.

Of outstanding strength are the targets set in **air quality** and **climate change** policies. Those include the air quality objective of *"not exceeding critical levels and loads..."* which is further specified by the 2010 national emission ceilings, the 2005/2010 EU ambient air quality standards. By 2008-12 an 8% CO_2 emission reduction from 1990 levels has to be achieved and a long term climate stabilisation of 2°C above pre-industrial temperatures is envisaged. These targets and objectives have been followed by a number of instruments including a number of pollutants emission limit values for specific industrial activities and a cap and trade system for CO_2. Climate change is high on the political agenda and further policy progress can be expected – although the emission trading system still needs to prove its ability to deliver. The further development of air legislation on the other hand is getting under increasing economic pressure. The achievement of the already existing targets is difficult and requires substantial changes in transport and energy consumption, for which Member States largely failed to develop plans. It will be important to emphasise the synergies of climate and air policies through a focus on energy demand and efficiency measures. There are many opportunities for ECOs to influence Air quality and Climate Change policies, especially through early participation in national implementation and application of the existing EU laws. Furthermore, the EU level stakeholder and expert processes in further developing legislation are an excellent opportunity for societal influence.

A further environmental media is **water**, which only recently complemented the former narrow and use specific protection against chemical pollution (in order to achieve water resources to be fit for drinking, fishing, bathing etc...), with the holistic objective of a *"good ecological status"* by 2015, comprising the protection of habitat conditions and biological quality elements. While this objective should present a key indicator for biodiversity protection and long-term environmental sustainability, its interpretation and standard setting is largely left to Member States. On the other hand, this objective is embedded in administrative and policy instruments (like water pricing) and an EU implementation strategy, which could help to avoid its dilution when integration with other policies, like agriculture, transport and land use planning, takes place. The mutual reinforcement of water and nature protection will be a key for driving structural change.

Furthermore, full exploitation of the environmental opportunities of the Common Agricultural Policy mid-term review 2003 will be essential. But finally, the EU wide and national interpretation of "good ecological status" must deliver water status maps, which tell the "ecological truth". Without public scrutiny and support this is unlikely to be achieved.

Completely lacking in Europe are protection objectives and targets for **soil and natural resources**. **Waste policies** can partly be understood as trying to fill this gap. Europe's waste policy is built on strong principles; for example, the waste hierarchy, but which are rarely implemented to the letter. The setting of enforceable overall targets and deadlines is incomplete, setting some umbrella targets – like landfill biodegradable waste reduction targets (25% by 2006, up to 60% by 2016), but still missing over-arching targets on recycling and for waste prevention. Some specific waste streams have been addressed with collection and recycling targets for batteries, packaging waste, cars and electrical and electronic waste and environmental performance standards, like land filling and incineration, to support air, soil and water protection. In order to better steer Europe's waste policy, a clear and enforceable target for the reduction of natural resource use would also need to be set. A growing awareness of Europe's global environmental footprint and impact in developing countries can be a key driver for this.

Besides this better information about chemicals in products and their impact on environmental media including soils, will sharpen (but not substitute) the use and enforcement of existing waste management principles- in particular prevention of hazardousness – through substance bans. At national level the scope for action is great: absence of or inadequate waste management plans at local level, subsidies for waste incinerators, illegal landfills and waste shipments to avoid sound waste treatment need to be challenged.

TOOLS AND MEASURES

Above all environmental "instruments" stands the framework for **public involvement, access to information and justice in environmental matters**. The UNECE Convention from 1998 has been setting the scene, unfortunately with its implementation in Europe lacking still behind. Nevertheless public participation depends on cultural and historical aspects, which means that in any case involvement, access to information and courts can and must be tested at EU and national level in order to make progress. The potential of achieving a structural change through successful participation and court cases is substantial.

To address the **environmental impacts of economic activities** the EU provides for a few market based and management provisions. In terms of market based instruments - aiming at internalising the costs of pollution - the EU presents the **Environmental Liability** and **Emission Trading Directives**. The first has a potential large and far-reaching scope, including biodiversity and nature conservation, water and soil protection. But its effectiveness is unsure, as the political compromise reached is rather weak with many loopholes – falling even behind some existing national systems. Its success at the end depends on whether the national application ensures that the polluter will have to pay environmental damage in practice, which means that financial guarantees are established and no permit defence is allowed. European Greenhouse Gas Emission Trading has a much smaller scope but a clear and enforceable target is set. Overall both instruments are an important first step towards market based instruments within the rather limited EU powers in designing market based policies.

In terms of management provisions the **Integrated Pollution and Prevention Control Directive** is the most important access to control industry activities and foster better environmental performance. Through the permitting procedure – to be concluded by 2007 for all covered activities (SMEs are excluded) - industrial installations are checked against EU "Best Available Techniques" reference documents. But as this is done at national or even local level and no level playing field is provided very different levels of performance can be observed in Europe. Further the quality of those reference documents is due to lack of data and balanced expert input not always satisfactory. Additionally public involvement in permit writing is yet very low. It can be concluded that the environmental effectiveness of the permits depends on the capacities of permit writing authorities and will be largely driven by EU environmental quality and emission standards to be met. Despite these shortcomings, the Commission is planning to defer emission controls and environmental media protection to the IPPC process as regards future, air, soil, waste and water policies. This should be prevented and instead EU protection standard should be set to avoid a huge gap between theory and practice.

The **Strategic Environmental Impact Assessment Directive** addresses governmental planning in order to integrate environmental considerations and allow better public participation. The Directive is met with reservation as its older sister Directive – on the assessment of projects, the **Environmental Impact Assessment Directive** - has been disappointing so far from a nature conservation point of view and rarely has a project been stopped or substantially altered. But the SEA Directive has been improved and both Directives have great potential especially in combination with pursuing the enforcement of other environmental targets and standards - like climate change, water and air protection - in order to achieve structural policy changes.

Instruments addressing the **environmental impacts of products** are crucial to move from end-of-pipe measures to tackling diffuse impacts, like the loss of chemicals or energy use. Only a few strong measures have been taken so far, including the **restriction of hazardous substances in electrical equipment, the market ban for carcinogenic substances, or the pesticides authorisation scheme**. But none of those measures are able to keep pace with their growing consumption and use and introduction of new substances.

Due to the complexity of addressing the multiple stages and environmental impacts of a product from cradle to grave , the discussion is rather theoretical and tends to imply expert follow-up work, as in the case of the **Integrated Product Policy** work. The EU attempted to promote voluntary instruments like the **European eco-label**, the **Environmental management and audit system (EMAS), technical standardisation** or **voluntary agreements**. These instruments suffer from the weakness that they leave all freedom to business to decide and little obligation to deliver. Nevertheless, EMAS and technical standardisation in conjunction with the **New Approach** are used increasingly as implementation tools for environmental legislation. Environmental citizens' organisations regard economic and fiscal incentives as more powerful tools and have advocated their application with little result yet at EU level. For the time being the only promising instrument is **public procurement** which allows the public sector to ask for high environmental performance of their purchases and thus to create outlets for eco-products. In the case of the **Energy**

Using Products Directive vague targets are set, but most decisions are deferred to Member State Committees setting performance levels and writing mandates for standardising body CEN. It will be a great challenge for societal groups to participate in such mandate writing and standardisation work, without which no real progress can be expected. Standardisation has so far failed to deliver in the case of packaging and electrical/electronic waste standards. In terms of the specific focus of reductions in impacts from hazardous substances there is an **EU chemicals policy** in place. But this policy managed to cover only a few percent of all chemicals and failed to prevent damage effectively. The Commission proposal for a new system - REACH - would for the first time make available safety data for 30,000 chemicals, make chemical producers responsible for the safe management of chemicals, and encourage the substitution of chemicals which have potential long-term or irreversible impacts.

OUTLOOK: THE ENFORCEMENT AND THE LEGISLATIVE CHALLENGE

These days attacks on environmental legislation as being bureaucratic, disproportionately costly and reducing business competitiveness (vis-a-vis China) is largely based on populism rather than driven by a careful analysis. Indeed Europe's environmental legislation has grown piece by piece and resulted in a respectable and sometimes complex body of legislation. But there is no empiric evidence of negative impacts on economy and, compared to EU's internal market, agriculture policies and national fiscal and subsidy policies, environmental legislation is rather simple. Nevertheless, enforcement and application of these laws is very poor partly because of their devolved character, because environment does not have its own voice or standing in courts, and because national and regional policy makers have not been involved in setting up EU environmental laws and administration is ill-equipped to deliver effectively. Instead of vehemently addressing this challenge at high level we observe a trend of delegating further environmental policy decision to lower levels – European Agencies, expert committees, regional expert networks, technical standardisation and even voluntary agreements. It is unlikely that those levels can provide solutions to persistent environmental problems. On the contrary, such "diffusion" of environmental decision-making may lead to "environmental dilution". This risks a further weakening of Europe's connection with its citizens – something needed more than ever when increasingly exposing Europe's peoples to globalisation pressures from increasingly open markets.

Existing environmental legislation represents a unique attainment for European values. These achievements need to be protected and their fruits harvested before they are rotten. The more EU environmental legislation is discussed, demanded and finally applied, the less the risk of watering down or repeal. This book can give practical assistance to citizens' organisations, environmental administrations and decision makers on both understanding and using EU environmental laws in order to achieve our environmental targets. Understanding interlinkages between different areas is important because targets are set at different levels and for different media and must be seen as supporting each other.

III.

EU Environmental Policies: A short history of the policy strategies

By Dr. Christian Hey[3]

18   III.1 Introduction

18   III.2 An idealistic start 1973 - 1982

19   III.3 Towards the Internal Market 1982 - 1987

20   III.4 Towards environmental policy integration 1987 – 1992
          (the Fourth EAP)

23   III.5 Roll-Back 1992 - 1995

25   III.6 1997 -2003: The last wave of environmental regulation?

26   III.7 The 6th EAP and the Thematic Strategies

27   III.8 Outlook

---

[3] Secretary General of German Advisory Council on the Environment

III.1   Introduction

Many issues under discussion today have a long history. Some issues which were already on the agenda during the 70s have recently returned to the forefront. A short history of policy strategies over the last 20 years therefore offers an insight into the current situation.

This analysis takes as its starting point the six Environmental Action Programmes. These are medium-term programmes and strategic policy documents which reflect the fundamental elements of contemporary environmental thinking and problem perceptions, as well as strategic policy orientation. New action programmes often reflect a change in the general political climate of their time.

But they are not binding programmes for action - even if they contain lists of planned activities. A short history of the Environmental Action Programmes puts the more specific industry legislation into the wider context of environmental policy strategies.

The following analysis of the six EAPs suggests, that in terms of principles, there has been much more continuity than change over the last 30 years. Yet in terms of more specific policy actions, one can observe a gradual learning process. It starts with hot-spot management, and gradually moves towards a more holistic and integrated approach, looking for synergies between business and environmental goals. Progress however has never been linear and European environmental policies over the last 30 years have always been very sensitive to wider economic and political cycles.

III.2   An idealistic start 1973 - 1982

The EC started its environmental policies with an ambitious programme. This contained many elements of today's ideas on "Sustainable Development". After the first United Nations Conference on the Environment in Stockholm in 1972 and growing public and scientific concerns on the limits to growth, the Commission became active in initiating an original Community policy. On the basis of European Council commitments in 1972 to establish a Community environmental policy, the first EAP was decided upon in November 1973[4].

This programme already established the argument that economic development, prosperity and the protection of the environment are mutually interdependent. It was argued, that "the protection of the environment belongs to the essential tasks of the Community" (ibid.). Among the most important objectives were:

- the prevention, reduction and containment of environmental damage
- the conservation of an ecological equilibrium
- the rational use of natural resources

---

[4] in: OJ C112/1 from 20.12.1973

The First EAP emphasised the need for a comprehensive assessment of the impacts of other policies, in an effort to avoid damaging activities. In this way, the First EAP already contained, in an embryonic form, many of the later ideas behind "sustainable development". Environmental policies in the EC originally had their own environmental justification, without this being subordinated to internal market objectives.

These ambitious targets were formulated in a spirit of optimism as regards the feasibility of far reaching policy change, which became frustrated during the following decades of environmental policy making.

Yet the first steps, as proposed by the First EAP, were more "down to earth". It proposed a gradual approach to defining environmental quality objectives. This started with research activities on the nuisance of pollutants, on the causes of pollution and on criteria for environmental objectives. At the end of this process a definition of product and environmental quality norms was put forward. The approach was based on the protection of single environmental media (water, air, soil etc.). The First EAP devoted most of its attention to water protection and waste, but it also contained a sectoral approach, with special reference to agriculture and spatial planning. Also mentioned were preparatory activities for emissions control.

The Second EAP (1977 - 1981)[5] was essentially a follow up to the first in terms of approach and objective, with simply a greater range of problems to be dealt with. Nature Protection received special attention.

In terms of a practical approach the First and the Second Programmes (1973-1981) advocated quality values for water and air. The quality objectives for drinking water were very strict – those for air could be achieved without strong policy intervention.

The evaluation of the practical success of this first period of environmental policy making is, in general, relatively critical. Initial enthusiasm declined considerably during the periods of economic recession (1975 – 1978, 1981 -1983). Nevertheless a number of framework directives, especially for water and waste, were decided during this period.

III.3 Towards the Internal Market 1982 - 1987

The Third EAP (1982 - 1986)[6] and partially the Fourth (1987 - 1992) reflect a considerable change in policy approach, being much more closely related to the completion of the Internal Market than their predecessors.

The Third EAP emphasised the potential risks and benefits of environmental policies to the Internal Market and issue linkage between the internal market and environmental policies became a key driver for programming and activities. Environmental emissions standards needed to be harmonised to avoid distortions to industry competitiveness. Product regulations had to be

---

[5] In: OJ C 139/1 from 13.6.1977

[6] In: OJ C 46/1 from 17.2.1983; Fourth EAP in: OJ C328/1 from 7.12. 1987.

harmonised to avoid non-tariff barriers emanating from different national product norms. On the other hand, the third EAP emphasised the economic benefits, especially the positive employment effects to be gained from environmental policies.

The environmental policy approach was also modified. The Third EAP shifted from a quality approach to an emission-oriented approach. It proposed formulating emission limit values for stationary, as well as mobile, sources. But beyond this new approach, in order to introduce better filter technologies for the reduction of emissions at the "end of the pipe", the objectives of the first and second EAPs were restated. The Third EAP also made positive reference to the first global strategy for "Sustainable Development" formulated by the IUCN in 1980. Waste avoidance, efficient resource use and integrated environmental technologies were some of the objectives of the third EAP.

The practice of environmental policies during the eighties was particularly concerned with clean-air policies, and noise and risk management for industrial sites.

This policy change came about partially as a result of strong German pressure. As a result of the discussion on the 'Waldsterben', the economic forest interests involved and emerging pressure from the Green Party, the German government decided on ambitious clean-air policies requiring emission reductions from large combustion plants and cars. During the 1980s, to avoid distortions in competition, German industries and the government successfully lobbied for a harmonised European emissions control policy. Other pioneering countries were also successful in exporting national policy innovations to EU level: a tradition of strategic environmental planning from the Netherlands, a culture of participatory environmental policies from many of the smaller countries, a focus on setting environmental quality objectives from the UK and more recently the reform of chemicals policies in Scandinavian countries.

III.4   Towards environmental policy integration 1987 – 1992 (the Fourth EAP)

1987 is often seen as a turning point in EC environmental policy, since environmental protection received its own chapter in the Treaty. Yet in terms of approach and practice, one finds much more continuity than change, with the Treaty codifying many principles, which can already be found in earlier policy documents. As with the Third EAP, the economics of European environmental policies remained central to the thinking behind the Fourth EAP, with an assumption of harmony between the objectives of the internal market and environmental protection. The harmonisation of environmental standards was to take place at a "high level". If this condition were met, national measures, which might distort free trade, would not be necessary. Harmonisation at a high level was justified as an essential component of the competitiveness of European industries in global markets.

The Fourth EAP marks a further change in the approach to environmental policy. The shortcomings of the earlier approaches (i.e. quality policy, emissions orientation) were recognised. An approach which relied entirely on environmental quality objectives was recognised as shifting problems to other media or other regions (the case of long range transboundary pollution). Likewise, it was acknowledged that, an approach which focussed on emission controls for stationary sources was unlikely to achieve certain ecosystem or health based quality objectives. The Fourth EAP instead proposed a more integrated approach. For the first time, environmental protection was not perceived as an additive, but rather as an integrated activity within the whole production process. Part of the integrated approach was to reduce energy or material inputs and to close cycles, so that waste streams could be minimised. Furthermore, pollution control was to systematically control all environmental media (water, air and soil) and involve an evaluation of the problem causing substances. Therefore the Fourth EAP started to discuss a "sectoral approach", analysing the impact of strategic economic sectors on the environment. For the first time ever, the evaluation of new, incentive based instruments, such as taxes, subsidies or tradable emission permits was announced.

This was an initial commitment for the strategic reorientation of environmental policies in the EC, which gradually took place between 1989 and 1994. The ideas of the Fourth EAP (integrated approach, sector analysis, new instruments) were further elaborated in the following years. This change is often characterised as a "paradigmatic change", a change from "trade orientation" to a "sustainability frame". Environmental policy is less perceived as an additive policy and more as an integrated part of economic decision-making.. "Sustainable development" gradually became a normative reference for environmental policy in the EU from the beginning of the 1990s onwards. The incorporation of the environmental dimension and the systematic search for "no regret strategies" were promoted. In other words, win-win situations were identified where both environmental and economic objectives could benefit. The White Paper on Growth, Competitiveness and Employment *(CEC 1993)* proposed a new development model, which tried to create employment and improve the efficiency of resource use by a shift in the relative prices of labour and energy. Sustainable development was perceived as a tool for improving the state of the environment, social efficiency and competitiveness simultaneously.

A number of external factors contributed to the further advancement and elaboration of the new policy approach. Among the most important were the emergence of new global threats (1), the respective preparations for the UNCED conference in 1992 (2), the wider support for economic instruments (3) and a new wave of environmentalism in Europe (4). :

(1) At the end of the 1980s, the debate on global environmental risks, especially that relating to climate change, reached the official agenda. Since 1985 a number of international conferences had been urging for dramatic policy changes. In 1988, following an initiative from the European Parliament taken two years earlier, the Commission reacted to those international changes with its first general communication to the Council. In its second communication to the Council in 1990, the Commission proposed a strategy to stabilise emissions by using a mix of efficiency standards, fiscal instruments and research.

The climate change debate has some inherent characteristics that require environmental policies beyond end-of-pipe technologies. It requires a long-term perspective, since both the impact of climate change and the redundancy of any effective policy measure can only be experienced in the long run. For CO_2 reduction, traditional end-of-pipe technologies are not yet available. That means that integrated efficient technologies, structural change in the economy, or even new production and consumption patterns are required. Furthermore, CO_2 reduction requires policy changes in several different sectors (such as energy, transport, agriculture and the chemical industry). To give long-term orientation to all those different sectoral actors a target led environmental policy approach, as established with the Kyoto-protocol and the EU commitments on reduction targets, is very helpful.

(2) Its global character required international action, where the Community could play a major role in "regime building" and as a "leader". This last characteristic made it attractive to the Commission because it could be used as a tool for strengthening European integration and the Commission's own role in international politics. Global leadership was an important incentive in drafting a proposal on an energy/CO2-tax, before the UNCED conference took place in June 1992 (ibid.)

(3) At the end of the 1980s, a new regulatory approach for environmental policies was promoted by the Commission -- especially the use of indirect, economic instruments. The Task Force Report on the Internal Market and the Environment (1989) was not the first document but the most explicit early document to propose environmental taxes. The different initiatives of the Commission became authorised by the "Dublin Declaration" of the European Council from June 1990, which asked the Commission to prepare a communication on "economic instruments". In the following years, several studies on the economic, social, and environmental impacts of these new instruments were edited. The pilot project for this new approach - the CO_2/Energy tax - was proposed in May 1992. The tax clearly focused on energy efficiency and fuel substitution, both of which were target-oriented approaches towards structural change. This shift towards economic and fiscal instruments took place in several OECD countries. During this period, the limits of the traditional approach to promote end-of-pipe solutions by regulations had become obvious, namely in the Scandinavian countries, Denmark, the Netherlands, and to a certain extent in Germany. Furthermore, the new regulatory approach fitted better into the "neo-liberal wave" rather than the previous command-and-control approaches, because it focused on market mechanisms, deregulation, and self-regulation.

(4) At the end of the 1980s, there was a mounting wave of environmentalism. Membership of environmental organisations increased considerably. Green parties were popular in several EU countries, and achieved good results at national level and in the European Parliament. Enquiries into environmental preferences confirm the rising level of public expectation between 1988 and 1992. At European level, a considerable number of new ECO (Environmental Citizens' Organisation) offices, mainly establishing access to the EU institutions for their national members, were set up between 1986 and 1992. Their capacity in terms of staff, professionalism and networking with members and experts increased considerably during the nineties. Thus, the new approach was greatly supported by increased public concern for the environment in general and strengthened capacities of "green" organisations and parties in particular.

This strategic reorientation was then explicitly formulated in the Fifth Environmental Action Programme (1992 - 1999).

Among the most interesting and innovative elements of the Fifth Environmental Action Programme were:

▶ The principal aim of sustainable development according to the definition of the Brundtland Report.

▶ Reference to the sectoral approach, which integrates an environmental dimension into the most polluting sectors (transport, energy, agriculture, etc.), and the limits of old end-of-pipe approaches. Instead, the action programme proposed structural change in favour of public transport, energy efficiency and waste prevention.

▶ The emphasis on new instruments, especially on market-oriented instruments such as fiscal incentives or voluntary instruments, which strengthen producers and consumers own interests in environmental decision-making.

▶ The new consensus-oriented approach taking into account the crucial role of non-governmental protagonists and local/regional authorities to represent the general interest of the environment. This may contribute to innovative concepts, raise public awareness, and enforce the implementation of EU directives.

▶ The setting of medium and long-term objectives for the reduction of some pollutants, and proposed instruments to achieve these objectives.

Hence, the Fifth Environmental Action Programme had in its philosophical part, all the necessary elements of a policy oriented towards "ecological structural change".

III.5 Roll-Back 1992 - 1995

The new approach of the Commission met considerable resistance from Member States. The period after the UNCED conference can be characterised as a downward cycle of environmental policies.

Unfortunately this bold initiative from the Commission did not find sufficient support amongst Member States. Shortly after the UNCED conference a new agenda was promoted by several Member States, which concentrated mainly on the competitiveness of industries and the decentralisation of environmental policies. This new agenda partly contradicted the ambitious ideas of the 5th EAP. Therefore little progress could be achieved on the more innovative projects of the 5th EAP - whereas decisions on others were taken relatively rapidly.

The proposal for an energy/CO_2 tax, a pilot project for using the new approach, was watered down during two years of negotiations and finally dropped as a Community tax in 1994. In May 1995 the Commission presented a new proposal which re-nationalised the responsibility for introducing such a tax for an interim period. Also, several other initiatives and ideas for reform came to a standstill because of strong opposition from certain industries, from other Directorates General of the European Commission, and from Member States.

Some of these have been well documented, including the watering down of the originally ambitious objectives of the packaging directive in 1994, the delays in the Strategic Environmental Impact Assessment directive proposal *(see chapter V.3.5.2)* and progress in environmental taxation in sectors other than the energy sector (e.g. transport,).

24

Member States reacted to the more ambitious elements of the 5th EAP with demands to re-nationalise environmental policies according to the subsidiarity principle. Upon the initiative of Germany and UK, a high level expert group was set up to analyse the potential to reduce regulations that impose excessive costs on the industrial sector. The so-called "Molitor-Group" systematically scrutinised environmental legislation that was only perceived in terms of its perceived cost dimension to the economy *(CEC 1995)*. Existing legislation was attacked particularly in the fields of water protection (the drinking water directive), waste (i.e. the principle of proximity; or the promotion reuse systems) and procedural law (Environmental Impact Assessment Directive).

As a response to the new agenda of several Member States a new regulatory approach emerged that focused on procedural requirements, framework directives, voluntary agreements and self-regulatory information and management tools. Such instruments are rather consensus-oriented and require the co-operation of industries. They are less demanding on European standards and are less substantive, but leave more flexibility and leeway to Member States.

As to the principle approach and the objectives, however, the Commission confirmed and further refined the approach of the Fifth EAP in its first and second progress reports on its implementation.

There are a number of reasons for the "roll-back of environmental policies":

(1) During this period, it became obvious that the Commission was overly optimistic on the willingness of Member States to follow "paradigmatic change". Some Member States were not willing to follow the new approach in substance. They were reluctant to support the new quality of European integration. The failure of the Fifth Environmental Action Programme was interpreted as a symptom for the limits to European integration in environmental policies.

(2) The pending economic crisis and difficulties in ratifying the Maastricht Treaty contributed to a more cautious attitude from the Commission as regards the promotion of innovative and far-reaching new proposals. Furthermore, the programmatic impetus met considerable resistance from both governments and interest groups -- especially from those who would have to bear the costs of such a new approach. Difficulties in getting the Energy/CO_2-tax proposal approved by the Council was just a symptom of the problem of implementing the "paradigmatic change" mentioned above. Evaluative reports concluded that progress on policies directed towards structural change "has been piecemeal and slow". The reports even observed a standstill in the efforts towards structural change.

(3) Due to reunification the preference structure completely changed in one of the potential leaders of EU environmental policies – Germany.. The discussion on the modernisation of environmental policies there came to a standstill, whereas the economic problems of reunification, especially high unemployment, became a primary concern.

III.6   1997 -2003: The last wave of environmental regulation?

At the end of the 90s one can observe a patchwork of different, partially contradictory trends, with different environmental policy approaches being promoted simultaneously. There was a certain revival of the "sustainability approach" New ambitious legislation – partly strengthening the emissions oriented policies of the eighties - can be observed, as well as continuing attempts at deregulation and diffusion of competencies.

Sustainability remains on the agenda. After it was strengthened as a Community target in the Amsterdam Treaty from 1997, the Commission and several Presidencies launched an initiative for environmental policy integration, called the Cardiff Process. Environmental policy integration and sustainable development became key elements of a complex architecture of strategy documents. The Commission shifted from its previous top-down approach and its instrumental focus of environmental policy integration towards a broader and less committed approach: basically sectoral Council formations were asked to identify the key problems of their sectors, to define objectives and to formulate activities in order to meet the objectives. Generally most of the sector strategies lacked committment, were vague and lacked innovation. The hopes of the Commission, that sectoral self-responsibility and voluntary action by transport, agriculture or economic ministers would work, were left frustrated.. Some progress was achieved on indicators, as were some sector targets, namely for transport and energy *(EEB 1999, 2001)*. A serious target setting process on some key environmental issues was not launched by the Commission *(see: SRU 2002)*.

However the revival of environmental legislation in the late 90s was impressive An unprecedented regulatory boom on many technical but also some very political issues started in 1996. This included :

◗ New complex and holistic framework legislation, such as the Ambient Air Quality Directive (96/62), the Water Framework Directive (2000/60) or the IPPC-Directive (1996/61), formulating an ambitious work programme for several decades, while delegating many decisions and tasks to member states, bureaucratic networks or to civil society and business. The reform of European Chemicals Policies launched in the late nineties and proposed in 2003 also belongs to this category. It is still to be seen if these new tools of cooperative governance mobilise sufficient resources and enthusiasm can drive environmental innovation.

◗ New target oriented legislation, setting maximum national emission ceilings for key pollutants, but leaving member states the freedom to choose how to achieve necessary reductions. The NEC-Directive (2001/81) is the most sophisticated example of this approach. Based upon long term targets and an assessment of a cost-effective reduction curve, the EU has set nationally differentiated emission ceilings for 4 pollutants. With the 2003 Emission Trading Directive, another target-oriented policy, setting nationally differentiated CO_2-targets – the so called burden-sharing agreement – became legally binding.

26

▶ The Completion, revision or modernisation of existing legislative programmes, such as the daughter directives on air quality (1999/30; 2000/69;2002/3), on emission control for cars (98/69) and lorries (99/96), fuel standards (in 1998) or the large combustion plants (2001)and the incineration directives (2000/76), the landfill directive /1999/31) or the revision of the Seveso, the Ecolabel and EMAS-directives generally lead to more ambitious standards and a more comprehensive system of protection.

▶ The introduction of many new environmental policy instruments - namely the establishment of producer responsibility, take back and recovery targets for some waste streams (End of Life Vehicles (2000/53) and WEEE (2002/96), Strategic Environmental Impact Assessment (2001/42) , Environmental Liability (2004), CO_2- Emission Trading (2003/87). All are incentive based instruments, internalising the external costs of giving feed-back to economic and public decisions.

▶ New procedural legislation or the revision of existing legislation strengthening civil society rights, notably the three Aarhus pillars: freedom to information, participation rights and access to justice (Directives 2003/4,2003/35 and CEC Directive proposal 2003/624).

Furthermore policy preparation at EU level became much more participatory, inviting environmental NGOs to play a role in committees, expert networks and numerous consultation processes and hence to slightly counterbalance influential industry lobbying at all levels of the Commission. .

Each of those pieces of legislation had more or less serious shortcomings and deficits. However the system of environmental programmes, duties, rights and incentives made impressive progress during that phase. New committed member states, the environmental Commissioners of that period, the strong and constructive support of "rainbow" coalitions in the European Parliament and of a majority of Green and Social Democrat Ministers in the Environmental Council all contributed to the unprecedented dynamics of that period. A "green triangle" of environmental policy making between Commission, Environmental Council and the European Parliament was able to successfully bypass traditional veto players, pursuing economic or institutional interests and succeeded in introducing new instruments, which would have politically failed politically even in so-called pioneering member states without European support. National environmental policies have become mainly EU driven.

III.7   The 6th EAP and the Thematic Strategies

The 6th EAP may fall within a secular change in support for ambitious environmental policies. The overall political agenda is driven by the development concerns of new member states, a new wave of deregulation linked with the debate on European Governance and the increasing relevance of economic considerations. All this is also reflected in a change of political majorities in Europe.

The 6th EAP does not share the ambitious goals of its predecessor. It is both more reluctant to set targets and to identify key instruments. The starting point of the 6th EAP is that so-called persistent environmental problems, such as climate change, the loss of biodiversity or the over-consumption of resources require a broader approach beyond environmental legislation. Furthermore the need for the consolidation of existing legislation is increasing, especially in the view of enlargement. Basically the 6th EAP formulates a framework of general principles and objectives, which will be more specified by so-called thematic strategies on key issues, such as pesticides, resources, recycling, soils, the urban environment, the marine environment, and clean air. The reform of chemicals policy and policies to reduce EU green house gas emissions also belongs to the key policy priorities for this first decade of the new millennium.

The 6th EAP adopts a very cautious approach. It identifies themes and principles. Specification takes place by strategies, which are partially frameworks for further frameworks. The political strategy of the 6th EAP is to postpone potentially contentious and controversial political decisions to later phases or to avoid them altogether by relying on cooperative approaches to environmental policy making. Cooperative approaches with industry, such as integrated product policies, the wider use of standardisation for environmental policies, voluntary agreements, cooperation with Member States' expert fora, or both (e.g. chemicals policy reform) rank high on the political agenda in order to manage complex risks, where knowledge both on the scale of the problem and on the availability of solutions is limited. It is evident that those new governance approaches relieve the legislator and strengthen the role of private and public professionals with specific technical skills. Furthermore the Commission is changing its key role from an initiator of legislation to a manager of policy processes. Environmental policy may hence lose its previous political profile and become more and more a theme for small specialist expert communities. Those communities are responsive to scientific evidence, but the selection criteria for representatives from civil society wanting to participate in those communities has also increased. The cooperative management of the policy processes is very demanding in terms of resources and staff and some processes simply fail to gain momentum because of insufficient public investment, Integrated product policy is certainly a case in point. So it is far from evident that cooperative arrangements deliver more than the traditional regulatory instruments. This applies especially to countries and situations where the negotiation capacity and expertise of public service and of environmental organisations is in the early stages of development. A further problem is that policy approaches become over complex. Holistic and integrated approaches promise to tackle and balance everything with everything at the same time. However the risk is that in the end they amount only to fine rhetoric on principles - and little action.

III.8 Outlook

Future environmental policies need to become refocussed. Persistent environmental problems are the challenge for the forthcoming phase of policy making and should be prioritised. Solving persistent environmental problems needs the involvement of other sectors, but environmental policy will have to play a key role. Setting quantitative and binding targets, which may be nationally differentiated but give direction to Europe's environment as a whole, and defining acceptable levels of risk and of environmental quality based upon the precautionary principle, will continue to be an environmental policy task. There is also a great deal of scope for improvement in emissions standards, and restrictions or incentives for further preventative behaviour from business and consumers. However, the behaviour of these two groups will not improve if the overall market signals are wrong. Implementation not only requires better cooperation and negotiation with other sectors and with industry, but there will also be a need in the future for someone who is entitled to negotiate on behalf of the environment, such as environmental citizens organisations and naturally the respective environmental authorities. Environmental legislation on targets and quality objectives are key starting points for negotiating with industry and member states. Clean air and climate change policies show that a target led approach can acts as driving force for improvement.

FURTHER READING:

ANDERSEN, M. S., LIEFFERINK, D. (1997): European Environmental Policy. The Pioneers. Manchester: Manchester University Press.

COLLIER, U. (Hrsg.): Deregulation in the European Union. Environmental Perspectives. London: Routledge, S. 3-22.

DE BRUIJN, T.J.N.M., NORBERG-BOHM, V. (Hrsg.) (2004): Sharing Responsibilities. Voluntary, Collaborative and Information-based Approaches in Environmental Policy in the US and Europe. Cambridge: MIT Press

DE CLERCQ, M. (Hrsg.) (2002): Negotiating Environmental Agreements in Europe. Critical Factors for Success. Cheltenham, Northampton: Edward Elgar.

DEMMKE, Ch., UNFRIED, M. (2001): European Environmental Policy: The Administrative Challenge for the Member States. Maastricht: European Institute of Public Administration.

EEAC 2003: European Governance for the Environment, Statement: download at http://www.eeac-network.org

EEB (1999): Do sector strategies work? An evaluation of four sector strategies on integrating environment and sustainable development: download at http://www.eeb.org

EEB (2001) Enviornmental Policy Integration. Proposals for a better institutional framework, based on an examination of the Cardiff Process, EEB document No. 2001/019; download at : http://www.eeb.org

EICHENER, V. (2000): Das Entscheidungssystem der Europäischen Union. Institutionelle Analyse und demokratietheoretische Bewertung. Opladen: Leske + Budrich.

CEC Commission of the European Communities (1993): Growth, Competitivity and Employment: White Paper

CEC Commission of the European Communities (1995): Report of the Group of Independent Experts on Legislative and Administrative simplification.

CEC Commission of the European Communities (2001): White Paper on Governance. Brussels.

CEC Commission of the European Communities (2002): Status of the Cardiff-Prozess. December 2002. – download at http://www.europa.eu.int/comm/environment/integration/cardiff_status.pdf

CEC Commission of the European Communities (2002a): Communication from the Commission – Action plan "Simplifying and improving the regulatory environment". Com (2002) 278 from 5.6.2002, 19 S.

GLASBERGEN, P. (Hrsg.) (1998): Co-operative Environmental Governance. Public-Private Agreements as a Policy Strategy. Dordrecht, Boston, London: Kluwer Academic Publishers.

GOLUB, J. (Ed.) (1998): New Instruments for Environmental Policy in the EU. London and New York: Routledge.

HEINELT, H. (Hrsg.) (2000): Prozedurale Umweltpolitik der EU. Umweltverträglichkeitsprüfungen und Öko-Audits im Ländervergleich. Opladen: Leske + Budrich.

HÉRITIER, A. (2001): New modes of governance in Europe: policy-making without legislating? Bonn: Max-Planck-Projektgruppe Recht der Gemeinschaftsgüter.

HEY, C. (1994): Die europäische Umweltpolitik. München, BeckVerlag.

HEY, C. (1998): Nachhaltige Mobilität in Europa: Akteure. Institutionen. Politische Strategien, Westdeutscher Verlag. Opladen, 298p.

HEY, C. (2001): From Result to Process-Orientation: The New Governance Approach of EU Environmental Policy. elni Review H. 2, S. 28-32.

HEY, C (2003). Industrylobbying in Brüssel: Einflussstrategien und –barrieren. In: ZUR, Sonderheft 2003. Nomos. Baden-Baden

HEY, C./ Brendle, U.(1994b): Towards a new renaissance: A new development model. Part A Reversing the rollback of environmental policies in the European Union. Brussels EEB, European Environmental Bureau.

HOLZINGER, K., KNILL, Ch., SCHÄFER, A. (2002): European Environmental Governance in Transition? Bonn: Max-Planck-Projektgruppe Recht der Gemeinschaftsgüter.

Liefferink, J.D./ Lowe, P.D./ Mol, A.P.J. (eds.) (1993): European Integration and Environmental Policy. London/ New York: Belhaven Press. .

KNILL, Ch. (2003): Europäische Umweltpolitik. Steuerungsprobleme und Regulierungsmuster im Mehrebenensystem. Opladen: Leske +Budrich

KNILL, Ch., LENSCHOW, A. (2000b): Implementing EU Environmental Policy – New Direction and old Problems. Manchester: Manchester University Press.

KNILL, Ch., LENSCHOW, A. (2003): Modes of Regulation in the Governance of the European Union: towards a Comprehensive Evaluation. European integration online papers 7 (203). Online im Internet: URL: **Http://eiop.or.at/eiop/texte/2003-001a.htm** [Stand 12.02.2003].

KRÄMER, L. (2002): Development of Environmental Policies in the United States and Europe: Convergence or Divergence? Florence: European University Institute. EUI Working Papers, RSC No. 2002/33.

LENSCHOW, A. (Hrsg.) (2002): Environmental Policy Integration. Greening Sectoral Policies in Europe. London: Earthscan.

Meuleman/Niestroy/Hey (eds.)(2003): Environmental Governance in Europe. RMNO Background Studies (V.02, 2003). Lemma, Utrecht.

OECD (2003): Voluntary Approaches for Environmental Policy. – Paris: OECD.

PALLEMAERTS, M. (1999): The Decline of Law as an Instrument of Community Environmental Policy. Revue des Affaires Européennes No. 3/4, S. 338-354.

Rehbinder, Eckart/ Stewart (1986): Integration through Law: Europe and the American Experience. Environmental Policy, Berlin: Walter de Gruyter.

RITTBERBER, B., RICHARDSON, J. (2001): (Mis-)Matching declarations and actions? Commission proposals in the light of the Fifth Environmental Action Programme. Paper presented to the Seventh Biennial International Conference of the ECSA, May 31-June 2, 2001. Online im Internet: URL: **Http://www.nuff.ox.ac.uk/Politics/ECSAA%RittbererRichardson.htm** [Stand 13.08.02].

SADELEER, N. de (2002): Environmental principles. From political slogans to legal rules. Oxford: Oxford University Press.

SCHARPF, F.W. (1999): Governing in Europe: effective and democratic. New York: Oxford University Press.

SRU (2002): Environmental Report 2002; unter: **http://www.umweltrat.de**

SRU (2004): Environmental Report 2004; unter: **http://www.umweltrat.de** (in June 2004)

WEALE, A., PRIDHAM, G., CINI, M., KONSTADAKOPULOS, D., PORTER, M., FLYNN, B. (2000): Environmental Governance in Europe. An ever Closer Ecological Union? Oxford: Oxford University Press.

WURZEL, R. K. W. (2002): Environmental policymaking in Britain, Germany and the European Union. Manchester: Manchester University Press.

IV.

Setting the objectives, targets and timetables: Sectoral legislation

32   IV.1   Introduction

36   IV.2   Nature

46   IV.3   Air

77   IV.4   Waste

125   IV.5   Water

IV.1 Introduction

By *Stefan Scheuer*[7]

In order to develop Europe's environmental policy, the general **EU objective of achieving** a *"high level of protection and improvement of the quality of the environment"* has to be defined. **Indicators** have to be developed, which allow a measurement of success and failure, and **specific targets with deadlines** set in EU legislation, to provide Europe's citizens with clear and enforceable rights to a clean environment.

The European Environment Agency EEA regularly publishes environmental indicators, which give an overview of the state of Europe's environment and the success of its environmental policy priorities (*EEA 2005*). Nevertheless, EEA indicators are dependent on data availability, which is strongly linked to monitoring and reporting requirements set out in EU environmental legislation. Many environmental problems not covered by legislation are therefore not addressed using indicators or benchmarking.

There are many different ways to establish environmental targets based on ecological carrying capacity and/or human health protection, the latter clearly dominant in Europe's environmental legislation. As the ecosystem approach encompasses all aspects of the biotic and non-biotic world it presents a long-term and thus abstract objective. Ecosystem variations should be separated into anthropogenic induced ones and those which are independent of human activities, in order to allow targeted policy action. For example Europe's objective "to halt biodiversity decline by 2010" (*see chapter IV.2*) is of such a general nature that without specific indicators or targets, which can be related to human activities, little direct policy action is likely. Nevertheless this objective is a key policy driver to developing and setting less integrated but more enforceable targets. So far, no overall indicator for biodiversity is available and only partial indicators - like farmland bird or threatened species, ecological water status, area and connectivity of nature protected land - have been or are being developed. This may now change with the SEBI2010 project which seeks to develop a European set of biodiversity indicators to assess and inform about progress towards the European 2010 goals but only started as recently as January 2005[8]. With regard to resource use – biotic and non-biotic – some indicators are available, including municipal waste volumes, water and energy consumption and land use (soil sealing, infrastructure), but are mostly of a very general nature and not used for establishing enforceable targets.

What is quite remarkable is the ignorance with which the European Union, one of the wealthiest regions in the world, treats its ecosystems. It seems unable to collect and make available sufficient information to allow us to form a clear picture of the true state of the environment, enabling us to set quantitative targets.

---

[7] EU Policy Director, European Environmental Bureau

[8] SEBI 2010: The Streamlining European Biodiversity Indicators 2010 project (SEBI 2010) is one of the more concrete outcomes from the Malahide Conference on biodiversity. This conference was held in Malahide, Ireland in 2004 and brought together stakeholder from all over the EU to discuss prioritiy for EU biodiversity policy. They agreed that indicators for biodiversity were badly needed in Europe. The project is led and run by the European Environment Agency and the coordination team consists of experts from a.o. the European Topic Centre, ECNC and the EEA. http://www.ecnc.nl/

Therefore, in terms of setting ecosystem related targets with deadlines in European legislation, rather little has happened. As indicated above, the "2010 biodiversity" target is neither enforceable nor easily related to specific policy action. However, this target can be, and is, linked to specific sectoral objectives - like the objective to "end eutrophication" from agriculture activities as set by the Nitrates Directive (*see chapter IV.5.7.4*) and the maximum NOx and NH3 air emission targets set by the National Emission Ceilings Directive (*see chapter IV.3.4*); achieving the Favourable Conservation Status of important flora, fauna and habitats as established by the Wild Birds and Habitats Directives (*see chapter IV.2.2*); or the achievement of "good ecological status" in rivers, lakes and coastal waters by 2015 as set in the Water Framework Directive (*see chapter IV.5.5.1*).

33

It must be noted that **environmental quality standards**[9] which are directly linked to the ecosystem approach have not been developed for any of the above examples. The most advanced ecosystem approach is represented by the Water Framework Directive, which describes the standards and parameters in a normative way, but leaves it up to Member States to set quantitative standards. EU wide water quality standards are to be set on the basis of an ecotoxicological approach for only a limited number of chemical substances. (*see chapter IV.5.5.2*). Most European environmental quality standards are developed on the basis of human health protection, e.g. drinking water, bathing water and air quality standards. Overall there are only a few dozen chemical quality standards (except the overall pesticide standard for drinking water) set at EU level for a few media (drinking water resources, rivers, lakes and air). This looks rather irrelevant when viewed in relation to the 100,000 plus chemicals which are allowed to be marketed and used in the EU and the other environmental compartments (soil, marine). Scientific uncertainties are high, specifically with regard to ecosystem protection. Uncertainty is mainly addressed through a "safety factor" in order to take into account the fact that the (eco)toxicological safe level depends on the species and that synergy effects with other pollutants occur regularly. Monitoring and compliance control is another peculiar problem, as it is often the most expensive aspect of setting (and enforcing) environmental quality standards. Therefore, the effectiveness of environmental quality standards in achieving overall ecosystem and human health protection is currently rather limited.

On the other hand **emission limit values** complement environmental quality standards. They regulate the source of pollution in order to restrict the level of permitted noise or pollutant emissions to the environment by means of general or abstract limit values. In principle, they are guided by state of the art technology or more economically driven by the best available technology. By focusing on the source of environmental pollution, emission limit values are to take account of principles to prevent pollution at source and to take precautionary action. Setting emission limit values should be seen as an optimisation obligation, essentially calling on the EU's legislative institutions to require those responsible to take all possible technical measures to limit and progressively reduce pollutant emissions. Unfortunately, in the process of agreeing on the concrete limit values, compromises are being made to keep unwilling governments on board. At

---

[9] *Environmental Quality Standards* are defined very differently. Sometimes they are generally described as rules related to environmental quality outlining a desirable quality level, concerned with individual aspects of the environment, such as a particular medium (soil, water, air) or a specific target (e.g. human beings, ecosystems, etc.). In this context they are understood as a quantitative and measurable level of pollution or deterioration, e.g. a chemical concentration in air or water.

EU level a few emission standards have been established for industrial discharges into water (*see chapter IV.5.7.2*), for air pollution from cars and combustion plants (*see chapter IV.3.5 and IV.3.6*) and for noise pollution from different machinery (e.g. road vehicles, lawn mowers, construction and home appliances). Furthermore, EU guidance for setting national, regional or local emission limit values for bigger industrial installations is provided via the IPPC Directive, based on Best Available Technologies BAT (*see chapter V.3.2*).

The **relation between emission limit values and quality standards** is historically a difficult one. Quality standards specify citizens' rights to certain physical environmental standards and should ensure the long-term stability of ecosystems. Emission limits impose restrictions on individual economic activities, should reduce overall pollution and as a minimum ensure the achievement of environmental quality standards. This so-called **"combined approach"** (*see chapters IV.5.5.2.2.2 on WFD and V.3.2.2.4 on IPPC*) should set minimum environmental quality standards, avoid filling levels of pollution up to these standards and thus ensure the implementation of the precautionary and prevention at source principles and prevent long-term or global pollution. But this is a "chicken and egg" dilemma. In many cases, emission limit values are set in relation to available environmental quality standards instead of best available technologies and thus allow the "filling up" of water or air to respective standards. This clear violation of precaution and pollution prevention might also be encouraged by the IPPC Directive (*see chapter V.3.2.3.2*), which allows authorities to take into account local environmental conditions when defining emission limit values. Different countries have chosen different approaches and have tried to impose them at EU level. Germany has traditionally applied a technologically driven emission limit approach while the UK has preferred setting emission limit values to determine what level of pollutant discharges is acceptable. Besides this being the result of geographic differences between the two countries (quick dilution or disposal of contamination on an island in contrast to a continental and transboundary situation) it is also the result of different types of national economic development and models of state intervention - which changed in the case of UK and Germany in the 1990s (*Wurzel 2002*).

The current tendency of EU environmental policies is to follow an environmental impact and thus environmental quality approach (*see Water Framework Directive IV.5, IPPC Directive V.3.2 and the Commission ideas for a revision of the Waste Framework Directive IV.4.6*) to the disadvantage of "state of the art" based emission limit values. This must not be seen as a negative development per se, as EU environmental quality standards can be very powerful policy drivers - as seen in Drinking Water or Air Policies[10] - potentially leading to substantial changes in e.g. transport and agriculture policies. It all depends on whether or not those standards are sufficient to prevent loss of biodiversity and maintain the ecological carrying capacity. Thus they need to be protected against legislative dilution or granting exemptions and whether more standards are finally set for more environmental media (soil, marine sediments) at EU level.

With respect to chemical standards, a lot is expected from Europe's chemicals policy reform (REACH), in delivering valuable information for many thousands of different chemicals, allowing the establishment of air, water and soil quality standards. On the other hand, there is strong

---

[10] The EU air quality standards (for SO2, Lead, PM10 and CO) to be met in 2005 will, on the whole, not be achieved by Member States and have triggered an intensive public debate in several countries, including legal challenges.

national resistance against setting new EU quality standards, as experienced in the ongoing process to establish new air and water quality standards. Finally, global and long-term pollution resulting from the emission of persistent and bioaccumulative chemicals can only be addressed through emission controls and limit values. But the boundaries of what classical emission controls can deliver - mainly dealing with point sources (industrial installations) - are quickly reached, bringing the diffuse sources (agriculture, consumer products, transport) into the spotlight. For these, BAT based concepts or product-by-product controls are rather ineffective and more generic approaches, such as the substitution of hazardous chemicals with safer alternatives throughout the value chain[11], have to be applied.

35

The following chapters explain and assess Europe's main legislation in setting environmental objectives, targets and deadlines for Nature Protection, Air, Water and Waste policies. Many deadlines set by those policies are due to be met in the very near future and a great number of Member States will fail to achieve them and instead seek exemptions or extended deadlines. At the same time, citizens and ECOs will have ample room to organise political and legal activities to challenge their government's failure to achieve these objectives and targets because of delayed or non-existent action.

In particular, Air and Waste policies will be revised in the coming years under the 6th Environmental Action Programme. At the time of writing, the tendency was to go for deregulation partly because of the pressure to simplify and reduce environmental legislation due to misconceptions of excessive cost burdens to industry. It is, however, also partly because of deficits in implementation resulting in Member States being susceptible to multiple infringement procedures resulting in little national appetite for further environmental legislation. In this area there is an important role for ECOs to ensure that Member States 'get it right' (correct transposition) and enforce transposed legislation themselves from the beginning, not only resulting in better environmental protection but also avoiding the future development of avoidable political antibodies.

BIBLIOGRAPHY AND FURTHER READING

EEA (2005) Technical report No 1/2005, EEA core set of indicators – Guide, European Environment Agency

WURZEL, R.K.W. (2002) Environmental Policy-making in Britain, Germany and the European Union, The Europeanisation of air and water pollution control, Manchester University Press

---

[11] A series of value-generating activities

IV.2　Nature

By Christine Falter[12] and Stefan Scheuer[13]
(in consultation with the EEB biodiversity working group)·

IV.2　Nature 36

　　IV.2.1　Introduction 36

　　IV.2.2　Birds Directive and Habitats Directive 37

　　　　IV.2.2.1　Introduction 37

　　　　IV.2.2.2　Objective, scope and tools of the Birds and the Habitats Directives 38

　　　　IV.2.2.3　Strengths and weaknesses 39

　　IV.2.3　Natura 2000...40

　　IV.2.4　Implementation deficits and link with other environmental legislation................41

　　IV.2.5　Way forward and the role of ECOs 43

IV.2.1　Introduction

Biodiversity is the most integrated indicator used to describe the well-being and functioning of eco-systems and thus presents the ultimate test of environmental sustainability. Surprisingly perhaps, little background information is available for this indicator and only a few proxies, like number of birds species or nature reserves, have been developed, leaving us with an immense knowledge gap and uncertainty about the most effective design of environmental policies. Precaution must therefore rule most of the decision-making. Conservation of pristine or stable semi-natural habitats is of crucial importance, as it allows the protection of highly valued – albeit sometimes little understood - ecosystems. Nature conservation has a long history in Europe and receives great public attention.

Pressure on habitats and subsequent loss of biodiversity as a result of population growth and industrialisation in Europe is intense and in the mid 1970s Europe started to put in place strategies to protect biodiversity – but with limited success. The low quality and lack of ambition with which Member States were implementing and enforcing EU laws became evident almost immediately. No other policy has been subject to as many complaints by citizens as the EU's nature protection laws. The deficit in integrating nature protection into other policies and providing coherent policies is also significant.

---

[12] former Policy Officer for agriculture and biodiversity at the European Environmental Bureau

[13] EU Policy Director, European Environmental Bureau

Europe's precious biodiversity continues to be under pressure on several fronts - including invasive species, wasteful land-use, poor planning and intensive farming practices. In addition there is now growing evidence of the serious effects of climate change on species distribution and survival.

The ambitious target to halt biodiversity decline by 2010

In 2001, at the EU Spring Summit in Gothenburg, the EU Heads of State and Government committed themselves to halting the decline in biodiversity in the EU by 2010, as part of the EU Sustainable Development Strategy. The target is also laid down in the legally binding Decision on the Sixth Environmental Action Programme (6th EAP) (EC 2002). The 6th EAP highlights nature and biodiversity as a top priority.

IV.2.2 Birds Directive and Habitats Directive

IV.2.2.1 INTRODUCTION

EU nature conservation policy is based on two main pieces of legislation – the 1979 Birds Directive (EC 1979) and the 1992 Habitats Directive (EC 1992).

The Habitats Directive focuses on the protection of wild species and their habitats. Each Member State is required to identify sites of European importance, so called Special Areas of Conservation (SACs), and to put in place management plans where necessary. Under the earlier Birds Directive Member States are required to classify so-called Special Protection Areas (SPAs), which are similar to SACs and may overlap at times. The sites identified under the two Directives make up Natura 2000, the EU's network of protected areas. This effectively is the cornerstone of EU policy for nature conservation.

The Birds Directive developed largely out of public concern about the annual killing of migratory birds, which was widespread in southern Europe, as well as a result of pressure from citizens' organisations and the European Parliament. As early as 1971, suggestions were mooted for Community legislation on bird conversation. Research undertaken by the European Commission which indicated a reduction in the number and population of migrating bird species due to hunting, agricultural intensification and the destruction of habitats resulted in the Commission proposing a directive in 1976. The initiative was controversial and many Member States were reluctant about the Community entering into this sphere raising questions about the legal basis of Community action on the grounds of remoteness from the functioning of the common market. The Directive was finally adopted in 1979, which is remarkable considering that it was eight years before the Single European Act (1987) gave the Community a clear legal basis for measures concerned with nature conservation[14].

---

[14] The Commission had argued that Europe's wild birds were part of the common heritage of the Community, that an effective protection demanded transboundary cooperation and that conserving wild birds was necessary to 'attain, within the operation of the common market, the Community's objectives regarding the improvement of living conditions, a harmonious development of economic activities throughout the Community, and a continuous and balanced expansion.'

The Second Action Programme on the Environment (1977) already hinted at a more ambitious proposal for the protection of habitats. The Commission mentioned in the programme that it would submit appropriate proposals for Community action, if necessary, to ensure that the Bern Convention, that was being developed by the Council of Europe at the time, could be satisfactorily applied. Many of the provisions of the Bern Convention on the Conservation of European Wildlife and Natural Habitats, signed in September 1979, served as a model for the Habitats Directive, which was agreed 13 years later. However, the 1992 Habitats Directive goes one step further by aiming to protect selected habitats for their own sake rather than because they were home to valuable species. The directive was proposed by the Commission in 1988 just after the Single European Act had extended the Community's competence to the environmental sphere. The directive was to a large extent a result of combined pressure from the European Parliament and the requirements of international law.

IV.2.2.2 OBJECTIVE, SCOPE AND TOOLS OF THE BIRDS AND THE HABITATS DIRECTIVES

The Birds Directive obliges Member States to preserve, maintain or re-establish a sufficient diversity and area of habitats for birds in order to maintain healthy populations of all species. This is to be done primarily by creating protected areas, managing habitats both inside and outside protected areas, re-establishing destroyed biotopes and creating new ones. Member States are to lay down a general system of protection for all species of wild birds, although exceptions are made for hunting and for certain other reasons.

Annex I of the directive lists particularly vulnerable species which are to be subject to special conservation measures concerning their habitat in order to ensure their survival and reproduction. Member States are to classify the most suitable territories (both land and sea) as Special Protection Areas (SPAs) for the conservation of these species and to prohibit the capture and hunting of these birds. Annex II lists species that can be hunted within certain limits.

The aim of the Habitats Directive is to contribute to the maintenance of biodiversity through the conservation of natural habitats and the protection of wild fauna and flora. The directive aims to establish a 'favourable conservation status' for both habitat types and wild species of Community interest. It requires Member States to designate sites as Special Areas of Conservation (SACs) and set up necessary conservation measures, including management plans, which will result in the establishment of a 'coherent European ecological network' of sites of Community importance to be known as Natura 2000.

Annex I lists habitat types whose conservation requires the designation of Special Areas of Conservation (SACs). A habitat type is defined as being of Community interest if it is in danger of disappearance within its natural range, where the Community has a special responsibility for their conservation because of the proportion of their natural range falling within EC territory, or which represents an outstanding example of one or more of the six biogeographical regions (Alpine, Atlantic, Continental, Macronesian, Mediterranean and Boreal). A seventh region, the Pannonian region, was added by the 2003 Accession Treaty.

Annex II contains a list of animal and plant species of Community interest, the conservation of which benefits from the special designation of their habitats. As with Annex I, a number of priority species are identified.

IV.2.2.3   STRENGTHS AND WEAKNESSES

One of the strengths of the Birds Directive is that it applies to *all* wild bird species in Europe, and that it introduced for the first time site-based conservation measures into EU policy (with the selection of SPAs to be based on ornithological criteria only). It has also provided a transboundary solution for the protection of migratory species (in this case birds), and tackles the problems of hunting and trade.

The directive is weak, however, insofar as it did not set a timeframe or threshold for the completion of the SPA network, merely referring to the obligation to create "sufficient sites". However, 25 years after the directive has been adopted, it is common knowledge that many Member States are still far behind fulfilling this obligation, and further judgments of the ECJ can be expected.

The margin of discretion available to Member States under the Birds Directive concerning the granting of derogations, and in particular to specify hunting seasons for migratory birds, has also led to problems, and has been interpreted differently in different Member States. A significant number of cases of alleged insufficient implementation have had to be resolved by the ECJ. The "Guidance document on hunting" under the Birds Directive developed by the European Commission in 2003 is based on case law by the ECJ and extensive stakeholder discussion and provides more clarity on many aspects of the directive, including criteria for determining the duration of the hunting season.

Compared to the Birds Directive, the Habitats Directive is a more modern legislative tool. It reinforces the basic principle of the Birds Directive, notably species and site protection measures, and extends some of the provisions to include, management and impact assessment provisions in article 6. This article also extends to the Birds Directive by means of Article 7, which amends the Birds Directive.

The creation of the biogeographical seminars which examine the ecological representativity of the national lists (amount, distribution, quality of the sites) has to be welcomed. The Habitats Directive lays down that the only criteria on which the selection of Natura 2000 sites can be based are ecological ones.

Article 8 offers the possibility of co-funding of the management plans (Article 8), which can help create jobs in the countryside and enable local people to benefit from conservation. Another strength of the directive is that Natura 2000 sites are protected even if the threats originate from outside the border of the conservation area. Compensation is obligatory if a Natura 2000 site is damaged. Natura 2000 contains a modern idea of developing networks and corridors (Article 10), which is especially important because of the impact of climate change.

Moreover the Habitats Directive has strengthened the importance of Environmental Impact Assessment (EIA) and Strategic Environmental Impact Assessment (SEA)*(see chapter V.3.5)*, because Article 6 of the directive provides real juridical impacts and the possibility to complain to the Commission. Another strength of the directive is that it foresees a strict timeframe; however, in spite of this a complete protection of sites will only be guaranteed six years after the adoption of the European list of sites.

A weakness of the Habitats Directive is, however, that at the time of its adoption it did not sufficiently provide for the protection of marine sites. Moreover there is no obligation for the designation of sites for the protection of migratory species. The definition of marine habitats is not as obvious as terrestrial ones, but at least, types of shores should be designated according to their importance to marine biodiversity to counter the deterioration of large coastal and marine areas throughout Europe.

In addition, there have been enormous delays in the Member States concerning the implementation of the Birds and the Habitats Directives, and both are amongst the most litigated directives of the EU. The assessments of plans and projects with a significant impact on a Natura 2000 site are often inadequate. Many of the provisions of the Habitats Directive, such as the definition of the concept 'favourable conservation status' are set out in guidelines which are not legally binding. Whether and in which way the concept of "favourable conservation status" also applies to the Birds Directive, is still under discussion.

IV.2.3   Natura 2000

The development of Natura 2000 is without doubt one of the most ambitious tasks in terms of nature conservation in the EU and presents an essential step towards the target of halting biodiversity loss by 2010. The Natura 2000 network now covers around 17% of the territories of the Member States before the EU's 2004 enlargement, an area about the size of Germany.

The Natura 2000 network will include three categories of sites: those hosting the habitat types of Community importance listed in Annex I of the Habitats Directive; secondly those sites comprising the habitats of certain animal and plant species of Community importance listed in Annex II of the Habitats Directive; and thirdly the 'Special Protection Areas' for birds classified by Member States under the Birds Directive.

Member Sates were asked to submit to the Commission by June 1995 a list of sites within their territories which are potentially of Community importance. The next step is for the Commission to draw up, in agreement with each Member State, a draft list of sites of Community importance (for each biogeographical region), drawing on the national lists. By 1998 the final list was supposed to be adopted by the Commission having been presented to a committee of Member State representatives where voting is by qualified majority. The third step is the designation of sites selected as being of Community importance as SACs by the Member States concerned. This is to be done as soon as possible and at the latest within six years. Member States have to set up the necessary conservation measures, including management plans, and to take all the steps necessary to avoid the deterioration of the habitats or disturbance of the species for which the 'areas' have been designated.

All plans or projects which individually or in combination with others are likely to have significant effects on sites are to be subject to an appropriate assessment of the implication for the conservation value of the site. Where an assessment of a plan or project indicates that it will damage the conservation interest of a site and there are no alternative solutions, but the plan or project must be carried out for 'imperative reasons of overriding public interest', including those of a 'social or economic nature', the Member State must take all compensating measures necessary to protect the overall coherence of Natura 2000. The Commission must be informed of the measures adopted[15].

The management and monitoring of Natura 2000 will require significant and continuous investment. However, it should not be forgotten that there are significant socio-economic and environmental benefits accrued from designating Natura 2000 areas (*IEEP and WWF 2002*).

In the exceptional case of this directive Community co-financing is to be provided to help Member States to meet their obligations as a huge financial burden will be imposed on some Member States because of the uneven distribution of habitats and species within the Community. Whereas the Natura 2000 network covers for instance 10% of the territory in France (at the moment), it covers 38% of the territory of Spain. On average the network covers 15 % of all European territory (*Le Monde 2005*).

The official deadline for the finalisation of Natura 2000 lists as set out in the directive was 1998. The first list of sites of EU conservation importance was approved by the European Commission for the Macronesian biogeographical region (which covers the Canary Isles, the Azores and Madeira) in December 2001. This was followed in 2004 by the adoption of the list of sites for the Alpine, Atlantic and Continental biogeographical region. Implementation in the new Member States is under way and when completed will make a sizeable contribution to the total area under protection.

IV.2.4 Implementation deficits and link with other environmental legislation

The implementation of the Birds Directive has been controversial and poor in several Member States. It has resulted in rulings against e.g. France and Sweden and many other countries have been referred to the ECJ. Likewise, the implementation of the Habitats Directive has been seriously delayed in certain regions and Member States. As a result the Commission has been warning since July 1999 that regional funding under the Structural Funds may be withdrawn from Member States where implementation of the Birds and Habitats Directives is particularly inadequate. The following year the Commission stated that operations financed by the Structural Funds

---

[15] In the case of sites hosting priority habitat types or species, the grounds for proceeding with damaging projects are restricted to those relating to human health or public safety, environmental improvements of primary importance or other imperative reasons of overriding public interest.

must not threaten areas likely to be included in the Natura 2000 network.. Otherwise funding would not be released. This is believed to be the first instance of the Commission threatening Member States to withhold funding as a way to make them apply environmental legislation.

The submission of national lists of proposed SAC sites as specified under Article 4 (1) of the Habitats Directive has been delayed in all Member States. In January 2000 the Commission took further action against several Member States who had failed to adequately transpose the Habitats Directive into national legislation and to submit complete lists of sites of potential importance for the establishment of the Natura 2000 network. In September 2001 the ECJ ruled that Ireland, France and Germany had contravened the Directive as they had failed to provide adequate lists of proposed SACs and to submit these before the 11 June 1995 deadline (*cases C-71/00, C-67/99 and C-220/99*). This was followed in March 2003 by a judgement of the ECJ against Italy for poor transposition of the Habitats Directive (*case C-378/01*).

In January 2005 the Commission sent a final warning to France for not having classified a sufficient number of SPAs under the Birds Directive, in spite of an ECJ ruling of 2002 (case C-202/01). This could lead to a second judgment of the ECJ with possible penalties for France.

The present state of the network still varies enormously between the Member States. The problem ranges from the lack of definitive lists in some Member States to a serious delay in the presentation and implementation of the site's management plans in others. This is particularly serious as the sites continue to be subject to projects that often reveal to be damaging. The European Commission has had difficulties in responding to all the complaints made by individuals and citizens' organisations in effective time, and has thus been unable to avoid the degradation of several sites.

Europe's Environmental Impact and Strategic Environmental Assessment Directives are crucial supporting instruments for the protection of designated sites from harmful developments (*see chapter V.3.5*).

Under the EU Water Framework Directive from 2000 Member States had to provide a list of all protected areas under Community legislation, including the ones for the conservation of habitats and species, by the end of 2004. The specific objectives for such sites have to be achieved at the latest by 2015 if no other deadlines are provided for (*see chapter IV.5.7*). It can be expected that the overall 2015 objective of "good ecological status" for all waters substantially supports and improves the status of nature protected sites, most of which are dependent on the functioning of their aquatic components (*see chapter IV.5.5.1*).

The new EU Directive on environmental liability from 2004 needs to be transposed and implemented at national level by 2007. This Directive ensures that companies (the ones covered by the Directive are the ones covered by other EU laws, e.g. IPPC) have to pay the costs for damages to the environment, including damages to protected habitats and species. Member States should encourage companies to take out insurance or other financial guarantees (*see chapter V.3.3.5*).

IV.2.5 Way forward and the role of ECOs (Environmental Citizens' Organisations)

National and European citizens' organisations play a key role in the implementation of the Habitats Directive on the ground and contribute to the management and monitoring of the designated sites:

▶ ECOs have proposed many Natura 2000 sites successfully. The European Habitats Forum[16] has taken part in biogeographical seminars and citizens' organisations have shown gaps in national lists. WWF has made shadow-lists, which have been extremely useful in these seminars. BirdLife has identified Important Bird Areas, and the ECJ has used these as a manual against e.g. Netherlands and Finland, which has led to the classification of more SPAs. Due to the work of Portuguese ECOs for instance, many projects and infrastructures which would have caused damage to natural areas have been stopped, postponed or more adequate minimisation measures have been imposed, like in the case of the Vasco da Gama Bridge in Lisbon or the South Motorway.

▶ ECOs have had the determining role in driving poor national implementation forward by launching national court cases, complaints with the European Commission, which eventually resulted in rulings of the European Court of Justice[17]. The EEB as made proposals to the Commission for improving the effectiveness of complaints (*EEB 2002*).

▶ ECOs help to ensure that both the EIA and SEA directives are properly applied (*see chapters V.3.5.1 and V.3.5.2*). The purpose of those directives is to ensure that the environmental consequences of development measures, plans and programmes on protected areas are identified and assessed during their preparation and before their adoption. The public and environmental authorities can give their opinion and all results are to be integrated and taken into account in the course of the planning procedure.

However, too many EU and national policies and financial incentives run counter to an adequate and ambitious implementation of the Habitats Directive to allow a positive outlook. The respect for nature protection and integration into structural and agriculture policies needs strengthening. Furthermore appropriate financing of nature protection must be ensured[18]. In 2004 the Commission estimated the costs for managing Natura 2000 at around 6.1 billion Euro per year. These finances are not at all guaranteed in the EU's proposed financial perspective 2007-2013. The Commission envisages financing Natura 2000 through existing funds, i.e. the Structural and

---

[16] The European Habitats Forum (EHF) is a group of ECO networks that provide input and advice to DG Environment, particularly relating to the implementation of the Birds and Habitats Directives and the establishment of the Natura 2000 network. The EHF meets with DG Environment twice a year after the Habitats Committee meetings and provides a coordinated means of communication between DG Environment and European nature conservation citizens' organisations.

[17] The decisions of the ECJ in the Marismas de Santona (C-355/90) and Lappel Bank (C-44/95) cases laid the basis for the Natura 2000 site selection rules, i.e. only ecological criteria.

[18] Although the responsibility for managing Natura 2000 lies with the Member States, Article 8 of the Habitats Directive provides for EU co-financing for the measures essential for maintaining or re-establishing priority habitats and species at a favourable conservation status. This requires a broad range of measures, including land acquisition, site rehabilitation, compensatory/incentive payments for (land) managers, administrative costs, legal fees etc.

Cohesion Funds, Rural Development and the proposed European Fisheries Fund. But this will need considerable awareness raising within the respective departments and remains rather uncertain and open to the reigning day by day political agenda.

First of all, for the Commission strategy to work, a clear conditionality in the allocation of funding would need to be established, including an automatic withholding of EU funding if Member States fail to deliver against the environmental acquis, particularly Natura 2000.

Although proper implementation of the directives will be essential for biodiversity protection, this will not be sufficient to achieve the EU's objective of halting the decline in biodiversity by 2010. Firstly most of Europe's biodiversity remains outside the scope of the NATURA 2000 network. Therefore it is crucial that nature protection is integrated in other EU (as well as national and regional) sectoral policies, including agriculture, transport, regional policy, fisheries, etc.

Furthermore environmental organisations should consider the opportunities from linkages with other EU laws. The implementation of the Water Framework Directive in particular bears strong potential improvement for nature and biodiversity protection outside designated areas. The Directive requires the achievement of a "good ecological status", which represents a aquatic biodiversity with minor disturbance, by 2015 for all rivers, lakes and coastal waters, including their floodplains and wetlands. In the case of Natura 2000 protected areas the relevant objectives must be achieved by 2015 at the latest. ECOs should highlight this link and participate in the ambitious implementation of the Water Framework Directive (*see chapter IV.5*).

Secondly a considerable negative impact on biodiversity is assumed to come from more subtle pressures, like chemical and genetic hazards. Therefore the 2004 Environmental Liability Directive provides another great opportunity for the prevention of damage to habitats and species and enforcing the polluter pays principle. Due to the flexibility left to Member States in implementing the Directive and the potential loopholes, its effectiveness will depend on the quality of transposition into national law by 2007. Most importantly ECOs should ensure that licensed industrial activities are not automatically excluded from strict liability and that financial guarantees are mandatory (*see chapter V.3.3.5*). Finally Europe's current chemicals policy reform could – if successfully adopted – end the accumulation of chemicals in the environment and wildlife, reducing the potential impact from hormone or neurological disruption or other irreversible effects on biodiversity.

BIBLIOGRAPHY AND FURTHER READING

EEA (2005) Technical Report No. 1, EEA core set of indicators - Guide

EEB (2002) EC complaints procedure: EEB's seven key recommendations for a change, Position Paper December 2002, Brussels

EC (1979) Council Directive 79/409/EEC of 2 April 1979 on the conservation of wild birds

EC (1992) Council Directive 92/43/EEC of 21 May 1992 on the conservation of natural habitats and of wild fauna and flora

EC (2002) Decision No 1600/2002/EC of the European Parliament and of the Council of 22 July 2002 laying down the Sixth Community Environment Action Programme, OJ L 242 of 10/9/2002

IEEP and WFF (2002), Promoting the Socio-Economic Benefits of Natura 2000, Background Report for European Conference, Brussels, 28–29 November 2002

Le Monde (2005) Natura 2000: comment la France veut rattraper son retard, 15 January

McCORMICK, J. (2001) (ed.) Environmental Policy in the European Union, Palgrave: London

RICHARTZ, S.(2004) Brussels in Brief: Finding the Funds for Natura 2000, IEEP, December 2004

WURZEL, R.K.W. (2002) Environmental Policy-making in Britain, Germany and the European Union, The Europeanisation of air and water pollution control, Manchester University Press

IV.3 Air

By *Kerstin Meyer* [19]

46

IV.3 Air 46

IV.3.1 Introduction 47

IV.3.2 The overarching EU environmental and health objectives for clean air 49

IV.3.3 The air quality Directives: the Framework Directive on ambient air quality
assessment and the four Daughter Directives 50

    IV.3.3.1 Introduction 50

    IV.3.3.2 Assessment 53

        IV.3.3.2.1 Strong points 53

        IV.3.3.2.2 Weak points 54

    IV.3.3.3 Implementation: 55

        IV.3.3.3.1 Will the objectives be achieved? 55

        IV.3.3.3.2 Have plans and programmes been made? 57

        IV.3.3.3.3 What is the quality of the plans and programmes? 57

        IV.3.3.3.4 How will the Directives be enforced by the Commission? 58

        IV.3.3.3.5 Room for ECO action 59

IV.3.4 National emission ceilings for certain atmospheric pollutants 60

    IV.3.4.1 Interim objectives and emission ceilings 60

    IV.3.4.2 Programmes and reporting 61

    IV.3.4.3 Review and revision 61

    IV.3.4.4 Assessment of the NEC Directive 62

        IV.3.4.4.1 Systematic overestimation of costs 62

        IV.3.4.4.2 Benefits to health and the environment 62

        IV.3.4.4.3 Level of ambition too low 63

    IV.3.4.5 Future developments 63

    IV.3.4.6 Room for ECO action 64

IV.3.5 Road vehicles 65

    IV.3.5.1 Why emission legislation for road vehicles? 65

    IV.3.5.2 Some aspects of regulating vehicle emissions 66

[19] Policy Officer, European Environmental Bureau. The author would like to thank Christer Ågren for his contributions to the text as well as for his many useful comments. Further thanks to Anette Hauer, for writing an earlier draft of the air quality chapter. I am also grateful to Duncan Laxen, Karsten Krause, Hugo Tente, Lesley James, Karola Taschner and Dragomira Raeva for their constructive comments.

IV.3.5.3 Current legislation 67

IV.3.5.4 Planned legislation...68

IV.3.5.5 Outlook – Room for ECO action 70

IV.3.6 Large Combustion Plants 70

IV.3.6.1 Why a Directive for large combustion plants (LCPs)? 70

IV.3.6.2 Requirements and emission standards 71

IV.3.6.3 Assessment 73

IV.3.6.3.1 Directive applies to existing plans.......................................73

IV.3.6.3.2 Country-wide "bubbles" 73

IV.3.6.3.3 Exemptions and derogations 74

IV.3.6.4 Review 74

IV.3.6.5 Outlook and room for ECO action...74

IV.3.1   Introduction

Air, like water, is an environmental medium, which circulates freely though the environment and together with soil/land and water forms the habitat for all flora and fauna. The different elements contained in the air we breathe every day and deposited on water and soil are influenced by an array of different factors such as weather patterns, complex atmospheric chemistry and anthropogenic activities leading to air pollution. Air pollution can both be a local and an international problem, because it can be transported over long distances. EU legislation on air pollution addresses two sides of the same coin: pollutant emissions as well as air quality *(see chapter IV.1)*. Emission legislation tries to reduce the amount of pollution emitted into the atmosphere. Air quality legislation aims at guaranteeing that the air we breathe is safe for human health as well as for the environment as a whole.

The four topics presented in this chapter make up the cornerstones of European air legislation. The air quality Framework and Daughter Directives set minimum quality standards for clean air that apply throughout the Union. The Directive on national emission ceilings (NECs) is the most important law regulating the emissions of air pollutants. It covers four air pollutants of crucial importance for human health, ground-level ozone, acidification and eutrophication. The Directive defines important interim objectives in order to protect the environment and human health and also spells out long-term environmental and health objectives with regard to air pollution, the latter based on the carrying capacity of the ecosystems *(see chapter IV.3.2)*. The two other sections in this chapter focus on large combustion plants and car emissions, thus addressing two of the most important sectors causing air pollution in the EU.

While most of the current air legislation in the EU was formulated during the 90s, air pollution has been in the political debate for much longer. Over time, different aspects of the problem became the focus of political attention. Strong environmental concerns brought air pollution onto the political agenda: the problem of acidification of Scandinavian lakes and rivers was discovered in the late 60s. This was followed by concerns over air pollution impacts on forests, including the acidification of forest soils in the 80s. Since the 90s, the debate has also focussed strongly on the health damage caused by air pollution, particularly with regard to urban air quality.

Clean air policy-making in the EU has been influenced by international negotiations on air pollution : under the 1979 Convention of Long-range Transboundary Air Pollution, and its various protocols, in particular the 1999 Gothenburg Protocol[20]. Important EU policy-goals relating to air pollution were laid down in the Fifth (1992) and Sixth (2002) Environment Action Programmes and the Community Strategy to Combat Acidification (1997) *(see chapter III)*.

Air pollution legislation in the European Union has so far been relatively effective. Over the past few decades it has reduced pollutant emissions and improved air quality substantially. However, combating air pollution is still a formidable challenge: we are far from reaching the EU's clean air objectives of not exceeding **critical loads** and levels and the effective protection of all people against recognised health risks from air pollution, which are the most important EU objectives with regard to air pollution *(see chapter IV.3.2)*.

> **DEFINITION**
> Critical loads specify the environmental carrying capacity for different ecosystems. They have been defined as: "The highest load that will not cause chemical changes leading to long-term harmful effects on the most sensitive ecological systems".
>
> *Source: Nilsson, J. (Ed) (1986): Critical loads for nitrogen and sulphur.*

Ground-level ozone and particulate matter (PM) are still important issues of concern for human health and effects on ecosystems, particularly as more and better evidence about the negative health effects caused by these pollutants[21] becomes available. Currently the life-expectancy of EU-citizens is shortened on average by about 9 months due to man-made PM, according to the most recent calculations under the Clean Air for Europe programme (CAFE) *(Amann et al. 2004)*[22]. This is comparable to the impact of traffic accidents *(ibid p. iv)*. In the year 2020, if all existing EU legislation concerning sources of air pollution is implemented, for large parts of the population life expectancy losses attributable to antropogenic PM are calculated to still exceed 6 months (and in countries like Belgium and the Netherlands it will still be about nine months) *(ibid, p.58)*[23]. Furthermore, the European Environment Agency (EEA) concludes that countries in Central and Eastern Europe still have problems with sulphur dioxide and nitrogen oxides *(EEA 2003)*[24].

---

[20] For further information, see: http://www.unece.org/env/lrtap/welcome.html and http://www.unece.org/env/lrtap/multi_h1.htm (official Convention website). From ECO point of view, see: http://www.acidrain.org/clrtap.htm.

[21] For example: WHO (2003) Health Aspects of Air Pollution with Particulate Matter, Ozone and Nitrogen Dioxide, Report on a WHO Working Group, Bonn, Germany, 13-15 January 2003, Copenhagen, http://www.euro.who.int/document/e79097.pdf , WHO: Health Aspects of Air Pollution – answers to follow-up questions from CAFE, Report on a WHO working group meeting, Bonn, Germany 15-16 January 2004, Copenhagen, http://www.euro.who.int/document/E82790.pdf. For overview, see: http://www.euro.who.int/eprise/main/WHO/Progs/AIQ/Activities/20020530_1).

[22] Report available at: http://www.iiasa.ac.at/rains/CAFE_files/Cafe-Lot1_FINAL(Oct).pdf

[23] Please note: These figures are probably underestimations. This computer model only calculates mortality for population above 30 years, it does not calculate infant mortality and thus underestimates overall effect. Also, for technical reasons, these calculations only reproduce a part of the total observed mass of PM, thus health effects are likely to be underestimated.

[24] http://reports.eea.eu.int/environmental_assessment_report_2003_10-sum/en/kiev_sum_en.pdf .

Acidification is still a prevailing problem in European forests and fresh waters. If no further measures are taken, in the year 2020, 150,000 km² of forests will continue to receive unsustainable amounts of acid deposition from the atmosphere and thousands of Scandinavian lakes will still not be able to recover from past acidification (*Amann et al. 2004, p.vi*). Furthermore, improved understanding of the nitrogen cycle reveals serious threats for biodiversity from excess nitrogen deposition from the atmosphere throughout Europe (*ibid, p. iv*). Current computer modelling under the CAFE programme shows that if no additional abatement measures are taken, biodiversity will remain threatened at more than 650,000 km2 (45% of European ecosystems) due to excessive nitrogen deposition (*ibid, p. vi.*).

Therefore it is crucial that further efforts are made to reduce air pollution and to ensure that existing air pollution related Directives (as well the related IPPC Directive − *see chapter V.3.2*) are implemented and enforced everywhere in the EU. When introducing further policies to reduce air pollution, it is also important to keep in mind that there are linkages between air pollution and climate change. Energy efficiency, renewable energy and sustainable mobility policies offer win-win solutions that simultaneously help to combat both air pollution and climate problems.

IV.3.2 The overarching EU environmental and health objectives for clean air

Protection of the environment and the health of EU citizens are important elements of the EU Treaty. It obliges the European Community to preserve, protect and improve the quality of the environment and to strive for a high level of environmental and human health protection in its policies (EC Treaty).

In the field of air legislation, the two most important Directives in this context are the air quality Framework 1996/62/EC and national emission ceilings (NECs) 2001/81/EC Directives.

The air quality Directives define minimum standards for the protection of health and the environment that are to be met everywhere. The two main aims of the air quality Framework Directive are:

▶ To "define and establish objectives for ambient air quality in the Community designed to avoid, prevent or reduce harmful effects on human health and the environment as a whole".

▶ To "maintain air quality where it is good and to improve it in other cases".

The first objective is rather broad. It is further specified in the four subsequent "Daughter Directives" that were agreed by the EU in the following years. The Directive requires the Daughter Directives to set effects-based limit values, aimed at safeguarding human health and the environment. The second objective complements this by indicating where the Directive applies. It clearly states that air quality should not deteriorate anywhere in the EU.

The NEC Directive further complements ambient air quality standards by setting the long-term environmental quality and health objective "of not exceeding critical levels and loads and of effective protection of all people against recognised health risks from air pollution"[25].

---

[25] Directive 2001/81/EC, Art. 1.

50

Critical loads specify the environmental carrying capacity for different ecosystems. They have been defined as: "The highest load that will not cause chemical changes leading to long-term harmful effects on the most sensitive ecological systems." (*Nilsson 1986*). It can be said that in a strict sense a critical load, according to that definition, is one that produces no effect on the most sensitive[26]. receptor, even in the long term. Receptors may be individual species, types of soil, ecosystems, etc With regard to human health, it is evident that in order to effectively protect all people against recognised health risks of air pollution, particular attention needs to be paid to the protection of vulnerable groups such as children or elderly people. These long-term objectives provide the benchmark for the success and ambition of EU air policy measures.

IV.3.3   The air quality Directives: the Framework Directive on ambient air quality assessment and the four Daughter Directives

IV.3.3.1   INTRODUCTION

The Framework Directive on ambient air quality assessment and management (*96/62/EC*) from 1996 lays down, for the first time, common rules and principles for setting effects-based air quality limit values to be met everywhere in the EU.

It lists 12 pollutants for which legislation, including limit values, measurement and assessment requirements, must be developed, and sets the timeframe for the development of the so-called "*Daughter Directives*". In the years 1999, 2000, 2002 and 2004 the EU subsequently adopted four Daughter Directives on ambient air quality, covering all the 12 pollutants.

The first Daughter Directive (*1999/30/EC*) sets limit values for sulphur dioxide, oxides of nitrogen, particulate matter, and lead in ambient air, and replaces the previous EU Directives adopted in the 80s. The limit values set in the Directive aim at protecting human health and are based mainly on the (1997) World Health Organisation's (WHO) guidelines. For sulphur dioxide (SO_2) and nitrogen dioxide (NO_2) additional environmental standards were introduced to be met outside built-up areas. The

DEFINITIONS

Limit value: a level fixed on the basis of scientific knowledge with the aim of avoiding, preventing or reducing harmful effects on human health and/or the environment as a whole, to be attained within a given period and not to be exceeded once attained.

Target value: a level, fixed with the aim of avoiding more long-term harmful effects on human health and/or the environment as a whole, to be attained where possible over a given period.

Alert threshold: a level beyond which there is a risk to human health from brief exposure and at which immediate steps shall be taken (...).

Source: Framework Directive (96/62/EC).

---

[26] For further information, see: Critical loads Environmental factsheet published by the Swedish NGO Secretariat on Acid Rain. Updated May 1998, http://www.acidrain.org/cl_fact.htm#Critical_loads

Directive further defines detailed measurement requirements, i.e. where to measure air quality (urban background, close to traffic, close to industry, etc.), how many monitoring stations per city, and which measurement technique should be used. Last but not least, a separate article defines the Member States' obligations regarding the dissemination of information to the public.

The second Daughter Directive (*2000/69/EC*) sets limit values for benzene and carbon monoxide (CO) for the first time in the EU. This is particularly important as it represents the introduction of an air quality standard for a carcinogenic pollutant – benzene – where no safe threshold can be defined. After the full implementation of this Directive it is estimated that emissions of benzene will drop 70% by 2010 and those of carbon monoxide by one-third by 2005.

51

The third Daughter Directive relating to ozone (*2002/3/EC*)[27] sets non-binding target values for ozone in ambient air to be attained "where possible" by 2010 as well as long-term objectives equivalent to the World Health Organisation's guideline values[28]. These targets correspond to the objectives set in Directive 2001/81/EC on national emission ceilings *(see chapter IV.3.4)*. The Directive also sets alert thresholds and requires Member States to take short-term action if these alert thresholds are exceeded and includes requirements to inform citizens about the actual pollution load.

The fourth Daughter Directive (*2004/107/EC*) also sets non-mandatory target values for arsenic, cadmium, nickel and polycyclic aromatic hydrocarbons (PAHs) as well as specifying monitoring requirements for mercury.

All Directives require the development of pollution reduction plans by Member State authorities. When a certain concentration of pollutants is exceeded (specified in detail in the different Directives), plans and programmes have to be made, listing the different policy actions (e.g. low emission zones, promotion of walking and cycling etc.) the authority plans to take for achieving the standard. These plans and programmes have to be reported to the Commission and to be made available to the public, in order to allow citizens to trace progress towards meeting the standards.

---

[27] This Directive replaces and older Ozone Directive (92/72/EC), which was repealed by that date.

[28] The possibility to set target instead of limit values in the case of ozone is already mentioned in the Framework Directive (Art. 4.1) taking *"account of the specific formation mechanisms of this pollutant"*. In the opinion of the EEB, this shouldn't, however, reduce the efforts to implement policies to reduce ozone pollution.

Table 1: Limit (LV) and target values (TV) in the air quality Directives.

POLLUTANT	ENTRY INTO FORCE	1 H. AVERAGE (HUMAN HEALTH)	24 H. AVERAGE (HUMAN HEALTH)	ANNUAL AVERAGE (HUMAN HEALTH)	8 H. MEAN VALUE (HUMAN HEALTH)	VEGETATION AND ECOSYSTEM
Sulphur Dioxide ($Sé_2$)	2005 (LV)	350µg/m3	125 µg/m3			20µg/m3 per year
Nitrogen Dioxide (NO_2)	2010 (LV)	200 µg/m3		40 µg/m3		30µg/m3 per year
PM_{10}	2005 (LV) 2010 (LV) [a]		50 µg/m3 [b] 50 µg/m3 [c]	40 µg/m3 20 µg/m3		
Lead (Pb)	2005 (LV)			0.5 µg/m3		
Benzene (C_6H_6)	2010 (LV)			5 µg/m3		
Carbon Monoxide (CO)	2005 (LV)				10 µg/m3	
Ozone (O_3)	2010 (iV)	180/240 µg/m3 [d]			120 µg/m3 [e]	AOT_{40} [f] = 18 000 µg/m3 hours
PAH	2012 (TV)				1 ng/ m³	
Cadmium (Cd)	2012 (TV)				5 ng/ m³	
Arsenic (As)	2012 (TV)				6 ng/ m³	
Nickel (Ni)	2012 (TV)				20 ng/ m³	
Mercury (Hg)	no LV or TV	-	-	-	-	-

(a) Indicative limit values to be reviewed in the light of further information on human health and environmental effects, technical feasibility and experience in the application of Stage 1 limit values in the Member States.

(b) Not to be exceeded more than 35 times a calendar year.

(c) Not to be exceeded more than 7 times a calendar year.

(d) At 180 (information threshold) the population should be informed, and at 240 (alert threshold) short-term action should be taken.

(e) Not to be exceeded more that 25 times per year.

(f) AOT_{40} = Accumulated exposure over the threshold 40 ppb.

IV.3.3.2   ASSESSMENT

IV.3.3.2.1   Strong points

Legally Binding Standards: The limit values in Directive 1999/30/EC and Directive 2000/69/EC are legally binding air quality standards. These are strong policy drivers for air quality management and policies for clean air. Citizens have the right to go to court if their authorities have not achieved the standards or not implemented appropriate measures to achieve them. This is a strong tool for environmental organisations which, together with affected citizens, can raise court cases about enforcing the air quality limit values (this has already been done in the UK and the Netherlands, and is likely to happen in Germany as well).

Simple: Limit values are relatively simple and easy to communicate. This makes it easy for environmental groups to use them in communicating with the general public. In Germany for example, EEB member organisation Bund für Umwelt und Naturschutz Deutschland (BUND) refers to the limit values when campaigning for better inner city transport policies and particulate filters for diesel cars.

More data: The Directives generate a wealth of air quality data. In the different Daughter Directives, measurement requirements (citing criteria, number of stations) have subsequently been laid down in more detail. This data will enable a better understanding of air quality problems throughout the EU and will thus help to identify gaps in the policy responses at local, national and EU levels. It will also facilitate better target policies and measures and could possibly be used to benchmark between Member States (However also *see the point on comparability under chapter IV.3.3.2.2*).

Public information: Public information clauses in the air quality Directives are quite good. Besides obligations to inform the general public, the second, third and fourth Daughter Directives include the explicit obligation to inform environmental and consumer organisations. Attention and pressure from citizens and environmental organisations are particularly important for the implementation of the ozone and heavy metal Directives, which set target values instead of limit values. Strong concerns from the general public can serve as an additional policy driver to ensure that countries and regions do everything they can to comply with the target values.

Systematic approach: Air quality plans and programmes can lead to a systematic approach to tackling the problem of air pollution. The implementation of the first Daughter Directive indicates that the obligation to make plans and programmes and to comply with limit values stimulates some regions and cities to deal with the problem of air pollution in a more systematic and integrated way. This helps to raise the issue of air pollution on the agenda of local and regional policy makers and encourages different sections within administrations to work together on air quality management.

Important related positive effects: If air pollution is tackled through integrated traffic management and sustainable mobility policies there will be other positive side-effects such as less noise and congestion, and better quality of life in cities. Findings from the Dutch National Institute for Public Health and the Environment identify transport-related air quality and noise problems as main factors contributing to the diseases caused by environmental problems in the EU (*National*

53

Institute for Public Health and the Environment 2004). Reducing both noise and pollution would yield considerable health benefits. Additionally, many measures reducing air pollution also reduce atmospheric levels of CO_2 the main greenhouse gas. This is a case for encouraging , for example, more stringent energy efficiency requirements in the industry and housing sectors as well as for measures promoting renewable energy.

54

IV.3.3.2.2 Weak points

Target values: Not all air quality Directives contain legally binding limit values. Directive 2002/3/EC on ozone, as well as Directive *(2004/107/EC)* on heavy metals and PAHs contain only non-mandatory target values. Particularly with regard to the latter, the EEB had demanded the setting of limit values, because of the carcinogenic properties of these pollutants. Cadmium and mercury are identified as priority hazardous substances in the Water Framework Directive *(see chapter IV.5).* Therefore the objective should be to phase out all anthropogenic emissions of these carcinogens. Unfortunately the air quality Directive does not make this link – it does not refer to the Water Framework Directive nor does it set limit values or ambitious long-term objectives. The target values require Member States to attain these levels, "where possible" *(2002/3/EC, Art. 2.9).* However, if they do not reach them, political and juridical action might be limited. Furthermore, without a clear EU level playing field Member States and their authorities are vulnerable to economic pressure groups. Specifically in the case of the Directive on heavy metals and PAHs, target values are explicitly excluded from being used as relevant environmental quality standards in the permitting procedures of industrial installations under the IPPC Directive *(see chapter V.3.2).*

Local versus national: Air quality Directives focus primarily on local pollution hotspots, but the substances addressed are not only of local origin, but also transported over long distances. Local authorities charged with implementing the Directives often claim that local policies and measures are not sufficient to bring down pollutant concentrations to the required levels. Often, additional measures at national or EU-level are needed to achieve air quality standards. However, there is not necessarily co-ordination between national and local

HOTSPOT AND BACKGROUND STATIONS
PM monitoring stations are located in three different categories of places: hotspots, urban background and rural background. Hotspot monitoring stations measure ambient air quality near busy roads or other highly polluted places, whereas urban and rural background stations give the concentration for the air quality in a city or region as a whole.

air quality policies. National policy objectives particularly in the field of transport, often run counter to local air quality objectives. Therefore, when additional EU measures are negotiated in the Council, Member States – because of short-term national interests - often block or water down legislative proposals that would help them achieve air quality standards (for example the currently negotiated revision of Directive 1999/32/EC on the sulphur content of ship fuel).

Comparability: Data generated by the Directives is not always comparable. Countries have some leeway in deciding where exactly to put pollution monitors, so in some countries hotspot measurement stations would be closer to the road centre than in others[29]. Countries that have prob-

---

[29] This is the case for instance with regard to the siting criteria for traffic-oriented PM10 samplers, which currently define a minimum distance from the kerbside, but no maximum distance (1999/30/EC).

lems in meeting the limit values in these hotspots may argue that other countries which appear not to have such problems only do so because when they take measurements they do not measure closely enough to the kerbside. Countries also have some freedom on deciding how to set up their mix of stations, i.e. the ratio between hotspot and background stations varies from country to country[30]. Furthermore, the two most commonly used types of measurement equipment used in the EU to monitor PM do not produce equivalent results. To provide comparable data, the countries that use non-standard measurement methods need to demonstrate to the Commission that the results of their measurement methods are equivalent to the standard method defined in the first Daughter Directive. If they are not, they need to apply a correction factor to the results they obtained. An expert working group on implementation has produced substantive guidance on how this equivalency can be demonstrated mathematically, but so far the guidance has not been used by Member States or correction factors are still missing for some monitoring stations. All these factors mean that EU-wide PM data is not fully comparable. This is impractical for establishing EU-wide trends and forecasts, which are politically relevant particularly if Member States are or are likely to be out of compliance with the limit values.

Ecosystem protection: Siting criteria for ecosystem monitoring stations are too lax. The current criteria defining where ecosystem monitoring stations need to be installed mean that it is not possible for example to apply ecosystem limit values at Natura2000 sites that are close to a road or relatively close to an agglomeration[31]. This means such areas would only be covered by the less stringent limit values for human health and thus could be exposed to levels of air pollution that are damaging to ecosystems and plants. This is in clear contradiction to the goals of EU nature conservation and biodiversity policies. In locations where specific EU legislation for ecosystem protection exists, ambient air quality legislation should complement this. The siting criteria in the Directive should be amended, making it possible to apply the stricter ecosystem limit values in these areas.

IV.3.3.3   IMPLEMENTATION:

IV.3.3.3.1   Will the objectives be achieved?

It is difficult to assess whether the limit values of EU air quality legislation will be achieved or not, as some of these limit values will enter into force in 2005 or 2010 respectively (i.e. after the time of writing of this article). Most of the following assessment will be focussed on the first Daughter Directive, as this is where most implementation experience has already been gathered.

With regard to PM_{10}, a wealth of monitoring data has been analysed by the CAFE expert working group on particulate matter. In summer 2004, this working group produced the Second

---

[30] PM monitoring stations are located in three different categories of place: hotspots, urban background and rural background. Hotspot monitoring stations measure ambient air quality near busy roads or other highly polluted places, whereas urban and rural background stations give the concentration for the air quality in a city or region as a whole.

[31] Sampling points for analysing air quality for ecosystems or vegetation should be sited more than 20 km form agglomerations or more than 5 km from other built-up areas including motorways, see: 1999/30/EC, Annex VI.

Position Paper on Particulate Matter, which will inform the Commission in preparation for the planned revision of the first Daughter Directive. It concluded that: "without additional policies and measures, there will be widespread non-attainment" of the PM limit values in the EU[32].

In hotspots, PM limit value exceedances can be quite substantial[33]. The highest concentrations - about twice the actual 24 hour limit value - were found at traffic hotspots and near industrial installations[34]. The high concentrations near industrial installations illustrate again, how air quality protection is linked to the IPPC Directive, as PM limit values have to be respected, when site-specific permits are issued. Permits for industrial installations are one way of achieving the proscribed level of air quality (see chapter V.3.2). Diffuse PM emissions are also important in the context of heavy metals which are regulated under the fourth air quality Daughter Directive, as some of the particulate mass can consist of heavy metal compounds.

In many countries PM_{10} levels frequently exceeded even the urban background measurement sites, and some countries also have very high rural background concentrations of PM. Rural background levels for example in the Netherlands are higher than levels at traffic hotspot stations in countries such as Finland, Norway or the United Kingdom[35]. This illustrates that current PM concentration is highly differentiated across Europe. Given these substantial exceedances in many locations, it is indeed unlikely that the limit values will be met in all places in the year 2005. This is particularly worrying, as several studies have clearly demonstrated the many health hazards caused by particulate matter[36].

Regarding ground-level ozone, the third Daughter Directive on ozone 2002/3/EC has only been in force since 9 September 2003. Based on ozone monitoring data, a significant reduction in ozone concentrations has yet to be achieved. The EEA concludes: "Under current legislation and with the rate of turnover of the vehicle fleet, further reductions will gradually occur towards 2010, and further reductions may be necessary to achieve the target values of the new ozone Directive"[37].

---

[32] http://europa.eu.int/comm/environment/air/cafe/pdf/working_groups/2nd_position_paper_pm.pdf , p. 12.

[33] Out of the number of stations exceeding the limit values plus margin of tolerance in 2001, several monitoring stations reported peak levels exceeding 100 µg/m3, which is more than twice the 24h limit value for PM10. CAFE Working Group on Particulate Matter (2004): p. 60.

[34] See: CAFE Working Group on Particulate Matter (2004): p. 60. Typical examples of industrial installations with diffuse dust emissions are steel mills, non-ferrous metal smelters, open mining, cookeries, cement production, large uncovered stock piles or loading and unloading of dusty goods, see ibid. p. 72.

[35] CAFE Working Group on Particulate Matter (2004): p. 67. When referring to any of these results it is important to keep in mind that there are comparability problems of PM10 levels in Europe (see weak points in this chapter).

[36] Medina, Sylvia, Elena Blodo, Michael Saklad (2004): Health Impact Assessment of Air Pollution in 26 European Cities and Communication Strategy – Latest Findings of the Apheis Programme, http://www.umweltdaten.de/whocc/Newsletter34.pdf, WHO (2003) http://www.euro.who.int/document/e79097.pdf .

[37] EEA (2003): Air pollution by ozone in Europe in summer 2003. Overview of exceedances of EC ozone threshold values during the summer season April–August 2003 and comparisons with previous years, Report to the European Commission by the European Environment Agency, European Topic Centre on Air and Climate Change based on data provided in the framework of Council Directive 92/72/EEC on air pollution by ozone by 15 September 2003, Executive Summary http://reports.eea.eu.int/topic_report_2003_3/en/tab_content_RLR

In the summer of 2003 exceptionally long-lasting and spatially extensive episodes of high ozone concentrations occurred, mainly in the first half of August. EEA states that these episodes appear to be associated with the extraordinarily high temperatures over wide areas of Europe. This demonstrates an important link between high ozone concentrations and climate change: if climate change were to result in warmer summers in Europe, ozone levels would increase, even if current emission levels of air pollution were to remain the same[38]. In order words, climate change would aggravate the problem of ground-level ozone, causing additional health and environmental impacts *(see chapter IV.2 on Nature Protection and Biodiversity)*.

57

IV.3.3.3.2   Have plans and programmes been made?

In summer 2004, the European Commission sent first warnings to nine out of the EU-15 Member States for failure to submit plans and programmes reducing NO_2 and PM. These formal notices were sent to Austria, France, Germany, Ireland, Italy, Luxembourg, Portugal, Spain and the UK[39]. None of these countries had submitted plans and programmes to the Commission by December 2003, even though there were quite significant exceedances of the limit values in the first Daughter Directive all over Europe.

Italy has the most zones reporting exceedances of the PM_{10} and NO_2 limit values **plus margin of tolerance**. No pollution-reduction plan or programme had been notified to the Commission by July 2004. This is particularly striking, as Italy is among those countries that – according to the PM position paper – have the highest exceedances for hotspots as well as urban background concentrations.

> **MARGIN OF TOLERANCE**
> Percentage by which the limit value may be exceeded in the years before it enters into force. The aim of this margin is to indicate, in which areas Member States need to implement policies to achieve the limit values. If the margin of tolerance is exceeded, plans and programmes must be made with the aim of meeting the limit value.

In France, lengthy implementation procedures seem to have contributed to the late start in making plans and programmes. Analysis by French EEB member France Nature Environment (FNE) also shows that out of 26 agglomerations that need to make air quality management plans, at the time of writing, only one has actually implemented an air quality management plan. 19 agglomerations are still beginning to design their plan, which means that the plans will not enter into force before the end of 2004 or beginning of 2005 respectively *(Roesch and Cambou 2004)*[40]. This is obviously too late, if they are intended achieve limit values in 2005.

IV.3.3.3.3   What is the quality of the plans and programmes?

It is difficult to assess the quality of the many plans and programmes that have been made and are still being made in the context of the first Daughter Directive. According to a recent

---

[38] EEA (2003): p. V and VI.

[39] Commission press release (2004) Outdoor air quality: Commission asks nine Member States to reduce pollution, reference: IP/04/872 Date: 08/07/2004.

[40] Available at , http://www.fne.asso.fr/PA/air/doc/GuidePPA.pdf . For further info: http://www.fne.asso.fr/PA/air/dos/sp_air_A9.htm#ppa

Commission workshop on plans and programmes[41] a frequent problem seems to be the lack of quantification of air quality impacts. Air quality impacts of certain measures are rarely modelled. Given this lack of quantification, the prioritisation of measures becomes difficult. Moreover, several cities and regions in the New Member States report the planned construction of ring-roads to solve inner-city air quality problems[42]. Obviously, this is a counterproductive way of managing urban transport policy and cannot be supported from an environmental point of view.

According to German EEB member (BUND) most plans and programmes drawn up by German cities are disappointing, because they are often quite noncommittal. Measures are not co-ordinated into an overall policy with concrete measures to reduce traffic at large. Effective measures like low emission zones with restricted access for high-emission vehicles (for example for diesel cars without particulate filters) have not yet been planned by any German city. However, as far as technical measures are concerned, some cities retrofit or intend to retrofit their bus fleets with particulate matter traps. All buses in Frankfurt and Lübeck use particulate filters, as do most of the buses in Berlin and Munich use them as well[43]. FNE shares the German analysis with regard to Dunkerque, the only French city to have already implemented an air quality management plan.. The measures specified in the plan are too general and there is a lack of targeted and quantified measures (*Roesch and Cambou 2004, p. 60*).

In the UK, there are around 130 local authorities required to make air quality management plans, and their quality varies. As there are so many plans and programmes, it is difficult for environmental citizens' organisations to assess the overall quality. An assessment for the UK Department of the Environment mentions the quantification of policy impacts as one of the first problems that come to attention in a comparative analysis of plans and programmes. It shows that in some cases it is extremely difficult to estimate the effect of certain measures on pollutant concentrations, because − again − some local authorities did not include quantified data regarding the emission sources. Cost-effectiveness assessment on a local scale has also proved difficult (*DEFRA 2004*)[44].

IV.3.3.3.4   How will the Directives be enforced by the Commission?

At the time of writing, it is not clear how the Commission will respond to the likelihood that many Member States will fail to comply with limit values. However, in response to the widespread lack of action plans so far and the indications that in the plans, effective measures aimed at reducing transport growth in city centres are often lacking, the EEB is calling for a strict enforcement of the

---

[41] Implementing Air Quality Legislation in the enlarged EU: Workshop on Plans and Programmes of Air Quality and National Emission Ceiling Directives, 1-2 September 2004, Brussels, **http://europa.eu.int/comm/environment/air/cafe/general/ workshop_on_plans_programmes.htm**.

[42] Prague reported a "speed-up the construction of new elements of communication system (inner and outer circuit)" as well as prioritising their street construction "according to its significance for air protection". The Silesia region in Poland as well as the Moravian-Silesian Region in the Czech republic presented "bypass building" and "continuation of road investment" as means to comply with the PM10 limit values. See: **http://europa.eu.int/comm/environment/air/cafe/general/ workshop_plans_presentations.htm**

[43] Interview with Martin Schlegel (29.11.2004), transport campaigner BUND.

[44] Available at http://www.defra.gov.uk/environment/airquality/laqm/eval/pdf/actionplan-report.pdf , p. 19 ff.

Directive. A high level of scrutiny from the Commission is needed with regard to the question of whether plans and programmes have been made in time and if the Directive has been properly implemented in EU Member States. Furthermore, the Commission should evaluate the effectiveness of policies and measures undertaken in action plans. In particular, it should investigate if and how local air quality management is complemented by supporting policies at national level or if it is in fact counteracted by national policy priorities particularly in the field of transport policy.

So far it is clear that air quality Directives have not achieved their double objective of avoiding, preventing or reducing harmful effects on human health and the environment as a whole while at the same time maintaining ambient air quality where it is good and improving it in other cases. This is mainly due to insufficient implementation and enforcement: plans and programmes have to contain effective measures to reduce air pollution and this needs to be supported by a coherent national policy. Given the dramatic health impact of PM specifically, but also continuing problems with high ground-level ozone concentrations, more action to reduce PM and ground-level ozone is urgently needed.

IV.3.3.3.5 Room for ECO action (Environmental Citizens' Organisation)

ECO action on these Directives can take place at local, national and at EU levels. At EU level, ECOs can participate in the Clean Air For Europe (CAFE) process, where current EU clean air policy is evaluated and an integrated strategy for the future of clean air policy is formulated. At the time of writing, it appears as if the Commission is likely to revise the first Daughter Directive and to introduce new limit values for smaller dust particles ($PM_{2.5}$), which are so far not regulated by air quality Directives. ECOs should pay close attention to this revision process to ensure that the revised directive is environmentally ambitious.

At national level, focus on implementation of the existing Directives will become increasingly important. While the onus of controlling Member States' compliance with EU Directives rests with the Commission, it remains limited and its effectiveness depends largely on ECOs making the Commission aware of specific problems. As air quality legislation has good provisions for informing the public, environmental organisations, citizens' groups as well as public health organisations can use these to assess local air quality. They can act as a watchdog, demanding that these standards are respected everywhere in the EU, particularly at traffic hotspots and in areas near industrial installations. Furthermore, the plans and programmes that need to be drawn up to reduce pollution offer a unique opportunity for ECOs to promote a sustainable urban transport policy that reduces traffic, traffic congestion and pollution. Local environmental groups can critically follow the development of these action plans and insist that the measures selected in them are sound from a longer-term sustainability perspective and that they are actually put into practice. This is important to ensure that these Directives are implemented in a meaningful way, leading to improved environmental and health conditions and a better quality of life in cities.

IV.3.4   National emission ceilings for certain atmospheric pollutants [45]

The Directive 2001/81/EC on national emission ceilings (NECs) covers four air pollutants, namely sulphur dioxide (SO_2), nitrogen oxides (NOx), volatile organic compounds (VOCs), and ammonia (NH_3). It sets long-term environmental and health objectives with regard to air pollution in the European Union as well as interim environmental objectives to be achieved by 2010 *(see chapter IV.3.4.1)*[46]. These objectives are strong political commitments defining the environmental quality objectives in the field of air pollution.

Air quality legislation and emission legislation are complementary. Air quality Directives set the minimum standards to be achieved in ambient air everywhere in the union (*see chapter IV.3.3*). Emission legislation tries to provide the tools to attain the reductions necessary to meet those standards as well as to achieve   long-term environmental quality objectives. This is a key Directive, both for defining long-term and interim environmental objectives as well as for defining country-by-country emission ceilings, which is the most important legally binding tool for meeting these objectives. Emission ceilings are also crucial for attaining the EU air quality standards for a number of pollutants, including SO_2, NO_2, fine particles (PM_{10}), and ozone.

IV.3.4.1   INTERIM OBJECTIVES AND EMISSION CEILINGS

The interim environmental objectives specified in this Directive are:

▶ **Acidification.** The areas where critical loads are exceeded to be reduced by at least 50% in all areas compared with the 1990 situation.

▶ **Health-related ozone exposure.** Ground-level ozone above the critical level for health to be reduced by two-thirds in all areas compared with the 1990 situation.. In addition, ground-level ozone load should not exceed a given absolute limit anywhere.

▶ **Vegetation-related ozone exposure.** Ground-level ozone above the critical level for vegetation to be reduced by one-third in all areas compared with the 1990 situation.. In addition, ground-level ozone load should not exceed a given absolute limit anywhere.

These interim objectives should be met by 2010. The Directive further specifies country-by country emission ceilings to be attained by 2010. The purpose of the emission ceilings is *"to meet broadly"* interim environmental objectives, given in Article 5 of the Directive. Member States must limit their annual national emissions so that these do not exceed ceilings laid down in Annex 1 of the Directive (*see Table 2 at the end of chapter IV.3.4.6*), and they must ensure that these emission ceilings are not exceeded in any year after 2010.

---

[45] This section of the text is largely reproduced from Environmental Factsheet No. 16, June 2004 by The Swedish NGO Secretariat on Acid Rain, see: http://www.acidrain.org/pages/publications/factsheet/factsheet16.pdf.

[46] To "move towards the long-term objectives of not exceeding critical levels and loads and of effective protection of all people against recognised health risks from air pollution", *2001/81/EC, Art. 1*. The long-term objective should be achieved "preferably by 2020", *Art 10.5 (b)*.

After the adoption of the NEC Directive, national emission ceilings for 2010 have also been agreed with the new Member States. These NECs are established in the accession treaties between the EU and each acceding country, and presented in Table 4 at the end of chapter IV.3.4.6.

IV.3.4.2 PROGRAMMES AND REPORTING

The Directive prescribes that by October 2002 at the latest, Member States were to have drawn up programmes for the progressive reduction of national emissions of the four pollutants and to have reported them to the Commission. These reports shall provide information on measures and action taken at national level to attain emission ceilings. National programmes are to be updated and revised by 1 October 2006 and Member States are obliged to make this information available to the public[47]. Member States must also make annual reports of their national emission inventories, and their emission projections for 2010, to the Commission. Methodologies for emission inventories and projections are specified in the Directive.

IV.3.4.3 REVIEW AND REVISION

Based on, among other elements, the information from Member States, the Commission was to report to the European Parliament and the Council in 2004 on progress made in the implementation of the national emission ceilings, on the extent to which interim environmental objectives are likely to be met by 2010, and on the extent to which long-term objectives could be met by 2020. It will have to report again in 2008.

The Directive's review was to be completed in 2004, including an evaluation of the indicative emission ceilings for the Community as a whole *(see next paragraph)*, and consideration of further cost-effective actions that might be taken in order to reduce emissions with the aim of attaining interim environmental objectives by 2010. The review report has not yet been delivered yet, but was to be part of the Thematic Strategy on air pollution to be published in July 2005. Furthermore, the Commission has launched a contract to evaluate national plans of the Member States, to provide technical input to the revision.[48]

The Commission intends to prepare and adopt by mid-2006 a legislative proposal to revise the national emission ceilings. This revision will build upon the work performed in the context of the Clean Air for Europe Programme and the Thematic Strategy on air pollution. The Commission may also propose *"further emission reductions with the aim of meeting, preferably by 2020, the long-term objectives"*.

---

[47] The programmes are available under: http://europa.eu.int/comm/environment/air/nationalprogr_dir200181.htm .

[48] The final reports are available at the Commission's website at: http://europa.eu.int/comm/environment/air/necr.htm

IV.3.4.4   ASSESSMENT OF THE NEC DIRECTIVE

IV.3.4.4.1   Systematic overestimation of costs

In essence the methodology used when developing the Directive is intended to attain agreed objectives for improving environmental and health protection. It should also bring about an equal relative environmental improvement everywhere in the EU, while at the same time ensuring extraordinary improvements in the worst affected areas.

A computer model for integrated assessment was used to carry out a so-called joint optimisation to find the most cost-effective way, for the EU as a whole, of achieving the environmental aims expressed in the NEC Directive. This enabled the Commission to propose differentiated national emission ceilings, which largely reflects the polluter-pays principle and should maximize the environmental benefits of emission reductions[49].

A drawback of this methodology is that it tends to overestimate the costs for emission reductions. The reason is partly that only technical emission abatement measures have been considered, no account having been taken of structural measures such as switching fuels from coal to gas, increasing energy efficiency, greater use of alternative energy sources, and changes in the transportation and agricultural sectors. Emissions could be reduced at much lower cost through some of these structural changes rather than by relying solely on technical end-of-pipe solutions.

Furthermore, a highly doubtful energy scenario has been used in the computer modelling. This is largely based on information submitted by the individual Member States, and would imply an *increase* in the EU emissions of carbon dioxide by about 8 per cent by 2010. Such an increase is in absolute disregard of the commitments made by the EU and its member countries under the Kyoto protocol, involving a reduction of 8 per cent in EU emissions of greenhouse gases (of which carbon dioxide is the most important). A computer model run simulating a low-CO_2 scenario that would roughly accord with the Kyoto agreement brought the extra cost down by more than 40 per cent.

IV.3.4.4.2   Benefits to health and the environment

The area of ecosystems where the depositions of acidifying air pollutants exceed critical loads will be diminished as a result of the Directive. There will also be reductions in the exposure to damaging levels of ozone, both for people and vegetation. By lowering the emissions of SO_2 and NOx, the Directive will help reduce exposure to health-damaging fine particles (PM_{10} and $PM_{2.5}$), as these two pollutants act as precursors to secondarily formed sulphate and nitrate particles. Thus the Directive will contribute to achieving air quality limit and target values for SO_2, NO_2, PM_{10} and ozone. Although no interim targets have been set for eutrophication, improvements can nevertheless be expected as result of lower emissions of NOx and ammonia. However, in all cases significant further reductions in emissions are needed in order to attain the long-term objectives for protecting health and the environment.

---

[49] A similar approach is now also being used in the Clean Air for Europe Programme (CAFE) generating the background data for the upcoming Thematic Strategy for Clean Air, which will outline the EU's priorities in air pollution policy until the year 2020.

The Commission also made an analysis of the quantifiable financial gains to be made from reducing emissions.. Account was taken chiefly of the effects on human health (morbidity and mortality), on farm crops and modern buildings and materials. Calculations showed the gains to be significant, and that the economically quantifiable benefits significantly outweighed the estimated costs. It should however be noted that a number of gains were not included, such as direct health effects of NO_2 and VOCs, less acidification of soil and water, less eutrophication, reduced effects on biological diversity, less long-term effect on forest productivity, and reduced damage to historical monuments.

63

IV.3.4.4.3 Level of ambition too low

A weakness of the Directive is that the country-by-country emission ceilings are not strong enough. Current NECs in the Directive will fail to reach even agreed interim environmental objectives for 2010 and will certainly not attain long-term objectives by 2020. This is because during the negotiations of the Directive, a political compromise between the Council and the Parliament was reached, which resulted in less demanding emission ceilings.

Initially the ceilings proposed by the Commission were relatively strict. These ceilings were largely also supported by the Parliament, but were firmly rejected by the Council. Another result of this compromise was that in the current Directive there is no strong legal link between the emission ceilings and interim objectives. The emission ceilings are only required to "meet broadly" interim objectives by 2010. In order to illustrate the gap between country-by-country emission ceilings and what needs to be achieved to meet interim environmental objectives, the Directive also contains so-called indicative emission ceilings (*set out in Annex II*). These are set for the EU as a whole (i.e. not for each Member State), and reflect the estimated emission reductions needed EU-wide to meet interim targets (*see Table 3 chapter IV.3.4.6*). In any case it is obvious that the attainment of long-term objectives will require significant further reductions in emissions of all four pollutants.

IV.3.4.5 FUTURE DEVELOPMENTS

The Directive was scheduled for review and revision by 2004. The revision would have been an opportunity both to strengthen the NECs for 2010, for setting new NECs for later target years (e.g. 2015 and/or 2020), and for deciding on a date for the attainment of the long-term environmental objectives.

As mentioned earlier, the first review and revision will be delayed by over a year, the reason being that the analysis and evaluation was to be co-ordinated with the ongoing Clean Air For Europe (CAFE) programme, initiated by the Commission in 2001[50]. The CAFE programme will result in a Thematic Strategy on air pollution due to be presented by the Commission by July 2005. The strategy is to be accompanied by proposals for revised and/or new Directives relating to air pollution. The review of the NEC-Directive will be one part of the Thematic Strategy, but legislative proposal revising the NEC-Directive and setting new national emission ceilings will be postponed until mid-2006. Current developments under CAFE indicate that the NEC-Directive may be extended to include national emission ceilings for fine particles (PM_{10} or $PM_{2.5}$, or both).

---

[50] For further information, see: http://europa.eu.int/comm/environment/air/cafe/index.htm

IV.3.4.6 ROOM FOR ECO ACTION (ENVIRONMENTAL CITIZENS' ORGA-NISATION)

As significant further reductions in emissions of all four pollutants are needed to attain the long-term objectives and the ambient air quality standards described in chapter IV.3.3, it is important that ECOs keep monitoring the implementation of this Directive as well as acting to influence new related policy developments. One possible way of influencing EU policy development in this context is for ECOs to become more engaged and involved in the CAFE programme. This could take place both at the national level (Member States provide input data to the analysis, as well as national views and opinions) and at the EU-level (by participating in CAFE working groups, evaluate ongoing work and provide opinions).

As regards the Member States' reporting of national programmes, only a minority of Member States had actually prepared and reported such programmes in accordance with the Directive obligations. Here, national organisations could play an important role by exerting national pressure on their governments. If prepared in accordance with the obligations, these programmes could provide useful information not only on future emission levels, but also on national forecasts regarding future activity levels in the sectors of energy, transport, industry, and agriculture. Moreover, if Member States produce and disseminate this type of information properly, the likelihood of compliance with other air quality legislation, such as the EU air quality standards, could be better evaluated.

Table 2: National emission ceilings for SO_2, NOx, VOCs and NH_3, to be attained by 2010 for EU15 Member States (kilotonnes).

COUNTRY	SO2	NOX	VOCS	NH3
Austria	39	103	159	66
Belgium	99	176	139	74
Denmark	55	127	85	69
Finland	110	170	130	31
France	375	810	1050	780
Germany	520	1051	995	550
Greece	523	344	261	73
Ireland	42	65	55	116
Italy	475	990	1159	419
Luxembourg	4	11	9	7
Netherlands	50	260	185	128
Portugal	160	250	180	90
Spain	746	847	662	353
Sweden	67	148	241	57
UK	585	1167	1200	297
EU15	*3850*	*6519*	*6510*	*3110*

Table 3: Indicative EU-wide emission ceilings for SO_2, NOx and VOCs (kilotonnes).

	SO2	NOX	VOCS
EU15	3634	5923	5581

Table 4: National emission ceilings for SO_2, NOx, VOCs and NH_3, to be attained by 2010 for the new Member States[i] and accession candidate[ii] countries (kilotonnes).

COUNTRY	SO2	NOX	VOCS	NH3
Bulgaria	856	266	185	108
Czech Republic	265	286	220	80
Cyprus	39	23	14	9
Estonia	100	60	49	29
Hungary	500	198	137	90
Latria	101	61	136	44
Lithuania	145	110	92	84
Malta	9	8	12	3
Poland	1397	879	800	468
Romania	918	437	523	210
Slovakia	110	130	140	39
Slovenia	27	45	40	20

[i] The NECs for the new Member States are not given in the NEC Directive (2001/81/EC), but in the accession treaty for each country.

[ii] The NECs for the two accession candidate countries Bulgaria and Romania have not yet been established. Therefore, the figures given in this table for these two countries are taken from the 1999 Gothenburg Protocol.

IV.3.5   Road vehicles

IV.3.5.1   WHY EMISSION LEGISLATION FOR ROAD VEHICLES?

In 2001, road traffic in the EU15 was responsible for nearly half of all emissions of nitrogen oxides (NOx) and one third of those of volatile organic compounds (VOCs). It also contributed significantly to the emissions of fine particulates (PM) and of the greenhouse gas, carbon dioxide (CO_2). Reducing vehicle emissions is therefore a crucial component of any strategy to combat pollution from PM and ozone, as well as climate change.

Emission requirements for light road vehicles have existed in the EU since the early 70s, while the first emission standards for heavy vehicles came at the end of the 80s. As a result of this legislation, over the years the air pollution rate per vehicle has decreased considerably. A new car or truck today may emit some 80-90 per cent less air pollutants, as compared to a vehicle produced in the 70s.

It is however important to note that while technological improvements have lead to significant improvements in the emission rates from new road vehicles, transport volumes in the EU have grown steadily, at about the same rate as the economy or above. Over some ten years since the early 90s, passenger transport has grown by almost 20% and freight transport by about 30% (*EEA 2004*). Furthermore, progress made with diesel cars was much slower than with petrol cars, where the introduction of the catalytic converter as exhaust gas treatment brought about the main improvements. Diesel engines have only been treated by an internal optimisation of the engine. This helps explain why, during the same time period, emissions of regulated air pollutants from the transport sector were reduced only by 24 to 35% (international aviation and marine shipping not included) (*EEA 2004*). In spite of these reductions in air pollutant emissions from road transport, serious air quality problems still exist, especially in urban areas.

Consequently, significant further emission reductions from motor vehicles are necessary for the EU to reach its environmental and health objectives as outlined in the two previous subchapters. Technical standards, which apply to all new vehicles in the EU, offer the possibility of achieving across-the-board emission cuts and can help to bring down background concentrations of PM and ground-level ozone. They are thus important measures that can help local authorities to achieve the air quality limit values (*see chapter IV.3.3.3*). Emission limits for vehicles have to be EU-wide, as internal market regulations require joint Community standards so that vehicles, like any other product, can circulate freely in the internal market of the European Union.

IV.3.5.2   SOME ASPECTS OF REGULATING VEHICLE EMISSIONS

The specific properties of different types of engine and fuel result in different emissions of the various pollutants (carbon monoxide, nitrogen oxides, particulate matter, hydrocarbons an carbon dioxide). Diesel-driven vehicles for example emit much more PM and NOx, but less CO_2 than petrol-driven ones. Moreover, the quality of the fuel plays an important role in determining pollutant emissions and also influences the functioning of some exhaust gas treatment equipment (*see also chapter V.4.7*). Lead in gasoline, for example poisons catalytic converters, which was one reason behind the introduction of lead-free petrol in the EU[51].

To test if vehicles comply with emission limit values, standard testing procedures are used -so-called test-cycles. They aim at creating repeatable emission measurement conditions and, at the same time, simulate real driving conditions. Typically any new engine model has to be emission certified before it is released onto the market.

---

[51] For a good and more detailed explanation of the different aspects of regulating vehicle emissions, also see: Taschner, Karola Dr. (1998): Auto-Oil I and II, in: EEB (1998): EEB Industry Handbook, Brussels, **http://www.eeb.org/publication/ INDUSTRYHANDBOOK.pdf.**

The reliability and credibility of the test procedures are important to ensure that all new vehicles comply with the legally binding emission limit values, but also to ensure that the vehicles maintain low emission rates in real-life driving conditions. The different test-cycles for cars and heavy-duty vehicles have come under critique, as they do not adequately reflect real-life driving conditions. This means that in reality the emissions of the existing car fleet are likely to be higher than calculated when the emission standards were set. ECOs have therefore argued that the current test-cycles should be reviewed and revised. Different test-cycles are used in the EU, North America and Asia, and emission standards from different regions may not always be directly comparable.

Furthermore, it is also important to ensure that vehicles maintain low emission rates, even after several years of use. Therefore durability requirements as well as in-use testing durability testing, road worthiness tests and on-board diagnostic systems are also important.

IV.3.5.3   CURRENT LEGISLATION

Emission standards for cars and light-duty vehicles are referred to as EURO 1-5 standards (the Commission is currently working on a proposal for EURO 5). The most important Directive in this field is Directive 70/220/EC, which has been amended a number of times[52]. Amendments include:

▶ Euro 1 standards (also known as EC 93): Directives 91/441/EEC (passenger cars only) or 93/59/EEC (passenger cars and light trucks)

▶ Euro 2 standards (*EC 96*): Directives 94/12/EC or 96/69/EC

▶ Euro 3/4 standards (*2000/2005*): Directive 98/69/EC as well as further amendments in 2002/80/EC

Table 5. EU emission standards for passenger cars, including UBA (Federal Environment Agency, Germany) proposal for 2008[53].

PASSENGER CARS	PM (MG/KM)		NOX (G/KM)		HC (G/KM)		HC+NOX (G/KM)	
	diesel	petrol	diesel	petrol	diesel	petrol	diesel	Petrol
Euro 1 (1992-93)	140	-	-	-	-	-	0.97	0.97
Euro 2 (1996)	80/100[i]	-	-	-	-	-	0.7/0.9[i]	0.5
Euro 3 (2000)	50	-	0.50	0.15	-	0.20	0.56	-
Euro 4 (2005)	25	25	0.25	0.08	-	0.10	0.30	-
Euro 5 – UBA Proposal (2008)	2.5	2.5	0.08	0.08	0.05	0.05	-	-

[i] Indirect Injection (IDI) and Direct Injection (DI) engines respectively.   *Source: ACID NEWS No.3, September 2004*

---

[52] For a list of all amendments, see: http://europa.eu.int/comm/enterprise/automotive/Directives/vehicles/dir70_220_cee.html .

[53] These standards currently regulate four groups of compounds: nitrogen oxides (NOx), hydrocarbons (HC), carbon monoxide (CO) and particulate matter (PM). Of these, carbon monoxide is less significant from the point of view of health and the environment.

The EURO 3 and EURO 4 standards were elaborated in the context of the Auto-Oil programme, in which information on the abatement potential and costs of vehicle technology and corresponding fuels were analysed. The data input for this programme was provided by the European Motor Industry Federations(ACEA) and the mineral oil industry (EUROPIA), as well as by a consultant[54].

EURO standards have also been set for heavy-duty vehicles. These are referred to with roman numbers, EURO I – V. The most important Directive is the Heavy Duty Diesel emissions Directive 88/77/EEC, which has subsequently been amended several times[55].

Table 6. EU emission standards for heavy vehicles and UBA (Federal Environment Agency, Germany) proposals (no EU legal obligation) for 2008 and 2010.

HEAVY DUTY VEHICLES	NOX (G/KWH)	(HC) G/KWH	PM (MG//KWH)
EURO I ('92 – '93)	9.0	1.23	400
EURO II ('95 – '96)	7.0	1.1	150
EURO III (2000)	5.0[i]	0.66[ii]	100/160[iii]
EURO IV (2005)	3.5[i]	0.46[ii]	20/30[iii]
EURO V (2008)	2.0[i]	0.46[ii]	20/30[iii]
EURO V UBA PROPOSAL (2008)	1.0[i]	0.46[ii]	2/3[iii]
EURO VI (UBA PROPOSAL 2010)	0.05[i]	0.46[ii]	2/3[iii]

[i] Both ESC and ETC test cycle.   [ii] ESC test cycle only.   [iii] ESC and ETC test cycle respectively.

It is expected that many engine manufacturers will have to fit heavy duty diesel vehicles with both particulate filters and NOx reduction technology to meet EURO IV requirements. However, some manufacturers are now able to meet the limit values without further exhaust gas treatment.

IV.3.5.4   PLANNED LEGISLATION[56]

A review of current emission standards for road vehicles in the EU began in autumn 2003. This work is being carried out by a subgroup of the Commission's Motor Vehicle Emissions Group (MVEG), with the participation of the member countries and various stakeholders.

---

[54] Environmentalists, consumer organisations, motoring and other citizen's organisations have raised concerns about the Auto-Oil process being intransparent, see: Taschner, Karola Dr. (1998): Auto-Oil I and II, in: EEB (1998): EEB Industry Handbook, Brussels, http://www.eeb.org/publication/1998/INDUSTRYHANDBOOK.pdf, p. 52-62. On costs of technology in the Auto–Oil programme see also: Stockholm Environment Institute (1999), Costs and strategies presented by indus- try during the negotiation of environmental regulations, Stockholm Environment Institute, Stockholm, Sweden, http://www.york.ac.uk/inst/sei/pubs/ministry.pdf, p. 19-23. On Auto-Oil in general, see: http://europa.eu.int/comm/ environment/air/autooil.htm.

[55] For a list of all amendments, see: http://europa.eu.int/comm/enterprise/automotive/Directives/vehicles/ dir88_77_cee.html.

[56] This section of the text is largely reproduced from Environmental Factsheet No. 17, September 2004 by The Swedish NGO Secretariat on Acid Rain, see: http://www.acidrain.org/factsheet17.pdf

On the basis of this work the Commission will present a Directive containing new standards. The proposed Directives for light and heavy vehicles are expected to be issued in spring and autumn 2005, respectively. The development of new technology in recent years, combined with new findings regarding harmful health effects, especially of PM, makes it likely that the Commission will propose significant strengthening of emission limit values, primarily for diesel vehicles.

69

In 2003, the German Federal Environment Agency (UBA) published a proposal for new emission standards for motor vehicles (*German Environment Agency 2003*)[57].

For passenger cars the UBA proposal include the following (*see Table 5, above*):

▶ Emission requirements should be fuel neutral, i.e. the same for all fuels.

▶ Emission limit values for PM should be strengthened by a factor of ten. This is likely to require the application of particulate filters, which can remove 90 per cent or more of particulates in the entire size range (*The Danish Ecological Council 2004*)[58]. The current Euro 4 standards for diesel cars can be met without such filters, at least by small cars. If petrol vehicles are also covered by the proposed new PM requirement it may mean that direct injection engines will have to be fitted with particulate filters.

▶ The NOx requirement for diesel cars should be strengthened by a factor of three, down to the same level as for petrol vehicles.

▶ The summation value for NOx + HC for diesel cars should be replaced with an HC limit value regardless of engine type.

The UBA proposal for heavy duty vehicles means (see Table 6, above):

▶ Fuel-neutral requirements.

▶ Agreed, but as yet, indicative PM standards for 2008 are lowered by a factor of ten, which is likely to require the application of particulate filters.

▶ Agreed, but as yet indicative, NOx requirements for 2008 are halved, and then halved again in 2010.

In its report, the UBA discusses whether emissions of particulates should also be counted by number, or whether simply regulating the weight would suffice.. The authors conclude that confining the limit to weight could lead to the engine makers concentrating primarily on eliminating the largest and heaviest particles, which have relatively little effect on health. They would therefore like to supplement the current weight-based standards with limits on the maximum number of particles within the size range that is inimical to health.

The extra cost for a diesel car to meet UBA EURO 5 standard proposals - compared with EURO 4 - is estimated to run to 200–400 euros. It would cost practically nothing, on the other hand, for a

---

[57] The report of the German Environment Agency can be downloaded from http://www.umweltdaten.de/uba-info-presse/hintergrund/FutureDiesel_e.pdf.

[58] On health effect of diesel particles and particulate filters, see the Danish Ecological Council at: http://www.ecocouncil.dk/download/dieselpjece_eng.pdf.

heavy vehicle to move from EURO V to EURO VI, since it would be enough in that case to improve the emission control equipment that is already needed to meet EURO V requirements.

The need to reach a relatively quick agreement on exhaust emission requirements is important not only to allow industry the time to prepare for the production of cleaner vehicles, but also to give member countries an opportunity to introduce tax incentives to favour vehicles that comply with the requirements early -such as diesel cars fitted with particulate filters.

IV.3.5.5 OUTLOOK – ROOM FOR ECO ACTION (ENVIRONMENTAL CITIZENS' ORGANISATION)

The EEB, together with the European Federation for Transport & Environment (T&E), has been calling on EU Member States and the Commission to support UBA proposals for strengthened emission standards as well as demanding that the introduction date be brought forward to 2008[59]. At the time of writing, technology that would easily meet the proposed emission limit values is already commercially available so there is no reason to delay widespread introduction until 2010.

Environmental citizens' organisations in several countries have been campaigning for the introduction of particulate filters for diesel vehicles as well as for national fiscal measures favouring earlier introduction of cars equipped with such filters. It is important that nationals continue to demand that their governments support ambitious new limit values both for PM and NOx.

With regard to urban air quality, the concept of "environmental zones" will probably gain more importance in the future. In the context of air quality plans and programmes *(see chapter IV.3.3.3)*, local authorities can create low-emission zones in urban areas. In these zones, it is possible to restrict the access of highly polluting vehicles. ECOs can call for the establishment of more low-emission zones in urban areas, with restricted access for diesel vehicles not equipped with particulate filters. This would put pressure on car manufacturers to provide incentives to retrofit existing vehicles.

IV.3.6 Large Combustion Plants

IV.3.6.1 WHY A DIRECTIVE FOR LARGE COMBUSTION PLANTS (LCPS)?

The Large Combustion Plant (LCP) Directive *(2001/80/EC)* applies to combustion plants with a thermal capacity greater than or equal to 50 megawatts (MW)[60]. The LCP-sector is dominated by power plants, but also includes other industrial combustion plants in sectors such as iron and

---

[59] See ECO letter to the Environment Council in October 04: http://www.t-e.nu/docs/Positionpapers/2004/2004-10-11_ngo_input_envi_council.pdf.

[60] The 50 MW thermal is related to energy input, not output, i.e. it is based on the amount of fuel the plants burn, not on the amount of (useful) energy they produce. Furthermore, the emission limit values in the Directive are set as milligrams of pollutant per cubic metre of air (mg/m³), which again has no relation to the amount of useful energy produced. ECOs have long been calling for changing this to unit (e.g. grams) of pollutant per gigajoule (g/GJ) of useful energy produced, as this would help to promote energy efficiency.

steel production and petroleum refineries. It is important to regulate emissions from these plants because they are the EU's largest source of SO_2 emissions, as well as contributing significantly to NOx and particulate matter emissions. According to the European Pollutant Emissions Register (EPER) these combustion installations contributed to 69 % of all sulphur emissions, 64 % of nitrogen oxide emissions and 41 % of particulate matter emissions in all sectors covered by EPER in 2001 (*for EU 15*)[61].

Substantial reductions in emissions from LCPs are necessary in order for the EU to meet its international and internal environmental aims, as laid down in the 1999 Gothenburg Protocol to the Convention on Long-range Transboundary Air Pollution, the EU's Fifth and Sixth Environmental Action Programmes. According to a recent study analysing the largest point sources of air pollutant emissions in Europe, in 2001 the 100 largest point sources of SO_2-emissions were still emitting 7.1 million tonnes of SO_2 a year, corresponding to 43 per cent of the total of 16.7 million tonnes from all sources on land in Europe (*Barrett 2004*)[62].

IV.3.6.2   REQUIREMENTS AND EMISSION STANDARDS

The first EU Directive on Large Combustion Plants entered into force in 1988 (*88/609/EEC*). It set emission limit values for plants built after 1987 (so-called new plants) as well as country-by-country ceilings for step-wise national reductions in SO_2- and NOx-emissions from plants built before that date (so-called existing plants). However, the emission reductions required by this Directive were not ambitious and did not at all reflect what technology could deliver at that time. In October 2001, it was replaced by the second LCP Directive (*2001/80/EC*), setting stricter standards for some categories of plant and including more plants within the scope of the Directive.

The Directive sets emission limit values for sulphur dioxide, nitrogen oxides and dust. These limit values are minimum standards, which mean that Member States are free to adopt emission limit values and compliance deadlines which are stricter than those of the Directive, as well as including other pollutants and laying down additional requirements. The Directive sets emission limit values for three categories of plants:

1) plants licensed before July 1987 (so-called existing installations)

2) plants licensed between July 1987 and November 2003 (so-called "old" new installations)

3) plants licensed after November 2002 (so-called "new" new installations)

The limit values vary according to the age and capacity of the plants, as well as the type of fuel they burn (see Tables 7 and 8).

---

[61] See: European Pollutant Emissions Register: http://www.eper.cec.eu.int. EPER holds emissions data on around 10,000 large and medium-sized industrial plants, which are listed in Annex I of the IPPC Directive and which exceed specified emission thresholds. This means the percentage figures given here reflect the share of LCP emissions from all installations covered by EPER, not from all sources in the EU.

[62] Available at http://www.acidrain.org/apc17.pdf.

Table 7: Emission limit values for SO$_2$ and NOx from plants licensed after November 2002 (mg/m^3)

	SO2			Nox		
Plant size (MW$_{th}$)	50-100	100-300	>300	50-100	100-300	>300
Solid fuels [i]	850	200	200	400	200	200
Liquid fuels	850	400-200 [iii]	200	400	200	200
Biomass	200	200	200	400	300	200
Natural gas [ii]	35	35	35	150	150	100

[i] NB. Where the emission limit values for SO$_2$ cannot be met due to the characteristics of the fuel, installations smaller than 300 MW$_{th}$ must either limit their emission levels to 300 mg SO$_2$/m^3, or achieve a desulphurisation rate of at least 92%. Larger plants must achieve a desulphurisation rate of at least 95% and limit emission levels to 400 mg SO$_2$/m^3

[ii] Specifically for gas turbines using natural gas, the limit value in most cases being 50 mg NOx/m^3.

[iii] Linear decrease

Source: Acid News no. 3/2001

Table 8: Emission limit values to be applied from 1 January, 2008 for SO$_2$ and NOx from existing plants (licensed before November 2002). Plant size in MW$_{th}$ and emission limits in mg/m^3.

	SO2			Nox	
Plant size	50-100	100-500	>500	50-500	>500
Solid fuels	2000 [i]	2000-400 [i,ii]	400[1]	600	500 [iii]
Plant size	50-300	300-500	>500	50-500	>500
Liquid fuels	1700	1700-400 [ii]	400	450	400
Plant size	>50	50-500	>500		
Natural gas	35	300	200		

[i] NB: Where the emission limits for SO$_2$ cannot be met due to the characteristics of the fuel, various rates of desulphurisation should be achieved (from 60 to 94%) - with the highest rate applicable for plants greater than 500 MW$_{th}$.

[ii] Linear decrease

[iii] From 1 January 2016 the emission limit value will be 200 mgNOx/m^3.

Source: Acid News no. 3/2001

This LCP Directive has important links with the IPPC Directive. The LCP Directive sets mandatory emission limit values for air pollutants, that have to be respected in the permitting procedure. The IPPC Directive also takes account of other environmental impacts of large combustion plants and enables permitting authorities to set additional obligations for individual plants *(see chapter V.3.2 on IPPC).*

IV.3.6.3   ASSESSMENT

IV.3.6.3.1   Directive applies to existing plans

The most important innovation of the new LCP Directive is that it sets emission limit values for *existing* plants, which was not the case in the 88/609/EC Directive, in which plants built before 1988 were exempt from such limit values. Under the new Directive, existing plants are subject to emission limit values as from 2008.

It was largely thanks to the efforts of the European Parliament as well as active lobbying by ECOs that the terms of the Directive now also include emission standards for pre-1987 plants. Additional analysis undertaken for the Commission in 2001 clearly showed that if no action were taken, existing (pre-1987) large combustion plants would remain important sources of SO_2- and NOx-emissions in the year 2010 (*Cofala and Amann 2001*)[63]. It was estimated that by applying the same emission limit values to the pre-1987 plants as those in Directive 88/609/EC for post-1987 plants, SO_2 emissions from these sources in the EU15 would be reduced by almost 70 per cent and NOx emissions by about nine per cent. In accession countries SO_2-emissions from existing LCPs would be 86 per cent lower and NOx emissions would be cut by eight per cent[64].

In addition to prescribing measures for pre-1987 plants, the new LCP-Directive also sets tighter emission requirements for new plants, i.e. those licensed after November 2002. In this context it should be noted that the emission limit values also for these post-2002 plants are not very strict, especially regarding the NOx limit values (*see chapter IV.3.6.5*).

IV.3.6.3.2   Country-wide "bubbles"

A weak point of the Directive is that it opens up the possibility for Member States to combine their emissions from existing LCPs in country-wide bubbles. This means that instead of applying emission limit values to each individual plant, Member States are allowed to reach the equal over-all reductions from existing plants via a so-called national plan for the country as a whole. National plans have to ensure that the overall total emission reductions achieved are the same as what would have been achieved by applying ELVs to individual plants.. This procedure thus allows highly polluting plants to exceed the emission limit values, provided this is compensated by additional emission reductions in other plants within the national bubble – which obviously may result in negative impacts on local air quality. Currently, eight Member States have submitted national plans under the LCP directive: the Czech Republic, Finland, France, Greece, Ireland, Netherlands, Slovenia, UK. However, Slovenia subsequently withdrew its plan and opted for the ELV approach.[65]

---

[63] Available at http://europa.eu.int/comm/environment/pollutants/combustion_report.pdf

[64] Ibid.

[65] Entec UK Ltd (2005): Preparation of the review relating to the Large Combustion Plant Directive – a report for the European Commission, DG Environment; Draft Final Report; April 2005; Entec UK Ltd

73

IV.3.6.3.3 Exemptions and derogations

Another weakness of the Directive is the possibility of avoiding emission limit values altogether for plants which will not be in operation for more than 20,000 hours between January 2008 and 2015. This means that the oldest and most inefficient plants – which are also usually the most polluting ones – are allowed to remain in operation until 2015 without restrictions on their emissions, provided they do not exceed the total amount of operating hours.

Furthermore, the Directive provides "relaxed" emission limit values for LCPs operating only during so-called peak-loads. Such relaxed limit values are allowed for peak-load plants that will be in operation for less than 2000 hours annually until 2015, and for less than 1500 hours annually as from 2016. As 2000 hours equals nearly three months of operation, this means in practice that old, inefficient and highly polluting plants are allowed to continue to operate at seasonal load, while still emitting up to four times as much pollutants as other LCPs.

IV.3.6.4   REVIEW

According to the Directive, the Commission was to have submitted a review report to the European Parliament by the end of 2004, which could be followed up with a proposal for revision. In the review it was to investigate possibilities for further emission reductions, analysing among other things the need for further measures, costs and advantages of further emission reductions in the power plant sector compared to other sectors, and the technical and economical feasibility of further emission reductions. It was to have included "as appropriate" a proposal of possible end dates or lower emission limit values for one of the derogations relating to the NOx limit values. Furthermore, the Commission is required to analyse the national plans provided by the Member States *(2001/80/EC, Art. 4.7)* However the Commission has not yet published a review report.

At the time of writing the Commission has appointed external consultants to support the review of the Directive, by providing additional data on large combustion plant emissions, abatement measures and costs, and addressing the points mentioned above. The final consultant report has not yet been published.[66] The Commission is likely to base its review report on the consultants' findings as well as on the results of the cost-effectiveness analysis conducted in the context of the Clean Air For Europe programme *(see chapter IV.3.1)*.

IV.3.6.5   OUTLOOK AND ROOM FOR ECO ACTION (ENVIRONMENTAL CITIZENS' ORGANISATION)

The Directive will result in further emission reductions from the LCP sector, especially regarding SO_2-emissions from old (pre-1987) plants, but these reductions will in most cases take place only after January 2008, the deadline set for these plants in the Directive. Expected NOx reductions from old plants are likely to be less significant, since strict NOx-standards for pre-1987 and pre-2002 plants will only apply from 2016 onwards.

---

[66] An interim report can be found under: http://europa.eu.int/comm/environment/air/future_stationary.htm.

However, the key problem with the Directive is that several derogations and late deadlines have meant that some old, inefficient and highly polluting plants will be allowed to remain in operation more or less unabated for many years.. A recent study, analysing the emissions from large point sources in Europe, found that around 90 per cent of the emissions of SO_2 from large coal-fired plants come from those that were commissioned before 1987 (Barrett 2004). These plants should be replaced or upgraded urgently not only because they cause air pollution, but also in order to combat climate change, as they are not only very polluting, but in many cases also very inefficient.

Another problem with the emission limit values set in this Directive is that they are technology-conserving, rather than technology-forcing. Emission limit values are set at levels that could be met with technologies that were commercially available by the time the Directive was drafted in the second half of the 1990s. Moreover, some of these limit values, i.e. those for NOx, will not come into force until 2016 - fifteen years after the Directive was agreed. By comparison, emission limit values for road vehicles were set at levels that were considered achievable within a few years from when the Directives were agreed. They have also been reviewed and successively tightened.

From an environmental point of view, there is a clear need to further strengthen the emission limit values for all three air pollutants, NOx, SO_2 and dust. The application of currently available abatement techniques can deliver emission reductions well below the limit values of the LCP Directive. Recent analysis has found that in some countries (e.g. Germany, Netherlands, Austria, Denmark and Sweden) there are already a number of existing plants which easily meet the emission limit values set in the LCP Directive for new post-2002 installations. There can therefore be no doubt that it is possible to achieve emission levels with conventional technology that are considerably lower than the current standards (*Barrett 2004, p.43*). The forthcoming review and possible revision of the LCP Directive provides an opportunity to strengthen emission limit values, remove the unnecessary derogations, and set stricter deadlines for implementation.

Demanding a change in the units used for measuring emission limit values, from pollution/m³ to pollution/GJ useful energy still remains a priority. Currently, the emission limit values in the Directive are set as milligrams of pollutant per cubic metre of air (mg/m³). ECOs have long been calling for changing this to grammes per gigajoule (g/GJ) of useful energy produced, as this would help to promote energy efficiency *(see footnote 60)*. At the national level, ECOs should pay special attention to the local air quality near large combustion plants. This is particularly important in Member States where national plans are being applied.

By 2007 at the latest all LCPs will need a permit according to the IPPC Directive (*see chapter V.3.2*). Such permits must consider Best Available Techniques as outlined in the EU guidance[67]. As authorities deal with the task of writing these permits ECOs should use this opportunity to have an input into the process, using it to insist on stringent emission limit values beyond the ones set by the LCP Directive.

---

[67] See for the guidance the European IPPC Bureau website at **http://eippcb.jrc.es/pages/FActivities.htm.**

BIBLIOGRAPHY AND FURTHER READING

AMANN, Markus; BERTOK, Imrich; COFALA Janusz; GYARFAS, Frantisek; HEYES, Chris; KLIMINONT, Zbigniew; SCHOPP, Wolfgang; WINIWARTER, Wilfried (2004) *Baseline Scenarios for the Clean Air for Europe (CAFE) Programme. Final Report*, October 2004, p.6, Luxemburg, Austria.

BARRETT, Mark (2004) *Atmospheric Emissions from Large Point Sources in Europe*, Air Pollution and Climate Series, The Swedish NGO Secretariat on Acid Rain, p. 5..

CAFE Working Group on Particulate Matter (2004) *Second Position Paper on Particulate Matter*, April 6th, 2004, p. 12

COFALA, Janusz; AMANN, Markus (2001) *Emission Reductions from Existing Large Combustion Plants Resulting from the Amendment of the Large Combustion Plants Directive Report to the European Commission*, DG ENV Contract No. B4-3040/2000/267962/MAR/D3, p. 30.

Department for Environment, Food and Rural Affairs (2004) *Evaluation of Local Authority Air Quality Action Planning through Local Air Quality Management*, Ref: CS/AQ/AD102126/RM/2131(Final)

EEA (2003) *Europe's Environment: the third assessment. Summary*, Copenhagen, Denmark, p. 35.

EEA (2004) Transport and Environment, EEA Briefing, Copenhagen http://reports.eea.eu.int/briefing_2004_3/en/Briefing-TERM2004web.pdf

German Federal Environment Agency (Umweltbundesamt) (2003) *Future Diesel*, July.

National Institute for Public Health and the Environment (2004) *Outstanding Environmnetal Issues. A review of the EU's environmental agenda*, Bilthoven, The Netherlands, p.41-42.

NILSSON, J. (Ed) (1986) *Critical loads for nitrogen and sulphur – report from a Nordic working group*, The Nordic Council of Ministers, Report 1986:11, Copenhagen, Denmark.

ROESCH, Alexandre, CAMBOU José (2004) *Les Plans de Protection de l' Athmosphère: soutenir et développer la contribution du monde associatif à la protection de l' air. Guide méthologique*, Réseau Santé Environment, France Nature Environnement, p. 44-46

The Danish Ecological Council (2004) *Diesel particles – a health hazard*, Copenhagen.

IV.4 Waste

By Melissa Shinn [68,69]

IV.4 Waste 77

IV.4.1 Introduction - Waste facts 79

IV.4.2 Legal and policy context of EU waste policy................................81

   IV.4.2.1 The EU Treaty and waste policy..81

   IV.4.2.2 EU Waste Strategies and key principles....................................82

      IV.4.2.2.1 EU Waste Strategies...82

      IV.4.2.2.2 Waste management principles 82

         IV.4.2.2.2.1 Polluter pays principle 83

         IV.4.2.2.2.2 Producer responsibility 83

         IV.4.2.2.2.3 Proximity principle 84

         IV.4.2.2.2.4 Rectification at source................................85

         IV.4.2.2.2.5 A Note on Definitions................................85

   IV.4.2.3 EU waste legislation...86

   IV.4.2.4 EU horizontal waste legislation...87

      IV.4.2.4.1 Waste framework Directive 75/442 amended by 91/156 87

         IV.4.2.4.1.1 The definition of waste 88

         IV.4.2.4.1.2 Waste hierarchy..89

         IV.4.2.4.1.3 Permits 94

         IV.4.2.4.1.4 The siting of waste installations and public participation 95

         IV.4.2.4.1.5 Waste Management plans 96

         IV.4.2.4.1.6 Reporting...97

      IV.4.2.4.2 Directive 91/689 on hazardous waste 98

      IV.4.2.4.3 Limiting the generation of hazardous waste through products 98

      IV.4.2.4.4 Shipment of waste, Regulation 259/93 99

   IV.4.2.5 EU legislation addressing Waste installations – limiting emissions 101

      IV.4.2.5.1 Directive 2000/76 on waste incinerators
         (including co-incineration) 101

      IV.4.2.5.2 Directive 1999/31 on landfills...102

      IV.4.2.5.3 Directive 2000/59 on port reception facilities 103

      IV.4.2.5.4 Other waste treatment or recycling installations...........103

---

[68] Senior Policy Officer at the European Environmental Bureau

[69] With the collaboration of Roberto Ferrigno, Christian Hey, Ludwig Kramer

IV.4.2.6   EU legislation addressing specific waste streams 103

    IV.4.2.6.1 Separate collection; recycling targets...103

    IV.4.2.6.2 Producer responsibility, individual producer responsibility...........104

IV.4.3 Voluntary Instruments 105

IV.4.4 Standardisation (see also chapter IV.4.5) 106

IV.4.5 Taxes and charges..107

IV.4.6 Waste Policy today and tomorrow - New approach in the 2005 Strategy?............108

IV.4.7 Implementation and decision-making 111

    IV.4.7.1   Responsibilities...111

    IV.4.7.2   Complaints 112

    IV.4.7.3   Use of Waste Management plans 112

    IV.4.7.4   The importance of working locally 112

IV.4.8 Links and crossovers with other EU legislation 113

    IV.4.8.1   Links with environmental media policies: 113

    IV.4.8.1.1 Links to Water Quality...113

    IV.4.8.1.2 Links to Air Quality 113

    IV.4.8.1.3 Links to Soil Protection 114

    IV.4.8.2   Links with information and participation legislation (see chapter V.2) 114

    IV.4.8.3   Techniques used by installations...115

    IV.4.8.4   Nature protection...116

IV.4.9 Tips for ECO action 116

IV.4.10 The Court of Justice on important waste cases 119

IV.4.1   Introduction - Waste facts

According to Eurostat data[70] each year in the European Union 1.3 billion tonnes of waste are produced (including manufacturing and construction and demolition waste but excluding mining and agricultural and forestry wastes which make almost a further 1 billion tonnes). This amounts to about 3.5 tonnes of solid waste for every man, woman and child (*European Environment Agency 2002*).

According to information published by the European Environment Agency (EEA), five major waste streams make up the bulk of total waste generation in the EU: manufacturing waste (26%), mining and quarrying waste (29%), construction and demolition waste (C&DW) (22%) and municipal solid waste (MSW) (14%), and agricultural and forestry waste the estimation of which is particularly difficult. 2% of this waste is hazardous waste, i.e. about 27 million tones[71].

Waste generation continues to grow practically everywhere in the EU. The OECD expects that by 2020 waste in the European Union might have increased by 45 percent with regard to 1995. The OECD emphasises specific trends associated with this growing **waste burden**[72]. In particular:

▶ Chemical products, and the wastes associated with their production and consumption, are substantially **increasing in both complexity and amount**, suggesting uncertain but seemingly growing risks to environmental and human health systems.

▶ **"Hidden flows"** materials that support economic activities but do not actually enter the market place, such as mining wastes and eroded soil, can represent as much as 75% of materials used by OECD countries.

▶ There is a **linked increase** of Gross Domestic Product (GDP) and municipal waste generation in the OECD area 40% growth in both factors since 1980.

▶ OECD wide recycling has been increasing, but without countervailing efforts toward waste prevention, a **near-doubling** of municipal waste within the next 20 years.

Municipal waste (MW) makes up approximately 14 % of total waste. Meeting the objective set in the fifth environmental action programme (5 EAP), to stabilise the generation of municipal waste per capita at 300kg/capita per year[73], has not been realised. In the 1960s, waste production in Europe was approximately 200 kg per capita per year. Today it is close to 500 kg.

---

[70] Waste generated in Europe, data 1990-2001, Eurostat, 2003. Eurostat states - *about 2.25 billion tons of waste has been generated in Western Europe between 1998 and 2001.* These numbers cover waste generated in the 18 Western European countries (the 15 EU Member States plus Switzerland, Iceland and Norway) between 1998 and 2001. During the same time period Eurostat estimates 550 million tons of waste have been produced in the 13 Candidate Countries located in Eastern Europe. For EUROSTAT information see **http://epp.eurostat.cec.eu.int/**   in the domain of ENVIRONMENT statistics under data category WASTE

[71] Waste generated in Europe, data 1985-1997, Eurostat, 2000, p.37. **Note:** It is still difficult to obtain precise breakdowns on waste quantities at the EU level (despite comprehensive data existing at National level). There is not yet a very complete EU database for waste, and the classification varies widely from one Member State to the other. It is hoped that the Regulation 2150/2002 on waste statistics (2002 OJ L 332 p.1.), adopted in 2002 will, as of 2006, establish a consolidated set of data on waste generation.

[72] ENV/EPOC/PPC(2000)5/FINAL - OECD Reference Manual on Strategic Waste Prevention, August 2000

[73] [1993] OJ C 138/59.

Most of the domestic waste produced is still either burnt in incinerators, or dumped into landfill sites[74]. However, both landfilling and incineration create environmental damage. In particular, land use, air pollution by the release of hazardous substances, and pollution of water and soils remain high. Waste landfilling is a contaminating activity of major relevance: on average, 65% of municipal waste generated in the EU (190 million tonnes in 1995) is still landfilled. In waste land- fills leachates can be emitted to the surrounding soils and subsequently enter groundwater and/or surface water. Of particular concern are those that operate, or have operated in the past, without complying with the minimum set of technical requirements set by the Landfill Directive. Landfills are also the biggest source of Green House Gas emission from Waste management activities – mainly due to their emission of methane, a GHG with more than 20 times the GWP of carbon dioxide.

Whilst the bigger volume waste streams are clearly not municipal waste, they pose the major challenge in terms of finding solutions for prevention, collection and recycling and treatment. This is mainly due to the fact that they arise from disperse sources such as households and similar installations whose municipal wastes are unpredictable in their composition and disperse in their many generation points. The majority of industrial waste streams are more uniform and predictable in their nature and so easier to handle (although not necessarily less hazardous). The same can be said for agricultural, mining and forestry waste streams. In most EU Member States industrial wastes are handled by dedicated waste handlers, often adapted to the nature of the specific waste stream in question and paid for directly by the industrial waste generator. Agricultural, mining and forestry wastes all had traditional sources for (mostly) reuse of the wastes – either straight back onto the land or in other sectors such as the energy sector. Mining wastes, mainly a problem of storage have however been a growing environmental problem – but the solution is mostly to be found in the proper provisions for mining waste facilities – also, in theory, funded by the producer of the waste. The organisation of the `communal' funding and collection infrastructure of the different municipal waste streams, mostly from urban sources with the multiplicity of producers and waste types, has been therefore the main challenge of waste authorities.

Some municipal waste streams are of particular relevance in terms of environmental impacts. One example of such a waste stream is waste electrical and electronic equipment. In 1998, 6 million tonnes of **waste electrical and electronic equipment** were generated (4% of the municipal waste stream). The volume of WEEE is expected to increase by at least 3-5% per annum. This means that in five years 16-28% more WEEE will be generated and in 12 years the amount will have doubled. The growth of WEEE is about three times higher than the growth of the average municipal waste. Because of its hazardous content, electrical and electronic equipment causes major environmental problems during the waste management phase if not properly pre-treated. As more than 90% of WEEE is landfilled, incinerated or recovered without any pre-treatment, a large proportion of various pollutants found in the municipal waste stream comes from WEEE. (CEC 2000).

---

[74] The EU Commission, Answer to Written Question E-1242/03, (2004) OJ C 33 E, p.100): estimated in 2003 that on average, 60 percent of domestic waste in the EU 15 Member States was landfilled (Belgium 32%, Denmark 15%, Germany 46%, Greece 93%, Spain 83%, France 47%, Ireland 92%, Italy 94%, Luxemburg 37%, Netherlands 15%, Austria 43%, Portugal 88%, Finland 57%, Sweden 38%, United Kingdom 83%). For other waste, no sufficient data were available.

Even comparatively small hazardous waste streams are of particular concern. It is estimated that in 2002 at EU level 2,044 tonnes of portable NiCd **batteries** were disposed of in the municipal solid waste stream. **The main disposal route for spent batteries and accumulators is landfilling.** It is estimated that 75% of disposed spent batteries are being sent to landfill sites.. The main environmental concerns associated with the landfilling of batteries are related to the generation and eventual discharges of leachate into the environment. However batteries in waste are also incinerated. In the case of incineration, metals such as cadmium, mercury, zinc, lead, nickel, lithium and manganese are found in the bottom-ashes and fly ashes. Incineration of batteries thus contributes to emissions of heavy metals in the air and to incineration residues, which themselves then have to be landfilled in hazardous landfills. According to risk assessments the EU cadmium emissions of portable nickel-cadmium batteries due to incineration was calculated to be 323 – 1,617 kg of cadmium per year to the air and 35-176 kg of cadmium per year into water. Total cadmium emissions of portable nickel-cadmium batteries due to landfill was calculated at 131-655 kg of cadmium per year (*CEC 2003*).

81

The European Environmental Agency stated in 2004: *"Trends in waste generation, a proxy for resource use intensity, are unsustainable. Treatment and disposal options are diminishing as quantitative increase and concerns about their potential impacts grow. Decisions on the location of incinerators have become very controversial in many countries. Landfill options are often limited by space as well as by fears of soil and groundwater contamination and their impacts on human health. The current policy tools for dealing with waste are inadequate and need to be complemented by approaches that promote smarter resource use by changing production and consumption patterns and through innovation"* (*EEA 2004, p.6*).

IV.4.2   Legal and policy context of EU waste policy

IV.4.2.1   THE EU TREATY AND WASTE POLICY

Normally, European waste legislation is based on **Article 175** of the EC Treaty and member states may maintain or introduce more stringent national provisions than those adopted at European level under article 176. The key **environmental objectives and principles** of waste legislation, which are laid down in Article 174 of the Treaty are essentially a high level of environmental protection, prevention, rectification of environmental damage at source, and the polluter-pays principle. Unfortunately, in practice these principles are not all fully implemented in EU waste legislation.

What is crucial to bear in mind in EU waste policy is that waste is movable and therefore in some aspects treated as a product. Legislation (e.g. on packaging waste or batteries) is based on the product-related provision of the EC Treaty **(Article 95).** and is bound also to Treaty provisions on allowing the free circulation of goods **(Article 28)**. The relationship between the Treaty provisions on environmental protection and free circulation is very controversial and often involves the European Court of Justice. The European Court of Justice decided that, in general, waste materials come under the provisions of **free circulation (Article 28)**[75]. This means that where no EU waste legislation exists, any national waste legislation will have to be assessed against criteria for the free circulation of goods. The only safeguard is the environmental protection provisions of the Treaty - in particular that waste management policy should favour a high level of environmental protection and that it should protect, preserve and improve the quality of the environment, provisions which apply also to article 95 of the EC Treaty.

IV.4.2.2   EU WASTE STRATEGIES AND KEY PRINCIPLES

IV.4.2.2.1   EU Waste Strategies

An **EU strategy for waste management** was first adopted, in the form of a communication, by the Commission in 1989 and reviewed in 1996 (*CEC 1996*). Both communications were followed by resolutions from the Council and from the European Parliament. Next to them, the EU environmental action programmes, in particular the most recent 6th action programme (6EAP) (*EC 2002*) and the thematic strategies on the management of resources and the prevention and recycling of waste (under preparation) (*CEC 2003b and CEC2003c),* contain elements for the EU's latest political approach in waste management. EU communications and resolutions are not legally binding[76], allowing them at any time to adopt new approaches, but they have political importance. An important exception is the final Decision on the Sixth Environment Action Programme, which is legally binding on the Commission and Member States.

IV.4.2.2.2   Waste management principles

EU waste management policy is based on a number of principles[77]. The most important are:

▶ Polluter pays principle;

▶ Producer responsibility;

▶ Proximity principle; and

▶ Rectification at source.

---

[75] Court of Justice, case C-2/90, Commission v.Belgium, (1992) ECR I-443, para 28: "it must be concluded that waste, whether recyclable or not, should be regarded as a product the movement of which must not in principle, pursuant to Article 30 EEC" (now Article 28 EC Treaty) "be impeded".

[76] See Article 249 EC Treaty, which enumerates the binding EU instruments, but does not mention communications.

[77] Article 174(2) EC Treaty: "Community policy on the environment shall aim at a high level of protection taking into account the diversity of situations in the various regions of the Community. It shall be based on the precautionary principle and on the principles that preventive action should be taken, that environmental damage should as a priority be rectified at source and that the polluter should pay". Note that there is a general problem with the environmental principles laid down in Article 174 EC Treaty which has an impact on the waste management principles. Indeed, it is not clear, what the legal meaning of these principles is. No decision by the Court of Justice has clarified, until now, this problem. In legal literature, some are of the opinion that principles are legal rules which must be respected in each specific case or, at least, in general policy. Others believe that these principles are more political guidelines or orientations. There seems to be consensus that the principles allow a certain measure to be taken, but do not require measures to be taken or legislation to be adopted which complies with this or that principle. The principles of Article 174 EC Treaty are not automatically part of national environmental law, though they may also exist in a rule or provision of national law. In contrast, principles which are laid down in EU Directives or regulations, are applicable in Member States.

In addition to these principles Waste hierarchy is a guiding steering tool, serving as a precaution-ary proxy for the greatest potential reduction in environmental impacts from waste generation and management.

IV.4.2.2.2.1 Polluter pays principle

This principle is found in article 174 of the treaty, enshrined in the waste Framework Directive (WFD)[78] – article 15 - and reflected in a number of other Directives (producers of waste must bear the costs of having licensed transporters and managers of waste handle their waste, especially hazardous waste). However, it is not frequently practised and there is no firm and detailed defi-nition of it. It is generally understood to mean that the costs of waste treatment and disposal shall not be born by the average taxpayer, but by the person that generated the waste. An exam-ple of the polluter-pays-principle is the charging of costs for waste collection: where municipali-ties charge citizens with the costs of household waste collection, treatment and disposal. These means, however, are at the discretion of local, regional or national authorities and are not regu-lated or mandated at EU level. Even EC legislation transposed or national legislation that does not comply with this principle is legal (for example there are legal acts in the UK that make the application of the polluter pays principle directly to the citizen illegal) (*Eunomia 2003*).

Fundamental to the scope of waste legislation is the fact that in the case of waste the interpretation of waste 'generator' can be far reaching and thus precautionary. The Court of Justice has ruled that the owner of the land where waste is (legally or illegally) deposited, can also be considered to be the hold-er of the waste and thereby responsible for ensuring its safe treatment or disposal[79]. A more recent judgment stated that a petrol company may, under certain conditions, be considered responsible for the contamination of the soil, where fuel leaked from a defective petrol tank at a service station[80].

IV.4.2.2.2.2 Producer responsibility

This principle is not defined in EU legislation. It originally stems from Swedish environmental pol-icy and law, but is used in a different way at EU level. It states the responsibility of the producer for the impact of his products, in the case of waste policy typically at the end of their useful life-time. However recently producer responsibility has been extended to other life-phases of prod-ucts, including design (e.g. Directive on restriction on hazardous substances in electrical and elec-tronic equipment (ROHS)[81], Directive on waste from electrical and electronic equipment (WEEE)[82] – for more information *see chapters IV.4.2.4.3 and IV.4.2.6.2*)[83]. Producer responsibility may take the form of obligations for the producer to recover products or to collect waste, to establish funds or deposit schemes for recovery or recycling, organise recycling or recovery or relate to the design and manufacture of the product in view of the later waste stage.

---

[78] Directive 75/442 on waste (1975) OJ L 194 p.39; amended by Directive 91/156 (1991) OJ L 78 p.32.

[79] Court of Justice, case C-365/97 Commission v.Italy, (1999) ECR I-7773.

[80] Court of Justice, case C-1/03 Van de Walle a.o., judgment of 7 September 2004

[81] Directive 2002/95/EC OJ L37 p 19

[82] Directive 2002/96 on waste electrical and electronic equipment (2003) OJ L 37 p.24.

[83] Note: Producer responsibility is different from producer liability: while producer liability deals with damage that is caused by a product which has to be compensated, producer responsibility aims at preventing environmental impairment and damage, but does not aim at compensation.

The Commission waste strategy of 1996 stated: "*Considering the life cycle of a product from man-ufacture until the end of its useful life, producers, material suppliers, trade, consumers and public authorities share specific waste management responsibilities. However, it is the product manufactur-er who has a predominant role. The manufacturer is the one to take key decisions concerning the waste management potential of his product, such as design, conception, use of specific materials, composition of the product and finally its marketing. The manufacturer is therefore able to provide the means not only to avoid waste by a considered utilisation of natural resources, renewable raw materials or non-hazardous materials, but also to conceive products in a way which facilitates prop-er re-use and recovery. Marking, labelling, the issue of instructions for use and of data sheets may contribute to this aim.*" In its resolution of 1997, the Council avoided the use of the notion, but mentioned that "*the producer of a product has a strategic role and responsibility in relation to the waste management potential of a product through its design, content and construction*".

Elements of producer responsibility are to be found in the Directive on end-of life vehicles[84] and on electrical and electronic waste WEEE. The 2002 Sixth environmental action programme called for further development of producer responsibility; this aspect was also identified for discussion in the Commission communication *Towards a future thematic strategy on waste recycling*, but unfortunately the debate was restricted to '*exploring the merits of complementing existing Directives with a new approach addressing materials rather than end-of-life products.*' Overall the communication towards the final Waste Thematic Strategy shows a clear lack of enthusiasm for advancing the producer responsibility approach to the remaining waste streams still not addressed[85].

IV.4.2.2.3 Proximity principle

This principle stipulates that waste should be disposed of as closely as possible to its place of gener-ation. The EU Court of Justice developed it from the principle in Article 174 of the Treaty that envi-ronmental damage should, if possible, be rectified at source. In the waste framework Directive 75/442, Article 5, it is established for the disposal - but not for the recycling and recovery[86]- of waste[87]. And the Regulation 259/93 on the shipment of waste allows EU Member States to oppose ship-ments to another Member State, if the shipment is intended for disposal, but only to raise some, explicitly enumerated objections if the shipment is intended for recovery.

---

[84] Directive 2000/53 on end-of life vehicles (2000) OJ L 269 p.14.

[85] Towards Communication – p 32 - 'addressing smaller waste streams through producer responsibility may involve signifi-cant resources (both in legislative/ administrative and financial terms) for a relatively limited environmental benefit.'

[86] Court of Justice, case C-203/96 (1998) ECR I-4075: "It.. follows from the provisions of the Directive (75/442) and the Regulation (259/93), and from the general scheme of the latter, that neither text provides for the application of the prin-ciples of self-sufficiency and proximity to waste for recovery.."

[87] Directive 75/442 (see also note 11 above), Article 5: "1. Member States shall take appropriate measures, in cooperation with other Member States where this is necessary or advisable, to establish an integrated and adequate network of dispos-al installations, taking account of the best available technology not involving excessive costs. The network must enable the Community as a whole to become self-sufficient in waste disposal and the member States to move towards that aim indi-vidually, taking into account geographical circumstances or the need for specialised installations for certain types of waste.

Inside EU Member States, the principle rarely plays a role[88]. However, at EU level, it is often made use of by Member States which want to implement a national waste management policy and keep waste within their borders, also in order to fully use the capacity of waste incinerators, land-fills or treatment installations.

IV.4.2.2.2.4 Rectification at source

This principle, stated in Article 174(2) of the Treaty, establishes "that environmental damage should as a priority be rectified at source". It was used once by the Court of Justice to justify a national ban for hazardous waste imports[89]. However, this judgment has remained isolated and must be considered to have been superseded by the provisions of Regulation 259/93 on the shipment of waste.

IV.4.2.2.2.5 A Note on Definitions

An important function of EU legislation is also harmonisation in interpretation of different terminology. So far the various definitions pertaining to waste management (waste itself, hazardous versus non-hazardous waste, disposal, recovery, recycling, reuse etc) have been established in a generic way in the framework Directive   - in particular the definition of waste itself (see chapter IV.4.2.4.1.1 for more detail) and the operations classified as recovery and disposal (listed in the Annexes IIA and IIB), with subsequent greater specificity of definition in the individual waste stream Directives. Especially for the definitions of recycling and reuse[90].

Clarification between waste recovery and disposal operations is critical for the implementation of EU waste management policies, drawing the line between recovery and disposal affects the interpretation of the waste shipments regime, the fulfilment of mandatory recycling and recovery targets (Packaging, End-of-Life Vehicles, Waste Electrical and Electronic Equipment Directives) and the planning and permitting of waste treatment facilities for recovery and final disposal.

The challenge of revising the outdated lists of Recovery and Disposal operations in the framework Directive is a complex one due to the interdependence with other Directives and the different uses they make of them (restrictions to movement, boundaries of targets, permit requirements). The Commission has recognised this in its preparation document Towards the Thematic Strategy on Waste from May 2003. However the Commissions proposed solution (at the time of writing) is causing the EEB (and other industry and local authority stakeholders) some concern as we fear it may be symptomatic of some deeper changes in policy approaches to waste (see chapter IV.4.6).

---

[88] A notable exception is Germany where the proximity principle is translated into the responsibility of municipalities to take care of the disposal of municipal waste, but recovery (as opposed to.disposal) of waste can be privatised/delegated.

[89] Court of Justice, case C-2/90 Commission v.Belgium para 34: "The principle that environmental damage should as a priority be rectified at source .means that it is for each region, commune or other local entity to take appropriate measures to receive, process or dispose of its own waste. Consequently waste should be disposed of as close as possible to the place where it is produced in order to keep the transport of waste to the minimum practicable".

[90] Sometimes however the different legal and political texts and their evolution over time can give rise to some inconsistencies. One example is the notion of re-use: In Article 3 of the framework Directive 75/442 on waste, , re-use is described as a form of recovery of waste. In contrast to that, Article 3(5) of Directive 94/62 on packaging and packaging waste, provides that only packaging that is no longer re-used, becomes waste.

IV.4.2.3   EU WASTE LEGISLATION

The overall scheme of EU waste policy and law is the following:

EU WASTE MANAGEMENT POLICY AND LAW

EU Treaty
Articles 28-30 and 174-176 of the Treaty

Community Waste Strategy
COM (96)399

1st level - Horizontal legislation

Waste Framework Directive	Hazardous Waste Directive	Shipment Regulation
75/442(91/156)	91/689	93/259

2nd level - Waste installations

Waste Incineration Directive	Landfill Directive	Port Directive Facilities Directive
2000/76	1999/31	2000/59

3rd level - Specific Waste Streams

Waste Oils	Sewage Sludge	Batteries	Packaging	PCBs	Cars	Waste EEE*	Animal waste
75/439	86/278	91/157	94/62	95/59	2000/53	2002/96	1774/2002

* EEE means electrical and electronic equipment

The three legislative levels of this scheme are interrelated, in the sense that, as soon as any of the specific waste streams of the third level address hazardous aspects, the Directive on hazardous waste of the first level comes into play; where batteries are landfilled, the corresponding Directive of the second level is to be examined, etc.

It is important to be aware of the hierarchy of the cascade of legislation from the EU to the national level. Where EU legislation is adopted, it prevails, in cases of conflict, over national law. In the area of waste management policy, the EU acted, until now, almost entirely by binding legislative instruments, Directives or regulations. Generally, these legislative measures were based on Article 175 EC Treaty. Exceptionally, where the measure principally concerned the product and not the waste, such as the Directives on packaging and packaging waste, on batteries, on the ban of dangerous substances in electrical and electronic products, the legal basis was Article 95 EC Treaty, in order to ensure the existence of uniform provisions within the EU internal market. Regulations were only adopted in the areas of transfer of wastes and of agricultural waste where it was deemed to be important to have uniform EU-wide provisions which could not (easily) be amended by national legislation[91].

---

[91] An EU Regulation is binding in its entirety and directly applicable in all Member States and, therefore, needs normally not be transposed into national law, while a Directive has to be transposed into national legislation. Furthermore, where EU legislation is based on Article 175, Member States may maintain or adopt more protective measures, see Article 176 EC Treaty. Where EU legislation is based on Article 95 EC Treaty, national legislation may only be maintained or new legislation introduced under the very restrictive conditions of Article 95(4 to 8). And where EU legislation is based on 37 EC Treaty, such as the Regulation on animal waste, Member States may not adopt national amending legislation at all.

IV.4.2.4   EU HORIZONTAL WASTE LEGISLATION

IV.4.2.4.1   Waste framework Directive 75/442 amended by 91/156

This Directive was adopted in 1975 and reviewed in 1991 by Directive 91/156[92]. It lays down the EU-wide **definition of waste** and provides for a European Waste List which lists the different cat-egories of waste and of hazardous waste. It introduces a so-called "**hierarchy for waste management**", according to which waste management should first of all aim at the prevention of waste generation. If that is not possible, material waste recycling and incineration of waste with ener-gy recovery should be pursued. At the end of this hierarchy ranks the landfill and incineration without energy recovery.- Note that this hierarchy has been interpreted as establishing a politi-cal, not a legally binding orientation of policies, as will be further explained below.

The Directive requires, furthermore, a **permit for waste treatment or disposal** activities, for which basic conditions are laid down. Waste treatment or disposal measures have to ensure that human health or the environment are not threatened; uncontrolled dumping of waste is prohib-ited. Member States have to draw up **waste management plans** and to **report on the application** of the Directive every three years.

The Directive introduces the proximity principle for waste disposal, and requires Members to apply the polluter pays principle. It also establishes a committee for adapting the Directive's pro-visions for scientific and technical progress. This Committee serves more and more as the steer-ing instrument on EU waste management policy.

The Directive's precise scope in terms of types of waste is still legally contested. In particular, it is unclear as to what extent those areas that are mentioned in Article 2(1.b) of the Directive come under its provisions[93]. De facto, the Directive does not apply to radioactive waste, agricultural waste and mining waste. For mining waste, the EU is presently discussing a proposal for a Directive (*CEC 2003d*). Central to operationalising the environmental objectives of the Directive (recital 4 and others) is Article 4 which provides for the general obligations for waste manage-ment[94], a provision against which activities of local, regional and national authorities and opera-tors should be assessed.

---

[92] Directive 75/442 on waste

[93] Directive 75/442, Article 2(1): "The following shall be excluded from the scope of this Directive: (a) gaseous effluents emit-ted into the atmosphere; (b) where they are already covered by other legislation (i) radioactive waste; (ii) waste resulting from prospecting, extraction, treatment and storage of mineral resources and the working of quarries; (iii) animal carcas-es and the following agricultural waste: faecal matter and other natural, non-dangerous substances used in farming; (iv) waste waters, with the exception of waste in liquid form; (v) decommissioned explosives". See for a discussion of this pro-vision Court of Justice, case C-114/01 Avesta Polarit, judgment of 11 September 2003.

[94] Directive 75/442 (note 11), Article 4: "Member States shall take the necessary measures to ensure that waste is recovered or disposed of without endangering human health and without using processes or methods which could harm the envi-ronment, and in particular: - without risk to water, air, soil and plants and animals, - without causing a nuisance through noise or odours, - without adversely affecting the countryside or places of special interest. Member States shall also take the necessary measures to prohibit the abandonment, dumping or uncontrolled disposal of waste".

IV.4.2.4.1.1 The definition of waste

The EU definition on waste is that waste is "any material which the holder discards, is obliged to discard or intends to discard". This definition is aligned to that of the UN Convention on the shipment of hazardous waste (Basel Convention). It has an objective element in the sense that a material becomes waste by virtue of a circumstance which is outside the control of the owner or holder of the material; namely the fact of abandoning the material or the fact that there are provisions which determine that certain material is to be classified as waste, for example because it may not be used any more; an example is a product which does not correspond to legal requirements.

It is important to note that the waste definition does not make the classification of waste dependant on the economic value of the material[95]. Where a person places a TV-set on the pavement in order to have it removed by the municipal waste collectors, the TV-set becomes waste, though it may still be in a state to be used.

Economic operators very persistently try to exclude from the notion of waste those materials that have an economic value, that are capable of being re-used, recycled or otherwise used economically. The reason for this is that the material would then be classified as a product and not be submitted to the provisions of waste legislation, regarding transport, export, handling, burning etc. If the legislator were following this request, this would mean that there would be no waste from precious metal production, from metals, wood, glass etc., as all this material is or could be re-used or recycled. Until now, the Court of Justice rejected all such attempts, arguing that this would contradict the EU definition of waste.

It may be asked - What is the problem, if something is considered a product and not waste? The answer is, that for waste materials the control regime is normally stricter than for products and adequate environmental standards for products rarely exist. In other words the waste regime can also be seen as a surrogate precautionary regime that has a function as long as product legislation is not yet developed and complete.

A good indication as to whether something is product or waste is the consideration of whether the generation of material in question would have been avoided altogether, if that had been technically possible. Therefore, the gold dust which is generated when a jeweller makes a gold ring, is waste, though it has a high economic value. Had the jeweller been able to avoid the generation of that dust, he would have done so.

A judgment in the Court of Justice in 2004 clarified that leaking substances - the case in question concerned a fuel leak from a petrol station – constitute waste. independently of whether the contaminated soil was removed or not[96]. This judgment might have far-reaching consequences on new cases of soil contamination, creating financial responsibility for those responsible for the original contamination.

---

[95] See, for example, Court of Justice, C-304/94 Tombesi, (1997) ECR I-3561.

[96] Court of Justice, case C-1/03 (note 38 above)

Where a material is classified as waste, it remains waste, until the recycling or re-use process is finished. Again, there are attempts from economic operators to have the waste character already finished at an earlier stage, for example after sorting (for paper), or at the first step of the recycling process. The problem is well illustrated by a Court judgment: a truck driver was stopped by the police, because he was transporting waste without a permit for the shipment. He defended himself with the argument that he was transporting the material to a specific installation, where it was to be recovered, that the transport was the first part of the recovery process and that therefore the material was a product, not a waste. The Court however disagreed and concluded that until the recovery process was complete the material should be classified as waste[97].

As mentioned previously, the definition of hazardous waste depends entirely on the **waste list for non-hazardous and hazardous waste**[98]. Thus, any waste that is listed in the waste list with an asterisk (*), is hazardous waste. Member States may, under certain conditions, classify other wastes as hazardous. The classification as hazardous waste has the consequence that stricter conditions for issuing a permit to handle such waste apply, that the mixing of such waste is not allowed and that the shipment and landfilling of such wastes is subject to more severe requirements.

Nuclear waste follows rules of its own. It is de facto not covered by EU waste legislation, but follows the rules of the Euratom-Treaty. Also, recyclable nuclear material is not considered waste, but "product", while under Directive 75/442 recyclable waste is waste. There are no attempts to develop specific rules on nuclear waste at EU level though there are some provisions on shipment of radioactive waste.

IV.4.2.4.1.2 Waste hierarchy

The EU waste hierarchy is meant to provide a ranking for environmental soundness in waste management and serve as a proxy for the most efficient options for reduction in environmental impacts - prevention being the most efficient, reuse the next etc... The hierarchy is established in article 3 of the waste framework Directive 75/442 and indicates that Member States should strive at ensuring first of all waste prevention, then material recycling, incineration with energy recovery and finally landfilling of waste. According to the 1996 Waste Strategy (*CEC 1996b*) the hierarchy is commonly understood as (starting with the first priority):

- prevention, then
- re-use of materials and products;
- material recycling;
- incineration with energy recovery;
- landfilling.

---

[97] Court of Justice, case C-359/88, Zanetti a.o. (1990) ECR I-1509.

[98] Commission Decision 2000/532 replacing Decision 94/3 establishing a list of wastes pursuant to Article 1(a) of Council Directive 75/442 on waste and Council Decision 94/904 establishing a list of hazardous waste pursuant to Article 1(4) of Council Directive 91/689 on hazardous waste (2000) OJ L 226 p.3.

One can also indirectly derive this hierarchy from a number of provisions. Article 174 of the Treaty suggests combatting environmental damage, at source if possible, and ensuring prudent use of natural resources. In addition - and this can be interpreted as implementation of the precautionary principle - if you don't know how big the threat is, it is better to avoid it in the first place. A preference for the recovery (material recycling and incineration with energy recovery) with regard to the disposal of waste (landfilling and incineration without energy recovery) can also be deduced from Regulation 259/93 on the transfer of waste. This Regulation allows Member States to completely prohibit the transfer of waste to another Member State, if the waste is to be disposed in that Member State. In contrast to that, if the transfer takes place to recover the waste, Member States may only raise a number of expressly enumerated objections against such a transfer.

Unfortunately, the waste hierarchy does not mean that a Member State is legally obliged to take measures in order to follow the higher level of the hierarchy, before recurring to a lower level. This means that waste prevention measures - in other words, product-related and production-process related measures - would have to be taken, before recycling measures could be adopted or even considered. This is obviously not (radically interpreted) very practical, as any economic activity creates residues (wastes) and intermediate measures are necessary. Consequently, the EU Commission has never considered the hierarchy in Article 3 of the waste framework Directive 75/442 as a legal rule, but rather as a political objective and guideline. However, the absence of a 'legal rule' does not allow the Commission to abandon waste prevention as a priority, to be implemented and stimulated to the best of its powers, despite its unarguable intricacy.

Built into this principle the hierarchy approach is the concept of waste as a form of pollution per se and prevention and reduction of the quantity of waste production (in general but especially for disposal) as a proxy for prevention and reduction of environmental impacts and the achievement of a high level of environmental protection.

The hierarchy and the quantitative proxy have been, and remain, a potentially effective and pragmatic tool towards the greatest reduction in environmental impacts. This approach is then reflected in the specific waste stream legislation and is the basis for the establishment of mandatory targets and the preference given in mandatory targets to recycling over other forms of energetic recovery and the phasing out or limits placed on specific hazardous substances[99]. There has, however, been no proper mandatory implementation of quantitative prevention at the EU level yet.

(a) Waste prevention

Waste prevention includes the minimising of waste generation (quantitative prevention), but also the minimising of hazardous properties of wastes (qualitative prevention), in order to facilitate recycling and disposal. While recycling (and indeed any other waste management process) is a production process and requires transport of the recyclable waste, energy use during the recycling process, and the disposal of the residues from the recycling production, waste prevention potentially avoids all these activities.

Waste prevention may be done by fixing requirements (limit values etc) for the manufacture or composition of products (**direct prevention measures**), or by influencing the behaviour of man-

---

[99] Examples are the reductions of hazardous substances in batteries, end of life vehicles and waste electrical and electronic products.

ufacturers or traders through taxes or charges on raw materials. Alternatively consumers/users may by influenced prices or by quality labels such as eco-labels (*see chapter V.4.2*) or other information (**indirect prevention measures**).

As mentioned previously, EU provisions on waste prevention are rare and despite the priority given in the policy framework driving prevention has not been given priority in implementation so far. The framework Directive 75/442 mentions clean technologies and the marketing of less hazardous products as a means to achieving waste prevention. It therefore alludes to one critical point: before a material is waste, it is a product. Thus, effective waste prevention policy must involve waste policy that steers **a strong product policy**. So far the EU has not developed a comprehensive product policy which systematically limits the use of materials for the production of goods, reduces the use of hazardous substances and increases reusability and recyclability. Practically no legislation exists that requires products to be made, in whole or in part, from recyclable materials or that they be re-usable and/or upgradable. For this reason, product based prevention measures at EU level are limited to voluntary tools such as the EU Ecolabel criteria scheme which has set criteria on some 22 products so far (*see chapter V.4.2*). The exceptions are a number of **Directives that prohibit or restrict certain dangerous substances** in products: e.g. mercury in batteries, heavy metals in packaging, in cars or in electrical and electronic equipment and brominated flame retardants in electronic equipment.

One of the waste prevention approaches that has been on the agenda for many years without much evolution is **EU level waste prevention targets**. The target in the 5th EU environmental action programme from 1993, to limit municipal waste generation to 300 kg per capita per year[100], remained without widespread effects, most likely because no implementation or control measures were laid down and the target itself was of a voluntary nature. Subsequently the 6EAP calls on the Commission, as one of the priority actions on sustainable resource and waste management, to develop (and implement) Waste Prevention targets to be achieved by 2010. Regrettably, bymid-2005, progress towards such concrete measures had still not been taken at EU level.

Several methods may be used to indirectly influence waste generation. Requiring that products **contain a certain percentage of recycled materials** reduces the overall amount of waste stemming from the production of such products (from extraction and processing of virgin materials). Examples are papers, textile products or glass that contain a percentage of recycled materials. Plastic recycling is not yet technically very far advanced, due partly to the diversity of different types of plastic materials. But it has undoubtedly a very important future role to play in waste management, given the predominance and increasing use of plastics in our products and packaging. Another example of recycling is food residues may be used for the making of animal feed or compost.

Waste generation may also be influenced via such products by tax incentives, financial charges, reduced VAT on eco labels or other instruments that give economic advantages or public recognition to producers that make such products, and to consumers or users that acquire such products. In this regard, **public procurement,** with the considerable market it offers (public procurement accounts for over 16% of the European Union's GDP) could play a leading role, requiring recycled, reusable or recyclable products - paper, office equipment etc - and thus promoting products that promote waste prevention.

---

[100] Fifth EU environmental action programme "Towards Sustainability" (1993) OJ C 138 p.5

A further very important economic incentive can also be given through Individual Producer Responsibility – bringing the end-of-life costs of the individual product to bear on the original producer (for more on this see chapter IV.4.2.6.2 on Producer responsibility below).

As regards household and municipal waste, the practice of municipalities of fixing waste fees according to the weight of the household waste (often called **Pay As You Throw**), has, when applied, proven to reduce waste quantities generated (especially Municipal Solid Waste) and simultaneously increase source separation of specific, recyclable waste items - glass, metal, cardboard and paper - and of composting kitchen and garden waste[101]. As EU law does not regulate waste collection fees, national, and very often also regional or local, authorities have been left the responsibility of organising waste separation and collection in a way which reduces waste quantities.

As concerns **reuse** there is also a scarcity of EU action. The closest the EU comes on reuse policy is to set standards for ´reusability´ in the context of the packaging and packaging waste Directive. Such standards (which took 10 years to develop) are not expected to deliver any significant support for reuse systems (*for more details see chapter IV.4.4*).

(b) Waste recycling

As mentioned previously, the Commission's waste management strategy of 1996 indicates that material recycling is generally preferable to waste incineration with energy recovery[102]. This statement found the cautious but unambiguous support of the Council[103]. Recycling can save resources: for example, the making of a private car which weighs about one tonne, but has been estimated to take about 60 tonnes of raw material. For each car, about 85 percent of the material by weight can be recycled (steel and other metals, plastics etc)[104].

Daughter Directives on specific waste streams on waste oils, batteries, packaging and packaging waste, end of life vehicles and electrical and electronic waste, have set targets on this at EU level with mixed success. In the case of waste oils, where no quantified targets have been set (just a

---

[101] For a comprehensive overview of PAYT in Europe and the results achieved see Report Waste Collection: To charge or not to charge? by Eunomia, March 2003 available from http://www.eunomia.co.uk.

[102] Commission COM(96) 399 para 42: "Material recovery implies the separation of wastes at the source. This involves end-users and consumers in the waste management chain and makes them more aware of the necessity and the ways to decrease the generation of waste. Indeed, it is the end-users and consumers who should carry out the separation of wastes before disposal in order to reintroduce recyclable wastes in the production cycle. Furthermore, energy strategies relying on waste supplies should not be detrimental to the principles of prevention and material recovery. Indeed, in many cases it can be assumed that by retaining the existing material structure of the recoverable waste, it will be possible to minimise the additional material and energy necessary to produce a new product. Also, material recovery addresses the concerns about emissions from waste incineration installations.

In view of the above, preference should be given, where environmentally sound, to the recovery of material over energy recovery operations. This reflects the greater effect on the prevention of waste produced by material recovery than by energy recovery."

[103] Council Resolution of 24 February 1997 (1997) OJ C 76 p.1 para.22: "(The Council) recognises, as regards recovery operations, that the choice of option in any particular case must have regard to environmental and economic effects, but considers that at present, and until scientific and technological progress is made and life-cycle analyses are further developed, reuse and material recovery should be considered preferable where and insofar as they are the best environmental option".

[104] Recovery Options for Plastic Parts from End-of- Life Vehicles: an Eco-Efficiency Assessment, Final Report, May 2003, by Okoinstitut e.V. for APME

general priority for recycling) there has been fragmented success. This is also strongly attributable to the conflict between the recycling policy which seeks to encourage recycling and the EU fuels tax Directive which grants tax relief on use of waste oils as fuels and thus encourages its incineration[105]. In the case of packaging waste however the Directive has had significant impact on EU wide levels of recycling, with all Member States achieving the obligatory 2005 recycling targets, contributing significant reductions in impacts of Packaging Waste management (around 1 million tonnes of oil equivalent and 3 million tonnes of CO_2 equivalent as a direct result of the Directive[106]). However, despite the apparent usefulness of recycling targets as a policy measure to stimulate EU wide progress on recycling, the Commission has lately shown preferences to abandoning such approaches (see chapter IV.4.2.6.1).

93

(c) Waste incineration

Industrial waste generators often prefer waste incineration to material recycling, as material recycling is often still more time - and cost - intensive. Citizens, though, do not like to have waste incinerators in their neighbourhoods. Once an incinerator is constructed - which is often very expensive - it needs to be fed continuously during its lifetime at a certain minimum volume for, on average, periods of 30 years or more. This inflexibility brings pressure on the operator to oblige waste managers to bring the waste to a specific incinerator, though the costs might over time eventually be higher than considering other options, such as material recycling or composting for example. In Germany and apparently also the Netherlands, this practice even led to some cases which oblige waste generators to deliver their wastes to a specific incinerator[107] or obliged them to deliver minimum quantities for incineration[108].

Public objections to installations which co-incinerate waste (steel works, cement kilns, power plants) are, so far, less evident (also due to the ambiguity of the activity concerned – ie such installations are not specifically built to incinerate waste so they are not such an obvious target), though the emission and pollution risks might be at least equivalent to those from dedicated waste incineration installations and probably greater. This, together with political pressure from the industries concerned to maintain less demanding emission requirements on co-incineration in relevant legislation (e.g. the waste incineration directive) has led to a considerable increase in co-incineration with an even greater increase potential in the future[109].

---

[105] As regards waste oil tax reliefs, Directive 2003/96 gives the following member States permission to apply, till end 2006, reduced tax rates for the burning of waste oils: DE ES, FR IRL IT LUX AUT PT FIN UK. After 2006, the Council will take a new decision (and given past experiences probably prolong these derogations).

[106] See Chapter 1.7.3 of the Study on the Implementation of the Directive on Packaging and Packaging Waste 94/62/EC and options to strengthen prevention and reuse of packaging, February 2005 carried out for the Commission

[107] Der Anschluss und Benutzungszwang law in Germany establishes specifically a public responsibility (as opposed to private) for taking care of non-recoverable municipal waste.

[108] Note: there is debate as to whether this can in fact be a positive policy in the sense that it may avoid the 'liberal market alternative' with different operators (private and public) collecting household waste, with the possible consequences of waste for disposal being transported throughout the country in search of the most economic solution...

[109] The German Advisory Council on the Environment (SRU) report of 2004 emphasises that waste meets 25% of the energy needs of the cement industry in average. The report notes that out of the total incineration and mechanical treatment capacity in Germany (about 20 Mi tonnes), co incineration is estimated at 1 Mio. tonnes (which means 5%). However there are many co incinerators with up to 90% of their energy needs met by waste and that the average waste share is growing.

(d) Landfill

EU waste policy tries to reduce the disposal of waste to landfills, as the materials could instead be recycled, incinerated or otherwise be brought back to the economic circuit, whereas they may constitute an environmental risk at the landfill, by the development of methane gases (a greenhouse gas), the leaking of polluted liquids to the soil and underground waters, or other risks. Within the EU, about two thirds of all waste continues to go to landfills, although there are wide variations between Member States. In almost all EU countries, numerous landfills are operating without a permit (recently over 8000 illegal waste dumps were identified in France[110]) whereby such landfills are normally neither placed on appropriate sites nor equipped with the necessary environmental protection and security mechanisms. Under EU law, any place where waste is stored for more than three years is a landfill. Legal cases concerning land-filling activities are a frequent cause of environmental legislative infringement cases brought to the EU level (mostly however linked to the requirements of the Waste Framework Directive and not the Landfill Directive itself).

IV.4.2.4.1.3 Permits

All persons who professionally treat or eliminate, trade or otherwise handle waste, need a permit to do so. For waste incinerators, the most relevant conditions are laid down in the waste incineration Directive; for larger installations the requirements of the Directive on integrated pollution prevention and control (*IPPC Directive, see chapter V.3.2*) apply in supplement.

The **incineration** Directive which also applies to co-incineration installations, does not apply to some specific incineration types; but nothing prevents a Member State applying the same or equivalent rules to those incinerators which are not covered by EC law. Applications for a permit for an incineration installation and decisions on that application must be made available to the public. Furthermore, every citizen has, under the legislation on access to information on the environment, the right to see at any moment the complete permit, including all conditions which are laid down in it. No administration is entitled to keep as confidential information on, for example, conditions on emissions into the air or the soil.

The permit must contain a detailed enumeration of all waste types which may be treated; this includes an indication as to whether the wastes are hazardous or not. It must also contain information on the emissions into air and water, the measuring methods and techniques and, where hazardous wastes are burnt, and information on the maximum content of hazardous substances. The permit must furthermore ensure that the legislative provisions on waste water, discharges of hazardous substances into water and air, and on landfills are respected. The permit must be regularly reviewed and eventually be adapted.

Where a waste incinerator also comes under the provisions of the Directive on integrated pollution prevention and control, the permit must achieve the emission levels required by the waste incineration Directives. This can be achieved by applying the "**B**est **A**vailable **T**echnology BAT". Those are described in the **B**est **Ref**erence (BREF) document for Waste Incineration elaborated by

---

[110] Commission Press release 13 July 2004 - IP/04/895 – " *The Commission has furthermore decided to refer France to the European Court of Justice over the existence of numerous illegal and uncontrolled landfills across the country .8,434 sites have been identified by the Commission in France's 95 departmental waste management plans.*

an EU body in Sevilla (Spain)[111]. These BREF documents only constitute guidelines for the permitting administration, but are not legally binding. Eventually the responsibility rests with the permit writers at local levels, how they balance environment and economic interests *(see chapter V.3.2)*.

Member States have to regularly control that the conditions laid down in the permit are respected. . Since the use of, or the requirements of, the BAT levels set in the BREFs are not always as ambitious as they should be *(see chapter IV.4.8.3 below on `Techniques used by installations´ for the example of the Waste Incineration BREF)* the scrutiny of such permits is important.

Note that they are free to apply the provisions of the Directive on integrated pollution prevention and control to all incinerators, also to those that are not listed in the incineration Directive itself.

Similar provisions apply to permits for **landfills**. The application for a permit must also contain a description of the hydrogeological and geological characteristics of the site. The landfill must conform to the waste management plan that was established for the area and in particular indicate, whether the landfill is for non hazardous, for hazardous or for inert waste and which is the landfill's permitted total capacity. The operator of the landfill must be a person who has the necessary technical knowledge to run the landfill operations. Again, the application and the permit are accessible to every citizen, at his request, and may not be kept confidential.

Other waste treatment **installations** are only submitted to general requirements for a permit, and Member States may even satisfy themselves with a registration instead of a permit requirement. This also applies to installations which treat hazardous waste; however, in this case, it is the European Commission which grants, under certain conditions, a derogation. In all these cases, applications, permits granted and decisions on derogations must be made publicly accessible by the relevant administrative authority. And once more, nothing prevents a Member State from applying EU provisions to those installations that are not regulated in detail by the incineration and landfill Directives at EU level.

IV.4.2.4.1.4   The siting of waste installations and public participation

Waste installations are traditionally not popular with the general public. While people recognise the necessity to have waste treated and disposed of, they are concerned about the health and environmental risks that waste treatment, in particular waste incinerators might bring to their neighbourhood.. These concerns can be met in three ways: first of all to plan incineration projects only when absolutely necessary (focussing first on prevention, re-use and material recycling), secondly, to choose the state-of-the art technology and accompanying robust control systems,

---

[111] "Best available techniques" is defined, in Article 2(11) of Directive 96/61 concerning integrated pollution prevention and control (1996) OJ L 257, p.26 as follows: "'best available techniques' shall mean the most effective and advanced stage in the development of activities and their methods of operation which indicate the practical suitability of particular techniques for providing in principle the basis for emission limit values designed to prevent and, where that is not practicable, generally to reduce emissions and the impact on the environment as a whole: - 'techniques' shall include both the technology used and the way in which the installation is designed, built, maintained, operated and decommissioned, - 'available' techniques shall mean those developed on a scale which allows implementation in the relevant industrial sector, under economically and technically viable conditions, taking into consideration the costs and advantages, whether or not the techniques are used or produced inside the Member State in question, as long as they are reasonably accessible to the operator, - 'best' shall mean most effective in achieving a high general level of protection of the environment as a whole".

and thirdly, to fully involve the public in the planning and decision making. EU legislation does in fact require public participation, which should include discussion about a proper evaluation of alternatives and the quality of the planned installation[112] (*see also chapter V.3.5*). Furthermore, for bigger waste management installations – such as incineration plants, an environment impact assessment is mandatory. Unfortunately, however, all other waste treatment or disposal installations have to undergo an environment impact assessment, only if, '*in view of their nature, size or localisation, they are likely to have significant effects on the environment*'.

The impact assessment procedure has the objective of avoiding or at least minimising negative impacts of the project on humans or on the environment[113]. The impact assessment procedure, if properly conducted, can grant the affected public – including those who live close by the future installations, as well as their associations - the right to participate in the elaboration of the impact assessment with a view to including their arguments and seeing their interests taken into consideration. The final decision, however, rests with the administration which must explain the decision taken.

IV.4.2.4.1.5 Waste Management plans

Practically all EU waste legislation – from the framework Directive to the individual waste streams Directives - requires Member States to set up and regularly review waste management plans. Some of these plans concern the general management; others deal with collection requirements or adapting existing landfills to EU requirements. Whether Member States establish national plans or leave such plans to regional or local authorities, is left to them, as long as the whole territory of the Member State is covered by the plans. Of course, for many sectors, it would make sense to obtain some transboundary planning, for example on incinerators, landfills or collection systems. However, general reluctance from the public to have waste from other Member States or even regions in their neighbourhood, has restricted transboundary planning to a large extent.

Normally, management plans have to be sent to the European Commission, the intention being to progressively achieve an EU-wide waste management overview and coordination. So far however, this coordination has not materialised effectively at EU level. It is also true that for some plans – for example the 'framework' plan required under article 7 of the waste framework Directive there is no periodicity of renewal required so some waste management plans submitted to the Commission have not been revisited or notified to the Commission for many years. Waste management coordination is thus almost entirely left to national administrations or to bilateral cooperation.

Under EU law, the notion of "plan" and "programme" is often used indistinctively. The Court of Justice stated that a plan/programme must constitute "an organised and coordinated system of objectives", that it must contain a timetable and must be reviewed at regular intervals[114]; this means that the adoption of legislation or of practical or ad hoc measures normally is not suffi-

---

[112] Directive 2003/35 providing for public participation in respect of the drawing up of certain plans and programmes relating to the environment and amending with regard to public participation and access to justice Council Directives 85/337 and 96/61 (2003) OJ L 156 p.17. See also Directive 2001/42 on the environment assessment of the effects of certain plans and programmes on the environment (2001) OJ L 197 p.30.

[113] Directive 85/337 (see previous footnote), Articles 2 and 4.

[114] Court of Justice, case C-347/97 Commission v.Belgium (1999) ECR I-309 (concerning a programme for batteries).

cient to comply with the requirement of establishing a plan but this legal interpretation is hardly enforced by the Commission.

The quality of waste management plans varies considerably across Europe. Some plans consist of hardly more than some lines. The EEA concluded recently in a Parliamentary briefing – "Although it is difficult to identify a general relationship between the quality of waste plans and trends in waste generation and management performance, there does seem to be a positive link in countries with very good coverage in their plans, where high levels and strong growth rates for recycling can be found. These countries are also good at limiting landfilling"[115]. Waste management plans are public and may not be kept confidential. Their elaboration must undergo an environment impact assessment (*see chapter IV.4.2.4.1.4*) and the public has a right to participate in the drawing up of such plans. This is important, as many plans contain provisions on the siting of waste treatment or disposal installations and even some details of the construction and operation of such installations.

Furthermore the EU framework Directive on waste provides that Member States may object to a shipment of waste that does not correspond to their waste management plans (recital 9). The Court of Justice has however decided that this provision is of no relevance, as the possibilities of objecting to such shipments was regulated by the EU Regulation on the shipment of waste[116]. Revision of this regulation, currently ongoing, appears to be going in the direction of reinstating this possibility (*see chapter IV.4.2.4.4*).

IV.4.2.4.1.6 Reporting

Reporting allows public authorities to assess whether the legislative requirements have been complied with in practice, and adapt, where necessary, the conditions for the specific waste installation or the waste stream.

Member States, based on commonly agreed questionnaires, are obliged to regularly report to the European Commission on the implementation of the Directives. The Commission is obliged to regularly establish an EU-wide feedback report, often specifically to the other EU institutions on the implementation of the Directives; sometimes reports on several Directives or over several years are assembled in one single report[117].

The national reports are accessible to the public, upon request. The same applies to the EU reports. The EU reports are often disappointing in terms of providing data that would allow verification of environmental protection on the ground, as they limit themselves – according to the information in the national reports - to describing the national legislative and administrative measures, without discussing their practical applications. Thus, the reports do not really constitute a yardstick to assess compliance, enforcement and real fates of waste streams and associated impacts/ improvements. The quality of the information in these reports in very much dependant on the level of detail and type of information required by the questionnaires established for

---

[115] Court of Justice, case C-347/97 Commission v.Belgium (1999) ECR I-309 (concerning a programme for batteries).

[116] Court of Justice, case C-203/96 (see note above).

[117] Article 16 of the waste framework Directive requires for example that on the basis of reports submitted to the Commission it shall publish a consolidated report every three years, and for the first time in April 1996 (ie in April 1999, 2002, 2005 and 2008 etc...)

these purposes. As these questionnaires are drawn up in committees with no scrutiny or access by stakeholders, excluding ECOs among others, it is difficult to influence their content.

Specifically on those waste incinerators with a capacity of two tonnes per hour the relevant authorities must furthermore establish an annual report on the functioning and surveillance of the installation, which also reports on air emissions and water discharges[118]. Again, this report is required to be accessible to the public. Similar reporting requirements unfortunately do not exist for landfills. Generally, however, it should be noted that all emissions into the air and the water from waste installations which are in the hands of public authorities, must be made available to the public, on request (e.g. article 12 of Waste Incineration Directive).

IV.4.2.4.2 Directive 91/689 on hazardous waste

This Directive of 1991[119] defines hazardous waste as those wastes which correspond to the criteria listed in annexes of this Directive and are actually laid down in Commission Decision 2000/532 of May 2000. The general requirements of Directive 75/442, such as permits or management plans are strengthened for hazardous wastes. Specific requirements restrict the mixing of wastes.

IV.4.2.4.3 Limiting the generation of hazardous waste through products

Limiting or even completely prohibiting the use of certain substances, because they would make the waste which is generated at the end of a product's life-span hazardous, is current legislative practice, though there is no systematic approach to the problem at all; it also occurs outside the waste sector. For example, ozone-depleting substances or lead in petrol.

The main substances which are affected by such restrictions are heavy metals and other toxic, bioaccumulative or persistent substances. There are numerous provisions in EU law that limit the presence of heavy metals, in particular of cadmium, lead, mercury and chromium in products for example vehicles. Other examples concern polybrominated biphenyls(PBB), polybrominated diphenyls (PBDE), commonly used as flame retardants, which are banned, in addition to heavy metals in the Directive on restriction on hazardous substances in electrical and electronic equipment (ROHS)[120], or asbestos restricted by the Directive on marketing and restriction of use Directive[121]. In view of the present stage of European integration and the statement in the 6EAP that *"chemicals that are dangerous should be substituted by safer chemicals or safer alternative technologies not entailing the use of chemicals, with the aim of reducing risks to man and the environment"*[122], Member States may prohibit the use of other substances in products, where otherwise the material would, at the waste stage, lead to the generation of hazardous waste; this possibility has hardly ever been used; an example is the ban of lead in hunting ammunition, adopted by Denmark and The Netherlands.

---

[118] Directive 2000/76 on waste incineration, Article 12(2).

[119] Directive 91/689 on hazardous waste (1991) OJ L 377 p.20.

[120] Directive 2002/95/EC OJ L37 p 19

[121] For more examples related to products see Directive 76/769 relating to restrictions on the marketing and use of certain dangerous substances and preparations, (1976) OJ L 262 p.1 and its subsequent amendments.

[122] Decision 1600/2002 laying down the sixth Community environment action programme, (2002)OJ L 242 p.1. Article 7(1).

In the future, with the adoption of the Commission proposed legislative revision on chemicals policy, REACH[123] (*see chapter V.4.6.4.6*) there would be additional tools to prevent hazardousness of waste by reducing and restricting the use of hazardous substances in products and articles. Whilst waste itself is not covered by REACH requirements it is foreseen that products resulting from recycled wastes will be, although there is strong pressure from the relevant industrial sectors to obtain exemptions for such products.

99

IV.4.2.4.4   Shipment of waste, Regulation 259/93

As waste materials are movable objects and Member States - inside the Member States local authorities or other public or private bodies - provide for waste treatment and disposal installations, there is a tendency from Member States to limit the movement of wastes. In contrast, waste holders normally want to bring waste to places where its treatment or disposal is cheapest; they find support in the EU Treaty which generally provides for the application of Article 28 to 30 EC Treaty on the free circulation of goods. The present ongoing EU discussion on reviewing EU shipment rules for waste is likely to give Member States more possibilities to oppose shipments to other Member States.

In order to control waste shipments, the EU introduced the principle of prior informed consent: a shipment may only take place, where the administration of the Member State of dispatch has informed the administration of the Member State of destination of the planned shipment and agreement to the shipment has been given. Details of the procedure, of the documentation to be established and the security measures to be taken are laid down in Regulation 259/93 (*EC 1993*) on the shipment of waste.

The regulation, in parts is aligned to international commitments (such as the UN Basel Convention on the transboundary movement of hazardous waste), deals with the movements (shipments) of waste within EU Member States and the export and import of waste into the EU. It divides waste into non-hazardous ("green"), hazardous ("amber") and very hazardous ("red") categories and determines, when a Member State may object to a shipment of waste from another Member State.

Member States may object to any waste shipment to another Member State, where the waste it to be disposed of in the other Member State. However, where the waste is to be recovered and not disposed of in that other Member State - recovering includes the recycling and the burning of waste with the recovery of energy - Member States only have few, specifically enumerated grounds for objecting to such shipments.

According to ECJ case law the shipment of municipal waste to a dedicated waste incinerator in another Member State is normally to be regarded as a shipment for disposal and may thus be prohibited by the Member State of dispatch[124]. In contrast to that, the dispatch of waste for incineration to a cement kiln of another Member State (for co incineration) normally constitutes a dis-

---

[123] COM(2003) 644 final, Brussels, 29.10.2003.

[124] Court of Justice, case C-458/00 Commission v. Luxembourg (2003) ECR I-1553. Note: Some claim that this is a bit to simplistic interpretation of the judgement of the ECJ. They would say, that the judgement has caused new legal uncertainty – and that therefore restrictions against the export of waste based upon the uncertain imperfect between recovery and disposal continue to be very contestable.

patch for recovery[125]. Non hazardous – 'green' - waste which is shipped for a recovery operation is however not subject to any restrictions.

The EU has also prohibited shipments of hazardous waste to non OECD countries[126]. In this case it is irrelevant whether the shipment is for recovery or for disposal. This legislation follows a decision of the Basel Convention, adopted in order to prevent waste shipments to developing countries, where there is often a complete lack of adequate waste management infrastructure. Whilst the EU has made these obligations binding it is still the fact that this BASEL decision is internationally not yet in force. There are frequent attempts to bypass this export ban[127], for example by passing waste shipment containers through less monitored shipping routes or by selling end of life cars or ships to third countries as second-hand vehicles, to which product, rather than waste, legislation applies (product legislation containing no restrictions on movements).

A major revision of the Regulation is currently being discussed[128]. The ongoing revision contains some important improvements notably as regards the conditions determining when and under which conditions export for "sham recovery" (disposal falsely passed off as recovery) can be forbidden. The proposal (under discussion in the European Parliament and Council) so far foresees new (strengthened) criteria for authorities raising objections based on , among others, the ratio of recoverable and non-recoverable waste, non use of BAT despite having an IPPC permit in the installation of destination, lack of treatment in accordance with *legally binding environmental protection standards in relation to recovery operations, or legally binding recovery or recycling obligations established in Community legislation* and possibilities for raising objections in cases where the planned shipment or recovery would *not be in accordance with national laws and regulations in the country of dispatch relating to the recovery of waste, including where the planned shipment would be destined for recovery in a facility which has lower treatment standards for the particular waste stream than those of the country of dispatch, respecting the need to ensure the proper functioning of the internal market*[129].

In addition it foresees a *new Article 3.5 which states that municipal solid waste must be managed as waste destined for disposal.*

Note that nuclear waste is not included under the waste shipment Directive, as a separate Directive exists for this (*EC 1992, p. 24*).

---

[125] Court of Justice, case C-228/00 Commission v. Germany (2003) ECR I-1439.

[126] Regulation 120/97 (1997) OJ L 22 p.14.

[127] A recent report by an EU enforcement pilot project IMPEL revealed significant rates of irregularities in containers opened at some EU ports, IMPEL report November 2004 . A conference held in sequence to this report reported back that recent research in the Netherlands suggests that 50% of all Dutch waste exports to other EU states are possibly illegal, as are over 70% of exports to non-OECD countries. See **http://europa.eu.int/comm/environment/impel/ tfs_notified_waste.htm** for more information.

[128] Commission proposal for a regulation on the shipment of waste, COM (2003) 379.

[129] Draft Council Common Position on 2003/0139 (COD) 30 June 2004 - Article 13

IV.4.2.5 EU LEGISLATION ADDRESSING WASTE INSTALLATIONS – LIMIT-ING EMISSIONS

EU legislation has laid down detailed provisions on landfills (*EC 1999*) and waste incinerators (*EC 2000b*), but not on installations which treat waste for example by composting or recycling. There is however a lack of provisions on the decommissioning of installations. For waste installations, general provisions of EU law also apply, such as Directive 85/337 on environment impact assessment for the construction of new waste installations (*EC 1985*), Directive 96/61 on integrated prevention and pollution control (*EC 1996c*) or Directive 96/82 on accident prevention (*EC 1996d*).

IV.4.2.5.1 Directive 2000/76 on waste incinerators (including co-incineration)

For waste incinerators and co-incinerators, EU legislation has established relatively strict rules concerning emission limits into air and water. The objective was to ensure the maximum protection of human health and the environment (and at the same time, increase the acceptability of incinerators). The waste incineration Directive, adopted in 2000, replaces earlier Directives on municipal waste incinerators and hazardous waste incinerators, although a number of transition provisions continue to apply until the end of 2005. It deals with all installations that burn waste, also those which burn waste together with other fuels, such as cement kiln or power plants (co-incineration); the Directive does not apply to installations which incinerate radioactive waste, animal carcasses or wood waste.

The Directive fixes detailed conditions for the construction and operation of incinerators, including the minimum temperature or burning waste, emission limit values for air emissions of a number of pollutants - including dust, halogenated gases, NOx, SOx, TOC, CO and heavy metals as well as dioxins and furans. It also regulates discharges to the aquatic environment of waste water resulting from the cleaning of exhaust in accordance with the emission limit values on TSS, Hg, Heavy metals, dioxins and furans set in Annex IV. It makes a link to other relevant community legislation concerning the release of polluting substances into soil, surface water and ground water.

Larger waste incinerators are obliged to apply the best available technology, though this notion, as laid down in the waste incineration Directive is very vague. Generally, the emission standards are relatively strict, though Member States may even apply more stringent standards (for example in the Netherlands). Due to high potential emission releases particular attention is focused on measures that are planned to be taken in the case of malfunctioning of an installation. EU law provides that the total amount of malfunctioning of an installation may not exceed 60 hours per year. However, it is not clear how this provision is being enforced in Member States.

Furthermore, it must be remembered that small incinerators are not expressly addressed by EU legislation and also that the requirements of using the best available techniques only apply to the operation of large installations. Particular care is therefore needed to control the permits and conditions which are granted to smaller incinerators. This also applies to incinerators which are part of hospitals and to production installations which burn their own waste during the production process etc.

The previous **loopholes for co incineration** have been bigger and have been partially closed by the revised EU Waste Incineration Directive but it still has a number of shortcomings. The Directive does not offer solutions for some areas of potential ecodumping - for example, the threshold values of co-incineration are established in accordance with the waste disposal requirements for only some of the relevant parameters. In particular, limits for particulates (dust), nitrogen oxides and sulphur dioxide are much less stringent for co-incineration. Thus for co-incineration some emission limit values are less strict than for waste incinerators. Moreover, the EU Incineration Directive makes no mention of so called `feedstock recycling´ (burning waste materials for use as chemical as well as energetic input to a process) in installations such as blast furnaces. This is an especially serious omission because these processes are increasingly promoted as 'recycling' (in order to reach recycling targets), especially by the (end-of-life) car plastics industry.

IV.4.2.5.2   Directive 1999/31 on landfills

Landfills for waste are subject to detailed EU legislation as regards their construction and the permits for their operation. Landfills emit gases, in particular methane, which is a powerful greenhouse gas; liquids (eluates) from landfills may leach into the ground and groundwater, transporting heavy metals or other harmful substances. Technically speaking, landfills that totally protect against this effect do not exist and it is only possible to greatly reduce and possibly delay such effects. Hence the importance of ensuring that the sites for landfills are chosen with caution, and operated with care and that wastes are not accepted at landfills for which the landfill is not designed (liquid or hazardous wastes).

This Directive applies to all new and existing landfills which are in operation. Landfills are divided into landfills for hazardous waste, landfill for non hazardous waste and landfills for inert waste. All waste must be treated before disposal. Landfills for inert waste may not be used for other wastes, landfills for hazardous waste only for hazardous wastes. Liquid waste, hospital waste, complete tyres and explosive, corrosive or inflammable wastes may not be landfilled at all. The Directive also regulates in detail procedures for accepting wastes, provisions on measuring and control procedures and after care measures.

For climate change reasons (methane emissions), the disposal of the biodegradable municipal waste fraction is to be progressively reduced – by 25% (by weight) until 2006, by 50% until 2010 and by 65% until 2016, compared with 1995 levels (with extensions in the timetables for some Member States that invoke the escape clause, claiming that they were dependent by over 80% on landfill in 1995 for municipal wastes). Some Member States have taken further action to restrict landfilling of biodegradable waste at the national level - Germany will apply an outright ban on all landfilling of untreated household waste (targeting in particular biodegradable wastes) in 2005 and Norway[130] will apply a similar ban, on landfilling of biodegradable wastes, by 2009[131].

Landfills must be designed to prevent groundwater pollution as laid down in Directive 1980/68/EC (*see chapter IV.5.7.5*), which requires prevention of input of all hazardous substances into groundwater (black list) and limiting all other pollutants to prevent negative impacts on legitimate uses (e.g. drinking water abstractions).

---

[130] Norway, although not a Member State of the EU has signed a European Economic Area Agreement that requires , among other things that Norway adopts EU environmental legislation and participates in EU environmental programmes.

[131] In parallel, a recycling target for all types of waste - currently 75% by 2010 is to be increased to 80% by 2009 (source: Environment Daily 1845, 21/03/05).

IV.4.2.5.3   Directive 2000/59 on port reception facilities

In order to prevent the dumping of waste at sea, this Directive (*EC 2000c*) requires all EU ports to establish waste reception facilities for waste from ships. Ships that enter the port have to pay a fee for these facilities, irrespective of the actual use of the facility.

IV.4.2.5.4   Other waste treatment or recycling installations

So far under EU law, waste treatment (or other recycling) or composting installations only have to comply with the general provisions of the waste framework Directive 75/442 and, where they exceed a certain capacity, with the requirements of the integrated pollution prevention and control Directive. Details of the permits and conditions of operation are therefore established by Member States, either by legislation or on a case-by-case basis for each installation separately. Permits and the conditions for operation should be publicly available on request. So far there are no EU requirements as to the regular publication of comparative data as regards the emissions, discharges and   waste generation of such installations. As concerns specific legislation on biowaste, despite various political calls and even clear commitments from the Commission itself and the existence of a widely discussed draft document (*CEC 2001*) there is a current political reluctance within the European Commission to come forward with a proposal for this waste stream[132].

IV.4.2.6   EU LEGISLATION ADDRESSING SPECIFIC WASTE STREAMS

There are specific EU Directives on waste oils (*EC 1975*), sewage sludge (*EC 1986*), batteries (*EC 1991*), packaging and packaging waste (*EC 1994*), PCBs and PCTs (*EC 1996f*), waste from the titanium dioxide industry (*EC 1992b*), end of life vehicles (*EC 2000b*) and waste from electrical and electronic equipment (*EC 2002b*). Animal waste is regulated by a separate regulation (*EC 2002c*), in the context of agricultural policy.

IV.4.2.6.1   Separate collection; recycling targets

A key element of EU waste policy and achieving progress in moving collective EU waste management environmental performance level up the waste hierarchy from disposal to recycling is the establishment of minimum EU recycling targets and the associated and necessary separate collection. Whilst there is no general EU obligation to separate waste, in some Member States, - for example Denmark, Sweden, Austria the obligation exists at the National level.

However, specific EU waste stream legislation requires Member States to set up *separate collection* schemes for some waste streams – in particular for batteries, packaging waste, cars and electrical and electronic waste. For some of these waste streams dedicated **collection targets** are set which mean that a certain quantity of a given material must be (separately) collected, (eg. waste from electrical and electronic production, where Member States shall ensure that on average four kilograms of such waste per citizen and per year is collected)[133]. For other waste streams the col-

103

---

[132] See Letter to Commisioner Dimas from EEB. ASSURE, FEAD, RREUSE, ISWA and ECN, 7 April 2005

[133] The EEB recommended higher targets, estimating that 4kg was equivalent to less that 30% of the potential collection possible – for more information see EEB argumentation paper *Towards Waste-Free Electrical and Electronic Equipment,* *March 2001.*

lection objective is more generic[134] or is related to the quantities generated. For example, for batteries (new Directive proposal still under development) it is proposed (by the European Parliament) as a percentage of the national sales of batteries in each country (as the quantities of total battery consumption differs greatly between the old and new Member States).

104

Recycling targets are percentages which must be reached for the recycling of certain materials. For example, the Directive on packaging and packaging waste requires that by the end of 2008, 60 percent by weight of glass packaging material must be recycled. Such recycling targets exist, in EU law, by now for packaging waste, cars, electrical and electronic equipment; they were not established for batteries or sewage sludge, where they might be less appropriate.

There is some discussion at EU level, whether such recycling targets for specific sectors should not be replaced by *recycling targets for materials* - for example a recycling target for all glass, not only glass from packaging waste. The problem here is how such a recycling target can be organised (how the responsibilities for achieving the target attributed) and enforced, and whether the efforts in a specific sector would not be diluted by spreading the responsibility over several sectors. (e.g. packaging paper targets if changed to just paper targets would be probably cause a shift to non-packaging sources of paper and possibly reverse progress made on separate collection of packaging).

What is clear is that recycling targets usefully fix objectives for authorities and the public to attain and are capable of mobilising efforts and means. They require better organised collection of data, well developed collection systems and accompanying measures to persuade the public to participate in collecting.

IV.4.2.6.2   Producer responsibility, individual producer responsibility

On the basis of the Extended Producer Responsibility principle, producers should be responsible for all costs relating to the management of some waste streams (including collection)[135]– under EU law this is most notably the case for waste electrical and electronic equipment (WEEE) where extended producer responsibility is specified to the level of individual producer responsibility (*article 8 of Directive 2002/96 on WEEE*). **Individual financial producer responsibility** means that each individual producer is liable regarding the costs relating to the waste management activities which are required for its own-brand products when they become waste. It does not however relate to the organisation and logistics of take-back systems, which is a common misunderstanding. Individual responsibility does not prohibit co-operation among producers – but should maintain clear-cut ultimate responsibility by an identifiable individual producer. This means that companies will need to be informed about the waste from their own products and the costs corresponding to them.

---

[134] For example for waste oils it is simply stated in article 1 - Member States shall take the necessary measures to ensure that waste oils are collected and disposed of without causing any avoidable damage to man and the environment. Which can be interpreted as all waste oils available for collection.

[135] For example for waste oils it is simply stated in article 1 - Member States shall take the necessary measures to ensure that waste oils are collected and disposed of without causing any avoidable damage to man and the environment. Which can be interpreted as all waste oils available for collection.

The added value of the concept of individual responsibility, is that through the **internalisation of external costs, it creates a direct upstream effect which should contribute to designing for the environment.** In this way, every improvement in design will have a direct effect on the costs the producer should have to bear for potential treatment at the end of the product's life and eventually on the price consumers have to pay for the product.

Improvements in terms of design which will have a direct effect on waste management costs can include improved ease of dismantling (requiring less time equipment), use of easy and positive post-waste value materials for recycling, the direct use of recyclate for the manufacturing of new products (cheaper materials) — which will lead in turn to the use of less raw materials and saving in resources[136].

An important caveat: individual financial responsibility only drives eco-innovation if a very demanding framework is set, thus preventing cost reduction strategies by ecodumping. This applies especially to recycling, collection and chemicals targets. Compromises on those points create incentives to reduce costs by choosing environmentally less performing waste management options that cause more damage to the environment..

The effectiveness of the individual producer responsibility in the WEEE directive will in the end depend greatly on the way the EU directive is transposed into National legislation. According to an industry platform (the Recycling Platform - created to follow-up on the producer responsibility requirements and their implementation) many member states are transposing the EU Directive into National legislation in such a way that producers are, in practice, pushed into systems of collective responsibility.

In contrast to individual financial responsibility, collective financial responsibility does not remunerate the innovating producer. Innovation in ecodesign by one producer will probably get diluted and lost in the collective financing model. Environmental improvement through a collective model may only be achieved via consensus, if the participants of a collective model agree on differentiated fees or if there is competition between individual and collective models. However, producers are unlikely to agree on an environmentally differentiated fee as there will always be some 'losers' that will have to pay a higher fee). So only allowing 'competition' between schemes drives innovation as it gives a company opportunities for choice and hence also for leaving a collective financial responsibility scheme, if it prefers to do so.

IV.4.3   Voluntary Instruments

The use of non binding instruments in waste policy so far has been exceptional. In 1981, a **recommendation** on the recycling of paper was adopted[137]. However, this recommendation was almost completely ignored by professionals and also by administrations, and had virtually no impact. In a number of cases, manufacturing lobbies have tried in the past to persuade EU insti-

---

[136] Note: Improvements at the level of waste management costs can also be achieved by a shift in company policy by trying to sell services rather than products. This will also contribute to the idea of sustainable consumption, less use of resources, of hazardous substances, etc. as mentioned earlier. These concepts are not recognised by the Directive.

[137] Council Recommendation concerning the re-use of waste paper and the use of recycled paper (1981) OJ L 355 p.56.

tutions to back up industrial commitments by establishing **voluntary agreements**, mainly in order to escape legislative measures. So far they have been unsuccessful.

IV.4.4 Standardisation (see also chapter V.4.5)

In the waste sector, few EU standards have been elaborated until now. The standardisation or New Approach[138] has only been adopted in the packaging waste sector. Following the New Approach the EU Directive on packaging and packaging waste, beyond the targets and mechanisms for waste collection and recycling, fixes general provisions in order to define common understanding of the composition and the reusable and recoverable (including recyclable) nature of packaging – these are called the **essential requirements**[139]. Promotion of prevention of packaging waste was also included. The Directive however only laid down very general guidelines and environmental objectives to be reached (in *Annex II of the Directive*), leaving it to industrial standardisation by CEN to elaborate European standards for prevention, reusables, recyclables and other aspects. The approach was a complete failure: though there was a detailed mandate given to CEN, the CEN standards were so poor that the Commission refused to accept them[140] and make them official by publication in the European Official Journal[141]. The EEB and the Consumer Standardisation watch dog ANEC followed these developments closely[142] and the EEB even provided alternative proposals[143]. Unfortunately this refusal did not stop the packaging standards drawn up by CEN: the only effect which the refusal had was that the presumption that packaging complying with the standards complied with the packaging Directive, did not come into effect and so Member States were left to find national solutions for enforcement. Pressure to find an EU solution (and certainly pressure from industry for formal adoption of the CEN standards) lead to a second attempt to define the standards in CEN, with a `new` mandate. The mandate however was not significantly different from the first (as a result even CEN acknowledged it could not achieve very different results from the first attempt). In February 2005 the Commission, after putting the standards quickly through Member State committees, performed a complete U-turn and published the standards (almost identical in content to the ones refused in 2000) in the Official Journal.

---

[138] The "New Approach" is a specific policy approach characterised by a clear separation of tasks and responsibilities between the European legislator and the European standards bodies (European Committee for Standardisation-CEN, European Committee for Electrotechnical Standardisation-CENELEC and European Telecommunication Standards Institute-ETSI) – for more information see chapter V.4.5 on Standardisation)

[139] The essential requirements only establish the results to be attained, without specifying the technical solutions to do so

[140] With the exception of EN 13432 "Requirements on packaging recoverable through composting and biodegradation" which was published in the OJ. Note that the standard on prevention was `published` but with an exception for the essential requirements listed under Annex II, point 1 third indent, which means this standard gives no presumption of conformity for the Packaging Directive.

[141] See NGO reaction paper - *Success for EEB and ANEC rejection of packaging standards March 2001*

[142] See EEB position paper *CEN at work_ How the requirements of the European Packaging and packaging waste Directive are by-passed by the CEN standards*, Sept 2000. All this was done from outside of the formal EU standardisation process as at that time the ECO EU level standardisation watch dog did not exist, only some German EEB members had, albeit restricted, participative access to the German national mirror body. See **http://www.eeb.org/activities/waste/ packaging/Index.htm**

[143] See EEB's own alternative proposal for credible standards (to prove that they could have been set) - *EEB on Drafting effective packaging standards*, June 2001.

The standards (as they now stand) still fail to promote packaging prevention, reuse and recyclability[144]. Some illustrative examples are:

▶ they allow companies to use vacuous arguments on packaging presentation, such as "it has to look bigger than it is", to justify voluminous and unnecessary packaging.

▶ they allow companies to produce so called 'refill pouches' (common for eg in laundry softeners) and claim that this is re-usable packaging while the pouch is in fact a one-way throwaway packet and no re-use in the true 'multiple trip' sense occurs. No minimum amount of 're-usage' is defined in the standard.

▶ they allow packaging with any percentage of recyclable materials to be called 'recyclable'. A packaging item with as much as 50% inert (non-burning) material, made of PVC or with unacceptably low energy content meets the standard on 'energy recovery'.

The reasons for the failure of the Packaging standards are primarily due to imbalances in the interests of the parties defining the standard. (*for more information see chapter V.4.5 on Standardisation*)

With the general trend of deregulation, simplification of legislation, making `better legislation` etc. it cannot be excluded that standards in the waste sector will become more frequent in future. In theory, they could replace much of the existing legislation, in particular where technical details are regulated, such as acceptance criteria for wastes in landfills or incinerators, emission standards, recycling and recovery targets etc.

IV.4.5   Taxes and charges

EU waste law does not, so far, provide for **economic instruments**, such as for taxes, charges or fees. Such measures are left to Member States, principally for reasons of subsidiarity: all attempts by the EU to introduce such measures at EU level have been rejected so far[145].

Economic considerations strongly influence the behaviour of waste generators and holders. A landfill tax increases the price for landfill disposals and may make forms of waste recovery economically more attractive. A tax on metal can packaging makes this form of packaging less attractive for producers and consumers. The example of collection charges for municipal waste graduated according to the weight of waste already mentioned above is another one.

Public authorities have the possibility of influencing waste pricing, as they frequently operate waste treatment and/or disposal operations, organise the collection of municipal waste, and occasionally industrial waste. They also sometimes have the possibility of initiating or even

---

[144] For more details on how the standards fail to achieve the objectives of the packaging Directive see *ANEC-ECOS Position Paper on the revised Packaging Standards prepared under the second standardisation mandate M317* , January 2001 and EEB press release *Commission encourages increases in packaging waste!*, 22 February 2005

[145] See, for example, Article 15 of Directive 94/62 on packaging and packaging waste, which provides: "Acting on the basis of the relevant provisions of the Treaty, the Council adopts economic instruments to promote the implementation of the objectives set by this Directive". No such measures were yet adopted.

adopting legislative measures. Influencing the pricing of waste treatment and disposal activities via taxes, charges and fees also requires from public authorities, a clear long-term strategy on waste prevention and recycling objectives to be reached, which treatment and/or disposal installations are to be favoured and considerations on what materials should be financially favoured or disadvantaged.

108

Another good example of the use of taxes is a tax on plastic shopping bags introduced in the Republic of Ireland which has cut their use by more than 90% and raised millions of euros in revenue. 3.5 million euros of the revenue raised is to be spent on environmental projects. The tax of 15 cents per bag was introduced in March 2002, in an attempt to curb litter, among other objectives.

IV.4.6 Waste Policy today and tomorrow - New approach in the 2005 Strategy?

(Note: This chapter is based on the state of discussion surrounding the elaboration of the EU Waste Strategy in June 2005.)

It is important to realise that waste management within the EU is traditionally marked by the **dispute on competence between Member States**, often also regions and local authorities, on the one hand, **and the EU** on the other hand. As Member States are responsible for constructing and operating waste installations, they can considerably influence waste flows. On the other hand the EU has the possibility of influencing waste management through the provisions on the free circulation of goods (and some wastes) which prevent Member States from closing their borders to waste imports or exports. This tension is a useful one for reinforcing such principles as the proximity principle.

The upcoming features of the Thematic Strategies generated by the 6EAP – in particular on waste prevention and recycling – which follow some new lines of thought, whilst not formalised (and so still subject to changes) are worth mentioning as they illustrate the political environment in which EU waste policy is evolving at this point in time.

The strongest evidence of changes in approach can be seen in the Commission's intentions for amending the waste Framework Directive (WFD). The Commission's proposals **for the WFD** revision involve at least 4 fundamental changes:

1. **undoing the current proxy approach to waste policy (waste hierarchy based on a quantitative approach to prevention > recycling > incineration > landfill)** by introducing a 'new' environmental objective and a lifecycle approach (as a new article 1 of the WFD). While nothing can be said against refining the quantity based approach in order to increase the environmental effectiveness of measures and ensure that environmental trade offs are better taken care off (e.g. climate versus water pollution), it would require a huge amount of data (e.g. chemical compositions of products and waste) and modelling capacity (e.g. predicting ecosystem responses). As this condition is not fulfilled and it is unlikely that Member States or industry

would be willing to invest in it, replacing the current waste volume approach with environmental and life-cycle assessments would

i. offer no further progress in terms of real implementation (no clarification of Member States' responsibilities/obligations towards implementation for example)

ii. potentially lead to increased environmental degradation and health impacts.

Further the 'lifecycle thinking' would need to be carried out at local authority level and their establishing of waste management plans[146], for which resources are currently not available. At the same time the attack on the credibility of the waste hierarchy gives increased justification for subsidiarity in waste policy - disengaging the EU from its responsibility of steerage in waste policy (as Life Cycle approaches and analyses are more robust at the national and local level) good examples of the risks of this approach can be seen in the already existing debate on the packaging and packaging waste Directive.

According to the current WFD the primary objective should be to restrict the production of packaging waste. As an avoidance technique the industry tries to steer the discussion in the direction of merely reducing the lifecycle impacts of each packaging unit. In practice this leads to the distraction of political pressure to reduce quantities of packaging generated and attempts to establish tools such as Packaging Environmental Indicators (PEIs). National pilot projects (for eg in The Netherlands) have revealed difficulties with PEI's ability to deliver good EU steerage ie steering to wards recycling/other forms of treatment (primarily due to lack of sufficient data, specific nature of different kinds of packaging and the high weight of the distribution (transport) factor in the lifecycle formula). In general the conclusion is that such indicators would be even less feasible at the EU level where data quality is an even greater challenge[147].

2. **introduce approaches to establish end-of-waste (fitness for use) criteria that may restrict the scope and precautionary nature of waste policy.** As stressed above the scope of the waste definition is the key to ensuring that the impacts of waste generation are reduced and most importantly that there is a strong (regulative pressure) incentive to generate LESS waste. A key argument for reducing the scope of waste policies is the promotion of recycling. But the actors behind this argument would instead suggest that the main interested actors are co incinerators who want to burn more waste – as a cheap fuel alternative - without too much administrative control, or material streams that would like to shed the material cycling responsibilities – eg plastics wastes. The EEB has explained in a study why it thinks that this is not beneficial to the environment[148].

---

[146] The life cycle approach risks asking Member Sates to go down something similar to the UK Best Practical Environmental Option (BPEO) route. This can lead to to endlessly complicated processes, requiring heavy resources (where citizens' organisations' interests often lose out) leading in turn to serious governance deficits and the lifecycle analyses are in turn highly vulnerability to distorting boundary and assumptions. The UK environment ministry has recognised this (*Changes to Waste Management Decision Making Principles in Waste Strategy 2000*, DEFRA, Dec 2004) and are now abandoning the BPEO approach. Sources indicate that it will probably go back to applying the waste the hierarchy and **strengthening the hierarchy approach** and consultation practices.

[147] The Study on the *Implementation of the Directive on Packaging and Packaging Waste 94/62/EC and options to strengthen prevention and reuse of packaging,* February 2005 done for the European Commission concludes that the Dutch experience suggests that *'a simple PEI is too simple for any other application than as a product development tool for industry'*

[148] See: EEB study: Ecodumping by Energy Recovery, 2001

3. **introducing pro-incineration approaches to the recovery definitions;** The Commission's proposed solution (at the time of writing) is to establish, through technical committees, a general definition of recovery based on the notion of substitution of resources, including a corrective mechanism to list specific processes in the disposal annex where it is considered that from an environmental perspective it is not appropriate to classify them as recovery (and so remove the recovery status maybe given to them by this very broad definition of recovery – eg landfill with biogas capture could be newly classified as recovery by this definition, and would need to be de-classified to disposal again). This correction mechanism may also make use of efficiency (eg energy efficiency) thresholds to clarify specific cases (ie waste incinerators). The proposal as it stands is unsatisfactory as, among other reasons:

   i. the wider interpretation of substitution of resources will mean that indirect power/heat generation achieved by energy in waste incinerators (substituting that in dedicated power generation installations) will classify all modern waste incinerators as recovery (effectively reversing ECJ ruling on the matter – for more details see below) and,

   ii. the use of efficiency thresholds is undefined, and if used only for energy efficiency, is insufficient to serve the multiple considerations for waste management (prevention hazardousness, separation of recyclables, proximity principle etc). The EEB has proposed an alternative approach for 'credible' recovery criteria.

There also serious concerns at this point in time on the part of the various stakeholders (ECOS and even industry interest groups) as to the democratic adequacy of the committees proposed to carry out such changes given the current poor track record of political decisions taken by the Commission using such committees[149].

4. **delegating setting of recycling standards to the IPPC Directive instead of setting up a dedicated horizontal daughter Directive** (as has been done for landfill and incineration). The EEB does not believe that the IPPC is the right place to address such waste policy aspects, for several reasons. One such reason is that the implementation of the IPPC Directive seems to be of low quality leading to huge national and regional differences in environmental protection (*see chapter V.3.2*)[150].

At the same time the Commission shows intentions of revoking (deleting) the Waste Oils Directive. Interestingly the Commission has failed to bring forward any environmental data justifying such a step and despite adequate evidence that the problem is not with the Waste Oils legislation but contradictory to EU economic policies (fuel tax exemptions encouraging the burning of waste oils).

---

[149] See EEB press release on case of ROHS technical adaptation committee - *Commission risks credibility in its implementation of electronic products safety rules*, 18 March 2005

[150] In July 2003 the Commission reported to Parliament and Council about the progress of IPPC implementation. The findings are alarming: huge delays, inadequate legal transposition and candidate countries facing serious capacity problems. It concludes that successful application of BAT by 2007 is essential to achieve more sustainable production patterns and therefore warns that if efforts of authorities are insufficient it might be necessary to establish a more harmonised approach, i.e. by setting EU wide emission limit values. But the Commission until now refrained from making any concrete proposals to revise or how to complement IPPC.

Similarly the Commission refuses to carry out its previous commitment to complement the waste legislative set-up with a much needed biowaste and composting Directive and prefers to consider product standards or IPPC run process standards as sufficient, leaving the legislative tools to Member States, with each one setting its own Strategies.

Both moves are difficult to place in an environmental impact approach, where there is the evidence for benefits of waste oil regeneration[151] and removing biodegradable waste from landfills and supplying clean compost to European Soils has widely documented benefits.

To conclude, a change from the existing approach to waste policy with a strongly precautionary (especially in approach to scope) and 'quantity as an impact proxy' approach, making full use of leadership from the EU level and using binding legislative tools to one that uses a complex lifecycle impacts approach, delegating leadership to the national level and preferring standards and non-legislative tools all point in the direction of a Commission that is deregulating on waste policy[152].

IV.4.7   Implementation and decision-making

IV.4.7.1   RESPONSIBILITIES

Implementation of EU legislative measures is, in principle, the obligation of Member States (*Article 175(4) of the Treaty*). In the area of waste management, this basic principle must not be forgotten, as EU waste legislation presupposes the adoption of national, regional or local legislative measures in order to be fully operational. It is furthermore most important that waste legislation is properly applied at all administrative levels, that the necessary monitoring and control measures are taken by the public authorities and that in particular the illegal shipment, illegal incineration, illegal operation of landfills and illegal disposal of wastes is exposed and stopped. In all Member States, the application of existing legislative provisions by waste operators and other persons continues to be a serious problem[153]. Generally, it must be said that in all Member States bad implementation, application and enforcement is one of the biggest problems for environmental law in general and for waste management law in particular.

---

[151] See EEB submission to Commission consultation on the revoking of the Waste Oils Directive - EEB response to Waste Oils consultation Feb 2005

[152] For a good and still relevant overview of the challenges waste policy faces see EEB publication – Towards a Low waste Europe – Ten Key Issues, April 2001.

[153] For example the Commission has recently applied to the European Court, because there are more than 8.800 unauthorised landfills in operation in France (see Commission Press release 13 July 2004 - IP/04/895). Though this practice is banned since more than a decade, unauthorised landfills operate not only in France, but in all Member States. A similar application is made against Italy, where nearly 5000 unauthorised landfills operate (Commission Press Release - IP/04/930 of 15/07/2004).

IV.4.7.2   COMPLAINTS

Individual persons and environmental organisations are entitled to complain to the European Commission, whenever they are of the opinion that waste legislation is not being applied correctly. Complaints to the European Commission on the infringement of EU waste legislation are frequent *These* complaints mainly concern the siting of waste incinerators or landfills, (some recent examples are *case 156/03 December 2004* Commission versus Italy, *case 420/02 Nov 2004* Commission versus Greece, *398/02 June 2004* Commission versus Spain etc) absence of environmental impact assessments, the operation of unauthorised landfills, and, more generally, the emissions or discharges from waste installations beyond the authorised limits.

IV.4.7.3   USE OF WASTE MANAGEMENT PLANS

As mentioned before, waste management plans have to be sent to the European Commission with the intention being to progressively achieve an EU-wide waste management overview and coordination. However, until now, the Commission has hardly ever looked at the quality or completeness of waste management plans or tried to integrate different plans into a greater planning structure, or looked at the details of national implementation reports. Waste management plans have a powerful potential for compelling Member States to implement and achieve waste management objectives. The current questionnaire that Member States have for reporting purposes[154] is probably too open and unspecific for these purposes – creating a tendency for Member States to just list what legislation has been adopted. Ideally the questionnaire should be revised and this should be done with proper stakeholder involvement and scrutiny through the use of a formal Waste Steering group (as exists for Air Policy containing both representatives of national competent authorities and stakeholders) and dedicated working groups.

IV.4.7.4   THE IMPORTANCE OF WORKING LOCALLY

The main challenges for local authorities is to set up waste management strategies and plans that are coherent with national waste management plans. Unfortunately, EU local authorities are generally not equipped to meet these obligations, at a time when strong political pressure for further devolution (of powers from EU level to national level) and more flexibility on waste policies is on the increase. There is an urgent need to fill the gap between those in charge of proposing new legislation, and those with the responsibility for implementing it. As it is often the local authorities that bear the burden of implementing the requirements of EU waste legislation, the need to improve their involvement in the preparation of legislation and to strengthen the support given for the exchange of best practices among them is vital. Working locally to achieve general environmental goals is thus critical for citizens' environmental organisations.

---

[154] The Questionnaires are laid down in **Commission Decision 94/741/EC** and **Commission Decision 97/622/EC** concerning questionnaires for Member States reports on the implementation of certain Directives in the waste sector. Available online at Commission Waste Policy website http://europa.eu.int/comm/environment/waste/legislation/d.htm

IV.4.8   Links and crossovers with other EU legislation

IV.4.8.1   LINKS WITH ENVIRONMENTAL MEDIA POLICIES:

Below are some of the links and crossovers from waste legislation to legislation on other environmental media and public participation requirements. The list is not exhaustive but gives some
examples of how, in particular other policies, can reinforce and even strengthen the ambition
level of waste policy. This in turn will assist us in achieving the necessary levels of pollutant reductions and reaching the objectives of clean air, water etc.

IV.4.8.1.1   Links to Water Quality

Landfills - see chapter IV.4.2.5.2. Annex I of the landfill Directive establishes requirements for
effluent (leachate) collection and treatment and foresees standards on leachate measurement
methodologies. The leachates collected must be 'treated to the appropriate standard required for
discharge'. Applicable standards can be found in Annex IX of the Water Framework Directive
2000/60/EC and further national and EU wide standards have to be developed following the procedures of the Water Framework Directive as outlined in *chapter IV.5.5.2.1*. Overall landfill
leachates must prevent pollution of soil, groundwater and surface water. Landfills must be
designed to prevent groundwater pollution as laid down in Directive 1980/68/EC (*see chapter
IV.5.7.5*), which requires prevention of input of all hazardous substances into groundwater (black
list) and limiting all other pollutants to prevent negative impacts on legitimate uses (e.g. drinking water abstractions).

Waste Incineration - *see chapter IV.4.2.5.1*. The regulation of incinerator effluents in the waste
incineration Directive has a direct link to other **relevant community legislation** as it concerns the
release of polluting substances into soil, surface water and ground water. This includes the 1980
Groundwater Directive which can be used to establish the level of environmental protection associated with emissions into water .

Waste oils – the waste oils Directive requires that Member States ensure the prohibition of any
discharge of waste oils into internal surface waters, ground water, coastal waters and drainage
systems.

On this point it is important to note that the concept of 'good water status' in the Water
Framework Directive (*see chapter IV.5*) is defined as good ecological as well as physio-chemical
status. This means that emissions to water bodies cannot result in a decline in the ecological
health of the water system and must allow the achievement of a close to natural state.

IV.4.8.1.2   Links to Air Quality

Waste Incineration – see chapter IV.4.2.5.1. The Directive on waste incinerators fixes emission limits for some air pollutants. These apply together with the general quality provisions on air specified by the Air Quality framework legislation and daughter Directives on specific pollutants (eg
on SOx, Particulate Matter (dust) and NOx) (see chapter IV.3.3).

Landfills – see chapter IV.4.2.5.2. The Directive on landfilling of waste sets emission controls on methane (a powerful global warming potential gas) and requires collection of methane (*Annex I paragraph 4 of the landfill Directive* – all methane must be collected and used or flared). Furthermore biodegradable waste targets are set for diversion of methane producing materials from deposition in the landfill – a 25% reduction (in weight) by 2006, 50% by 2010 and 65% by 2016, compared with 1995 levels (with extensions in the timetables for some Member States that make use of the escape clause that they were dependant by over 80% on landfill in 1995 for municipal wastes).

Waste oils – the waste oils Directive requires that Member States ensure the prohibition of: *any processing of waste oils causing air pollution which exceeds the level prescribed by existing provisions* (i.e. community waste incineration and presumably also therefore air quality legislation)

IV.4.8.1.3   Links to Soil Protection

Waste oils – the waste oils Directive requires that Member States ensure the prohibition of deposit and/or discharge of waste oils harmful to the soil.

ECJ Case 1/03 Van de Walle[155] on contaminated soils: This case ruled that where fuel is leaked from a petrol station, the leaked fuel and the contaminated soil are both waste, even before the soil is excavated. Under certain conditions, also the petrol company may be obliged to remove the contaminated soil. This has important implications for enforcement of the polluter pays principle in contaminated soils clean-up situations.

IV.4.8.2   LINKS WITH INFORMATION AND PARTICIPATION LEGISLATION (see chapter V.2)

National, regional and local waste policies are heavily influenced by grassroots action (for example in the UK the UK government recognises that due to public resistance large increases in dedicated incineration capacities is not a viable political option, leading them to more actively explore not only recycling strategies   but also other incineration strategies such as co incineration). In this context the right of access to information is crucial. Applications for a permit and the content of the permit granted, including all conditions on the operation and emissions of an installation, must be made available to the public according to the EU Directive 2003/35 providing for public participation[156]. Administrations are not entitled to restrict access to any data on emissions into the air, the soil or the water that stem from waste installations or the shipment, handling or treatment of wastes.

Citizens and environmental organisations have a right to comment on **applications for a permit, on applications for the construction** or a significant change of a waste installation. Where they might be affected by a waste installation, they are entitled to participate in any environment impact assessment which the administration undertakes. Furthermore, they are entitled to par-

---

[155] Judgment of 7 September 2004.

[156] Directive 2003/35 providing for public participation in respect of the drawing up of certain plans and programmes relating to the environment and amending with regard to public participation and access to justice Council Directives 85/337 and 96/61 (2003) OJ L 156 p.17.

ticipate in the making of waste management plans which often determines the siting of waste installations[157].

Another useful source of information is the recent establishment of a European Pollutant Emission Register (EPER)[158]. The first set of emissions data, covering the year 2001, was published in February 2004. The second set, covering 2004, will be reported in 2006. EPER contains data on the main pollutant emissions to air and water reported by around 10,000 large and medium-sized industrial facilities (including waste incinerators and waste disposals sites) in the 15 EU Member States and Norway. Fifty pollutants are covered. All reported emissions data are publicly accessible through the EPER website, which is hosted by the European Environment Agency[159] in Copenhagen. The website also provides descriptions of each of the substances, their uses, major emission sources and their impacts on human health and the environment.

IV.4.8.3   TECHNIQUES USED BY INSTALLATIONS

Larger waste installations (notably Incineration, Mechanical biological Treatment (MBT), waste storage, some metal and chemical liquid and sludge physio-chemical treatments, immobilising ashes, preparing fuel from liquid waste, preparing waste as fuel) also come under the Directive on integrated prevention and pollution control **(IPPC)** and must thus comply with best available techniques BAT (*see chapter V.3.2*). This technique is normally laid down in individual permits and Best Available Technique Reference documents **(BREFs)** are established at EU level (see chapter on IPPC for more information.) Note that BAT should be used to achieve waste management objectives and emission levels stricter than the specific emission levels set in Waste Directives.

The environmental performance benchmarks contained in the BREFs are not always satisfactory (*for information on this see chapter on IPPC V.3.2.3*). In the case of the final draft document for the Waste Incineration BREF from May 2005 the BAT associated emission levels were clearly under ambitious. The EEB, which participates in the BREF development process, therefore demanded (in November 2004) that so called 'split views' be recorded in the final text of the Waste Incineration BREF. This means the EEB is not convinced that the emission levels corresponding with BAT in the revised BREF were indeed a good representation of EU BAT. The split views, if maintained in the final BREF document, will show alternative, more ambitious benchmarks that authorities emitting

---

[157] See Annex I of Directive 2001/42 on the environment assessment of the effects of certain plans and programmes on the environment (2001) for specific criteria OJ L 197 p.30. Annex I includes among other elements ..

(f) the likely significant effects (these effects should include secondary, cumulative, synergistic, short, medium and long-term permanent and temporary, positive and negative effects) on the environment, including on issues such as biodiversity, population, human health, fauna, flora, soil, water, air, climatic factors, material assets, cultural heritage including architectural and archaeological heritage, landscape and the interrelationship between the above factors;

(g) the measures envisaged to prevent, reduce and as fully as possible offset any significant adverse effects on the environment of implementing the plan or programme;

(h) an outline of the reasons for selecting the alternatives dealt with, and a description of how the assessment was undertaken including any difficulties (such as technical deficiencies or lack of know-how) encountered in compiling the required information;

[158] 2000/479/EC: Commission Decision of 17 July 2000 on the implementation of a European pollutant emission register (EPER) according to Article 15 of Council Directive 96/61/EC concerning integrated pollution prevention and control (IPPC) (notified under document number C(2000) 2004)

[159] See http://eper.cec.eu.int/eper/

Waste Incinerator permits at the local level can require. The split views demanded by the EEB concerned the emission level ranges for releases to air and water and are in the range of 1/2 to 1/3 of the presented ranges. More specifically concerning emissions to air of total dust (as NO_2), Mercury (Hg) and its compounds, Cadmium (Cd) and dioxins and furans (PCDD/F), and emissions to water of Hg; Cd; Arsenium (As); Tin (Sb);Vanadium (V) and PCDD/F.

IV.4.8.4   NATURE PROTECTION

The siting of a waste installation, for example an incinerator or a landfill, within a designated European habitat (*see chapter IV.2*), is additionally subject to assessment under Directives on the protection of birds and the protection of habitats[160]. Only exceptionally do such installations not constitute a significant disturbance to habitats, due to emissions, traffic increase etc.

IV.4.9   Tips for ECO action (Environmental Citizens' Organisation)

There are a number of ways in which environmental organisations at local, regional, national and European level could consider taking action.

(1) **Use the complaints procedures (*see chapter VI.2*)**. Make sure that the existing provisions of EU waste law and of national, regional and local waste management law are actually applied. EU law is binding and, in the case of diverging rules, prevails over national, regional or local provisions. Where this is contested, the case should be submitted to the EU Court of Justice as the final arbitrator..

This applies to all areas of waste legislation that were mentioned above: the classification of materials as waste and not as used products, by-products, secondary raw materials etc.; the push towards a policy and concrete measures to increase and improve waste prevention and waste recycling and to get away from landfilling; the development and putting into practice separate collection and recycling of waste fractions; the achievement of recycling targets, where such targets have been fixed etc

(2) There is an accumulating body of **experience on waste prevention** in the EU Community. Good cases and examples on waste prevention initiatives can be found in for example the Belgian (Flemish region) Municipal waste authorities (OVAM) who have a report on their waste prevention planning and indicators[161], the Spanish (Catalan regional) authorities who recently established the Catalan Foundation for Waste Prevention and Responsible Consumption[162], the Waste Prevention project in Vienna (Austria)[163], the EEA EU Topic Centre on waste has a database with examples and

---

[160] Under COUNCIL DIRECTIVE 92/43/EEC of 21 May 1992 on the conservation of natural habitats and of wild fauna and flora

[161] See OVAM publication Flemish National Waste prevention indicators (September 2002), translation to English available from EEB

[162] See www.pangea.org/cepa

[163] See www.abfallvermeidungwien.at

case studies[164] and the OECD has recently developed waste prevention indicator documents. ACCRR (the Association of Cities and Regions for Recycling-Brussels[165]) is undertaking a study on prevention experiences in local authorities and ICLEI[166] has many useful local authority links. The EEB has also drafted a working document with proposals on EU level Waste Prevention tools and indicators which includes some information on experiences with waste prevention[167].

(3) Bear in mind that **material recycling** is frequently - and for the great majority of waste fractions - preferable to the burning of waste, even where energy is recovered from the burning process. An incinerator has a lifetime of more than thirty years and requires a constant flow of waste to keep it going. This might stifle innovative techniques for waste recycling. **Do a close analysis of real costs**. Incineration projects are also frequently favoured for subsidies – direct and indirect which reduces their real costs. Investigation of their real costs compared to alternative options – such as flexible combinations of mechanical-biological treatment, composting and source separation for recycling can be very revealing[168]. Do local authorities pay enough attention to the different waste streams? Is collection and sale of economic valuable waste items ensured, in order to lessen the public authorities' burden for waste management? Is composting promoted, with educational and informative campaigns, accompanying that organisation?

(4) **Insist on having waste management plans** being developed and regularly updated which contain details as to the precise quantities of waste that is generated, the places where this waste goes for treatment or disposal and, if any possible, targets and time-plans which indicate how and when the present waste management will be improved. Insist in timely and wide-spread public consultation on the draft plans. The looser such plans are, the higher the risk for the individual citizen to see the bill for waste collection go up at regular intervals. Furthermore insist on these plans being published and made public!

Local authorities must set up waste management strategies and plans that are coherent with national waste management plans. As a basis for this, data on waste amounts and composition have to be gathered and an inventory of existing waste management facilities, including the various dumpsites, has to be drawn up. Based on the future integrated waste management strategy envisaged by the national authorities and in line with European policy, implementation steps to meet aims need to be well planned and effectively executed. This could include the closure of dumpsites, the setting up of waste collection systems with separate waste stream collection and sorting, the building of recycling and treatment facilities, and the building of landfills according to EU standards. A clear division of physical and financial responsibilities between the actors (levels of administrations, producers and importers of waste and packaging, citizens) is essential. A robust financing strategy needs to be included into the planning process.

---

[164] The long established European database on waste: Wastebase http://waste.eionet.eu.int/wastebase has opened a new section counting a total of 90 European waste prevention success stories. Cases from 14 European countries from Norway in the North to Greece in the South, to Belgium in the West and Bulgaria in the East are represented in the new section (European Topic Centre newsletter April 2004)

[165] See http://www.acrr.org/resourcities/dematerialisation/dematerialisation.htm

[166] See http://www3.iclei.org/implementationguide/humansettlements/provision_services_infra.htm

[167] See EEB Working Document: Elaboration of EEB PROPOSALS FOR CONCRETE EU LEVEL WASTE PREVENTION MEASURES TO BE COMMITTED TO IN THE THEMATIC STRATEGY ON WASTE PREVENTION AND RECYCLING, April 2004

[168] For examples of incineration being subsidised see FOE Report on perverse subsidies for incineration in the UK - *Money to burn*, December 2003.

An option as yet unexplored is for ECOs to make use of EU waste management **reporting questionnaires**, either directly – providing alternative implementation information to the EU level authorities or on the basis of ECO developed questionnaires. This has been done with great success (good impacts on National and EU level responsible authorities) by EEB member organisations in the case of implementation challenges of the Water Framework Directive and will be repeated now with the Air Quality daughter Directives.

(5) All **waste installations** are required to have a **permit for operation**. The permit and the conditions of operation laid down in it, including emission limits, and discharges into water must be laid down in the permit. Even where waste installations exceptionally do not come under the requirements of EU legislation on integrated pollution prevention and control, they should apply the best available techniques for operation, in order to minimise damage to citizens' health and to the environment[169]. Where more precise provisions for the operation of installations are laid down in national or regional legislation, insist in comparing the content of the permit with this legislation. It should not be forgotten that EU waste incineration legislation also applies to those installations which principally generate products and only exceptionally use waste for incineration. Note that all provisions of the waste **framework** directive apply to waste incinerators of all sizes.

(6) **Waste installations normally have to report to the permitting authority** on the operation of the installation, the quantity of waste handled, malfunctioning and other incidents. The administration has the obligation to control that permits are complied with and to report its findings. All such reports are accessible to the public under **EU law on access to information on the environment**[170]. Ask to see these reports; compare the data with the data in the permit, in the waste management plan and with other available data. Clarify contradictions between different data and draw the attention of the administration to their obligation to provide information that is accurate, usable and useful.

(7) Be particularly attentive to **the generation of hazardous waste**. Where does it come from and where does it go? What do industrial installations do with their waste: do they separate waste fractions and have them collected separately? Do the figures in waste management plans for hazardous waste correlate with local or regional data? How do waste generators bear their producer responsibility; are there strategies to prevent waste generation, to recycle waste instead of burning or landfilling it?

(8) **Investigate waste shipments**: is waste, as a rule, disposed of as closely as possible to its point of generation? Are the shipment provisions respected? Is there any control of shipments by local or other public authorities? Is waste - for example electronic goods, end of life cars or ships, including pleasure boats – being sold as second hand goods and thereby circumventing waste legislation? A recent enforcement pilot project in a sample of EU Member States carried out under the European Union Network for the Implementation and Enforcement of Environmental

---

[169] This approach has been successfully demanded by the Dutch EEB member Stichting Natur en Mileu with reference to Dutch waste incineration installation permits

[170] Under Directive 2003/4/EC of the European Parliament and of the Council of 28 January 2003 on public access to environmental information – see http://europa.eu.int/comm/environment/aarhus/index.htm for more information.

Law (IMPEL) which is an informal Network of the environmental authorities of the Member States revealed high levels of illegal shipments including many so called 'recyclable' wastes destined for Asia[171].

(9) A lot of **waste disappears** from public attention; it is dumped at sea or in less accessible places on land, or , more and more frequently, **(co) incinerated** by private persons or companies, mixed with other items and sold as products etc. Try to get clarity on the quantities of wastes that are generated, by installation, by geographical entity or by other criteria. Clarify the differences in data on waste management and ask public bodies, such as prosecutors, administrations or the police to investigate the matter. Public opinion is always a strong ally in clearing up abuses and misuses.

(10) **Voluntary agreements are rarely suitable for the waste sector** which is marked by the situation that, as a rule, everybody wants to get rid of waste; this is the decisive difference to products which people want to acquire. Furthermore, there are a large number of competing small and medium sized companies in waste management and the fact that waste collection and disposal is also a public interest issue (for reasons of public health) means that it must be ensured that waste is regularly collected and disposed of at carefully selected places or installations. In view of the very different administrative waste management structures in Member States and the difficulties of monitoring the application of non binding instruments, it does not make sense to adopt such instruments (certainly not at EU level). Frequently professionals in the waste sector and often also producers of products prefer the level playing field that is reached by legislative instruments rather than having to deal with free-riders. The OECD has published a report investigating voluntary approaches, reaching the conclusion that they rarely offer much beyond 'business as usual' (OECD 2003).

(11) **Keep a close eye on standards** – at National level and through the European watchdog ECOS. Examples of EU standards that are currently under development that could have potentially important implications for national waste policy implementation are : the EU Solid Recovered Fuel standard (sets – environmentally insufficient - limit values on waste for use as fuels), packaging standards (insufficient to shift producers from business as usual except for compostability standard), methodology for characterisation of leaching from landfills, characterisation waste etc...More detailed information can be found at the ECOS website **www.ecostandard.org/**

IV.4.10 The Court of Justice on important waste cases

The waste sector is the environmental sector where most Court judgments were made : between 1976 and end 2003, the Court of Justice handed out 91 judgments, 27 of them alone in 2002/2003. This number is high, because EU waste management legislation is often poorly applied. On the other hand, there is the double conflict between Member States and the EU over

---

[171] A recent report by an EU enforcement pilot project IMPEL revealed significant rates of irregularities in containers opened at some EU ports, IMPEL report November 2004 . A conference held in sequence to this report reported back that recent research in the Netherlands suggests that 50% of all Dutch waste exports to other EU states are possibly illegal, as are over 70% of exports to non-OECD countries. See **http://europa.eu.int/comm/environment/impel/tfs_ notified_waste.htm** for more information.

119

responsibility for waste management (Member States in general wish to keep the waste 'at home' and manage it the way they think fit, the EU prefers that it be subject more to free movement considerations) and between economic operators and EU legislature. Economic operators try to escape waste legislation by trying to have their materials not classified as wastes but as products, by having them classified as non hazardous or by not having restricting shipment rules applied.

Relevant judgments of the Court of Justice include the following:

(Details on these cases can be found by using the 'Case Law' and 'Search Form' links at the European Court of Justice website: http://curia.eu.int/index.htm)

Shipment of waste

C-118/86 Nertsfoeder[172]. A Member State may not completely prohibit the shipment of waste which is to be recovered in another Member State.

C-203/96 Dusseldorp[173]. Exporting waste for recovery is regulated by Reg.259/93. The proximity principle does not apply.

C-209/98 FEAD[174]. Under certain conditions, Member States may provide for local monopolies for waste recovery.

C-228/00 Commission v.Germany[175]. The shipment of waste to a cement kiln for use as a fuel is a shipment for recovery.

C-458/00 Commission v.Luxembourg[176]. The shipment of municipal waste to a waste incinerator normally constitutes a shipment for disposal, even where energy is recovered.

Packaging

C-302/86 Commission v.Denmark[177]. A Member State may set up a take back system for packaging, even if this makes it more difficult for one-way packaging from other Member States to accede to the market.

C-380/87 Enichem v.Cinisello-Balsamo[178]. Local bans of certain types of packaging are not prohibited.

C-463/01 Commission v.Germany[179]. Deposit and return systems are compatible with the free circulation of goods, but may not discriminate against imports.

Notion of waste

C-418/97 Arco Chemie[180].

C-9/00 Palin[181]. Remaining stones in a stone quarry are wastes.

---

[172] (1987) ECR I-3883.

[173] (1998) ECR I-4075.

[174] (2000) ECR I-3473

[175] (2003) ECR I-1439

[176] (2003) ECR I-1553.

[177] (1988) ECR I-4607.

[178] (1989) ECR I-2491.

[178] (2000) ECR I-4475.

[180] (2000) ECR I-3473.

[181] (2002) ECR I-3533.

C-444/00 Mayer Parry[182]. Production residues for consumer products are wastes.

C-1/03 Van de Walle[183]. Contaminated land is waste, even before excavation.

Recoverable material - recovery

C-359/88 Zanetti[184]. Also recoverable and reusable waste comes under the notion of waste.

121

C-422/92 Commission v.Germany[185]. Recoverable materials are not secondary raw materials, but wastes to which EC law applies.

C-304/94 Tombesi[186]. Recoverable items are waste.

C-102/97 Commission v. Germany[187]. Directive 75/439 requires that priority be given to the recycling of waste oils.

C-6/00 ASA Austria[188]. The disposal of waste in old mines may, under certain conditions, constitute a recovery operation.

Waste and EC Treaty

C-2/90 Commission v.Belgium[189]. Wastes are covered by Articles 28 and 29 EC Treaty. The principle of rectification at source allows import bans.

C-155/91 Commission v.Council[190]. Article 175 EC Treaty is the correct legal basis for Directive 75/442

C-209/94P Buralux[191]. Individuals cannot tackle in court Regulation 259/93.

C-102/97 Commission v.Germany[192]. Directive 75/439 requires that priority be given to the recycling of waste oils.

Removal of waste - soil contamination

C-365/97 Commission v.Italy[193]. The closing of an illegal landfill is not sufficient. The waste must also be removed.

C-1/03 Van de Walle[194]. Where fuel leaks from a petrol station, the leaked fuel and the contaminated soil are both "waste, even before the soil is excavated. Under certain conditions, the petrol company may also be obliged to remove the contaminated soil.

---

[182] (2003) ECR I-6163.

[183] Judgment of 14 December 2004.

[184] (1990) ECR I-1509.

[185] (1995) ECR I-1097.

[186] (1997) ECR I-3561.

[187] (1999) ECR I-5051.

[188] (2002) ECR I-1961.

[189] (1992) ECR I-4431.

[190] (1993) ECR I-939.

[191] (1996) ECR I-615

[192] (1999) ECR I-5051

[193] (1999) ECR I-7773.

[194] Judgment of 7 September 2004.

FURTHER READING AND BIBLIOGRAPHY

CEC, European Commission Waste Policy website **http://europa.eu.int/comm/environment/waste/legislation/**

CEC (1996) Communication from the European Commission COM(96) 399final on the review of the Community Strategy for Waste Management of 30 July 1996

CEC (1996b) COMMUNICATION FROM THE COMMISSION on the review of the Community Strategy for Waste Management, COM(96) 399final.

CEC (2000) Proposal for a DIRECTIVE OF THE EUROPEAN PARLIAMENT AND OF THE COUNCIL on waste electrical and electronic equipment, Explanatory memorandum · COM(2000) 347 final, Brussels, 13.6.2000

CEC (2001) Second Draft Commission working document on Biological Treatment of waste, DG Environment, February.

CEC (2003) Proposal for a DIRECTIVE OF THE EUROPEAN PARLIAMENT AND OF THE COUNCIL ON BATTERIES AND ACCUMULATORS AND SPENT BATTERIES AND ACCUMULATORS, Explanatory memorandum of COM(2003) 723 final, Brussels, 21.11.2003

CEC (2003b) *Towards a thematic strategy on the prevention and recycling of waste*, COM(2003) 301 of 27 May 2003.

CEC (2003c) *Towards a thematic strategy on the sustainable use of natural resources*, COM(2003) 572 of 1 October 2003.

CEC (2003d) Proposal for a Directive on the management of waste from the extractive industries, COM(2003) 319 of 2 June 2003.

EC (2002) Decision of the European Parliament and Council 1600/2002 laying down the sixth Community environment action programme, OJ L 242 p.1.

EC (2002b) Directive 2002/96 on waste electrical and electronic equipment, OJ L 37 p.24.

EC (2002c) Regulation 1774/2002 laying down health rules concerning animal by-products not intended for human consumption, OJ L 273 p.1.

EC (1975) Directive 75/439 on the disposal of waste oils OJ L 194, p.23.

EC (1985) Directive 85/337 on the assessment of the effects of certain public and private projects on the environment OJ L 175 p.40.

EC (1986) Directive 86/278 on the protection of the environment, and in particular of the soil, when sewage sludge is used in agriculture OJ L 181, p.6.

EC (1991) Directive 91/157 on batteries and accumulators containing certain dangerous substances, OJ L 78, p.38.

EC (1992) Directive 92/3 on the shipment of radioactive waste, OJ L 35, p.24.

EC (1992b) Directive 92/112 on procedures for harmonising the programmes for the reduction and eventual elimination of pollution caused by waste from the titanium dioxide industry, OJ L 409 p.11.

EC (1993) Regulation 259/93 on the supervision and control of shipments of waste within, into and out of the European Community, OJ L 30 p.1.

EC (1994) Directive 94/62 on packaging and packaging waste, OJ L 365 p.10.

EC (1996c) Directive 96/61 concerning integrated pollution prevention and control , OJ L 257 p.26.

EC (1996d) Directive 96/82 on the control of major-accident hazards involving dangerous substances, OJ L 10, p.13.

EC (1996f) Directive 96/59 on the disposal of polychlorinated biphenyls and polychlorinated terphenyls (PCB/PCT), OJ L 243 p.31.

EC (1999) Directive 1999/31on the landfill of waste, OJ L 182, p.1.

EC (2000b) Directive 2000/76 on the incineration of waste , OJ L 332, p.91.

EC (2000c) Directive 2000/59 on port reception facilities for ship-generated waste and cargo residues OJ L 352 p.81.

EC (2000d) Directive 2000/53 on end-of life vehicles, OJ L 269 p.14.

EC (2001b) Directive 91/157 on batteries and accumulators containing certain dangerous substances, OJ L 78 p.38.

EEA (2002) European Environment Agency statistics from Environmental signals report 2002 - http://reports.eea.eu.int/environmental_assessment_report_2002_9/en/signals2002-chap12.pdf

EEA (2004) EEA Signals 2004, Copenhagen, Denmark, p.6.

EEB (2001) Towards a Low waste Europe – Ten Key Issues, Brussels, April 2001.

EEB (2001) argumentation paper Towards Waste-Free Electrical and Electronic Equipment, Brussels, March 2001

EEB (2003) Position Paper "Towards Thematic Strategy on Waste Prevention and Recycling", second edition, Brussels, December 2003

EEB (2004) Working Document: Elaboration of EEB proposals for concrete EU level waste prevention measures to be committed to in the Thematic Strategy on waste prevention and recycling, Brussels, April 2004

EEB, Further EEB position papers on waste streams and waste management – http://www.eeb.org/activities/waste/Index.htm

Eunomia (2003) *WASTE COLLECTION: TO CHARGE OR NOT TO CHARGE?*, A Final Report to IWM (EB) Chartered Institution of Wastes Management Environmental Body, March 2003.

European Council (1997) Council Resolution of 24 February 1997 (1997) OJ C 76

European database on waste: Wastebase http://waste.eionet.eu.int/wastebase - new section on European waste prevention success stories.

EUROSTAT information see **http://epp.eurostat.cec.eu.int/** in the domain of ENVIRONMENT statistics under data category WASTE

OECD (2000) Reference Manual on Strategic Waste Prevention, ENV/EPOC/PPC(2000)5/FINAL, August 2000

OECD (2003) *Voluntary Approaches for Environmental Policy: Effectiveness, Efficiency and Usage in Policy Mixes*, OECD, Paris.

IV.5   Water

By Stefan Scheuer[195]

IV.5 Water 125

   IV.5.1 Introduction and Summary 126

   IV.5.2 State of the aquatic environment in Europe..128

   IV.5.3 From 1975 to the WFD in 2000 (see also chapter IV.5.7) 129

   IV.5.4 Implementation deficits and way forward...130

   IV.5.5 Environmental objectives under the Water Framework Directive WFD...............131

      IV.5.5.1 Good Ecological Status for surface waters - Article 4.1(a) 132

      IV.5.5.2 Good Chemical Status – Article 4.1 and 16...................................133

         IV.5.5.2.1 EU quality standards and emission controls...................................134

            IV.5.5.2.1.1 Priority substances and identification of hazardous substances...134

            IV.5.5.2.1.2 Upcoming Daughter Directives on quality standards and control measures...136

            IV.5.5.2.1.3 Article 16 of the WFD – a precedent for EU chemicals legislation?....................................137

         IV.5.5.2.2 National Standards and Emission Controls.......................................138

            IV.5.5.2.2.1 Setting EQS..138

            IV.5.5.2.2.2 Emission Controls and Combined Approach.......................138

      IV.5.5.3 Groundwater Objectives and Protection - Article 4.1(b) 139

      IV.5.5.4 The no-deterioration objective – Article 4.1...................................140

      IV.5.5.5 Derogations - Article 4.3 – 4.7 141

   IV.5.6 Tools and measures under the WFD 142

      IV.5.6.1 River Basin Planning – Article 3 and 13 142

      IV.5.6.2 Public Participation – Article 14 143

      IV.5.6.3 Water Pricing - Article 9..144

   IV.5.7 The WFD - a new umbrella for environmental laws..................................146

      IV.5.7.1 Surface Water 75/440, Fish Water 78/659 and Shellfish Water 79/923 EEC Directives 147

      IV.5.7.2 Dangerous Substances Directive 76/464/EEC and its 'Daughter Directives'...148

---

[195] EU Policy Director, European Environmental Bureau

IV.5.7.3 Urban Waste Water Treatment Directive 91/271/EEC....................................149

IV.5.7.4 Nitrates Directive 91/676/EEC...149

IV.5.7.5 Groundwater Directive 80/86/EEC..150

IV.5.8 Assessment of the WFD 151

IV.5.9 Outlook and ECO action..152

IV.5.1   Introduction and Summary

Despite 30 years of extensive EU water protection legislation, the general state of waters has not improved, with small rivers and groundwater in particular having deteriorated further, and the true state of the aquatic environment in terms of key hydrological and ecological parameters remaining largely unknown. This failure is partly due to extremely low quality implementation and enforcement, growing consumption and use of chemicals and partly due to the patchwork nature of European legislation. A patchwork which has failed to establish an overall objective for sustainable water management and lacks integration in relevant policy sectors – mainly agriculture, transport, energy, product policies and land use planning.

In 2000 the EU adopted a comprehensive new water law, the Water Framework Directive WFD, which for the first time established an overall objective for all surface, groundwater and coastal waters in the EU to be achieved by 2015. The WFD provides an umbrella for all relevant water policies, repeals a number of Directives, including the Freshwater, Shellfish Water, Groundwater and Dangerous Substances Directives by 2013[196], establishes close links with nature conservation and provides a wide range of management tools, including public involvement, long-term and integrative planning and water pricing. The Directives on Nitrate pollution form agriculture, Urban Waste Water Treatment and Bathing Water protection remain in place.

The overall objective of the WFD – "good status" - is described for all surface waters (rivers, lakes and coastal waters) in a normative way as a "slight deviation" from the aquatic biodiversity found or estimated to exist under conditions where there has been only very minor human impact. For groundwater, good status means that groundwater quality and quantity does not negatively impact surface water status or the ecology of terrestrial ecosystems which depend on groundwater. The strong ecological orientation of this European environmental objective is a novelty in European law. For the first time ever, it provides a truly environmental objective to be respected by all human activities and therefore asks for strong integration in all other policies at EU as well as at state level. However, the objective needs to be defined in quantitative terms largely at national or eco-regional level, which could lead to very different ambition levels in different countries. The quality of this national ecological standard setting process will be largely determined by the quality of the political and technical debate, administrative culture and civil society

[196] Council Directive 78/659/EEC on the quality of freshwaters to support fish life; Council Directive 79/923/EEC on the quality of shellfish waters; Council Directive 80/68/EEC on the protection of groundwater against pollution; and Directive 76/464/EEC on the pollution by dangerous substances.

involvement. Only if the benefits of a functioning and healthy aquatic ecosystem are made part of the wider public debate and the well-trodden paths of current water management are challenged by the wide range of already existing alternatives will it be possible to maintain the integrity of the WFD objective. In theory the WFD requires ecological objectives to be set irrespective of their socio-economic impact. Derogations from these objectives are possible afterwards if given conditions are met. However, in practice this purist separation rarely happens and decision makers will always want to know the "costs" of achieving a certain objective before agreeing to it out of opportunistic political reasons.

127

Besides the overall ecological standards to be set at national level the WFD carries on the long tradition of establishing chemical quality standards. Those standards must support the ecological objective. The methodology for setting such individual standards is well developed although synergistic effects of the total pollution load are not taken into account and thus might be inappropriate for achieving the good ecological status. The WFD provides a harmonised EU approach to setting standards and for the first time implements the precautionary principle for chemicals that persist in the environment or could cause irreversible or long-term damage. This is probably one of the outstanding successes of ECO campaigns. Emissions of chemicals identified with such properties need to be phased out in order to achieve zero or natural background concentrations. Unfortunately the WFD does not provide the necessary control mechanisms (e.g. a ban on the use or marketing of a certain chemical) but relies mainly on a combined pollution control approach. This means Member States establish emission controls and limit values based on best available techniques or best environmental practices as developed under the Integrated Pollution and Prevention Control (IPPC) (see chapter V.3.2), Urban Waste Water Treatment, Nitrates, Dangerous Substances (see chapters IV.5.7.2, IV.5.7.3 and IV.5.7.4) or other relevant Directives. Together with general emission control measures (e.g. prior authorisation, licensing of activities etc...) those measures must lead to achieving the chemical quality standard. Otherwise emission controls must be strengthened. For many pollutants this approach will be inefficient as it fails to address the main emission pathway – their widespread use in every-day products. The future adoption and implementation of an ambitious new EU Chemicals Policy (Registration, Evaluation and Authorisation of all Chemicals, REACH, see chapter V.4.6.4.6) could fill this gap.

The largest part of the WFD is devoted to planning and management measures in order to achieve the objectives. These measures include the setting up of river basin authorities, environmental and economical stock taking, a risk analysis of failing to achieve the objectives, public consultation and information, a programme of measures and finally the establishment of river basin management plans, which need to be periodically revised. The many administrative arrangements to be undertaken and EU reporting requirements are complex and pose a huge task for national governments to implement them. Considering the flexibility the Directive gives to Member States in setting the overall objective and the negative experience with implementation of old EU water laws, it is without doubt very important to have tight EU control and as much transparency as possible. On the other hand the real danger is that in the jungle of EU reporting requirements water managers and politicians will lose sight of what really matters– achieving a truly sustainable and transparent water management that restores and maintains the carrying capacity of aquatic ecosystems and contributes to halting further biodiversity decline.

IV.5.2 State of the aquatic environment in Europe

Rivers are sometimes described as the veins of the earth. They transport all water and all materials from or added to the land which they drain, and their volume, speed and dynamic are the result of the characteristics of the basin. This means that all human activities within a certain river basin will leave their mark on the rivers, the lakes and finally the coastal waters – even if such activities seem to be far away from the actual water course. Therefore the state of the aquatic environment cannot be described via a single line running through the land or a single parameter, but includes all groundwater, the river bed, the riparian zone, wetlands and flood plains and requires a wide range of parameters, of which the biological ones allow for the most integrated assessment.

The true state of European water resources and aquatic ecosystems is unknown. Monitoring programmes are inadequate or non-existent in many Member States, and where they are in place, their results often remain inaccessible to the public. According to the European Environment Agency, there is a large gap between what is required by the WFD in terms of ecological assessment and monitoring and what is undertaken by Member States and it is currently not possible to assess the ecological quality of European waters due to lack of information (*EEA 2003, p.5*). There are no dependable assessments of the status of rivers, lakes, coastal waters, wetlands and groundwater. The WFD will require an assessment system for the first time, delivering reliable and comparable ecological status data for all waters, regardless of the European region concerned. It is astonishing in a rich continent like Europe that the EU is currently unable to accurately indicate the extent of pollution and disruption of its aquatic resources.

The limited information available at EU level is the result of EU legislation dealing mainly with specific chemical pollution, e.g. BOD loads, pesticides, heavy metals, nutrients or specific activities, e.g. drinking water provision, waste water discharge. This data suggests that over the last two decades there has been some improvement in reducing phosphate and organic pollution, mainly through waste water treatment. But nitrate levels in groundwater and surface water remain high – in many places above the drinking water standard and almost everywhere leading to significant negative impacts on biodiversity. Pollution from heavily regulated chemicals and metals is decreasing, while pesticide contamination of drinking water resources is identified as a problem in many countries. But for most hazardous chemicals no information is available.

While the more obvious signs of river pollution -dead fish and foam floating on the surface - are fortunately seldom encountered in Europe today, more subtle biological effects have been detected. Trace quantities of endocrine disrupting chemicals, for instance, have been shown to interfere with the hormone regulation of fish, leaving them infertile. Hazardous chemicals such as these may be responsible for declining fish catches in several European countries, yet the exact mechanisms of action are extremely difficult to identify.

Natural rivers with intact riparian land have become a rarity in both northern and southern Europe and 57% of Europe's 89,000 km long coastal zones are used for agricultural, industrial or urban purposes. (EEA 2003)

At least two thirds of all drinking water in the EU depend on groundwater reserves. These valuable underground resources are not only jeopardised by pollutants, but also by excessive abstraction. 60% of European cities overexploit their groundwater resources (*EEA 1995*). Along the coastlines in Southern Europe and on many islands, seawater is already intruding into the depleted underground aquifers, making them unusable as drinking water. The main causes of this unsustainable use, apart from city supply, are irrigation and tourism. The water exploitation index in southern Europe has not improved since 1980 and the irrigated area has increased by 20% since 1985 (*EEA 2000*). Allocated irrigation water per hectare increased between 1993-1999 by 19% (EEA *2003b*). This leads to the promotion of unsustainable water management solutions, like big inter-basin water transfers and dam constructions, as in the Spanish National Hydrological Plan published in 2000.

Alarming conclusions must be drawn from the European Environment Agency's reports. Despite an improvement in some seriously polluted rivers and lakes, the general pollution situation of European waters has not markedly improved since the 1980s. Groundwater as well as smaller water resources are specially threatened with further deterioration and many water resources are over-exploited, especially for agriculture in the South.

IV.5.3   From 1975 to the WFD in 2000 (see also chapter IV.5.7)

Water legislation was one of the first sectors to be covered by EU environmental policy and comprises more than 25 water-related directives and decisions. The first wave of legislation took place from 1975 to 1980, resulting in a number of directives and decisions which either lay down environmental quality standards (EQS) for specific types of water, like the Surface Water, Fish Water, Shellfish Water, Bathing Water and Drinking Water Directives, or established emission controls and emission limit values (ELV) for specific water uses, like the Dangerous Substances Directive and the Groundwater Directive. These directives were mainly based on the first Environmental Action Programme (1973), which called for both approaches to be used. In practice, however, the dual approach not only led to highly fragmented water legislation, but also to huge implementation problems. It proved less successful than expected in its environmental outcome.

The second wave of water legislation from 1980 to 1991 was less comprehensive. Apart from the introduction of two new instruments, the Nitrates and Urban Waste Water Treatment Directives, several 'daughter directives' implementing the Dangerous Substances Directive were adopted.

Due to this patchwork of legislation from the 1970s onwards, new and more co-ordinated water legislation was demanded by both Council and Parliament. During the 1990s, a major revision of the EU water policy was prepared, finally resulting in the Water Framework Directive 2000/60/EC[197]. Not only will this directive repeal six earlier water directives and one regulation and effect a number of other pieces of water legislation, but it will also provide the basis for subsequent legislative initiatives.

---

[197] OJ L 327/1, 22.12.2000, p.1-72

The WFD tries in particular to reconcile the traditionally conflicting approaches of ELVs and EQSs. The crucial Directives to be repealed by the WFD in this respect are the Dangerous Substances and Groundwater Directives. While their emission limit approaches are in principle taken over by the WFD, it is doubtful whether this will result in an equivalent level of water protection. The EQSs set by the Surface Water, Shellfish and Fish Water Directives are a lot more easily taken up than the WFD 'good status' objective to be achieved by Dec 2015 for all EU waters.

The Drinking Water and Bathing Water Directives remain as free-standing directives, yet Member States are required to co-ordinate the protection of these waters under the scope of the WFD. The Directives stemming from the 'second wave' of water legislation, Urban Waste Water Treatment and Nitrates Directives, will not be repealed by the WFD, but some of them will be revised. Several of their requirements will have to be co-ordinated via the River Basin Management Plans. In addition, the achievement of the objectives for the 'protected areas' designated under the Nitrates and Urban Waste Water Treatment Directives is required by 2015 (Article 4).

IV.5.4   Implementation deficits and way forward

Hardly any of the water protection Directives has been fully implemented and enforced in the prescribed way or by the prescribed deadline (*Demmke 2000*), nor have its objectives been achieved. 13 Member States were found guilty by the European Court of Justice for non-compliance with water legislation in 54 cases concerning 10 Directives in the period 1998-2004. There are no reliable figures with regard to the number of infringements proceedings launched by the Commission and ECO complaints submitted. But from the Commission's 2003 report on the implementation of the Urban Waste Water Treatment Directive it can be assumed that a considerable number of infringement proceedings are resolved before a court judgement. Furthermore, from the experience with ECO complaints in the field of nature conservation, a very low percentage of complaints ever lead to infringement proceeding by the Commission.

The implementation situation may be called disastrous, and in terms of EU-wide common water protection standards, the Community is far from its goal. To add to the problem, during the first wave of water legislation, Member States were not obliged to report in detail on any progress in implementing and transposing EU water legislation. As a result, a lot of cases never came before the Commission and a huge number of infringements is likely not to have been the subject of legal proceedings. With Council Directive 91/692/EEC of 23 Dec 1991 on Standardising and Rationalising Reports on the Implementation of Certain Directives Relating to the Environment, Member States are obliged to report in detail on the implementation of environmental directives. Consequently, the number of cases against Member States that are brought before the European Court of Justice by the Commission because of implementation shortcomings has risen sharply.

What does the WFD do differently to overcome this serious problem? Firstly it establishes extensive reporting obligations and secondly the European Commission and Member States are engaged in a Common Implementation Strategy, which has already produced 13 thematic guidance documents[198]. Despite this encouraging start, two EEB surveys from January 2004 (*EEB 2004*) and November 2004 (*EEB and WWF 2005*) revealed that the quality of the national transposition laws tends to be poor. Half of all Member States have not clearly stated the overall WFD objectives in their laws and most countries have established rather weak competent authorities to deliver the WFD. With regard to the quality of implementation, the picture is not very bright either. Almost none of the Member States have taken early public involvement in the WFD implementation seriously although the situation improved in 2004. Ecological data which Member States should have collected already to ensure a smooth running of the Intercalibration exercise is missing (*see chapter* IV.5.5.1).

IV.5.5   Environmental objectives under the Water Framework Directive WFD

The WFD aims to protect the physical and biological integrity of all aquatic ecosystems and hence establish a basis for moving towards sustainable human water use. Environmental protection is thus one of the main objectives of the Directive. The integrated and ecologically oriented assessment of the surface water status with its corresponding objectives are central instruments.

The environmental objectives are set out in article 4 of the WFD. The overall objective is a 'good status' to be achieved for all waters by Dec 2015.

For surface waters, 'good status' comprises a 'good ecological' and a 'good chemical status'. Ecological status is determined by biological, hydro-morphological (e.g. the habitat conditions) and physico-chemical quality elements. The point of reference is given by the biological parameters of undisturbed waters. These are waters with only 'very minor' human impacts (*see chapter* IV.5.5.1). The good chemical status is determined by existing surface water quality standards, still to be developed new EU legislation setting standards for EU relevant pollutants ("priority substances") and national standards for national or regional relevant pollutants following a prescribed methodology in the WFD (*see chapter* IV.5.5.2).

For groundwater, 'good status' is determined by a good 'quantitative' and a 'good chemical' status. Good quantitative status is achieved when the abstraction is less than the natural recharge and additionally sufficient water is available for surface waters and groundwater dependent terrestrial ecosystems to avoid any significant damage to them. Good chemical status is achieved when existing quality standards are met, e.g. Nitrates and Pesticides, and the chemical contamination has no significant negative impacts on surface waters or dependent terrestrial ecosystems and allows safe drinking water supply (*see chapter* IV.5.5.3).

This 'integrative' approach, which extends current chemical water quality targets to water quantity, habitat quality and biological targets, is an improvement in the protection of our aquatic environment. Under the WFD, waters have to be protected and enhanced in a more 'holistic' manner. This will require rehabilitation measures such as the provision of ecologically oriented

---

[198] All guidance documents developed under the Common Implementation Strategy for the Implementation of the Water Framework Directive can be found at http://europe.eu.int/comm/environment/water/water-framework/implementation.html. A comprehensive and critical assessment of those guidance documents has been published by EEB and WWF in 2004.

water flows to support natural biodiversity. The risks stemming from chemical pollution not covered by traditional monitoring (because of its complexity or synergies) should now be detected through the required ecological assessment. As soon as the biological system in a given water body reacts negatively on chemical contamination, the causes should be identified and controlled in order to achieve 'good ecological' status.

In parallel to the good status objective, article 4 of the WFD also calls for i) prevention any further deterioration in the status of surface and groundwater, ii) a progressive reduction in pollution from priority substances and a phasing out of emissions from priority hazardous substances for surface waters and iii) a progressive reduction in groundwater pollution, in order to reverse any significant and sustained upward trend in the concentration of any pollutant. The first clause is clearly a stand-still clause to avoid the repetition of past water management errors and is logically the first measure needed to ensure that the objectives can be achieved Clauses ii) and iii) reiterate existing EU chemical water quality objectives and should in a precautionary way support the achievement of the ecological objectives, recognising that our knowledge of the causal relationship between chemical contamination and biological reaction is limited. The stated aim of phasing out the emission of priority hazardous substances which is independent of whether there is scientific proof of specific environmental damage or not is novel and is the legal transposition of a European Commission and Member States commitment under the 1998 Convention for the protection of the North-East Atlantic[199].

IV.5.5.1   GOOD ECOLOGICAL STATUS FOR SURFACE WATERS - ARTICLE 4.1(A)

The application of ecological assessments and the exact definition of the 'good ecological status' objective need further clarification. In its Annexes II and V, the WFD gives a first 'guideline' with working instructions and normative definitions, but specific numerical values, e.g. to characterise 'good ecological status', still need to be developed.

The WFD divides ecological status into five classes as following:

ECOLOGICAL STATUS CLASSIFICATION	NORMATIVE DESCRIPTION	COLOUR CODE
High	*No, or only very minor, anthropogenic alterations* to the values of the physico-chemical and hydromorphological quality. The biological quality elements show *no, or only very minor, evidence of distortion*. These are the *reference* conditions and communities.	Blue
Good	*Physico-chemical conditions ensure ecosystem functioning and biological quality elements deviate only slightly from the reference.*	Green
Moderate	*Biological quality elements deviate moderately from the reference.*	Yellow
Poor	*To be defined by Member States*	Orange
Bad	*To be defined by Member States*	Red

[199] Sintra Statement, OSPAR Convention.

The first step is to group all surface waters, rivers, lakes and coastal waters in their respective ecological types. Secondly for each ecological type biological reference values have to be derived.
Therefore one needs to find waters under undisturbed or only very minor impacted conditions –
this means surface waters the ecosystem of which is unlikely to be impacted by human activities,
like agriculture, air pollution, infrastructure works, water abstractions or diversions. The reference values for relevant biological parameters are derived from these reference waters. In many
parts of Europe it will not be easy to identify such surface waters and thus historical or modelled
values might be used. Thirdly, the biological boundary which the WFD uses as a reference point
to determine which elements 'deviate slightly' has to be set. This reference point will in effect be
an ecological standard as everything below it will be deemed as failing to meet WFD objectives.

What sounds like a rather straight forward process is highly complex and would result in very different standards across Europe – meaning that 'good status' in Sweden would probably not be
comparable to 'good status' in Italy. The WFD therefore provides several safety-nets. Firstly, it
does not rely solely on biological quality elements but requires the setting of quality standards
for specific pollutants and general physico-chemical parameters (e.g. temperature, oxygen, nutrients) in a legally binding way and requires assessment of hydro morphological (e.g. flow dynamics, substrate, tide) elements to support the ecological standard (*CEC 2003*). Secondly a so-called
"one out – all out principle" is applied concerning all the different biological, physico-chemicals
and hydromorphologial quality elements (like oxygenation condition or composition and abun-
dance of benthic invertebrate fauna) defining "good status", which means that the worst quality
element determines the overall status.

And finally, the WFD establishes a joint European process – the so-called "Intercalibration" - to
ensure that the national ecological assessment systems are in line with the WFD normative definitions and that the standard values are comparable between all states. This Intercalibration exercise is ongoing and has to end in 2006 (*CEC 2003*). It has already become clear that Member
States have failed to collect the relevant data and lack the political ambition to make this
Intercalibration exercise a success. It is likely that a repetition of the exercise will be necessary,
once data is available from the monitoring required to be in place by 2006 under the WFD.
Otherwise there is a danger that the WFD will do no more than confirm existing insufficient
national assessment systems, making a nonsense of the entire process.

IV.5.5.2   GOOD CHEMICAL STATUS – ARTICLE 4.1 AND 16

The WFD's objective is to achieve "good chemicals status", which is part of the overall "good status", to progressively reduce pollution and to phase out emissions of priority hazardous substances. Therefore, Member States and the EU are required to set quality standards as well as
emission controls for specific chemicals, in particular substances or groups of substances on the
indicative list in Annex VIII[200] and the priority list in Annex X to the Directive. It is important to
note here that action at Community level always involves a legislative process. As there is no legal
hierarchy between secondary legislation, e.g. between the WFD and its Daughter Directives, any
new law can change or override the WFD requirements.

___

[200] This indicative list comprises virtually all known pollutants or groups of pollutants.

IV.5.5.2.1 EU quality standards and emission controls

For specific substances, the EU sets Community-wide standards which have to be met as part of the objective of achieving 'good chemical status'. Community standards existing at the time of entry into force of the WFD (inter alia the Daughter Directives to the Dangerous Substances Directive (76/464/EEC) as listed in Annex IX, and other water-relevant directives) have to be continually observed. For bodies of surface water, environmental objectives established under the first River Basin Management Plan required by the WFD shall, as a minimum, require quality standards at least as stringent as the existing ones.

However, the relevant Daughter Directives of the Dangerous Substances Directive (76/464/EEC) will have to be reviewed by the Commission, and revised control measures proposed, including the possible repeal of controls on the substances covered by these directives but not included in the Water Framework Directive's list of priority substances (see below).

IV.5.5.2.1.1 Priority substances and identification of hazardous substances

In November 2001 a list of 33 priority substances, which establishes Annex X of WFD, was adopted as a Decision by the European Parliament and Council under the procedures laid down in Article 16 of the WFD. The list identifies 11 priority hazardous substances and 14 substances for review as potentially hazardous substances.

For the substance selection the Commission has chosen a simplified risk based assessment procedure, which is based on the intrinsic hazards of a substance and available monitoring data about the occurrence of the substance in water (COMMPS procedure[201]). This is an important improvement from the classical and flawed risk assessment procedure which requires a theoretical exposure assessment based on emission pathway models. This classical procedure takes a long time, puts the burden of proof on public authorities and undermines the application of the precautionary principle. Nevertheless, a number of shortcomings can be reported, requiring improvements under COMMPS for its future application. For example, a great number of substances for which no data were available at Community level from the national monitoring programmes or business confidentiality applies were left out. As a consequence, the COMMPS procedure only covered 95 substances on the basis of monitoring data and 123 substances on the basis of modelling data.

Amongst the finally adopted 33 WFD priority substances, 3 are classified as UNECE POPs[202], 13 as hazardous by OSPAR, which means that they are either POP-like substances or highly toxic, persistent and bio accumulative. Another 16 were selected under the OSPAR Convention in 1998 and 2000 for priority action, with a view to ceasing their release by 2020. It is obvious that priority substances targeted by other international agreements should be identified as priority hazardous substances. The Community has internationally committed itself to cease emissions of these substances by 2020.

The identification of priority hazardous substances was problematic. 14 substances have been left for a re-evaluation whether they are hazardous. 8 of which are pesticide active ingredients which have been detected to be ubiquitous in European waters (Atrazine, Chlorpyrifos, Diuron, Endosulfan, Isoproturon, Pentachlorophenol, Simazine, Trifluralin).

---

[201] http://europe.eu.int/comm/environment/water/water-framework/preparation_priority_list.htm

[202] 1998 Protocol on Persistent Organic Pollutants (POPs) to the UNECE Convention on long-range Transboundary Air Pollution on POPs

Three chemical manufacturers (Dow Chemical, Finchimica, Makteshim Agar) challenged the European Union at the European Court of Justice in February 2002 over inclusion of these substances in the WFD list of priority substances. The case was rejected. The companies claimed that the exclusive basis for selection of substances under the WFD should be full risk assessments in accordance with regulation No. 793/93 (*see chapter V.4.6.4.2*) and not the chosen COMMPS procedure. In other words, that there is no room for the application of the precautionary principle. In addition, it is claimed that regulating pesticides under the WFD is in contradiction to their marketing authorisation under the Pesticide Authorisation Directive (91/414/EC) (*see chapter V.4.6.5.1*).

CAS Number	Name of priority substance	X priority hazardous(X)*** under review	Chemical for Priority Action under OSPAR Convention	UNECE 1998 POPs
15972-60-8	Alachlor			
120-12-7	Anthracene	(X)***	+	
1912-24-9	Atrazine	(X)***		
71-43-2	Benzene			
n.a.	Brominateddiphenylethers	X	+	
7440-43-9	Cadmium and its compounds	X	+	
85535-84-8	C10-13-chloroalkanes (**)	X	+	
470-90-6	Chlorfenvinphos			
2921-88-2	Chlorpyrifos	(X)***		
107-06-2	1,2-Dichloroethane			
75-09-2	Dichloromethane			
117-81-7	Di(2-ethylhexyl)phthalate (DEHP)	(X)***	+	
330-54-1	Diuron	(X)***		
115-29-7	Endosulfan	(X)***	+	
959-98-8	(alpha-endosulfan)			
206-44-0	Flouranthene			
118-74-1	Hexachlorobenzene	X		+
87-68-3	Hexachlorobutadiene	X		
608-73-1	Hexachlorocyclohexane	X	+	+
58-89-9	(gamma-isomer, Lindane)			
34123-59-6	Isoproturon	(X)***		
7439-92-1	Lead and its compounds	(X)***	+	
7439-97-6	Mercury and its compounds	X	+	
91-20-3	Naphthalene	(X)***	+	
7440-02-0	Nickel and its compounds			
25154-52-3	Nonylphenols	X	+	
104-40-5	(4-(para)-nonylphenol)			
1806-26-4	Octylphenols	(X)***	+	
140-66-9	(para-tert-octylphenol)			
608-93-5	Pentachlorobenzene	X		

87-86-5	Pentachlorophenol	(X)***	+	
n.a.	Polyaromatic hydrocarbons	X	+	+
50-32-8	(Benzo(a)pyrene),			
205-99-2	(Benzo(b)fluoroanthene),			
191-24-2	(Benzo(g,h,i)perylene),			
207-08-9	(Benzo(k)fluoroanthene),			
193-39-5	(Indeno(1,2,3-cd)pyrene)			
122-34-9	Simazine	(X)***		
688-73-3	Tributyltin compounds	X	+	
36643-28-4	(Tributyltin-cation)			
12002-48-1	Trichlorobenzenes	(X)***	+	
120-82-1	(1,2,4-Trichlorobenzene)			
67-66-3	Trichloromethane (Chloroform)			
1582-09-8	Trifluralin	(X)***		

IV.5.5.2.1.2 Upcoming Daughter Directives on quality standards and control measures

For priority substances a progressive reduction in pollution is to be achieved by establishing Community-wide environmental quality standards and source controls by the procedure laid out in Article 16. For so-called priority hazardous substances, the cessation of discharges, emissions and losses shall be achieved within 20 years at the latest. There is no derogation provided in the Directive from these obligations.

For the first list of 33 priority substances the Commission should have proposed standards and measures by 20 November 2003 but is running late and announced its proposals for March 2005. Those standards and measures need to be adopted by the European Parliament and Council in a full legislative process. If no agreement is reached Member States have to set national quality standards and control measures by 2007, which can be seen as a safety net procedure. The work on setting quality standards is already well advanced at the technical level. The main questions remaining are:

▶ How to use maximum allowable concentrations (MACs) to avoid environmentally damaging peak concentrations, which would not be identified through annual averages (e.g. pesticide peaks during summer)?

▶ Setting standards for sediments or organic matter in case of bio accumulative substances?

▶ How to interpret standards for priority hazardous substances, as their concentrations should finally go towards zero or background concentrations, as required by the WFD?

▶ How to interpret the WFD requirement to reduce the level of drinking water treatment, e.g. could the environmental standard value be higher than the drinking water standard taking into account certain treatment technologies?

The development of control measures poses major problems. Much work has been devoted to identifying and quantifying the sources and pollution pathways. In many cases the diffuse sources are significant and would need to be addressed in order to achieve the quality standards or phase out all emissions. This would in most cases require EU wide market and use restrictions, which are under the auspice of DG Enterprise and not DG Environment, the body is in charge of the WFD. Therefore DG Environment is trying to use 'its' legislative tools including IPPC Directive (see chapter V.3.2) and delegate the main measures back to Member States (see also chapter IV.5.5.2.2.2 about WFD required emission control actions by Member States), who have already reacted negatively about this workload being "dumped" on them.

The cessation or phase out of emissions, discharges and losses of specific chemicals or groups of chemicals has been part of national and international legislation for a long time, both in the European Union and in the US and Canada. However, actually achieving cessation has so far proven politically difficult. Some stakeholders argue that if the principal emission sources of a substance have been stopped and major reductions achieved, further reductions are economically difficult and unnecessary for the protection of human health and the environment. This is a doubtful argument especially for substances exerting potential negative effects at very low quantities. This is for example the case for highly persistent and highly bio accumulative chemicals (vPvBs) which once released even in low quantities, accumulate in human breast milk and are atmospherically transported even to the most remote regions on earth. Continued emissions of such substances – even in low quantities – pose a non-quantifiable risk and present a real life experiment with human health and the environment. History shows that often damage has only become apparent when it was too late to adequately react (see cases of asbestos, DDT, etc...).

Hence the standard emission control measures developed in the seventies, the classical end of pipe solutions, are clearly inadequate for such substances. Instead, effective control measures for priority hazardous substances need to be precautionary and look at all stages of a chemical's life – production, transport, application, use in articles/products, waste. Each of these stages has different release patterns, and recent studies suggest that in most cases the later life stages (use in products or waste) pose the biggest problems (CEC 2002a)[203]. Thus, the use of the IPPC Directive or similar instruments is inadequate to effectively phase out emissions of hazardous chemicals (see chapter V.3.2.4.1).

The cessation of emissions, discharges and losses of priority hazardous substances can only be reliably achieved by consequently applying the principle of substitution, i.e. either by replacing them with safer substances or by non-chemical alternatives. The EU committed itself to such an approach in the preparation of a new EU chemicals policy REACH (see chapter V.4.6.4.6).

IV.5.5.2.1.3 Article 16 of the WFD – a precedent for EU chemicals legislation?

In brief, for priority hazardous substances, the WFD prescribes EU-wide measures which would lead to the cessation of all discharges, emissions and losses no later than 20 years after the adoption of the respective measures. In several respects, the WFD is hence a legislative novelty:

---

[203] EU Risk Assessment for Decbromodiphenylether. According to that study releases into the environment from use and disposal of articles containing the substance account for more than 70% of all releases.

▶ for the first time in EU law, it introduces a legally binding obligation on Member States and the European Union to phase out emissions of certain chemicals by a certain deadline;

▶ again for the first time, the WFD fully recognises and applies the precautionary principle to the control of chemicals, by declaring hazard based assessments (i.e. evaluation of chemicals by their intrinsic properties) a valid instrument for prioritising substances for action. In other words, full risk assessments are no longer the only mechanism to regulate chemicals under EU law. This is a major step forward in dealing with the environmental legacy of one hundred years of virtually uncontrolled chemical technology.

The pending revision of EU chemicals legislation should take these developments into account. The Commission's 2001 White Paper on Chemical Policy indeed applies a corresponding method-ology. Whatever the legislative tools emerging from the imminent EU chemical policy reform, it would be a huge failure to ignore the precedent set by the Water Framework Directive's provi-sions on the cessation of discharges, emissions and loss of hazardous substances, or the legisla-tive principles enshrined therein.

IV.5.5.2.2   National Standards and Emission Controls

IV.5.5.2.2.1   Setting EQS

Environmental quality standards (EQSs) for all pollutants 'identified as being discharged in signif-icant quantities' into bodies of surface water have to be set *at Member State level* according to the procedure laid out in Annex V, 1.2.6. (an indicative list of the main pollutants is provided in Annex VIII). In setting an EQS, detailed data on biological toxicity and the aquatic ecosystem need to be taken into account. Hence, environmental quality standards (EQSs) are likely to differ from region to region and from water type to water type. The environmental quality standards (EQSs) are subject to peer review and public consultation.

Compliance with these EQSs is required for the achievement of the objective of 'good ecological status' and 'good chemical status' (defined in Annex V) by Dec 2015.

For 'High Status' surface water bodies, Member States, with regard to the non-deterioration pro-vision (article 4.1.a.i), must:

▶ prevent non-synthetic pollutants discharged in significant quantities from reaching concentra-tions in the water body above the range normally associated with undisturbed conditions;

▶ prevent synthetic pollutants discharged in significant quantities from reaching concentrations above the limits of detection.

IV.5.5.2.2.2   Emission Controls and Combined Approach

The WFD's Article 10 requires Member States to control, by 2012 at the latest, all substance dis-charges and process covered by the IPPC, UWWT, Nitrates and Dangerous Substance Directives as well as all discharges of the priority substances of Annex X of WFD following a *combined approach*. That means firstly establishing or implementing

a) emission controls based on best available techniques, or

b) relevant emission limit values, or

c) in the case of diffuse impacts the controls including, as appropriate, best environmental practices.

Second, in case an EQS is not met via the above mentioned emission controls more stringent controls have to be set.

As a basic measure, Member States are required to include in the programme of measures to be
established by 2009 and be made operational by 2012, all sources liable to cause pollution prior
to regulation, such as prohibition, prior authorisation, or registration.

Furthermore, Member States have to establish emission controls and quality standards for the
priority substances in Annex X by Nov 2006, in case the EU fails to adopt such measures in time.

IV.5.5.3 GROUNDWATER OBJECTIVES AND PROTECTION - ARTICLE 4.1(B)

The big dilemma with effective protection of this most important drinking water resource is that
we do not know much about it. What can be done for surface water - using biology to derive the
carrying capacity of the ecosystem – cannot be replicated in the case of groundwater as there is
as yet little or no information about its ecosystem available. Groundwater biology is estimated to
play a big role in its self-cleaning capacity, and its reaction to pollution is rather different to that
of surface water ecosystems[204] (*Griebler et al. 2001*).

Therefore traditionally precautionary action, e.g. preventing the entry of all pollutants, is applied.
The 1980 Groundwater Directive, which will be repealed by the WFD in 2013, followed such an
approach, by setting up a "black" list of substances, which are persistent, bio accumulative and
toxic or of similar concern, which have to be prevented from entering the groundwater, and a
"grey" list of substances, which have to be limited from entering groundwater.

Unfortunately the WFD did not take up this approach and instead merely calls for the prevention or limitation of the entry of substances, without specifying what should be prevented and
what should be limited. During the WFD negotiations much effort was devoted to establishing
groundwater quality standards, e.g. based on drinking water standards. These efforts were eventually fruitless and resulted in a compromise article 17, which gives the Commission a mandate
to develop a new Groundwater Directive to fill the gap. In September 2003 the Commission published its proposal which, according to ECOs, represented a serious weakening in the existing
protection levels under the WFD and the 1980 Groundwater Directive. The Commission tried to
establish quality standards, but these are arbitrary and impractical to monitor in a harmonised
way, and would not prevent increased pollution, thus undermining the necessary precautionary
action of preventing the input of pollutants.

---

[204] Groundwater is the habitat for numerous microorganisms and animals. In Europe more than 2000 lager species, such as
worms, shrimps or even amphibians, are identified. These species, protozoans and a rich spectrum of bacteria are crucial factors of in situ water purification.

140

What does the WFD set as groundwater objectives?

Firstly, it requires the achievement of sustainable quantitative water management by 2015. This means that groundwater abstractions must be less than the natural recharge rate and sufficient water flow must be provided to avoid any significant negative impact on water ecosystems and groundwater dependent terrestrial ecosystems. Saltwater or other intrusions due to groundwater level changes must be prevented. This water quantity objective is novel in EU law and a major step towards sustainable water management.

As regards chemical quality, the WFD requires that the input of pollutants be prevented or limited, although without specifying which substances should be limited and which should be prevented altogether. Finally, groundwater chemical quality must not lead to a deterioration of the status of surface waters or terrestrial ecosystems depending on groundwater. A statistically significant increase in the concentration of pollutants in groundwater has to be prevented and ultimately reversed.

In line with the WFD approach to surface water chemical objectives, one could interpret this objective similarly as being that all "hazardous" substances need to be prevented from entering groundwater and that all other substances be limited in to prevent rising concentrations and damage to surface water and terrestrial ecosystems.

In order to put this interpretation into practice the new Groundwater Directive could extend the strategy to prevent surface water pollution (see chapter IV.5.5.2.1) to groundwater.

Finally the WFD prohibits direct discharges of pollutants into groundwater, as one of the measures required to achieve the Directive's objectives. However, Member States may exempt certain direct discharges from specified activities from this general prohibition, provided that the objectives set for the groundwater body are not compromised and the dischargers have obtained authorisation. But some of the exempted activities concern very toxic substances (e.g. injecting mining waste back into certain groundwater aquifers).

The WFD has not solved the groundwater protection problem and a great deal of work still needs to be done. But most importantly it requires Member States to develop a better understanding about their groundwater and its strong – but usually underestimated – interrelation with surface waters.

IV.5.5.4   THE NO-DETERIORATION OBJECTIVE – ARTICLE 4.1

Preventing the deterioration in the status of surface and groundwater has been identified as a crucial objective in protecting the environment from ongoing or planned human developments Any further deterioration would make the achievement of a good status more difficult or even impossible. There has been a long discussion about this no-deterioration objective and when and how it has to be applied.

The first difficulty the WFD poses is that the objective is linked to the status of the respective water bodies. Deterioration within one status class is allowed, only when the water changes the class from e.g. moderate to good or from good to high, the objective of no-deterioration is breached. This means that until appropriate monitoring is in place in 2006 and the status assessment can be finalised there will be uncertainty about the exact status of certain water. On the other hand it also depends on the size of the impact of a specific project how certain or uncertain one can judge a deterioration in water's status, e.g. the building of a dam in a free flowing river means the total disruption of the river continuum and certainly means the deterioration of status even without awareness of all biological parameters.

141

But there have been other problems. Some Member States have claimed that the no-deterioration objective only becomes legally binding in 2009 or 2012, when the programme of measures is established or becomes operational. The Commission rejected this claim and told the European Parliament in March 2001 that 'since the Directive entered into force on 22 December 2000 a strict no-deterioration clause has applied, which should prevent a repetition of past errors'.

The no-deterioration objective could in theory be one of the most powerful WFD tools against unsustainable human developments and would have a direct impact on most land use planning, infrastructure, transport and energy projects. However, ECOs have so found it very difficult to make use of it, as was the case with the Spanish National Hydrological Plan from 2000. For the time being Europe's nature conservation legislation provides the better tested and thus more successful instrument for preventing ecological deterioration resulting from human activities.

IV.5.5.5   DEROGATIONS - ARTICLE 4.3 – 4.7

Good status (good ecological and good chemical status for surface waters, good quantitative and chemical status for groundwater) has to be achieved by 2015 - with the possibility of two 6 years extension periods, in the case of technical infeasibility, disproportionate costs or natural conditions, which need to be reported (Article 4.4). Lower objectives (Art 4.5) may be set for the same reason and if the additional following conditions are met:

▶ No other significantly better environmental option to serve the socio-economic or environmental needs are available;

▶ remediation measures are taken to achieve the highest quality status under the unavoidable impacts

▶ no further deterioration occurs in the water status;

▶ and the reasons for setting lower objectives are reported and reviewed every six years.

A special derogation for surface waters has been introduced in article 4.3 in case of heavy infrastructure, which changes the character of a water, e.g. harbours, flood defence in cities. Those waters can be identified as heavily modified or artificial. For such waters a new objective is determined taking into account the physical and unavoidable changes. This means the biological standards will be lower than for "natural waters", but the chemical quality standards remain unaffected.

In order to designate a water as heavily modified or artificial, Member States have to prove that the following conditions are fulfilled:

(a) the physical restoration, e.g. removal ,of dams or dykes, to achieve good ecological status would have significant adverse effects on:

- the wider environment
- navigation, including port facilities, or recreation
- activities for the purposes of which water is stored, such as drinking water supply, power generation or irrigation

- water regulation, flood protection, land drainage, or
- other equally important sustainable human development activities.

(b) the objectives delivered by the physical modification cannot be achieved by a significantly better environmental option.

There is a high that Member States will make extensive use of these derogations. In some countries, e.g. Germany, The Netherlands, it is expected that the majority of waters will be designated as heavily modified.

Finally the WFD accepts the failure to prevent the deterioration or achievement of good status (Article 4.6) provided the following conditions are met:

- If there is no better environmental alternative.
- If the activity is a new physical modification or groundwater abstraction or presents new sustainable human development.
- If the activity is of overriding public interest or delivers human health or safety benefits, which outweigh the environmental benefits.
- If all practicable measures are taken to mitigate its effects.

For all three above described derogations, the WFD requires that in addition the applied derogation must not make achieving WFD objectives impossible in other water bodies, or contradict EU environmental policies, or lower the environmental protection set out in other EU legislation, such as the Habitats Directive.

IV.5.6   Tools and measures under the WFD

IV.5.6.1   RIVER BASIN PLANNING – ARTICLE 3 AND 13

The WFD requires a detailed, long-term and iterative planning process and the setting up of adequate administrative arrangements including the designation of competent river basin authorities (Article 3). The success of achieving the ambitious environmental objectives in an effective way and whether they will be achieved at all, depends to a large extend on the establishment of a competent authority able to oversee a river basin district and equipped with sufficient competencies and capacities to integrate different policies, including land use planning, navigation, fishing, tourism and agriculture.

The central tool is the River Basin Management Plan (RBMP) under article 13 for each river basin district, which includes inter alia:

▶ description and characterisation of the river basin, including environmental assessment of human activities, economic assessment of water uses, description of pollution sources and risk analysis of failing to achieve the objectives (Article 5);

▶ a list of environmental objectives and exemptions established for surface and groundwater (Article 4);

▶ a list of protected areas (Article 6);

▶ a map of the monitoring stations (Article 8);

▶ measures to achieve cost recovery for water services and to implement the polluter pays principle (Article 9);

▶ programme of measures (prior-authorisations, prohibitions or other general binding rules) and specific additional measures to achieve the environmental objectives (Article 11).

The initial RBMP for each river basin district will have to be completed by Dec 2009 and reviewed and updated every six years thereafter (2015, 2021 etc). The preparation of RBMPs is a most important area of influence for ECOs, since this is where all relevant issues for the achievement of the WFD objectives are negotiated. Notably, Member States are required to ensure a full and comprehensive public consultation on all the issues covered by the plans (article 14).

In the case of international river basins – whether they fall entirely within the European Union or extend beyond the boundaries of the Community – Member States are asked to ensure co-ordination and co-operation with the aim of producing one single international River Basin Management Plan. If such an international RBMP cannot be produced for some reason or other, Member States are still responsible for producing River Basin Management Plans for the parts of the international river basin district which within their territory.

IV.5.6.2   PUBLIC PARTICIPATION – ARTICLE 14

Public Participation has a prominent place in the WFD. Firstly the European Union has developed a strong agenda of increasing transparency and public participation. Secondly water management is historically a public service a large public administration behind it. Civil servants working in the sector tend to have a background in civil or building engineering, which without doubt leads to impressive advancements inhuman development. But such development has often been pursued at the expense aquatic ecosystems, leaving nearly no rivers without a channel, dam or concrete enforcement. The WFD takes a strong line in tackling this problem and shifting water management from working against towards working with nature. A very important tool in making such a shift is the constant pressure to explain to the public what is happening and to present alternative management solutions.

If the WFD is to be truly successful it requires not only repair measures but a joint societal effort. Again transparency, access to all documents and open discourse with the public are essential in preparing such a joint effort involving different actors and across society as a whole.

The WFD thus requires Member States to encourage the active involvement of interested parties in WFD implementation. In specific article 14 requires a six month consultation on the timetable and work programme for the RBMPs in 2006 on the main water management issues in 2007 and on the draft RBMP in 2008. All background documents must be made available on request.

The EEB together with WWF carried out a survey of the WFD implementation quality end of 2004, which provides insight into the state of play of respective public participation in different Member States (*EEB and WWF 2005*). The overall starting point is quite discouraging, but in several countries, public authorities' attitudes towards involving ECOs in WFD implementation improved between the end of 2003 and the end of 2004. However, the survey shows that in many countries, public participation and active involvement still have a long way to go to reach WFD objectives. A lot of time has passed without public authorities having developed appropriate strategies. But there are also very encouraging examples, such as the case of Schleswig-Holstein, where ECOs are reimbursed for participating in the implementation exercise, and Scotland, where the Scottish Executive is actively gathering information held by societal stakeholders.

IV.5.6.3   WATER PRICING - ARTICLE 9

The use of economic instruments is of growing importance for environmental policies. One of the key priorities of the 5th Environmental Action Programme (1992-1999, see chapter III) was the widening of the range of environmental instruments, which has been reconfirmed by the 6th Environmental Action Programme (2002-2012). For several years the use of economic and fiscal instruments has been promoted in EU environmental policy with mixed success - like phasing out environmental harmful subsidies, energy taxes and emissions trading. The EU has so far only managed to successfully introduce emissions trading for climate change purposes. Taxation and subsidies largely rest within the sole competence of Member States. Indeed a number of countries have introduced environmental based taxes and charges. But overall environmental taxation remains at a stable low level of 5-10% of total tax revenues in the EU (*EEA 2002*). On the other hand environmentally harmful subsidies are conservatively estimated to be 5% of GDP in OECD countries (*OECD 2004*), with subsidies in the field of water management worth some 15 billion US Dollars per year.

Member States are now obliged under the WFD to develop water pricing policies for water uses in order to support the achievement of the environmental objectives and to implement a cost recovery for water services, taking environmental and resource costs into account.

The potential of using water pricing for improving current inefficient water management and of moving towards sustainable water use is great. Fair water pricing could - especially for the supply-driven management in the agricultural sector - reduce the level of harmful subsidies, lead to more efficient water use and finally to demand management (*EEB 2000*).

Nevertheless, as regards full cost recovery, including environmental costs, water pricing is quite a sophisticated question. Water is not just a commercial good, market forces are not easily applicable and the economically based calculation of environmental costs is of course complicated. But if the Polluter Pays Principle is to be taken into account, then the inclusion of environmental (damage) or resource costs is a crucial precondition (*EEB 2000*). Simple and straight forward action, such as identifying and reducing subsidies, charges or levies for water abstraction and use and earmarking them etc., is needed. Ecological and holistic oriented water status objectives, strategies against pollution from dangerous substances, river basin management and public participation are the main tools and objectives within which water pricing should be set.

The introduction of the full cost recovery principle for water services in the WFD has not been an easy task and during the WFD legislative process, opposition, especially from Member States, where the potential of financial instruments is greatest, hindered EU-wide clear and binding obligations on water pricing.

Article 9 of the WFD obliges Member States to take into account the principle of recovery of the costs for water services and specifically includes environmental and resource costs. The Article specifies that Member States have to ensure by 2010:

▶ that their water pricing policy is an incentive for efficient water use and thereby contributes to the environmental objectives;

▶ an adequate contribution for the different water uses to the recovery of the cost of water services.

The first obligation is very important, because it makes clear that water pricing has to be seen and used within the frame of the environmental objectives of the WFD.

The second obligation gives wide room for interpretation. What is meant by an "adequate contribution"? Further to that, full cost recovery is limited to water services, which are defined in Article 2(3) as "all services (abstraction, impoundment, storage, treatment and distribution of surface water or groundwater, waste water collection and treatment facilities) which provide, for households, public institutions or any economic activity". Whether self services or hydropower are considered to be water services is open to interpretation.

Due to the above-mentioned strong opposition from Member States, with low level of cost recovery for water services (particular Spain and Ireland), exemptions from the application of the full cost recovery principle are wide.

When establishing water pricing policies under Article 9, Member States can take into account social, environmental and economic effects as well as geographic and climatic conditions.

Further to that, Member States can simply decide not to establish any water pricing policy for a specific water use activity (e.g. irrigation) under the condition that this does not compromise the achievement of the Directive's objectives.

The economic analysis for each river basin district (Article 5), which had to be concluded by 2004, will be the basis for establishing water pricing policies and for implementing the full cost recovery and the polluter pays principle.

According to Annex IV of the Directive, the economic analysis should therefore contain sufficient information including the estimation of environmental costs. Annex IV further specifies that long-term forecasts of supply and demand for water have to be accounted for, which is very important in the light of the possible changes in the water cycle due to climate changes.

The economic analysis and the steps to implement the full cost recovery obligation have to be reported, which allows stakeholders to have an influence on how far their government is willing to move towards cost recovery and internalisation of external (environmental) costs.

IV.5.7 The WFD - a new umbrella for environmental laws

The WFD not only repeals a number of EU water laws but also provides a general coordination framework for many others, which includes setting deadlines for achieving their respective objectives. Furthermore, it feeds into important political debates with regard to Europe's chemicals management and product policy as well as relevant other sectoral policies. The following table tries to give an overview on this network of legislation and policies.

WATER FRAMEWORK DIRECTIVE IN LEGAL TERMS	
Repeals by 2013	• Fish Water Directive 78/659/EEC • Shellfish Water Directive 79/923/EEC • Groundwater Directive 80/68/EEC • Dangerous Substance Directive 76/464/EEC
Requires Member States to register and monitor all protected areas and achieve their respective objectives by 2015 if not specified otherwise	Protected areas under: • Drinking Water Directive 98/83/EC • Bathing Water Directive 76/160/EEC • Nitrates Directive 91/676/EEC • Urban Waster Water Treatment Directive 91/271/EEC • Wild Birds and Habitats Directives 92/43/EC and 79/409/EEC
Requires Member States to develop or implement Best Available Technique BAT or Best Environmental Practice BEP by 2012 if not specified otherwise *And apply combined approach (first implement BAT or BEP, than use EQS to check whether it is sufficient, if not strengthen BAT or BEP)*	For pollutants and activities under • IPPC Directive 96/61/EC • Dangerous Substances Directive 76/464/EEC • Nitrates Directive 91/676/EEC • Urban Waster Water Treatment Directive 91/271/EEC
Requests a review of biocides or pesticides marketing authorisations in order to met quality standards set according to article 16 of WFD	• Pesticides Directive 91/414/EEC • Biocides Directive 98/8/EC

WATER FRAMEWORK DIRECTIVE IN GENERAL POLICY AND POLITICAL TERMS

Develops a key indicator	For Europe's objective to halt biodiversity decline in 2010
Greatly expands nature protection and conservation	Especially including wetlands and floodplains outside Europe's nature legislation
Encourages better integration of water protection	into other EU policy areas such as energy, transport, agriculture, fisheries, regional policy and tourism. The Commission's proposed EU funding instruments for regional and rural development and social policies for 2007-2013 foresee funding of WFD measures (WWF 2005).
Requires Member States to ensure appropriate administrative arrangements to implement WFD (including the setting up of an competent authority)	This means as a logical consequence that water protection must be integrated with all land use planning policies and other sectoral policies.
Requires the identification of priority and priority hazardous substances and the phase out of their emissions	First step to implement Europe's commitment under several international conventions For the first time a more efficient methodology for the selection of substances of concern selection has been used. Politically calls for: • filling the knowledge gap about the safety of most chemicals marketed and used in the EU. • efficient substitution of such chemicals and/or market and use controls Both could be delivered by the new Chemicals Policy REACH

147

IV.5.7.1 SURFACE WATER 75/440, FISH WATER 78/659 AND SHELLFISH WATER 79/923 EEC DIRECTIVES

These three directives require Member States to establish Environmental Quality Standards for specific water bodies and water uses.

The 1975 Surface Water Directive aims at protecting relevant surface waters intended to be used for drinking water purposes, such as lakes, rivers and reservoirs. Member States have to designate those waters and have to take all necessary measures to comply with the standards set in the directive. Most of the requirements of the directive have been integrated into the 1980 Drinking Water Directive. Consequently, the Surface Water Directive will be repealed by the WFD in Dec 2007.

The objective of the 1978 Fish Water Directive is to protect and improve the quality of fresh waters that support, or could support, certain species of fish. Similarly, the 1979 Shellfish Water

148

Directive aims to protect and improve the quality of coastal and brackish water bodies, in order to contribute to the quality of edible shellfish products. In order to achieve the objectives of both directives, Member States have to designate the relevant water bodies, to monitor the quality of these water bodies and to take measures to ensure compliance with the minimum standards set by the Directives ('guide' as well as 'imperative' values are laid down).

The Fish Water and Shellfish Water Directive will be repealed by the WFD in Dec 2013. The achievement of a good ecological and chemical status for all waters through the WFD should imply the achievement of quality standards to support fish and shellfish life. Nevertheless, nothing in the WFD explicitly prevents the lowering of standards from these Directives once they are repealed.

IV.5.7.2   DANGEROUS SUBSTANCES DIRECTIVE 76/464/EEC AND ITS 'DAUGHTER DIRECTIVES'

The 1976 Dangerous Substances Directive is an important component of EU water legislation and provides the framework for subsequent regulation to control the discharge of specific dangerous substances. It applies horizontally to all surface water in the EU and, by setting emission standards, partially follows the second approach of the 1973 Environmental Action Programme.

The objectives are the elimination of pollution by the dangerous substances listed in Annex I ('Black List') and the reduction of pollution by Annex II substances ('Grey List'). The regulatory measures to be used by Member States to achieve these objectives are prior authorisations for any discharge of List I substances, which can be granted only for a limited time period. List I substances are identified on the basis of their toxic, persistent and bioaccumulative properties.

Due to differing viewpoints between the UK and other Member States, the directive follows a parallel approach and makes provisions (under Article 6.1 and 6.2) for the EU Council to adopt both Emission Limit Values ELVs and Environmental Quality Standards EQSs for each 'black list' substance. Member States are free to choose which approach to use. For the time being, only the UK has followed the EQS approach.

In 1982 the Commission identified 129 'candidate' substances[205] which qualify as List I substances according to Directive 76/464/EEC. However, only 18 out of the 129 substances have been regulated up to now through daughter directives as List I substances (Mercury Discharges 82/176/EEC, Cadmium Discharges 83/513/EEC, Mercury Discharges 84/156/EEC, Hexachlorocyclohexane Discharges 84/491/EEC, and Dangerous Substance Discharges 86/280/EEC).

The WFD will repeal the 1976 Dangerous Substances Directive in 2013, except for Article 6, the List I substances, which was replaced immediately by the WFD. In addition, the 1982 'candidate' list of 129 substances is replaced by the 33 priority substances in Annex X of the WFD (*see chapter IV.5.5.2.1.1*). The WFD specifically mentions that the EQSs established under the first River Basin Management Plan have to be at least as strong as the ones established under Directive 76/464/EEC.

---

[205] OJ C 176, 14.7.1982, p. 3.

IV.5.7.3   URBAN WASTE WATER TREATMENT DIRECTIVE 91/271/EEC

The Directive's objective is to protect the environment from urban waste water discharges through collection and treatment of such discharges, which according to the WFD has to be achieved by 2015 at the latest. The setting of Emission Limit Values ELVs for organic biodegradable substances, nitrates and phosphates from urban wastewater treatment plants is dependent on the local aquatic environment, i.e. how sensitive it is with regard to eutrophication or whether it is protected for drinking water abstraction. Some countries identified their whole territory as sensitive[206], some identified parts of their territory[207], and finally a number of countries do not identify sensitive areas at all, but apply the most stringent ELVs instead[208]. In a tiered approach, Member States have to establish waste water collection for agglomerations with more than 2000 p.e. (population equivalents) by the end of 2005 starting with agglomerations above 150,000 p.e. by the end of 1998.

The Directive has delivered the single biggest improvement in the quality of European rivers and lakes through reduced pollution with organic biodegradable substances. Nevertheless, 40% of European waters still show euthrophication symptoms and up to 50% of this impact can be attributed to urban waste water discharges (*EEA 1999*).

The Directive has been often cited as the most expensive European law ever. Indeed waste water collection networks and treatment plants are costly infrastructures, but its human health and environmental benefits are likely to be much higher. Additionally, it is often forgotten that the Directive does not prescribe technical standards, but sets emission limit values, which can be achieved often better and cheaper through decentralised waste water treatment systems and the separate collection of waste water of different qualities.

Implementation of the Directive has been slow and the Commission launched 34 infringement procedures against Member States between 1994 and 2003 mainly on transposition, reporting, designation of sensitive areas and collection systems (CEC 2004). Whether ELVs are met is subject to an ongoing evaluation by the Commission.

IV.5.7.4   NITRATES DIRECTIVE 91/676/EEC

The Directive's objective is to reduce water pollution caused by nitrates from agriculture sources, which according to the WFD has to be achieved by 2015 at the latest.. Similar to the Urban Waste Water Treatment Directive the Nitrates Directive requires Member States to identify specific problem areas (Nitrate Vulnerable Zones NZVs) and apply emission control measures to meet the objectives. Beyond that, Member States have to establish codes of good agricultural practice, to be applied by farmers on a voluntary basis, in order to provide general protection for all waters.

By the end of 1993 Member States had to designate as NZVs all surface or ground waters which would either fail to meet the drinking water standard of 50mg nitrates/l or suffer from eutrophication if no pollution control measures are taken.

---

[206] Belgium and Sweden

[207] Germany, Spain, France, Greece, Ireland, Italy, Portugal and the United Kingdom

[208] Denmark, Luxembourg, the Netherlands, Finland and Austria

For the NZVs, Member States have to establish action programmes within two years, which ensure inter alia that a quasi "emission limit value" of 170 kg of Nitrogen applied per hectare per year is achieved. This limit value can be adjusted according to local soil, crop and climate conditions after approval by the Commission and with the condition that the environmental objectives are achieved.

150

Considering that eutrophication of European waters remains today's biggest threat to our aquatic ecosystems, a lot more must be done to correctly apply the Nitrates Directive. Member States have been rather reluctant to regulate the agricultural industry. By end of 2001 for all countries but Denmark the Commission has launched infringement proceedings for failure to correctly apply the Directive (CEC 2002b).

The Commission launched 56 infringement procedures against Member States between 1994 and 2001. This indicates the high level of disrespect for this important European law and is clearly a huge waste of money: "The cost of nitrate reduction lies in the range of EUR 50-150 per hectare per year, but this is estimated to be 5 to 10 times cheaper than removing nitrate from polluted water. A 2002 study estimates that denitrification of UK drinking water costs £19 million a year and projects the total UK cost of achieving the European Union nitrate standard for potable water at £199 million over the next 20 years. Consumers, rather than the polluters (i.e. farmers), pay almost all of the bill." (EEA, 2004)

IV.5.7.5 GROUNDWATER DIRECTIVE 80/86/EEC

In principle, all discharges of pollutants into groundwater were regulated by the 1976 Dangerous Substances Directive which in its Article 4 explicitly obliged Member States to apply a zero emission regime for discharges of List I pollutants into groundwater. This article also referred to a future directive on groundwater and ceased to be applicable with the adoption of the 1980 Groundwater Directive.

In a similar way to the Dangerous Substances Directive, the Groundwater Directive divides pollutants into two categories – a 'black list' and a 'grey list'. However, the objectives are to prevent 'black list' substances from entering groundwater and to limit 'grey list' substances introduced into groundwater. To reach these objectives, the national competent authorities have to prohibit any direct discharges and to take all necessary measures to prevent indirect discharges with regard to 'black list' substances. All discharges of 'grey list' pollutants are subject to prior investigation and authorisation.

The Groundwater Directive has not been able to meet the challenge of effectively preventing long-term and diffuse groundwater pollution. A lack of instruments and of integration into other policies is the main reason for this. Nevertheless, the zero-emission obligation applies to all sources of groundwater pollution and represents a precautionary principle for certain substances, which are identified on the basis of their toxic, persistent and bioaccumulative properties. The Commission proposal for a new Groundwater Directive from October 2003 following article 17 of the WFD is less explicit with regard to prevention and limitation of input of substances (EEB 2003).

The 1980 Groundwater Directive will be repealed by the WFD in Dec 2013.

IV.5.8   Assessment of the WFD

Strengths

Above all the WFD sets and describes an ecological objective in detail and provides a common process to achieve some degree of EU harmonisation. Compared to the so far rather generic ecological objectives (halt biodiversity decline; protection of certain individual species and habitats) this is a major step forward. If successful in terms of technical and political ambition the result would be the first ever substantial indicator to test and challenge human development against ecological sustainability.

Further to this overarching objective and in recognition of its vulnerability for political pressure due to its complexity, the WFD provides several "safety nets" and environmental principles, including the no-deterioration of status, the one out – all out principle (worst quality element determines overall status), setting of chemical quality standards, phase out of emission of hazardous chemicals and the Intercalibration for achieving a EU wide comparable and WFD consistent interpretation of the ecological objective. The legal requirement to phase out emission of hazardous chemicals is novel in EU law and a major step towards implementing the precautionary principle for chemical threats which are unacceptable due to their potential long-term, global and irreversible negative impacts. International commitments for the protection of the marine environment can now be achieved.

Like no other environmental Directive before it, the WFD provides a great number of policy instruments and administrative arrangements, which in summary should allow transparency, early control and proper enforcement, and ensure sufficient integration with other environmental as well as sectoral policies to achieve these objectives. If those reforms are achieved this would mark a new era in water management and governance.

Weaknesses

Many of the above mentioned strengths suffer in one way or another from legal vagueness and complexity. The WFD was developed under strong re-nationalisation pressure – with the aim of strengthening national competencies, while still believing in the necessity to set high environmental standards at the same time. As a result of this the WFD is ambitious but open ended, granting each Member State a multitude of opt-out and exemption possibilities, which in the worst case scenario would lead to 25 different national ambition levels in defining the objectives. This would – depending on economic, employment and internal market pressures – easily result in a race to the bottom and environmental dumping, which would most likely result in no more than confirming existing and insufficient water protection standards. Such a stand still would leave all the heavy administrative and reporting requirements of the WFD without a clear new objective and eventually would make the WFD completely redundant.

Whether such a negative scenario can be avoided depends on the strength of the European Common Implementation Strategy (see chapter IV.5.4) and more specifically, on the Intercalibration process, to deliver a harmonised and ambitious objective. This process must be well resourced and at the same time resist political pressure to reduce perceived costs from setting ambitious objectives. Such pressure depends partly on the national administrative set-up and its ability to leave the well trodden path of "technological solutions only" and instead engage in an open dialogue with other governmental departments and stakeholders, convincing them of the benefits of integrated and long term water management. Most governments and water management authorities are still very far away from this "ideal" and are reluctant to seek stakeholder input or to communicate in a transparent way. Nevertheless positive changes in attitude towards ECOs have been observed during the year 2004.

Agriculture policies will have a particularly important role to play. Bad communication in Denmark and the Netherlands over the WFD's allegedly dramatic consequences for farming in Europe has already lead to huge resistance towards implementing the WFD.

IV.5.9 Outlook and ECO action (Environmental Citizens' Organisation)

The WFD can be seen as a major contribution to Europe's sustainable development agenda. When correctly implemented, the WFD should deliver the environmentally sustainable conditions needed for aquatic ecosystems through defining a good status for all waters in Europe. But there is still a long way to go and several drawbacks can be observed – in terms of lack of political commitment to fulfil the letter and spirit of the WFD. Often the exercise is seen as too costly and potential benefits are largely ignored. Public participation in implementation is still in the early stages of development and needs a major effort from public authorities as well as environmental organisations. If this is not taken seriously the WFD could turn out to be a "white elephant".

But the WFD is also an integrative policy which links up with many existing environmental policies and will rely on the successful delivery and implementation of specific policies like the new chemicals policy, the greening of agriculture, transport, energy and regional policies.

ECOs are provided with ample opportunities to play a role and shape new water management under the WFD. It is not easy to find in the jungle of paragraphs, ambiguous formulations and flexible wording, where such opportunities lie. The following paragraphs should help ECOs and other stakeholders to better identify what they can do make the WFD work.

> For more detailed advice the reader is referred to the following publications: Lanz & Scheuer, *EEB Handbook on EU Water Policy under the Water Framework Directive*, EEB 2001; and *Tips and Tricks for the Water Framework Directive Implementation*, EEB and WWF 2004 (see: http://www.eeb.org/activities/water/publications.htm)

1. Highlight the many environmental and socio-economic benefits, including:

▶ *Increasing water security* – catastrophic water events are becoming more frequent due to climate change, but also to regional water mismanagement. Long-term integrated river basin management and protecting/ restoring the environment's natural capacity to balance extreme weather conditions should reduce local impacts of droughts and floods.

▶ *Only paying once and once only for the right measures to tackle water problems* - Replacing the current uncoordinated water management approaches by a single integrated and coordinated one can largely increase the efficiency of measures and reduce overall costs.

▶ *Paying less in the future for water treatment* – Today households pay the bill for cleaning up industrial and agricultural water pollution. Making those polluters pay and reducing pollution at source reduces water supply prices and increases water quality. According to the EEA reducing nitrate pollution from agriculture is 5-10 times cheaper than removing nitrate from polluted drinking waters (*EEA Signals 2004*).

153

▶ *Having high quality water available for all as needed* – water shortages are expensive to business and pose health hazards. Appropriate water pricing and sustainable resource management would lead to a more reliable water supply.

▶ *River, lake, wetland, estuarine and coastal habitats, and species available for all to enjoy* – functioning ecosystems, which improve recreation and leisure are still underestimated values. They can be an important source of prospering local economies.

▶ *Fishing and tourism* – once good water status is reached higher revenues can be achieved.

2. Public Participation and access to documents:

ECOs need to request that their authorities develop a structured and financed participation framework. Often this only happens after ECOs persistently ask pressing questions about the state of implementation and request access to information. Many ECOs are afraid of getting involved due to the technical complexity of the issues and the capacities involved in tackling them. But one does not need always technical knowledge in order to be a constructive player in the process. In the first instance authorities need to be able to explain to non-experts what they are doing. Inability to do so constitutes bad administration. In many cases the pressure on authorities to make them explain in plain language what they are doing leads to more accountable and open management, which allows new paths to be explored.

3. Encourage and get involved in a public discourse about "good ecological status" and setting specific objectives:

So far nobody can claim to know exactly what is meant by "good ecological status". Of course, many biologists working on the issues have developed ambitious ideas, but the general tendency of authorities is to change their existing ecological assessment systems as little as possible. Together with the constant pressure from politicians to design a cheap and easily achievable objective and to make sure that their own region and country does not look bad in comparison with others, this drives down the ambition to provide a truly holistic ecological assessment as required by the Directive. e.g. there is almost no national system which consistently uses fish as a key indicator for assessing the state of waters. The WFD clearly requires to monitor and use fish as a key biological indicator. So far Member States have not been able to bring forward sites with sufficient fish monitoring data in order to carry out a joint Intercalibration exercise which achieves WFD conformity and EU wide comparable interpretation of what constitutes "good ecological status". Following the scientific driven development of a definition of "good ecological status", specific objectives for each body of water have to be set, which may take socio-economic conditions into account and thus resulting in extending deadlines or lower environmental ambition. Blanket exemptions, like defining all flood defences or agriculture activities as overriding public interests, should be avoided and must be challenged.

154

4. Explore synergies with other water legislation and nature conservation:

Water pollution from agriculture activities and households remains a major challenge. The Nitrates and Waster Water Treatment Directive can help a lot but are far from being applied and enforced. In specific the waste water treatment in the 10 New Member States has still a long way to go. Alternatives to the inflexible and large scale solutions mainly used in Europe 15 should be promoted, including waste water separation systems and reed bed infiltration systems. Infrastructure projects threaten remaining natural water courses, which are often lying in designated nature conservation areas. Nature conservation obligations can thus be used in synergy to avoid further deterioration.

5. Integration with other Common Agriculture Policy CAP and Rural Development:

The mid-term reviewed CAP provides better opportunities to support WFD implementation and achievement of its objective. Specifically the Rural Development plans for the period after 2006 can serve to finance and steer agriculture management plans which conform to the WFD.

6. Challenge new infrastructure, like inland navigation, flood defences and hydropower:

With the WFD's no-deterioration principle any new infrastructure work has to undergo a detailed environmental impact assessment, the lack of better alternatives to provide the service has to be proven and mitigation measures have to be established.

7. Insist on phasing out the emissions of priority hazardous substances:

Firstly the Commission has to propose emission controls for priority hazardous substances which achieve cessation of all emission within 20 years. ECOs will have to make their voice heard in the European Parliament and at government level to avoid attempts by different industry sectors to weaken ambitions and undermine the precautionary principle. If the EU fails to adopt appropriate measures Member States have a duty to achieve cessation of emissions themselves. It will be necessary to remind governments of this duty.

BIBLIOGRAPHY AND FURTHER READING

CEC (2002a) EU Risk Assessment for Decbromodiphenylether.

CEC (2002b) *The Implementation of Council Directive 91/676/EEC, Synthesis from year 2000 member States reports*, Commission report COM(2002)407.

CEC (2003) Common Implementation Strategy for the WFD - Guidance Document No 6 Towards a guidance on the establishment of the Intercalibration network and the process on the Intercalibration exercise.

CEC (2004) *Implementation of Council Directive 91/271/EEC,* Commission report COM(2004) 248 final Brussels, 23.4.2004.

DANIELPOL, D.L.; GIBERT, J.; GRIEBLER, C.; NACHTNEBEL, H.P. & NOTENBOOM, J. (2001) Groundwater Ecology, a tool for management of water resources. Office for publications of the EC. Luxembourg.

DANIELPOL, D.L.; GIBERT, J.; GRIEBLER, C.; GUNATILAKA, A.; HAHN, H.J.; MESSANA, G.; NOTENBOOM, J. & SKET, B. (2004): The importance of incorporating ecological perspectives in groundwater management policy. *In: Environmental Conservation* (2004, Issue 3)

DEMMKE, C (2000) *Towards effective environmental regulation: Innovative approaches in Implementing and Enforcing European Environmental Law and Policy*, European Institute of Public Administration.

EEA (1995) Europe's Environment - The Dobris Assessment.

EEA (1999) *Report no 4: Nutrients in European Ecosystems*.

EEA (2002) *Environmental Signals 2002*.

EEA (2003) *Europe's water: An indicator-based assessment*, p. 5.

EEA (2003b) *Indicator Fact Sheet: Mean Water Allocation for Irrigation in Europe*.

EEA (2004) *Environmental Signals 2004*

EEB (2000) A Review of Water Services in the EU under liberalization and privatisation pressures. A Special Report, Brussels, July 2002

EEB (2003) *Future EU Groundwater Protection*, EEB Position on Commission's proposal for a Groundwater Directive, November.

EEB (2004) *The Quality of National Transposition and implementation - A snapshot, Results of an ECO Questionnaire*, Brussels.

EEB/WWF (2004): *Tips and Tricks for the Water Framework Directive Implementation*

EEB and WWF (2005) *The Quality of National Transposition and Implementation of the WFD at the end of 2004, A second "Snapshot " Report - Assessment of results from an ECO questionnaire,* Brussels.

KAMPA E. / HANSEN W. (2004): *Heavily Modified Water Bodies*, Springer-Verlag Berlin Heidelberg

LANZ K. / SCHEUER S. (2001): *EEB Handbook on EU Water Policy under the Water Framework Directive*, Brussels, January 2001

OECD 2004: *Environmentally harmful subsidies and international instruments*

WURZEL, R.K.W. (2002): *Environmental Policy-making in Britain, Germany and the European Union, The Europeanisation of air and water pollution control,* Manchester University Press

WWF (2005) *EU Funding for Environment, A handbook for the 2007–13 programming period,* Brussels April 2005, page 31.

V.

Horizontal legislation – providing tools to achieve environmental objectives

158 V.1 Introduction

160 V.2 Public Involvement in Environmental Decision-Making

172 V.3 Environmental Measures on Production, Planning and Management

258 V.4 Product Policy

V.1   Introduction

By *Stefan Scheuer*[209]

Europe's environmental policy is based on the European Union's general objective of achieving a *"high level of protection and improvement of the quality of the environment"* and applying the general principles of *integrating environmental protection* into all policies, *precaution, prevention, damage control at source* and *polluter pays*[210]. In chapter IV of his book a number of quantitative and qualitative objectives for air, nature protection and water are explained which specify and thus make operational the *"high level protection"* objective. Soil protection objectives are still missing and have only marginally been taken into account through the air, water nature and waste policies. Alongside this, in order to implement the general principles, specific objectives and controls have been established. These include specific media objectives and standards sector specific emission controls, for air and water emissions in particular - and management provisions - especially for waste management. All of these are explained in chapter IV.

The steady economic growth and rising 'standards' of living that the EU has experienced over the last fifty years have left us with environmental challenges that are well known. Emissions of greenhouse gases and some air pollutants are still increasing, and the EU will find it difficult to meet its commitments under the Kyoto protocol and Air Quality Ceilings. At the same time our biodiversity and food chains are under threat from habitat fragmentation, hazardous chemicals, poor land-use planning and insufficient protection of our waterways and seas. There are increasing problems disposing of municipal waste, the volume of which increases at roughly 3 % per year. At the same time the Total Material Requirement of the EU is estimated at around 50t per capita of which non-renewable resources form 88% and 1/3 of the total is being met through imports from non-EU countries (*Bringezu, Moll and Schütz 2003*).

So, despite a comprehensive set of environmental legislation the EU has failed to achieve key environmental objectives. This is largely due to the failure of Member States to put in place the right measures in time, but also because the European Union promotes policies which contradict environmental objectives. The "environmental management" principles - above all the integration principle - established by the Treaty have largely not been applied. Economic growth based on increasing material and natural resource use and consumption has been quicker than environmental efficiency gains.

Increasingly, therefore, environmental policies have been focussing on setting generic environmental standards for products, production, planning and management, which are applicable for a wide range of sectors or products. Thus they can be described as horizontal environmental policy tools, although sector specific environmental legislation increasingly includes management and planning provisions (e.g. Water Framework Directive – *see chapter IV.5*). Some follow a market based approach, like Greenhouse Gas Emission Trading or Environmental Liability; others require public authorisations for production or products, like IPPC or Pesticides Authorisation

---

[209] EU Policy Director, European Environmental Bureau

[210] Treaty of the European Community - Articles 2, 6 and 174.

Directives. Some are voluntary, like the EU Flower, others set mandatory requirements for planning processes, like the Strategic Environmental Impact Assessment.

This development has to be welcomed in general but i) it remains far from completion and ii) raises concerns about the division of powers in the European Union and the application of the "subsidiarity principle". The most powerful "horizontal tools" are fiscal and subsidies policies, which, if they reduced the pressure on labour and increased it on environmental resource use, could strengthen economic performance, societal welfare and environmental protection (Jørgensen 2003). But EU fiscal policies must be adopted by unanimity. Until today, no harmonised environmental taxes or charges were ever decided at EU level on the basis of environmental objectives. There have been some 'minimum taxes' set for internal market purposes (e.g. taxes on fuels)[211]. This instrument is thus largely left to the discretion of Member States.

159

The "subsidiarity principle" established in the EU Treaty suggests that the EU only takes action where Member States fail to achieve EU objectives and that decisions are taken as closely as possible at citizen level. As a lot of "horizontal instruments" directly address this lowest decision making level – the local level – a number of complications arise and add to national resistance against perceived "bureaucratic over-regulation" coming from Brussels. Nevertheless one has to argue that obviously Member States have failed to achieve the objectives and therefore should either make a stronger effort or accept interference from the EU. Environmental quality standards cannot be withdrawn just because one fails to achieve them! Before that, one should prove that all efforts have been made to take action.

The following chapters explain a selected number of those "horizontal instruments", assess their strengths and weaknesses and make suggestions for action to make best use of those instruments in order to achieve environmental objectives. Therefore, specific attention is given to establishing the necessary cross-references and links to the environmental objectives described in chapter IV. For the purpose of allowing easier access to the different "horizontal tools" they are divided in two categories: i) Measures on Production, Planning and Management and ii) Product Policies.

The chapter starts with the most important environmental instrument – public participation in environmental policies. Although dealing with the environment, the three pillars of participation - public involvement, access to information and justice - touch upon a much wider field and must be understood as supporting the European Union's founding principles of liberty, democracy and the rule of law.

BIBLIOGRAPHY AND FURTHER READING

BRINEGZU, S. AND SCHÜTZ, H. (2003): Zero Study: Resource Use in European Countries - An estimate of materials and waste streams in the Community, including imports and exports using the instrument of material flow analysis. March, 2003.

JØRGENSEN, C.E. (2003): Environmental Fiscal Reform: Perspectives for Progress in the European Union. EEB, June 2003.

---

[211] Note: there are no taxes on fuels as such, only a harmonised range for taxes on fuels. It remains the sole responsibility of Member States to decide whether they wish to introduce such taxes or not.

V.2 Public Involvement in Environmental Decision-Making

By Mary Taylor [212]

V.2   Public Involvement in Environmental Decision-Making.....................................160

  V.2.1 Introduction 160

  V.2.2 The Aarhus "Public Participation" Convention 161

     V.2.2.1 Ratification..162

     V.2.2.2 The Three Pillars...162

        V.2.2.2.1 Access to Information 162

        V.2.2.2.2 Public Participation..163

        V.2.2.2.3 Access to Justice..164

     V.2.2.3 The EC institutions..165

     V.2.2.4 The Aarhus Convention Compliance Committee.........................166

     V.2.2.5 The Outlook 166

  V.2.3 Public Information on Pollution - Pollutant Release and Transfer Registers (PRTRs)..167

     V.2.3.1 The European Pollutant Emission Register 168

     V.2.3.2 The PRTR Protocol 168

     V.2.3.3 Outlook 170

V.2.1   Introduction

Those who seek to protect the environment, at whatever scale of activity, are usually quite disadvantaged as compared with the economic forces at work around us. In recent years there has been some recognition of the need to address the balance of power and allow citizens a greater role in environmental decision-making. For that role to be effective it follows that citizens must be able to access relevant information and have opportunities for comment. Decision-makers also need to be accountable which means their decisions, and the basis for those decisions, should be open to public scrutiny. In cases where citizens' rights are not upheld, or there is lack of enforcement of laws, then the public should be able to act in order to ensure that such laws are respected.

---

[212] Campaigner at Friends of the Earth, England, Wales & Northern Ireland.

EU environmental laws (*see especially chapter IV*) increasingly provide their own public participation provisions. They are considered to be essential for good application and enforcement of the laws, a very weak part of EU environmental policies. The list of Member States condemned by the European Court of Justice in Strasbourg for violating environmental laws is very long. Yet the list of infractions is likely to be only the tip of the iceberg. Water, soil, air and nature have no voice of their own and citizens are usually dependent on the European Commission to take up their complaints since they have practically no access to the Court in environmental matters.

Public participation rights, therefore, have to be seen as one of the most important horizontal tools in environmental policy, and the Aarhus "Public Participation" Convention and its implementation in Europe is a major step forward. However, access to justice remains weak and needs urgent underpinning to strengthen the ability of citizens to use the courts in defence of the environment.

In this section a brief overview of some tools that enable access to information and public participation - the Aarhus Convention, and its offspring, the Protocol on Pollutant Release and Transfer Registers - is given.

V.2.2 The Aarhus "Public Participation" Convention

The UNECE Convention on Access to Information, Public Participation in Decision-making and Access to Justice in Environmental Matters (Aarhus, 1998) is a milestone in environmental democracy, granting procedural rights to individuals with respect to access to environmental information held by public authorities and public participation in decision-making[213]. This transparency and participation is, according to the objective of the Convention, "to contribute to the protection of the right of every person of present and future generations to live in an environment adequate to his or her health and well-being...". So here is an international instrument clearly making a link between environmental protection, health, human rights and democracy. The Convention has far-reaching consequences for the conduct of government and businesses and involvement of the public in matters related to environmental protection.

The Convention does not set standards for environmental quality in itself, but sets out rules for the public's right of access to environmental information held by public authorities and a right of public participation in permitting procedures (e.g. granting a pollution licence to a power plant) and, if somewhat less enthusiastically, in the development of plans, programmes, policies and laws. It also provides for increased access to justice in a number of circumstances, such as wrongful decisions under the terms of the Convention or even to challenge more general contraventions of environmental laws.

The Aarhus Convention was negotiated with much input from the ECO community, and, despite its imperfections, it is a profound instrument providing a number of tools for campaigners to greatly improve government accountability, environmental decision-making and the involvement of stakeholders.

---

[213] The UNECE pages on the Aarhus Convention provide very useful material on the Convention and ongoing processes: www.unece.org/env/pp. "The Aarhus Convention: An Implementation Guide" has helpful commentary on each of the articles of the Convention. It is at: http://www.unece.org/env/pp/acig.htm

V.2.2.1 · RATIFICATION

In the EU, implementation of the Convention is somewhat complex — a number of individual countries have signed and gone on to ratify the Convention with their own national laws, others have only signed and are waiting for Community legislation[214].

The European Community itself has ratified the Convention in 2005[215], and the Community has to develop legal instruments that will apply the rules of the Convention to Community institutions as well as to the Member States. At the time of writing some of the legal instruments are not completed, although the ratification means that we should be able to use the provisions of the Convention.

The Aarhus Convention provides for minimum standards and any country, or indeed the European Community, can go further than the Convention. Conversely, it is also possible that some laws ostensibly implementing the Convention do not fully respect its intent or find them-selves not quite in line with the overarching Community legislation or even the Convention itself. Given the room for interpretation and development of concepts in this area there is considerable work to do — including by ECOs - to ensure the coherence and consistency of national and Community legislation.

Reference is often made to the three "pillars" of the Convention, as reflected in its title, each of which deserves particular consideration. This brief introduction cannot really do justice to the Aarhus Convention or describe its full scope and ongoing debates at EU or MS level, but we give a number of references for those who require more detail.

V.2.2.2 THE THREE PILLARS

V.2.2.2.1 Access to Information

An earlier EC Directive on access to environmental information (90/313/EEC) had already been in force for some years in EU Member States at the time the pan-European Aarhus Convention was being developed. The Directive itself inevitably influenced the negotiations but, since it was also being used in practice and subject to review under its own terms, its shortcomings were also exposed. The final Convention did make some advances beyond 90/313/EEC, and in its turn was transposed in 2003 into a new information Directive, 2003/4/EC. Member States should now have developed their own national laws to reflect this Directive by 14 February 2005.

Progress includes the definition of environmental information, which is very broad and includes cost-benefit and other economic analyses and measures such as policies likely to affect the environment. It is also much clearer that health and safety information is within the scope. Environmental information should NOT be interpreted as simply about the state of the environ-

---

[214] Of the 25 EU countries, currently only Slovakia has not signed the Convention; 19 countries have ratified, 5 have not yet ratified (Germany, Greece, Ireland, Luxembourg and Sweden).

[215] The ratification decision (in French) is at http://register.consilium.eu.int/pdf/fr/05/st05/st05457.fr05.pdf; further aspects of the dossier can be tracked at: http://europa.eu.int/prelex/detail_dossier_real.cfm?CL=en&DosId=186271

ment or just about information held by environment ministries. The definition of public authorities is now much improved. Where public functions have been devolved or delegated or even
privatised, it is clear that these secondary bodies cannot escape obligations to provide access to
information. However, the big exception still is that MS may exclude bodies or institutions that
act in a judicial or legislative capacity. This includes bodies such as parliaments, courts and tri-
bunals, and can also apply to authorities which may have both legislative and administrative functions. Note that in this latter case the authority should not have a blanket exemption - only those
areas which are legislative in nature should be exempted[216].

There are a number of exemptions that may be applied to justify refusal of access to information.
These include ; for reasons of national security, commercial confidentiality, unfinished documents
and personal data. However, all are discretionary and have to be weighed against the public interest in releasing the information. Information on emissions into the environment shall be disclosed.

One area that is still problematic occasionally is that of charges for information requests.
Examination of public registers or environmental information *in situ* must be free of charge, but
supply of information may be subject to charges. These have to be a "reasonable amount", but
this is somewhat open to definition and charges can cause difficulties for cash-strapped ECOs.
However, it is also the case that the Courts have noted that high charges would be perverse if
they restricted access to information[217]. If the cost of information is prohibitively high this creates
a serious inequality between, for example, corporations with access to large budgets and much
poorer ECOs.

V.2.2.2.2   Public Participation

The Aarhus Convention places considerable emphasis on the procedures for public participation
in decision-making on specific activities (such as IPPC permitting, *see chapter V.3.2*). It also provides for public participation with respect to the drawing up of plans, programmes and policies.
These are not defined in the Convention itself but would be expected to include environmental
strategies and planning policies.

Whilst some Community legislation already conformed to the Aarhus principles, and future environmental legislation will incorporate public participation as it is drafted, it has been necessary
to amend some existing legislation. Thus the new Directive on public participation, 2003/35/EC,
is not a perfectly horizontal piece of legislation applicable across the board. In particular, it
amends the Integrated Pollution Prevention and Control (IPPC) Directive (96/61/EC) (*see chapter
V.3.2*) and Environmental Impact Assessment Directive (85/337/EEC) (*see chapter V.3.5.1*), improving the public participation provisions, ensuring that they also apply to neighbouring countries
and providing for access to justice. Additionally it also brings into line public participation provisions for plans and programmes drawn up under the following Directives:

---

[216] Note that Article 8 of the Convention promotes public participation in the preparation of drafts of legislation, this being
regarded as an executive function.

[217] ECJ (1999). Commission v. Germany (C-217/97)

- 75/442/EEC on waste
- 91/157/EEC on batteries and accumulators
- 91/676/EEC on nitrate pollution
- 91/689/EEC on hazardous waste
- 94/62/EC on packaging
- 96/62/EC on air quality.

The essence of the process is that the public must receive early notification of intent to draw up (or modify or review) a plan or programme, with details of how to obtain information and how to participate and submit comments. The relevant authorities shall take account of any such comments, and in turn inform the public of their decision. Parties must also make public the reasons and considerations on which the decision is based so we believe that decision-makers should respond to each specific point raised by the consultation responses.

This Directive has to be transposed into national laws by June 25 2005.

An important loophole in the Convention was the exemption of public participation obligations for GMO-related decisions. With an amendment to the Convention, agreed by the Meeting of Parties on the 27th of May 2005, this loophole is partially addressed. The main weakness of this amendment is that it does not guarantee access to justice in cases where public participation requirements are violated or when contributions from the public are ignored without clarification. The amendment will only enter into force when some 30 countries have ratified it.

V.2.2.2.3   Access to Justice

The Commission has produced a draft Directive on access to justice for implementation by Member States (*CEC 2003*). This is going through the co-decision procedure with Parliament and is not yet final at the time of writing. The draft Directive leaves much of the details to Member States, so (assuming that any flaws are not amended prior to adoption) there should be possibilities for improving on the draft at national level.

Access to the courts in pursuit of environmental protection or exercise of rights is still not well developed. One of the most contentious issues is that of legal standing for ECOs (and others), i.e. the recognition of sufficient status to bring legal proceedings. A totally open system providing for "*actio popularis*" was opposed by the Commission on the grounds that this goes further than the Convention. Others have objected that enhanced access to the courts would create a huge number of actions, flooding the system. However analysis of environmental cases brought by ECOs in a number of countries where ECOs have broader standing indicates that this view is not well-founded. Such actions formed a minuscule proportion of the overall case load (0.0148% in one study) and indeed were far more successful on average than other cases, emphasising the highly focussed and targeted nature of the legal cases fought by ECOs (*SRU 2005 and de Sadeleer 2005*).

Member States can of course instate such a provision if they see fit. However, it seems likely that countries will set some hurdles to standing, and so some ECOs may find themselves with insufficient power to pursue cases at times. An automatic right of standing for environmental cases of broad public interest would be very helpful.

There is however a possibility to become a "qualified entity", a concept which is not found in the Aarhus Convention and which is opposed by environmental citizens' organisations. Legally established ECOs which have the aim of protecting the environment and have an organisational structure (including audited accounts) which enable them to pursue this objective are likely to qualify. Thus larger, longer established ECOs might pass the tests, but a local neighbourhood group set up to challenge a recent development proposal will have difficulty. This does not exclude the possibility of court action through some other route though, such as by a member of the public who can show "sufficient interest" or whose rights are impaired.

The potential costs of legal action are also a huge barrier to environmental justice. Not only does the losing side face its own costs, but often those of the opponent, despite the fact that an environmental case is likely to be very much in the public interest. One idea is to further the notion of an "Aarhus Certificate", whereby environmental cases are recognised as such early on and may then be funded out of more general funds, rather than one brave ECO facing the entire burden (*Coalition for Access to Justice for the Environment 2004*).

The draft Directive also excludes the Aarhus provision for challenge of acts and omissions by private persons which contravene national law relating to the environment (but covers those by public authorities). This seems to have pleased industry and ECOs have been urging amendment of this.

V.2.2.3   THE EC INSTITUTIONS

The Commission has also drafted a proposal for application of the Aarhus Convention principles to EC institutions and bodies[218]. The first important point to note is that these bodies are not limited to the European Parliament, Council, Commission, Court of Justice and Court of Auditors. It is interpreted much more broadly as *"any public institution, body, office or agency established by, or on the basis of, the Treaty establishing the European Community and performing public functions except when and to the extent to which they act in a judicial or legislatve capacity[219]"*. Thus EU agencies set up under Community legislation are caught within the net.

Much of the Regulation aims at parallel provisions for the Community bodies as for authorities in Member States. However there are a number of weaknesses and the overall effect is that the Aarhus provisions are less firmly embedded here than they are in the legislation for Member States.

Regulation 1049/2001 on access to Community documents will remain in force, but it is somewhat amended by the new draft Regulation. As noted above, the definition of Community institutions and bodies is clarified and extended, and any person, not just an EU citizen, may make a request for information.

---

[218] Proposal for a Regulation on the application of the provisions of the Aarhus Convention on Access to Information, Public Participation in Decision-making and Access to Justice in Environmental Matters to EC institutions and bodies; http://europa.eu.int/eur-lex/en/com/pdf/2003/com2003_0622en01.pdf

[219] From the explanatory memorandum accompanying the draft Regulation

Whilst the public participation provisions cover "plans and programmes relating to the environment", the draft stops short of including policies, legislation and financial programmes. Thus the public is rather excluded from the decision-making on high level EU policies.

The concept of "qualified entities" for legal standing appears again, as discussed above.

Since the Regulation is still in draft form, details may be amended before final adoption, but its progress can be tracked at the Commission's website[220].

The Regulation is likely to enter into force in 2006, after negotiations between the Commission, Environment Council and the European Parliament over the final text. In its first reading, the European Parliament demanded some important improvements to the Commission text. The Council however, in its "Common Position" of December 2004, went in the opposite direction. Effectively it wants to remove the "access to justice" pillar from the Regulation, and it also wants to exempt the European Investment Bank from its scope. In the autumn of 2005, the EEB will try to convince the European Parliament, during the second reading, not to accept such a weakening of the text.

V.2.2.4   THE AARHUS CONVENTION COMPLIANCE COMMITTEE

One potentially very useful mechanism for promoting better implementation of the Aarhus Convention is the international Compliance Committee, established under its auspices. This operates on a UNECE-wide basis for Parties to the Convention. The Committee can receive complaints directly (through the Aarhus Convention secretariat) from members of the public, although domestic remedies should be considered first. However, if these prove inadequate for some reason, ECOs or individuals can approach the Compliance Committee[221].

The compliance procedure is designed to promote compliance with the Convention, rather than provide redress for infringement of an individual's rights. A number of complaints have already been received, dealing with issues such as rights of legal standing for ECOs, failure of national laws to respect the Convention and failure of public participation procedures. Reports of the Committee meetings are available at the UNECE web site. As of writing no final reports have been filed on particular cases, but these promise to be of very great value in setting standards for implementation.

V.2.2.5   THE OUTLOOK

Whatever the content of the laws themselves, in practice freedom of information, public participation and access to justice are invariably also affected by cultural attitudes, historical perspectives, national implementation programmes, training for officials and the amount of resources given to the efforts. Actual practice will inevitably vary considerably from one country to another and it is vital that the ECO community uses the Convention and its legal offspring, not only to further particular environmental campaigns, but to test the effectiveness of national laws and

---

[220] http://www.europa.eu.int/prelex/detail_dossier_real.cfm?CL=en&DosId=186297

[221] Complaints can be received after a country has been a Party for one year.

check them against the standards and aspirations of the Aarhus Convention. Making use of the relevant laws for practical purposes is probably the best way to test and improve the implementation of the Convention, through exposing problems and promoting change. The documentation of the limitations of 90/313/EEC for example was very influential in shaping the Aarhus Convention, and this collection of practical experiences and review needs to continue. Various review procedures are even built into the legislation in fact (for example, Commission must review the Directive on public participation by 25 June 2009), providing excellent opportunities for ECOs to highlight weaknesses and press for improvement.

ECOs should also seek to push their own country's implementation as far as possible beyond the Convention and Directives. As mentioned above, legal standing and the costs of court action are one area where there is a critical need to make progress. The use of exemptions to withhold information (all of which are discretionary) should also be monitored closely. Charges for information should also be challenged.

A number of task forces and working groups established under the Convention continue to develop aspects of its application, again with ECO involvement. This is largely organised through the European ECO Forum's Public Participation Campaigns Committee, including an email discussion group. See www.participate.org for more details

V.2.3   Public Information on Pollution - Pollutant Release and Transfer Registers (PRTRs)

PRTRs are public databases showing quite detailed information on quantities of pollutants released or transferred to other sites, the specific sources responsible and their exact locations. Because they 'name and shame' individual factories and other pollution sources, they are widely acknowledged as a tool for increasing corporate accountability, driving down pollution and improving the management of chemicals. Many successful websites have presented PRTR data using interactive maps and helpful background information and are highly engaging tools for public use.

Having information about exactly how much pollution is coming from a factory stack or what dangerous chemicals might be stored in your neighbourhood might seem to be only fair. But it is only relatively recently that citizens have had systematic access to some of this information across the EU.

PRTRs have a longer history though. The first public PRTR was established in 1986 in the US, known as the Toxics Release Inventory. This was at least in part catalysed by the catastrophic accident at Bhopal, India, in 1984 when thousands of people were killed by a release of methyl isocyanate from a Union Carbide pesticide plant.

A number of initiatives are under way in the EU. The first EU-wide instrument, the **European Pollutant Emission Register** (*CEC 2000*), EPER, was adopted in 2000. But a number of countries already had their own **national systems** in place and some of these continue to be more extensive than EPER. A pan-European (UNECE) **Protocol on Pollutant Release and Transfer Registers** was agreed under the Aarhus Convention in May 2003 in Kiev. The Commission has proposed replacing the existing EPER with a PRTR Regulation that incorporates the requirements of the Protocol. This proposal is currently under discussion.

168

We describe very briefly here features of EPER and the Protocol, but should emphasise that each Member State should set up a national system, with public participation, so there is much scope for ECOs to influence these national systems and take them beyond the scope of the international instruments.

More detailed information on PRTR systems is available in the European ECO Forum booklet: "Your Right to Know About Sources of Pollution - A brief introduction to the protocol on pollutant release and transfer registers", December 2003.

V.2.3.1 THE EUROPEAN POLLUTANT EMISSION REGISTER

EPER requires major industries (those that report under the Integrated Pollution and Prevention Control Directive, *see chapter V.3.2*) to report on the annual releases of 37 air pollutants and 26 water pollutants, including major greenhouse gases and acid rain gases[222]. The information is available online[223] and allows users to view pollution data for around 10,000 individual industrial sites (including larger livestock farms and landfill sites) across the EU. Search facilities allow the user to explore for particular substances or by country or even company name. It is quite simple to find out who released, for example, the most sulphur dioxide in 2001 in any country. Since the data are only reported, for the time being, on a three-yearly cycle the next set of data published will be for 2004.

As much as EPER is a major step forward for the right-to-know movement, it is a very limited type of pollution inventory and gives only a partial view of all the releases and transfers[224] of pollutants from an industrial site. A major step forward will be implementation of the new Aarhus Convention **Protocol on Pollutant Release and Transfer Registers,** the PRTR Protocol[225].

V.2.3.2 THE PRTR PROTOCOL

The Aarhus Convention itself is very much concerned with access to information held by public authorities rather than private entities. But one section gave scope for the development of pollution inventories in order to channel more information about the environmental impact of corporate activities into the public domain. It required the establishment of national systems cataloguing the annual releases of pollutants from major industries.

---

[222] Some of the air and water pollutants are the same substances, making a total of 50 different substances in all.

[223] www.eper.cec.eu.int

[224] A release is an emission directly into the environment (such as to the air, into water or onto land); a transfer is usually used to describe wastes that are sent to another site for treatment or disposal, e.g. wastes sent to an incinerator.

[225] www.unece.org/env/pp/prtr.htm

The negotiations on the PRTR Protocol were finalised in 2003, and at the time of writing all EU countries except Slovakia and Malta had signed. It is also open to countries outside the UNECE (so globally) and a total of 36 countries plus the European Community are now signatories. The European Commission has prepared a draft Regulation in order to transpose the Protocol into EU law[226]. Once this becomes law, it is expected that data will be collected for the year 2007, which will be published in 2009. National data sets should be available ahead of the EU-wide web version.

Although the current EPER heavily influenced the negotiations and final form of the PRTR Protocol, the Protocol makes some significant advances. We do not have space here for much detail, but gains include:

▶ A total of 86 substances (up from 50, although a number are obsolete pesticides, at least in the EU)

▶ Capturing a wider range of reporting industry, including parts of the mining sector, large ship-yards, major sewage treatment works

▶ Releases to land are included for some pollutants

▶ Waste transfers are included if they are off-site (eg to a landfill or incinerator)

▶ Annual reporting (rather than three-yearly)

▶ Reports on diffuse sources of pollution such as traffic are to be developed

▶ Explicit language on public accessibility and user-friendliness

Also note that the EU's PRTR draft Regulation includes four further substances to ensure coherence with the Water Framework Directive (*see chapter IV.5*) and the Stockholm Convention on POPS. The timeframe for data collection and subsequent publication is also tighter as compared with the Protocol.

Some uses of PRTR data

▶ Identifying major polluters

▶ Showing decreases (or increases) in one factory's pollution over time

▶ Augmenting pollution prevention programmes and targets

▶ Locating "hot-spots" with particularly high emissions

▶ Helping emergency planning

▶ Analysing pollution and socio-economic factors eg the location of factories amongst populations of lower income or ethnic minorities

▶ Analysing investment and insurance risk

[226] The draft Regulation is at: http://europa.eu.int/eur-lex/lex/LexUriServ/site/en/com/2004/com2004_0634en01.doc

V.2.3.3 OUTLOOK

There are a number of areas where ECOs can be active. The PRTR Protocol is quite specific about public involvement, stating that *"Each Party shall ensure appropriate opportunities for public participation in the development of its national pollutant release and transfer register, within the framework of its national law"*. Member States should be challenged to put this into effect, involving ECOs and indeed other stakeholders. We have a role to play in ensuring that our countries embrace the Protocol/European EPER, preparing to meet the obligations as a very minimum but increasing the scope where possible.

It seems unlikely that the EU instrument itself will expand much beyond the PRTR Protocol requirements (except as noted above) in the short term. Many countries feel that implementing EPER and expanding it to be compatible with the Aarhus PRTR is already quite demanding. However, in the longer term there is huge potential for evolution and this will be greatly helped by "consumer demand". Industry is making it quite clear it dislikes the sort of transparency that we seek, and does not shy from using the "anti-competitive" arguments.

With respect to the scope of national systems, one obvious area for expansion is in the numbers of pollutants. The US system, the Toxics Release Inventory, publishes information on over 600 toxic pollutants which makes the 86 of the PRTR look very unambitious. There are also some industrial sectors that have escaped reporting requirements. Rather ludicrously the nuclear industry and petrochemical storage facilities were excluded from the Protocol for example, much to the dismay of the ECO representatives. There is scope for including these in national systems, although political and corporate opposition may have to be overcome of course[227].

ECOs should also use the information generated by the current EPER or existing national systems to help raise awareness and to increase the pressure for lower pollution levels. Publish those league tables of polluters! PRTRs need to be used, and useful analysis can lead the way for improvements. ECOs have in the past lead the way in developing internet access and undertaking analysis of pollution data[228]. This needs to continue if we are to keep the spotlight on pollution and waste disposal by industry and achieve further improvements in reporting.

BIBLIOGRAPHY AND FURTHER READING

DE SADELEER, ROLLER & DROSS (2005). Access to Justice in Environmental Matters and the Role of ECOs, Empirical Findings and Legal Appraisal. Europa Law Publishing, Netherlands.

EC (2000) Commission Decision (2000/479/EEC) of 17 July 2000 on the implementation of a European pollutant emission register (EPER) according to Article 15 of Council Directive 96/61\EC concerning integrated pollution prevention and control

---

[227] For example, the national system for England and Wales, the Pollution Inventory, includes radionuclides and the nuclear industry.

[228] For examples, visit: www.scorecard.org; www.foe.co.uk/campaigns/sustainable_development/pollution_and_poverty/

CEC (2003) Proposal for a Directive of the European Parliament and of the Council COM(2003) 624 final on access to justice in environmental matters, Brussels, 24.10.2003

Coalition for Access to Justice for the Environment (2004). Briefing at http://www.elflaw.org/files/CAJE%20General%20briefing.pdf

European ECO Forum (2003): Your Right to Know About Sources of Pollution - A brief introduction to the protocol on pollutant release and transfer registers.

SRU (2005): German Advisory Council on the Environment (2005). Access to Justice in Environmental Matters: The Crucial Role of Legal Standing for Non-Governmental Organisations. SRU, Germany;

UNECE pages on the Aarhus Convention provide very useful material on the Convention and ongoing processes: www.unece.org/env/pp. "The Aarhus Convention: An Implementation Guide" has helpful commentary on each of the articles of the Convention. It is at: http://www.unece.org/env/pp/acig.htm

171

V.3 Environmental Measures on Production, Planning and Management

172

V.3 Environmental Measures on Production, Planning and Management..........................**172**

   V.3.1 Introduction 174

   V.3.2 IPPC-Directive...175

      V.3.2.1 Introduction 175

      V.3.2.2 Key elements of the IPPC Directive 177

         V.3.2.2.1 The permits 177

         V.3.2.2.2 Conditions of the permit 177

         V.3.2.2.3 BAT 177

         V.3.2.2.4 Environmental Quality Standards as the basic approach of the directive 179

         V.3.2.2.5 How are ELVs defined? 179

         V.3.2.2.6 Public participation and pollutant emission register (Art. 15) 180

         V.3.2.2.7 Transboundary effects (Art. 17) 180

         V.3.2.2.8 Community ELV's (Art. 18)................................180

         V.3.2.2.9 Prevention of accident 181

      V.3.2.3 Strengths and Weaknesses 181

         V.3.2.3.1 Strong Points...181

         V.3.2.3.1 Weak Points 181

         V.3.2.3.3 IPPC implementation: Does it deliver improvements? 182

            V.3.2.3.3.1 *Quality of European BAT Reference documents* 183

            V.3.2.3.3.2 *Quality of national permits and use of BAT* 186

      V.3.2.4 IPPC in the wider policy context: An effective tool? 186

         V.3.2.4.1 Phase out of hazardous chemical emissions 188

         V.3.2.4.2 Preventing and Recycling Waste.........................189

         V.3.2.4.3 IPPC Permits as an Industry Defence....................190

            V.3.2.4.3.1 *Environmental Liability*....................190

            V.3.2.4.3.2 *Future Chemicals Policy (REACH)* 191

         V.3.2.4.4 Air and Climate Policies................................192

      V.3.2.5 Outlook and follow-up actions...193

   V.3.3 Environmental Liability - first step towards "Making the polluters pay"...............197

      V.3.3.1 Introduction: From Seveso to Baia Mare 197

      V.3.3.2 Making the Polluters Pay: a long way to go.........................197

V.3.3.3   Objectives, Scope and exemptions 198

        V.3.3.3.1 Objective..198

        V.3.3.3.2 Scope: Timing and Type of damage 199

        V.3.3.3.3 Scope: Type of activities 200

        V.3.3.3.4 Exemptions..200

        V.3.3.3.5 Financial security 201

V.3.3.4   Strengths and weaknesses of the new liability regime 202

V.3.3.5   Conclusions and ECO action (environmental citizens' organisation)...........205

V.3.4 Greenhouse Gas Emissions Trading Directive...208

  V.3.4.1   Introduction: Climate Change Policy in the EU 208

  V.3.4.2   Emissions trading: theory and practice..210

        V.3.4.2.1 The concept: cap and trade 210

                V.3.4.2.1.1   How it works in theory...210

                V.3.4.2.1.2   Environmental effectiveness....................................211

        V.3.4.2.2 The practice: the EU Emissions Trading System (ETS) 212

                V.3.4.2.2.1   Design features of the EU ETS 213

                V.3.4.2.2.2   Why the success of the ETS is important 214

                V.3.4.2.2.3   The Review 215

  V.3.4.3   Assessment of the EU Emissions Trading System 216

        V.3.4.3.1 Strong points..216

                V.3.4.3.1.1   General..216

                V.3.4.3.1.2   Coverage 217

                V.3.4.3.1.3   Compliance system...217

                V.3.4.3.1.4   Public participation and transparency....................218

        V.3.4.3.2 Weak points 218

                V.3.4.3.2.1   Setting national targets and rules 218

                V.3.4.3.2.2   Link to external credits from JI and CDM...............219

                V.3.4.3.2.3   Impact on other policies and sectors 221

  V.3.4.4   Outlook and Prospects for ECO action (environmental citizens' organisation) 223

        V.3.4.4.1 Monitoring implementation 223

        V.3.4.4.2 Targets and rules for 2008-12 225

        V.3.4.4.3 The review process 226

V.3.5 Environmental Impact Assessment..228

  V.3.5.1   The EIA Directive 228

        V.3.5.1.1 Introduction and key elements..228

        V.3.5.1.2 Positive effects for the environment 230

        V.3.5.1.3 Weaknesses and constraints..231

V.3.5.1.4 Opportunities for action 233

V.3.5.2   **The SEA Directive**...**237**

         V.3.5.2.1 Introduction 237

         V.3.5.2.2 Objective and Tools 238

                 V.3.5.2.2.1   Public consultation, the environmental report and monitoring

                 ...240

                 V.3.5.2.2.2   The SEA Directive and Biodiversity 242

                 V.3.5.2.2.3   Strategic Environmental Assessment and Climate Change.242

         V.3.5.2.3 Transposition and Implementation 243

         V.3.5.2.4 Strength and Weaknesses of the SEA Directive...............................244

         V.3.5.2.5 Outlook and action...245

V.3.6   **Environmental Management and Audit Scheme**.....................................**247**

V.3.6.1   Introduction...**247**

V.3.6.2   Potential of EMAS **249**

V.3.6.3   Third party audit...**249**

V.3.6.4   EMAS and environmental management standards...**250**

V.3.6.4   Shortcomings of EMAS and ISO 14001:1996...**252**

V.3.6.8   Promotion of EMAS..**254**

V.3.6.9   **Opportunities for environmental citizens' organisations and Demands for**

V.3.1   Introduction

Environmental Measures on Production, Planning and Management can be roughly understood as provisions to integrate environmental considerations into production and planning of economic activities.

Those measures aim at internalising the costs of pollution, which is otherwise paid for by the general public or consumer, thereby implementing the "polluter pays principle". This is very explicit in the case of Environmental Liability and Emission Trading (*see chapters V.3.3 and V.3.4*), where either the potential polluter will be made financially responsible when the damage happened and therefore needs to prepare for this event in advance or the polluter has to pay for "emission rights" immediately. Both systems should lead to a market-based, and therefore the most economic, solution.

Environmental Liability has the biggest environmental protection potential as it can cover all economic activities, but largely depends on its national transposition and the strength of enforcing authorities to set a "price" for environmental deterioration. The Directive allows for a number of loopholes, like permits and state of the art defence, which dramatically reduce the scope of the Directive and help to enforce traditional command and control legislation rather than providing an additional market based instrument[229]. How far this tool will thus contribute to environmental protection remains unclear.

European Greenhouse Gas Emission Trading on the contrary has a much smaller scope, mainly addressing one pollutant and a limited range of economic activities, but the target is set. It is clear how much will be achieved with this instrument.

In case of the Integrated Pollution and Prevention Control (IPPC) Directive *(see chapter V.3.2)* – the centre piece of EU environmental industrial policies - environmental consideration should become part of the business management culture and pollution be prevented at source. It covers in principle all potential pollution of air, water and soil in an integrated way, but only a limited number of economic activities. The IPPC is strongly linked to most other environmental policies, follows a local (decentralised) approach and its effectiveness depends on European environmental quality and emission standards.

The Strategic Environmental Impact Assessment (SEA) Directive *(see chapter V.3.5.2)* addresses most governmental planning in order to integrate environmental considerations and allow better public participation. It complements the older Environmental Impact Assessment Directive dealing with single projects *(see chapter V.3.5.1)*. The experience with the implementation of the latter was disappointing – rarely has a project been significantly changed or been stopped because of the impact assessment. Environmental impact assessments press for a cultural and mentality change in public administrations, which can take a long time. Meanwhile the SEA Directive provides excellent opportunities for enforcing the correct application of existing environmental protection standards and requirements as set out in chapter IV of this book.

Finally there is the voluntary Environmental Management and Audit Scheme (EMAS, *see chapter V.3.6*), which can be an important internal business system for increasing environmental awareness and efficiency. However, because of lack of transparency and indicators, the environmental outcome remains difficult to assess and any interference with legislation or replacing it should therefore be rejected.

V.3.2   IPPC-Directive

By *Stefan Scheuer* [230]

V.3.2.1   INTRODUCTION

The Council Directive 96/61/EC on "Integrated Pollution Prevention and Control" from 1996 (IPPC) is at the heart of the environmental regulation of industrial activities in Europe. It introduced an integrated permitting procedure for 30 industrial sectors covering the pollution of water, soil and air and bases emission controls and applying a "best available technique" (BAT) concept.

---

[229] Wennerås, P. and Permi Dedfence (2004) *Environmental Liability Regimes – Subsidizing Environmental Damage in the EC?*, The Yearbook of European Environmental Law, Volume 4, December.

[230] EU Policy Director, European Environmental Bureau. With contributions from Andrew Farmer (IEEP), Dirk Jepsen (Ökopol), Lesley James (FoE); Chapters 2 and 3 based on Industry Handbook chapter 4 by Christian Hey, 1998.

IPPC relies on centralised (EU level steering) as well as decentralised (national/local permitting) elements– the latter clearly dominating. As one of the core elements at European level, reference documents (BREFs) are established, which describe BATs and provide associated Emission Limit Values. For pollutants for which "the need of Community action has been identified" the Commission is mandated to propose EU wide obligatory emission limit values. This it has so far failed to do. On the decentralised side Member States have the possibility of providing general binding rules for applying the IPPC Directive, including national emission limit values. Furthermore, the BREFs are "only" to be considered by local authorities in granting permits to and setting Emission Limit Values for industrial installations and can specifically be adjusted to the local conditions – environmental as well as socio-economical. These permit conditions apply to all new installations. Existing installations must have a permit by 2007. There are close links between IPPC and nearly all other EU environmental policies, including the EU's waste, air, climate, water, environmental liability and management and pollution emission register policies (see chapters IV and V), where the IPPC permit and/or BAT concept is utilised to achieve their objectives. To be able to understand the above-mentioned policies it is essential to have some knowledge of the IPPC Directive.

Other links are also important – such as those with water, chemicals and waste policies as well as the increase in the use of market based environmental policy instruments. This creates a highly demanding and complex network of legislation, which needs careful scrutiny, especially on whether IPPC's "best available technique" concept merits such an increasingly important role in environmental policies. At the same time, some industries and Member States have identified IPPC as too high a regulatory burden for industry's competitiveness, launching debate on whether IPPC should be revised to allow for more flexibility (e.g. to introduce emission trading) (ENAP 2004).

The EEB has cautiously welcomed IPPC, but warns of the danger of too much regulatory devolution, leaving local permitting authorities vulnerable to business pressure and competition, and reliance on other environmental policies setting harmonised standards. The EEB has participated since 1997 in the development of BREFs and encouraged its Members to play a role in the national implementation process. Based on its experience, the EEB regards the quality of BREFs and granted permits as rather mixed. The quality of national transposition legislation and implementation quality is alarming (CEC 2003)[231].

Given this, and the above mentioned high expectations of IPPC to address more environmental problems on the one hand, whilst at the same time providing more flexibility in its implementation on the other, requires a a carefully assessment of the Directive's implications.

This chapter will begin by explaining the IPPC's key elements, before addressing its strengths and weaknesses, the quality of BREFs and national implementation and finally its standing within the wider environmental policy framework. Conclusions are drawn which outline EEB's views on what IPPC can and what it cannot deliver in order to make best use of the IPPC as an effective instrument in environmental policies.

---

[231] European Commission report about IPPC implementation, July 2003 finds: huge delays, inadequate legal transposition and candidate countries facing serious capacity problems. It concludes that successful application of BAT by 2007 is essential to achieve more sustainable production patterns and therefore warns that if efforts of authorities are insufficient it might be necessary to establish a more harmonised approach, i.e. by setting EU wide emission limit values. But the Commission until now refrained from making any concrete proposals to revise or how to complement IPPC.

V.3.2.2   KEY ELEMENTS OF THE IPPC DIRECTIVE

The IPPC-directive defines an integrated permitting procedure for 30 industrial sectors, such as the metal, mineral, chemical and waste-treatment industries. Permits are intended to, above all, prevent pollution in air, water and land and where prevention is not possible, to control this pollution (Preamble 8 and Art. 1). Industrial operators who want to receive a permit, must comply with several ambitious principles. They must apply preventive measures against pollution, not cause significant pollution, avoid waste production, and recycle or deposit unavoidable waste , use energy sources efficiently and take preventive action against accident risk and aftercare after economic activity has stopped (Art. 3).

V.3.2.2.1   The permits

The IPPC directive prescribes that all industries require a permit. This applies to new (Art. 4) as well as to existing ones.[232] (Art. 5, Art. 13). Existing installations must meet the requirements of the directive and have a permit by 2007. According to Art. 13 Member States should reconsider periodically the operating conditions of existing installations.. However, they are free to define the periods for reconsideration. In any case, any major changes require a revised permit.

A company asking for a permit has to provide the permitting authority with a set of basic information, especially on its activities, the energy and raw material requirements, its emissions, and the techniques which it shall apply to reduce emissions (Art. 6). Applications must be sent to the competent authorities, which have to develop a co-ordination mechanism so that all aspects of the permitting procedure can be effectively managed (Art. 7). On the basis of available information and verification, if they meet the essential requirements, the authorities may grant or refuse a permit (Art. 8). The IPPC directive does not define timetables for the delay between application and issuing a permit.

V.3.2.2.2   Conditions of the permit

The core Article of the IPPC is art. 9, which defines the conditions of the permit. A permit will only be given if : a company provides complete information, it meets the emission limit values defined by the competent authority, monitoring requirements are fulfilled, accident prevention measures are carried out and some additional conditions are met. All the above conditions have to be defined by the competent authorities, which means normally that they will be defined at national or regional level. The competent authorities also have to consider all other relevant information which it receives about the new installation (Art. 9.2).

V.3.2.2.3   BAT

The key concept, which aims to achieve the principles of art. 1 and 3, is that of "Best Available Technique". This concept is much more restrictive than the far-reaching principles might suggest. ,Each of the words themselves – 'best', 'available' and 'technique' – require definition.

"Best available techniques" are, according to the directive, the "most effective and advanced stage in the development of activities." It must be proven that such activities are practically suitable in order to derive emission limit values. ELVs are founded upon the basis of BAT – but "without prescribing

---

[232] "Existing" means an installation which was in operation before November 1996

the use of any technique" (Art. 9.4). It is not absolutely clear how this confusing cross-reference may be interpreted. An open interpretation is that even if one finds an ELV which corresponds to a BAT, there might also be other techniques which lead to the same values. Such an interpretation empha- sises the flexibility and technological openness which eventually may lead to innovation. Another interpretation is that this formulation reflects the opposition of many Member States and that of industry to identifying what a BAT really is. Industry is pressing for a wide range of techniques and corresponding BATs, so that the concept will lose its binding character (*UNICE 1998*).

"Best" is related to the most effective way of achieving a "high general level of protection of the environment as a whole". What this high general level may mean is not defined in this directive, but it is defined in several other directives (e.g. Habitats Directive, Air Quality Framework Directive and the respective daughter directives, Water Framework Directive). Favourable Conservation Status of important flora, fauna and habitats is established by the Birds and Habitats Directive (*see chapter IV.2*). As regards clean air policies, a programme to define European air quality standards is quite advanced (*see chapter IV.3*), as regards water a "good status" is set in a normative way and quality standards for 33 chemicals are to be expected soon (*see chapter IV.5*). With regard to soil quality, no EU policy has yet been developed and the future EU Soil Protection Strategy as outlined by the 6EAP is hardly likely to set standards in the near future. So the quality of "Best" depends on the quality of legislative work to be carried out in the course of the coming years.

"Available" implies that several conditions be met: scale, economic viability, efficiency and acces- sibility. Available techniques are already developed in terms of scale; they are hence "ripe" tech- niques and capable of being applied widely across Europe. They must have proven the market test – that means they must have been applied under normal market conditions ("economically and technically viable conditions"). They must be efficient, meaning that their cost-benefit rela- tionship has to be considered. Establishing the external financial costs of pollution and hence the benefits of its reduction is still an ongoing methodological problem, which has produced many scientific controversies. In determining whether or not a technology is really "efficient", the crucial factor is the choice of methodology used to identify the external costs being defined., Finally – despite the existence of an internal market without barriers to trade – the directive explicitly stresses "accessibility" as criteria: the technique must be locally accessible.

"Techniques" encompass a wide range of actions: from better technology to better operation and design of factories. This new term also initiated a controversial discussion on its interpreta- tion. For some, this definition is a leap forward from end-of-pipe solutions to an integrated view on production processes, making the concept innovative and far-reaching. Others fear that the focus on "software", on organisation and operation, may lead to a neglect of hardware, such as filters and other end-of-pipe technologies. If this interpretation becomes reality, the concept would narrow down rather than widen the range of actions to be considered.

The key concept of the IPPC-directive, the BAT, is open to a wide range of interpretations, despite – or perhaps because of - its elaborated definition. It contains restrictions, which raise doubts that the concept as such will reach the level of aspirations defined in Art. 3. However, it certain- ly also offers opportunities for policy innovation, if it is interpreted as a new requirement to apply integrated plus end-of-pipe technologies to reduce pollution.

V.3.2.2.4 Environmental Quality Standards as the basic approach of the directive

According to Art. 9.4 Member States have to define ELV's "based on the best available techniques" (see above). Furthermore they have to take into account the technical characteristics of the plant, its geographical location and local environmental conditions. In other words, the concept of BAT is further restricted and a fully decentralised approach is applied to define the specific conditions for a permit. The concept of "local environmental conditions" can only be appropriately understood by looking at the background to the directive's history. It is a weaker formulation of the original proposal of the Commission that BAT is not required, where given quality values may also be achieved with less effort. In other words, the idea of "local environmental conditions" implies a right to pollute until given Quality Values are reached. It is in essence a "filling-up" clause – even if "filling-up" might be interpreted as being contradictory to the basic principles of EU-environmental policies and the principles mentioned in the directive itself. However a pure filling-up policy is also restricted by several safeguards. The respect of local considerations should not lead to the neglect of long-distance and transboundary pollution (Art. 9.4). So the traditional policy of trying to achieve high local quality standards by the long-range diffusion of substances is not allowed according to the directive.

The primacy of quality standards also applies, if defined levels cannot be achieved by BAT alone. Where BAT falls short of achieving EU environmental quality standards, additional measures will be required (Art. 10). This is reconfirmed by the "combined approach" set out in Article 10 of the Water Framework Directive 2000/60/EC (*see chapter IV.5.5.2.2.2*).

The conditions for a permit hence are defined decentrally. National authorities have considerable discretion on how to strike the balance between environmental and economic considerations, which are included in the directive *(Breuer 1997)*. Without explicitly mentioning it, the IPPC-directive is based upon a quality-oriented approach. IPPC follows a combined approach, but with EQSs as the primary basis. ELVs have to consider those values, either by becoming lenient in clean regions or by becoming stricter in heavily polluted regions. The ELV-elements are defined as qualitative side-conditions to be respected (for instance: transboundary pollution, meeting the objectives of this directive). Further work on ELV is reduced either to an information exchange activity without legally binding character (Art. 16) or to an optional activity (Art. 18). Both elements are interlinked and require further description.

V.3.2.2.5 How are ELVs defined?

According to Art. 16 the Commission organises an exchange of information on the different national ELVs and on the BAT. It has to publish a regular report - every three years - on how the IPPC-directive is interpreted in Member States.

The information exchange is presently delegated to the IPPC-Bureau in Seville. This is a technical bureau linked to a research institute of the European Commission (IPTS). Member States and industry participate in this process. While ECOs are not explicitly mentioned in the directive, the internal mandate especially refers to the EEB as one of the participants in the process. As a result of this information exchange BAT-Reference Documents (BREFs) are elaborated, which serve as an information pool for the permitting authorities and the applying industries. The whole process of information exchange is delegated to Technical Working Groups, with the participation of the European Commission, Member States, the respective industries and the EEB. The general methodology of the elaboration of the BREFs and its results are discussed during an Information Exchange Forum in which the same participants meet on a regular basis. The information exchange has become a highly political exercise, because there are fundamentally different inter-pretations of the use of BAT. Industry strongly emphasises the "subsidiarity" principle and the need to define BAT at the local level. The reference documents should contain a wide range of possible techniques and focus more on the qualitative description of them, than on resulting ELVs. Some governments and ECOs perceive BAT, and the resulting ELVs, as a reference to permit-ting authorities. The list of selected techniques should be short and only contain those which are most innovative.. Nor should the cover of ELVs be too wide-ranging. BAT, as such, can be seen as a European concept which should eventually lead to harmonised minimum standards, according to Art. 18. However, this point of view is held by a minority of those in the relevant committees.

V.3.2.2.6   Public participation and pollutant emission register (Art. 15)

Public access and participation is required by Art. 15 and related articles and has been amended by the Directive 2003/35/EC on Public Participation. In general, ECOs are defined as "the public concerned" and therefore have the right to participate in the permitting process and to chal-lenge the permitting decision before a court of law. Permitting authorities must make publicly available the permit and the reasons and considerations for the decision, including how public opinions have been dealt with in the process, and the monitoring results.

Nevertheless it has to be kept in mind that, in practice especially, access to courts remains very difficult for citizens or ECOs because of high court costs. Further care has to be taken with respect to the format of consultation, sometimes incomprehensible documents or short dead-lines, all of which hamper effective involvement.

Based on article 15.3 the Commission established the European Pollutant Emission Register 2000/479/EC, which requires Member States to report to the Commission, in a specific format, all emission from installations covered by the IPPC. This information can be accessed at www.eper.cec.eu.int (see chapter V.2.3).

V.3.2.2.7   Transboundary effects (Art. 17)

The directive also requires permitting authorities to inform and eventually consult the authori-ties of neighbouring countries, if significant transboundary effects can be expected. In such cases the information should also be made available to the public.

V.3.2.2.8   Community ELV's (Art. 18)

Art. 18 is the only article which gives the Commission the mandate to formulate harmonised emission limit values, if "the need for Community action has been identified" – in particular on the basis of the exchange of information provided in Art. 16. Art. 18 also contains an old-direc-tives clause, which requires the application of the ELVs of older directives.

V.3.2.2.9   Prevention of accident

The directive expands the scope of accident prevention, as established by the 'Seveso' Directive 96/82/EC. The Seveso Directive requires notification from installations keeping or handling specific dangerous substances above a certain volume and obliges such installations to draw up and implement major-accident policies.

The IPPC Directive requires consideration of accident prevention for defining BAT and obliges installation operators to put in place accident prevention measures. (Article 3). The IPPC Directive is applying to a much wider range of installations than the Seveso Directive and therefore presents an important complementary EU instrument for the prevention of industrial accidents.

V.3.2.3   STRENGTHS AND WEAKNESSES

V.3.2.3.1   Strong Points

The strongest point of the IPPC directive is certainly its integrated and preventive approach to pollution from industrial installations. It is a major shift from the narrow end-of-pipe type of environmental policies towards wider issues, taking the minimisation of waste or energy efficiency into account (*Cleary et al. 1997*). This also implies a wider approach towards pollution reduction activities, including management and organisation issues. The integrated approach of the IPPC directive also applies to the joint view on all environmental problems, from global to local damage - including the protection of water, air and soils. Certainly the implementation of such an integrated approach raises a lot of methodological problems – such as assessing a technique which might have a strong record in one area of environmental problems and a very weak record in another (*Geldermann et. al. 1998*). Available studies suggest expert judgement to set priorities in the case of cross-media trade-offs. Nevertheless as regards its integrated approach, the IPPC directive has incorporated some essential elements of sustainable development – which need follow-up and proper implementation.

Other relatively strong points are the information and participation requirements according to Art. 15. They give environmental organisations and local citizens groups the possibility of influencing the permit granting procedure. The European Pollutant Emission Register, which allows easy and comparable access to information on installation emissions offers new opportunities in public control and benchmarking. This despite the fact the directive does not make any indications about the quality of information and is relatively restrictive on the number of substances to be reported. It falls far behind the state of art of Pollution Emission Registers as applied in the US (*see chapter V.2.3*).

V.3.2.3.1   Weak Points

The weakest point of the directive is that it has dropped one of the fundamental rationale of European environmental policies: that a harmonisation of standards is needed to avoid the unfair competition of regulatory systems or unfair economic competition and to solve international environmental problems. The IPPC-directive offers the possibility of doing this in Art. 18 and also refers to it in Art. 9.4. However, the directive basically follows a decentralised approach of norm setting. So Emission Limit Values ELVs are set at national or even local level (eg by individual permitting officials), taking into account local conditions. As this leaves considerable scope for interpretation, it can mean ambitious control policies in one country and a filling-up of clean areas in others. There is a risk that precautionary action which aims at maintaining good environmental quality - where it still exists - will no longer be taken.. On the other hand, heavily polluted areas may have to make considerable efforts to meet environmental quality objectives - if they are defined on a strict level in other directives.

Another weak point is the definition of BAT in practice. In theory, the.BAT information exchange process could be a source of innovation and fast diffusion of the most modern solutions throughout Europe. However, this would require a well financed process, with dynamic and highly qualified staff and a free flow of information on the real performance of average installations and pioneers in the respective sectors. The Commission does not have the money and the political will to support pluralistic participation in the process of information exchange. So, even the simplest figures on the respective sectors under discussion – such as the number, types, and size of installations - are not available. Stakeholders, including environmental organisations, have a formal right to participate, but missing rules for the process and the fundamental lack of finance make it very difficult for their experts to follow and influence discussions.

The information exchange process lacks even the finance necessary for the linguistic translation of respective reference documents on BAT. ,Conversely, the information exchange process in Seville where the EIPPC Bureau is based, is strongly influenced by participating industries, which offer expertise and even staff. But industry's position is clear: as a result of the information exchange process no clear ELVs should be recommended, but rather a wide range of BATs with their respective ELVs. Industry rejects even the European information exchange on innovative BAT using the "subsidiarity" argument. So much depends on the willingness and the enthusiasm of Member States on whether the information exchange focuses on the performance data of the "pioneers" or on some average existing permit conditions, already applied by a few Member States. The latter would imply a lost opportunity and would raise doubts on the usefulness of the whole information exchange process.

V.3.2.3.3   IPPC implementation: Does it deliver improvements?

The Directive is currently being implemented and so progress and quality is uncertain. Member State authorities interpret the Directive in their own context, industry submits permit applications, etc. At this stage it is difficult to have an overview of the progress being made and whether implementation is in line with the Directive's requirements[233]. Of particular concern is progress on ensuring that permits are issued to existing installations and their conditions met prior to the October 2007 deadline. Some Member States (eg Italy) seem to be establishing the conditions for a last minute rush that authorities will not be able to cope with. Even some Member States with programmes for dealing with these installations (eg the UK) have been criticised by the Commission as a 2007 bottleneck still seems likely.

---

[233] An examination of these issues is given in: http://www.europarl.eu.int/comparl/envi/pdf/externalexpertise/ ieep/ippc_brief.pdf

Ultimately, responsibility for checking whether Member States are complying rests with the European Commission. Process elements are easier for the Commission to address, ie whether the right installations have been.given permits. However, to check whether the emission limit values in the permits comply with the Directive requires an assessment of whether BATs were correctly determined on a case by case basis. Even where an installation is similar to many others across the EU, such checking is an enormous task. The regulator has the task of ensuring that permits are complied with. Unfortunately the European Pollutant Emission Register does not aid the Commission in compliance checking in this instance as the Register has a different scope and records total emissions (mass), as opposed to compliance with concentration values.

Therefore ECOs have an important role in informing the Commission and providing complaints. Nevertheless, it is important for the Commission to undertake sample audits to check conformity with the Directive. Where particular Member States demonstrate poor compliance, then these should be examined in more detail.

Since 1997, the EEB has participated in the development of the BAT- Reference documents (BREFs) and encouraged its Members to play a role in the national implementation process. From this experience the quality of BREFs and permits must be judged to be of rather mixed quality and ECO participation in the permitting procedures is low. The quality of national transposition legislation and implementation quality is alarming according to a European Commission report (*CEC 2003*).

V.3.2.3.3.1   *Quality of European BAT Reference documents*

To date, 21 BREFs have been finalised through the information exchange process facilitated by the European Commission and its Research Bureau in Seville. A further 12 BREFs are in preparation, which altogether largely cover the different sectors within the scope of the IPPC Directive. The BREFs development started with the "big" sectors - pulp and paper, steel and cements - and is now concluding with chemicals, ceramics, waste incineration etc.. In addition to the sector-specific BREFs, some "horizontal" BREFs have been or will be developed. These include monitoring systems, storage of material, economic and cross-media, and energy efficiency.

The EEB has participated in 13 of the 21 finished BREFs and is continuing its involvement in 4 of the 12 open BREF development processes.

Based on this experience, the following general process conclusion can be drawn:

1. **Industry bias:** There are not enough national experts to steer the process from the EIPPC Bureau in Seville and insufficient capacities from Member States (in many cases only Germany provides detailed and structured input). Participation from Member States has decreased over the years, leaving the majority of participants from industry. This situation allowed industry to commission the writing of the BREF on "Storage of bulk or dangerous materials" and to develop a "shadow" BREF for "Large Combustion Plants" where they did not like the official BREF. In other cases, industry just halts the process by not providing sufficient data.

2. **Lack of data to determine BAT and associated pollution emission values**: Compared to the knowledge that exists throughout Europe about specific types of industry, about their environmental effects and approaches to their regulation, the amount of information that is made available by participating industries and Member States is relatively poor. Even very basic information about numbers, size and types of installations is missing in most cases. Overall, this lack of data makes focussing on the task in hand very difficult Experience has shown that especially in the more heterogenic sectors (like e.g. Waste Treatment, Surface Treatment of Metals or Surface Treatment using Solvents) this has led to mistakes in the selection of the types of installations covered by the BREF. Industry participants often claim business confidentiality on real life environmental performance data. If they are forced to provide data, they provide data with a range of one or even two orders of magnitude. Sometimes, EEB experts find themselves in the situation of being the only party in the process ready to bring in ambitious performance parameters from existing plants – either by screening independent studies or by utilising the knowledge of sector experts or interviews with sector suppliers and others.

3. **Political instead of technical debates**: the result of points 1 and 2 above is that lobby strategies dominate and hamper any rational exchange of information.. In the case of the BREF on "Large Combustion Plants", industry "experts" did not even try to argue on technical grounds but rather preferred to "negotiate" over the outcome, in order to dilute any strong statements about BAT associated emission levels based on economic arguments. At the end of the process, industry even offered an alternative BREF, threatening to promote amongst its members.

The outcome of this process - the BREFs - are publications of approximately 500 pages long, of mixed quality, which makes any overall assessment very difficult. The following general observations can be made:

1. Good overall description of available installation techniques and technologies. This is undoubtedly valuable information for permitting or controlling authorities as well as for ECOs.

2. Description of environmental performance is of low quality : often substantive tables with emission data are provided, but this on an erratic basis without further explanation as to why some aspects are left out and others clouded with detail.

3. In general only the "classic" air pollutants from point sources are well documented. Quantified emission data even within one single BREF is often not comparable because of the use of different measurement units (daily, yearly averages). Finally, emission data is presented in terms of concentration values, rather than as emission per production or input unit, which means that no comparison can be made of the efficiency of different installations.

4. Emission levels associated with the selected BATs are rare and often provided as ranges, employing a factor of up to ten.

As a result the value of the BREFs for competent authorities in writing permits is seriously hampered in many cases. The main objective of BREFs - to provide emission levels in relation to the overall performance of an installation associated with BAT - is inadequate.

Table 9: BREFS and their current status

BREFS	STATUS
Pulp and Paper manufacture	Adopted Dec 2001
Iron and Steel production	Adopted Dec 2001
Cement and Lime production	Adopted Dec 2001
Cooling Systems	Adopted Dec 2001
Chlor-Alkali manufacture	Adopted Dec 2001
Ferrous Metal processing	Adopted Dec 2001
Non-Ferrous Metal processes	Adopted Dec 2001
Glass manufacture	Adopted Dec 2001
Tanning of hides and skins	Adopted Feb 2003
Refineries	Adopted Feb 2003
Large Volume Organic Chemicals	Adopted Feb 2003
Common waste water and waste gas treatment and management systems in the chemical sector	Adopted Feb 2003
Textile processing	Adopted Jul 2003
Monitoring systems	Adopted Jul 2003
Intensive Livestock Farming	Adopted Jul 2003
Smitheries and Foundries	Adopted May 2005
Slaughterhouses and Animal By-products	Adopted May 2005
Management of Tailings and Waste-Rock in Mining Activities	Finalised Jul 2004
Emissions from storage of bulk or dangerous materials	Finalised Jan 2005
Economic and cross media issues under IPPC	Finalised May 2005
Large Combustion Plant	Finalised May 2005
Food, Drink and Milk processes	Final Draft Jun 2005
Surface treatment of metals	Final Draft May 2005
Waste Incineration	Final Draft May 2005
Waste Treatments [Previously Waste Recovery/Disposal activities]	Final Draft May 2005
Large Volume Inorganic Chemicals - Ammonia, Acids & Fertilisers	Draft Mar 2004
Organic fine chemicals	Draft Dec 2004
Polymers	Draft Apr 2005
Surface treatments using solvents	Draft May 2005
Speciality inorganic chemicals	Draft May 2005
Large Volume Inorganic Chemicals - Solid & Others	Draft June 2005
Ceramics	Draft June 2005
Energy Efficiency	Started 2003

V.3.2.3.3.2 Quality of national permits and use of BAT

The EEB has not made an assessment of the involvement of ECOs in the national permit granting process, but so far feedback from EEB Membership remains very limited suggesting little practical involvement and low level of encouragement from the responsible authorities.

Because this work is still in progress and because there are many uncertainties attached to it, an assessment of the quality of national implementation is impossible. Nevertheless some theoretical conclusions can be drawn from an EU perspective. As outlined in chapter V.3.2.2.4, national environmental quality standards are likely to be the main driver for setting emission controls under the IPPC Directive and filling up to such standards is permitted. Based on the quality of the BREFs – as outlined above - this judgement seems to be true and therefore huge differences in the application of the IPPC Directive across Europe are very likely. **It is not BAT which seems to drive emission controls, but rather environmental quality standards set by pioneering countries or EU legislation.** This also means that the perceived hierarchy of a combined approach – preventing pollution first through BAT controls and secondly strengthening controls if environmental objectives are not achieved - as it is used in other legislation, like the Water Framework Directive 2000/60/EC - is likely to be invalid in practice.

V.3.2.4   IPPC IN THE WIDER POLICY CONTEXT: AN EFFECTIVE TOOL?

Regulating industrial environmental performance is far from simple, especially when attempted at EU level. The simplest way is to establish standard levels of environmental performance through setting emission limit values – an approach which has been done followed in the past and is still appropriate in some cases. However, this is only practical where there are clear classes of industrial activity across the EU which are similar enough in their activity to enable a standardised regulatory approach. An alternative method is to set out principles of industrial regulation which must be followed. IPPC does this and can, therefore, cover a wide range of industrial sectors and complex installations. The downside is that neither industry nor the community can be certain about what is to be required of an installation and who should be held accountable.

Even though precise outcomes are difficult to predict, one can expect the IPPC Directive to deliver environmental improvements in industrial performance, most obviously with regard to the many installations across the EU which are operating well below what could reasonably be considered to be BAT.

It is generally agreed that IPPC introduced a radical change to industrial regulation in the EU. Until this time EU regulation was based either on the setting of common emission limit values or environmental quality standards. IPPC complements these approaches. Rather than standard operating conditions, it has standard principles of operation. This is a flexible approach which can take local conditions into account and can incorporate a provision for ongoing improvement. It can also be applied to a wide variety of activities, giving it a dynamic quality. The downside - unfair regulatory competition and vulnerability to industry pressure, dependency on the ability to set environmental quality standards and lack of regulatory and thus lack of pollution prevention - is explained in chapter V.3.2.3.2.

It is important to stress that the IPPC Directive does not produce a 'level playing field' in the sense usually used by business in its approach to the single market. IPPC itself does not impose specific conditions on the same types of enterprises across the EU. Regulatory authorities consider the range of recommendations in a BREF, take account of local conditions and make their own decisions on what emission limit values are appropriate. One can, therefore, consider that while the IPPC will help practice across the EU to converge, it will not become standardised. Where the resulting differences in practices have no environmental consequences, this is not a problem. However, this flexibility could be abused and it will be important to ensure that comparative assessments are made of the emission limit values established in permits so that regulators in one or more Member States are seen to be implementing the spirit of the Directive as well as its letter.

187

The implementation of IPPC is being carried out during a period of debate over the nature (and future) of environmental regulation at EU and national level. This includes a consideration of the nature of environmental policy formulation as well as the types of instruments which might be appropriate.

IPPC is, generally, flexible. However, it does impose certain obligations which act as limits on what other instruments can be used by regulators. The first of these is a requirement to set emission limit values in a permit. The second set of requirements includes boundaries as to what is, and what is not, an installation and, therefore, what is the 'object' of the regulation. The former obligation means that instruments which rely on not setting ELVs (such as emissions trading) are difficult or impossible to operate alongside IPPC. This problem is given further consideration below. The second set of obligations are less clear and actual interpretation (and practice) in Member States varies. Indeed some restrictions derive from Member State law (eg a permit can only be issued to a single legal entity). However, there are innovative approaches which seek to make wider regulatory decisions to groups of installations (eg on an industrial estate or within a company). Other instruments are also in place such as negotiated agreements (well established in the Netherlands, for example). These have their place as long as they are seen to deliver beyond legal requirements and can be effective for some issues. However, the issue of enforcement is problematic and for most Member States such instruments remain an occasional useful tool for specific companies and issues. This is likely to remain the case and such instruments should not be allowed to interfere with the effective implementation of IPPC.

There is a debate within the EU and some individual Member States on 'better regulation'. Everybody wants *better* regulation. However, what is meant by 'better' depends upon whom one asks. For many it is a smokescreen for deregulation. IPPC is criticised as being burdensome and bureaucratic. The Dutch government, based on an unpublished survey of Member States, claimed[234] that the IPPC Directive is the most burdensome of all environmental legislation. However, this position is not explained. In fact, when Member States consider the implications of IPPC they are actually looking at national implementation. For some Member States (eg Sweden) IPPC is implemented through existing laws and has practically no discernable additional consequence. For some others it has been a major change to the way that regulation is undertaken, and they obviously view it as a burden. However, one must separate EU obligations from the

---

[234] ENAP conferences organised in 2002-2004 by Ministry of Housing, Spatial Planning and the Environment of The Netherlands. ENAP stands for Exploring New Approaches in regulating industrial installations

Member State requirements and when one does so one finds that industry is generally not concerned with the burdens of the Directive. Basic requirements of IPPC — analysis of installation activity and performance, permit application, monitoring, inspection and review — are essential and must remain in place as they are needed to ensure the protection of health and the wider environment. These requirements must be implemented in a way that does not introduce unnecessary costs to industry. But the experience many years of interaction with industry suggests that any slackening of such requirements will not maintain environmental performance.

Therefore, there is currently no justification for extracting selected industrial activities from the obligations of the IPPC Directive. There is possibly a case for examining the exact thresholds of what is covered by the Directive (eg excluding 'trivial' activities) as well as bringing new activities into the scope of the Directive (such as intensive cattle units, more waste management facilities, airports, construction sector etc).

There are close links between IPPC and nearly all other EU environmental policies - including EU's waste, air, climate, water, environmental liability and management and pollution emission register policies (see chapters IV and V) - where the IPPC's permitting and/or BAT concept is utilised for achieving their objectives. More links are discussed. In view of the specific problems identified in *chapter V.3.2.3.3* those possible links as well as some existing ones will be assessed.

V.3.2.4.1 Phase out of hazardous chemical emissions

The *Water Framework Directive* 2000/60/EC (WFD is innovative in implementing procedures to honour Europe's commitment to cease emissions of hazardous[235] chemicals by 2020. The WFD mandates the Commission to establish a list of priority substances (completed at the end of 2001) and identify the hazardous ones, for which the Commission has to propose controls to cease their emissions within 20 years (still to be carried out) (*see also chapter IV.5.5.2*). In its draft deliberations for control measures the Commission heavily relies on the IPPC permitting procedure and BAT concept. It suggests that for IPPC covered industrial installations Member States have to ensure that pollution reduction programmes are established including the cessation of emissions (no detectable concentrations). Furthermore, Member States should identify BATs for the specific emission sources.

> *Hazardous chemicals are substances, which are able to accumulate and persist in the environment or potentially can cause irreversible negative effects (like infertility or cancer). For such chemicals no safe standards can be set and in applying the precautionary principle Europe has committed itself to end their emission into the environment by 2020 (OSPAR Convention 1998).*

While at first sight it might seem appropriate to attach a new condition to the granting of IPPC permits, e.g. a programme to achieve the cessation of an emission, this is made immediately redundant for cases where cessation is technically or economically not feasible. This is likely to be a regular scenario, as local authorities are unlikely to be in a position to prove the availability of safer alternative techniques or technologies. The further suggestion of identifying within the IPPC BREFs process BAT for the relevant emissions also seems to be the wrong approach. IPPC is

---

[235] The term "hazardous" is used here as in the context of the Marine Conventions and the Water Framework Directive 2000/60/EC. It addresses substances of very high concern for the environment (and human health via the environment) due to their persistence, liability to bioaccumulate and liability to cause irreversible effects an low or unknown dose.

not designed to deal with the long-term and irreversible effects of hazardous chemicals and does not address the cumulative effect of several emission sources, which, especially in the case of persistent chemicals, plays a crucial role. In practice no BREF has yet dealt in any great detail with the emissions of a particular substance because of its toxicity.

V.3.2.4.2   Preventing and Recycling Waste

In the discussions on the Thematic Strategy on waste (*see chapter IV.4.6*) the Commission states that *'Another main challenge facing Community waste policy is to move towards a level playing field across the EU for recycling guaranteeing a high level of environmental protection with recycling being supported by an efficient internal market'.*

To reach this objective the Commission proposes to adopt measures in order to fill the gap on waste recycling standards, as follows:

◗ The extension of the IPPC directive to cover waste treatment activities not yet covered; and

◗ setting EU reference standards for the operation of waste treatment facilities. This is intended to be linked to the determination of quality standards for recycled products in an Annex of the directive 75/442/EC on waste. This would complement the technical requirements set at facility level through extension of the IPPC directive.

The IPPC directive already covers a limited number[236] of waste treatment operations – mostly pre-treatment operations and not recycling as such. Best available technique reference documents (BREFs) are currently being developed for these. It also covers waste incinerators with existing BREFs. The Commission claims that the extension of this directive would initiate a bottom-up process by which BREFs would be developed for all waste management operations and the permits of recycling installations would be updated by allowing authorities to include emission limit values and other appropriate permit conditions based on best available technology. The idea is to move *towards the harmonisation of environmental standards at the level of waste management installations.*

This approach gives rise to several practical concerns:

1. **Deficiencies in ability to produce real harmonisation or Best level of environmental standards**: Quality of BREFs so far is mixed and lacks guidance on ELVs, leading to a strong decentralisation of setting emission controls based on local environmental impacts. The BREFs are subject to national interpretation, which raises questions as to how harmonised the application of the levels set in any BREFs will be in reality. Additionally there is a risk that the BREFs do not really represent the best levels of available (and economically) feasible technology. This leads to restrictions in the opportunities that Member States' might have for objecting to waste movements; if, for example, their national standards are better than the BREF guidance and IPPC BAT is taken as a benchmark for 'environmentally sound treatment' under the EU Waste Shipment regulation.

---

[236] IPPC covers: MBT, waste storage, some metal and chemical liquid and sludge physio-chemical treatments, immobilising ashes, prep fuel from liquid waste, prep waste as fuel. IPPC does not cover: composting, battery recycling, Waste electronics and End of life Vehicles recycling/dismantling, oils, spent catalysts, other types recycling – except paper, steel indirectly

2. **Steering of waste streams not possible – emission standards alone not enough:** The IPPC Directive deals with individual installations and has almost no focus on how the waste should be treated at other stages of the waste treatment chain. For instance, if mobile phones are treated in an installation before actual recycling takes place, a relatively high portion of the precious metals could be lost (simple shredding in a big shredder). The alternative is disman- tling, which separates parts with high content of precious metals for recycling. Thus, the approach to steering waste for recycling must be **waste treatment chain orientated** and not installation oriented as it is the case in the IPPC Directive. This implies that other measures, in addition to IPPC standards, are needed to steer the choice between different processes (shredding or dismantling) and different kinds of installations.. The risk is that waste policy goes in the direction of merely setting installation standards and abdicates responsibility for the steering mechanism, with the excuse that standards are sufficient. This is a particular risk for waste streams such as biodegrabable wastes.

3. **IPPC is not very suitable for the majority of waste recycling facilities, which are SMEs** (the Council has on this basis previously rejected the idea of applying IPPC to SMEs[237]). The Commission talks about possible 'IPPC light' approaches for SMEs. This may further dilute an already fragile level of environmental ambition (given national flexibilities in interpretation of BREFs).

4. **Recycling standards needed urgently** – it is not clear how soon and to what extent IPPC will be revised at all to incorporate the proposed waste recycling activities. There is considerable resistance from Member States to revising the IPPC framework before implementation experi- ence has been shown to bring positive results. (post 2008). There is urgent need for recycling standards in some recycling/pre-treatment activities to reduce the risk of eco-dumping (move- ments of wastes to less performing and therefore cheaper installations).

V.3.2.4.3 IPPC Permits as an Industry Defence

Industry federations regularly argue that once a company has obtained an IPPC permit for an activity it has fulfilled all its duties, no further permits or specific authorisations should be neces- sary and that responsibility lies subsequently with public authorities. Considering the IPPC weak- nesses in theory and practice (*see chapter V.3.2.3*) this argument is very problematic and would lead to a weakening in companies' environmental responsibility and undermine the implemen- tation of the precautionary principle to be applied in cases of scientific uncertainties or lack of data.

V.3.2.4.3.1 Environmental Liability

The new environmental liability Directive 2004/35/EC from 2004 is a first step towards the appli- cation of a general market based instrument, moving towards the "polluter pays principle" *(see chapter V.3.3)*. It is built on the assumption that the risk of liability for potential polluters would result in more careful and preventive behaviour. Making businesses that damage the environ- ment legally and financially accountable for that damage is but a way to internalise external envi- ronmental costs and urge industry to seek for the most effective strategy for avoiding environ- mental damage. One element which provoked controversial discussion during the Directive's

---

[237] Council Resolution, Amsterdam, 18-20 April 1997

development was the "permit defence" - meaning that a company holding a permit would no longer be liable for environmental damage caused via the permitted activities. The final Directive now allows industrial installations holding IPPC permits to be exempted from paying the costs of remedial action if the installation's operator can prove that the environmental damage had been caused by an emission or an event authorised in the IPPC permit.

This exemption is likely to dramatically reduce in practice the effectiveness of environmental liability by referring responsibility back to public authorities, for the following reasons :

Firstly European BAT reference documents are rather limited when it comes to addressing all possible significant environmental damages and therefore provide little help for authorities issuing permits.

Secondly, permits are granted on the basis of local environmental and socio-economic considerations. So in cases where weak emission control conditions appear in the permit because of local economic considerations, the operator would, in addition to this benefit, have the possibility of defending himself against being held liable for paying the clean up of any resulting environmental damage. It is, of course, possible for the responsible authority to reject such a defence claim. Obviously, however, the pressure on the authority to grant the exemption will be rather high, as it is the authority which has set the permit conditions in the first place, and therefore shares a certain responsibility.

V.3.2.4.3.2 Future Chemicals Policy (REACH)

In 2003 the Commission presented its proposals for a regulation on the Registration, Evaluation and Authorisation of Chemicals (REACH) in order to reform Europe's chemicals policies (see chapter V.4.6.4.6). One of the key elements of this reform has been the identification of hazardous chemicals[238] and substituting their uses with safer alternatives. The process established for achieving this objective is called 'authorisation'. For identified chemicals a deadline is set after which all un-authorised uses of that chemical are banned. In order to obtain an authorisation industry has to proof adequate control or the lack of safer alternatives plus over-riding societal interest. The main spirit of this process can be clearly identified as achieving the 2020 cessation of emission of hazardous chemicals (OSPAR Convention 1998 and Water Framework Directive 2000). However, the Commission's proposals exclude emission from those industrial installations which hold IPPC permits, exempting them from needing an authorisation in order to continue using hazardous chemicals.

This case was rather simple. Industry argued on the basis of "legal security" that once an industry received a permit it should not have to seek further authorisation. This argument was accepted. From an environmental perspective, however, the IPPC Directive is unsuited to addressing the specific concerns of hazardous chemicals.

Firstly, only a very limited number of substances are specifically addressed in the BREFs with very limited information. This means that emissions of hazardous substances, even if they are not addressed in the BREFs and subsequent permits, would be completely exempted from any emission control.

---

[238] The term "hazardous" is used here as in the context of the Marine Conventions and the Water Framework Directive 2000/60/EC. It addresses substances of very high concern for the environment (and human health via the environment) due to their persistence, liability to bioaccumulate and liability to cause irreversible effects an low or unknown dose.

Secondly, hazardous chemicals represent a category of substances which include persistent substances and are thus liable to long-range transport, causing pollution in remote areas - like groundwater, marine sediments or Polar Regions. Europe's commitment is therefore to phase out the emission of such hazardous substances altogether. As the IPPC permitting procedure is largely driven by local environmental objectives, it is completely inadequate for achieving such "global" environmental objectives.

Thirdly, effective phase out of emissions can only be achieved if their widespread use is substituted with safer alternatives. This substitution test is currently very weakly addressed — if at all — in setting BATs.

V.3.2.4.4 Air and Climate Policies

When considering IPPC in the context of wider air and climate policy, three measures are particularly relevant: the Air Quality Framework and Daughter Directives; the Large Combustion Plant Directive; and the Greenhouse Gas Trading Directive.

By focusing on individual installations, IPPC takes no direct account of the number of installations in any particular area. It is therefore complemented by the local air quality standards set in Air Quality Framework and Daughter Directives (see chapter IV.3); indeed, Article 10 of the IPPC Directive specifically requires measures beyond BAT if these are necessary to meet local environmental quality standards. However, the 4th Daughter Directive - like the 3rd which did not set any legally binding air quality standards on ozone — sets no legally binding air quality standards.

The Large Combustion Plant Directive (LCPD) sets emission limit values for LCPs for SO_2, NOx and dust. It is therefore tempting to make comparisons with those set under IPPC, but this would be flawed on two counts. Firstly, the LCPD applies only to air emissions, whereas IPPC takes an integrated approach that considers emissions to all media — air, land and water. Secondly, the LCPD standards are legally binding, whereas the IPPC benchmark standards are not. IPPC only becomes legally binding when it has been applied to an individual installation, taking account of local environmental, technical and geographical factors. However, the two directives have to be applied without prejudice to each other, so the LCPD does provide an initial basis for the sort of binding minimum ELVs that the EEB would like to see. It is important to stress, however, that the ELVs under the LCPD are not necessarily BAT and it is likely that there will be good reasons in many cases where stricter ELVs should be applied as IPPC permits are issued.

The relationship between IPPC and the Greenhouse Gas Trading Directive (GHGTD) (see chapter V.3.4) is particularly problematic. The installation-specific nature of IPPC significantly conflicts with the geographical flexibility required for trading, whether it is for GHGs or acid gases within the LCPD National Plan option. The GHGTD contains an amendment to IPPC that supposedly resolves this by exploiting synergies between the two approaches, but in reality this is fundamentally flawed on three counts. Firstly, the amendment facilitates trade by removing any requirement to set site-specific ELVs for CO_2 under IPPC, and it is anticipated that this will be extended to other GHGs, provided that they do not cause significant local pollution. However, despite an initial statement that GHG emissions under trading must be no higher than they would have been under IPPC, the GHGTD contains no requirement for actually checking this . Therefore GHG emissions are being driven by the politically determined Kyoto targets rather than the technically determined BAT standards. Secondly, releasing the constraints of site specific GHG ELVs did not provide sufficient flexibility for trading whilst installation-specific energy efficiency requirements remained in place. Therefore, the setting of energy efficiency requirements has now been made optional under IPPC, again without any guarantee that the resulting standards will be at least equivalent to what would have been achieved under BAT. Lastly, the removal of GHGs and energy efficiency from IPPC and their consideration in relative isolation is not compatible with the integrated approach required under IPPC.

As in all trading systems, a distinction has to be made between the level of ambition (cap, emission standard etc) and the means of achieving that level of ambition (command-and-control, trading, taxes etc). In the case of trading, this means of compliance in theory offers the potential advantages of a simpler management system and cost savings, although the example of the US suggests that these savings have usually been significantly overestimated. However, trading itself is environmentally neutral because it is simply a means of compliance , and it is the nature of trading that an overachievement of the standard in one part of the system is balanced by an underachievement in another part. Any cost savings achieved by trading only get reflected in the level of ambition if there is a political will to do so, and without that will, the lower compliance costs simply result in higher industrial profit. With the removal of CO_2 from the IPPC under GHG trading, the level of ambition is set entirely on the basis of the political Kyoto negotiations, and lacks valuable information on the environmental potential and economic costs that occur in BAT assessments. It is this information that is necessary to inform any political will to properly reflect the lower costs of trading systems in the level of ambition. This combines with the disadvantages of the potential downgrading of IPPC energy efficiency requirements, the undermining of an integrated approach to industrial management, and the danger that Member States will downgrade IPPC for other pollutants to facilitate trading.

V.3.2.5 OUTLOOK AND FOLLOW-UP ACTIONS

9 years of IPPC implementation have shown that its decentralised elements are by far the central dominating aspects and that the main drivers for protecting the environment are environmental quality standards and objectives rather than ambitious and pioneering technologies. The EU level BREFs development is biased towards industry interests. Permitting authorities are exposed to economic pressures with little help from national or EU level. ECOs have received great opportunities through the public participation and information provisions to expose "bad" companies and influence the permitting conditions. But active participation in the permitting processes seems to be very limited for the time being. As a result huge differences in the environmental performances of industrial installations and eco-dumping are possible. At the same time there are signs of positive developments with a gradual improvement in the environmental performance of installations, especially with the approach of the 2007 deadline – when all installations will finally need to be in possession of a permit and to meet its conditions.

In order to promote the positive aspects and avoid negative development under the IPPC, the following action is recommended:

1. **Increase quality of European BAT reference documents:** Member State authorities and ECOs have to have increased capacity to balance industry domination. More and comparable emission levels associated with BATs have to be provided to guide and strengthen permitting authorities. Therefore industry must be obliged to provide real life emission values overriding unjustified confidentiality claims.

2. **Setting of EU wide emission limit values:** based on the BREFs and persistent environmental problems in the EU the Commission should finally use its mandate under the IPPC Directive to propose mandatory ELVs.

3. **New and ambitious air, soil and water quality objectives:** The environmental success of the IPPC Directive will depend on emission controls being steered by stringent air, water and soil quality objectives. Therefore the list of 33 priority substances under the Water Framework Directive has to be extended and quality standards adopted rapidly. The new EU soil protection policy should from its start foresee the setting of strategic quality objectives substantiated, as far as possible, by quantitative parameters.

4. **Refute attempts to weaken (increase flexibility of) the IPPC Directive**: In 2007 a revision of the Directive is foreseen. Industry - as well as Member States - has expressed its concerns about the costs and administrative burden – especially in light of the approaching 2007 deadline. Further attempts to extend emission trading beyond. Greenhouse gases are in conflict with the local protection approach of the IPPC. In light of the still to be concluded quality assessment of the IPPC implementation process , it is very important to provide regulatory stability in the first place and only propose changes once environmental outcomes are properly assessed.

5. **Highlight boundaries of what IPPC can and cannot deliver:** the many attempts to use the IPPC permitting and BAT elements as the principle means to address long-term and global environmental problems should be rejected. IPPC is not designed to phase out emissions of hazardous chemicals or steer waste policies towards prevention and recycling.

6. **Improve quality of national implementation:** ELVs and quality standards will be set at national levels as long as no European ones exist. That means the discussion on implementation must be used for a wide and ambitious interpretation of the requirements of the IPPC directive. Only if some pioneer countries formulate demanding standards, do we get a new chance for harmonised standards at the European level. On the national level ECOs may also insist on a regular review of the permits in a period of within a maximum of 10 years.

7. **Local participation and control:** The IPPC-directive offers new opportunities to monitor the permitting process at the local level. Local groups should actively use this new information and opportunity for consultation. A well-informed local public may exert much more effective social control on industries than a vaguely defined European directive may ever do. In this sense – the shifted responsibility of the IPPC directive to the local level – should also be interpreted as a challenge for local groups. They should be helped by practical advice on how to use the new information and consultation rights from their national organisations. Nevertheless it has to be kept in mind that in practice especially access to courts remains very

difficult for citizens or ECOs due to high court costs. Furthermore, care has to be taken with respect to the format of consultation, sometimes incomprehensible documents or short deadlines, which hamper effective involvement.

BIBLIOGRAPHY AND FURTHER READING

BREUER, Rüdiger (1997) Zunehmende Vielgestaltigkeit der Instrumente im deutschen und europäischen Umweltrecht – Probleme der Stimmigkeit und des Zusammenwirkens, in : Neue Zeitschrift für Verwaltungsrecht, 9-1997, p. 837-839

BUND 1997: Anforderungen an die Umsetzung der VU-Richtlinie aus Sicht des BUND,

CEC: Council Directive 96/61 EC of 24 September 1996 concerning integrated pollution prevention and control, in: OJ L 257 from 10.10. 1996, p. 26-40

CEC (2003) Communication for the Commission to the European Parliament and to the Council COM(2003) 354 final On the Road to Sustainable Production, Brussels, 19.6.2003

CLEARY, T.V./ O'BRIEN A./ O CLEIGIGH, L. (1997) Integrated Environmental Protection Based on BAT, Study for the European Commission, April 1997

ENAP (2004) High Level ENAP Conference "Exploring New Approaches in regulating industrial installations" organised by Dutch EU Presidency 17/18 October 2004

FARMER, A. and BRINK, T.P. (2004) Better and simpler, Simplification of regulation for a more competitive Europe. A background report for the conference on the 1996 IPPC Directive. A report for the Ministry of Housing, Spaitla Planning and the Environment of The Netherlands

GELDERMANN, J./ JAHN. CH./ SPENGLER,TH. (1998) Proposal for an intergrated approach for the assessment of cross-media aspects relevant for the determination of "Best Available Techniques" BAT in the European Union, DFIU, Karlsruhe 1998

HERITIER, A. /MINGERS, S./ KNILL, C. /BECKA, M. (1994) Die Veränderung der Staatlichkeit in Europa – Ein regulativer Wettbewerb: Deutschland, Großbritannien, Frankreich, Leske + Budrich, 1994

HEY, Christian/ BRENDLE, Uwe (1994) Reversing the Roll-Back of EU Environmental Policies, EEB, Brussels, 1994

IEEP and Ecologic (2003): Implementation of the IPPC Directive (96/61): Analysis and progress of issues, Policy Brief for the EP Environment Committee, EP/V/A/2003/09/01, Brief number 01/2003

PALLEMAERTS, Marc (1997) The proposed IPPC Directive: Re-regulation or De-Regulation, in: European Environmental Law Review, Vol. 5 NO. 6, June 1996, p. 174 - 179

PALLEMAERTS, Marc (1998) The Decline of Law as an instrument of Community Environmental

Policy; Proceedings of the Conference on: "The Community' s Environmental Policy at 25: Stocktaking and Future Prospects"; Université Libre de Bruxelles, 1998.

SCHAFER, Erich (1997) Stellungnahme zu den Bestrebungen, die Definition des Begriffes Stand der Technik als Maßstab für die Emissionsbegrenzung in der österreichischen Rechtsordnung zu ändern, Wien, Febr. 1997

ROSCHER, Heike (1998) The devolution is going to eat its children – Reflections to the Fourth meeting of the IPPC BAT Information Exchange Forum, 16-17-1 1998; also in: DNR-Rundbrief 2/1998

TASCHNER, Karola (1996) Directive on Integrated Pollution Prevention and Control – Perspective from environmental NGOs, Paper to the International Conference on Environment and Industry, Sevilla, 4/5 November 1996

UNICE (1998) UNICE's Implementation of IPPC Working Group Opinion on the Implementation of the IPPC Directive, 26.1.1998

WENNERAS, P. and PERMI DEDFENCE (2004) *Environmental Liability Regimes – Subsidizing Environmental Damage in the EC?*, The Yearbook of European Environmental Law, Volume 4, December.

V.3.3   Environmental Liability - first step towards "Making the polluters pay"

By *Rosanna Miccichè*[239]

V.3.3.1   INTRODUCTION: FROM SEVESO TO BAIA MARE

The list of accidents with lasting, catastrophic consequences for the environment is regrettably long. The disaster of Seveso back in July 1976, the fire in the Sandoz factory in Basel in 1986, the burst of a containment dam and the consequent spill of toxic waters and mud into Doñana National Park in southern Spain in 1998, and the accidents in Baia Mare and Baia Borsa, in January and March 2000 respectively, constitute the most emblematic and notorious cases. In addition to episodes of exceptional character, problems of soil and water pollution are becoming more and more pressing, due to the gradual release of pollutants mainly from industrial and agricultural activities (*EEA 2000*)[240].

As a direct consequence of public outrage spurred by those events, many EU Member States have adopted specific regulations on liability for environmental damage. These measures are very diverse in scope and nature (*CMS Cameron McKenna 1995*). Contextually, the European Institutions have set a debate in motion about the opportunity for a Europe-wide regime of liability for damages to the environment that aims at implementing the "polluter-pays" principle, enshrined in the Treaty establishing the European Communities (*Krämer 1997*)[241].

This subchapter is divided into four parts. The first part describes the rocky road that led to the adoption of a European Directive on environmental liability. The second outlines the main tenets of the environmental liability regime as envisaged in the final text of the Directive and its most controversial provisions. Thirdly, the provisions are examined in more detail, identifying strengths but also deficits, which may jeopardise the proper functioning of the Directive. In the last part, conclusions are drawn concerning the future role of the Community's environmental liability regime in changing the behaviour of potential polluters, and suggestions are made as to what ECO action can be taken to make the best out of the Directive.

V.3.3.2   MAKING THE POLLUTERS PAY: A LONG WAY TO GO

A first proposal for a European liability regime covering damages caused by waste was made in 1989, and quickly abandoned. In 1993, the Commission revitalised the debate around a possible future EC liability regime in its Green Paper on Remedying Environmental Damage[242]. Meanwhile,

---

[239] Followed the legislative process of the Environmental Liability Directive and worked actively throughout its key phases for the European Policy Unit of Greenpeace.

[240] A study carried out by the European Environment Agency has revealed that no less than 300.000 sites have been qualified as actually or potentially contaminated. The Agency estimates that the cost of partially recovering those is between 55 and 106 bn ?, equivalent to 0.6-1.25% of EU's GDP

[241] Art. 174, paragraph 2 of the Treaty establishing the EC (ex Art. 130 R), introduced by the Single European Act (1986)

[242] COM (1993) 47 final, 14 May.

on June 21 1993, the Council of Europe adopted a Convention on Civil Liability for Damage result-ing from Activities Dangerous to the Environment[243].

After consultation and studies carried out (*Brans and Uilhoorn 1997, Deloddere and, Donatienne 1997*), the Commission published a White Paper in 2000 aiming at exploring the different options for Community action in the field of environmental liability (*CEC 2000*)[244]. Adhesion to the Lugano Convention was quickly ruled out because of strong opposition from some Member States and industry to its far-reaching and broad scope. The White Paper on Environmental Liability conclud-ed in favour of a horizontal directive providing for strict liability for damage caused by EC regu-lated dangerous activities, covering both traditional damage (to health and property) and pure-ly environmental damage (contamination of sites and damage to biodiversity), and fault-based liability for damage to biodiversity caused by non-dangerous activities.

On 23 January 2002, the Commission finally published a Proposal for a Directive on environmen-tal liability with regard to the prevention and remedying of environmental damage. Unfortunately, following heavy pressure from some national governments and industry, the final proposal was weaker than earlier drafts and the White Paper in numerous respects, and con-tained a number of shortfalls seriously undermining the regime as a whole.

On 21 April 2004, after more than 2 years of heated discussions, the European Parliament and Council adopted the Directive (*2004/35/EC*) on Environmental Liability with Regard to the Prevention and Remedying of Environmental Damage[245]. Notwithstanding the joint advocacy effort by ECOs, the final text is very disappointing from an environmental point of view. Dangerously, it also leaves wide discretion to Member States in several key aspects of the Directive. The Directive entered into force on 30 April 2004, and gives Member States 3 years to transpose its provisions into national law. A revision of the functioning of the Directive, including proposals for amendments, is not envisaged until 30 April 2014.

V.3.3.3 OBJECTIVES, SCOPE AND EXEMPTIONS

V.3.3.3.1 Objective

The underlying objective of the Directive is to apply the ***polluter-pays principle*** throughout the European Union. Liability rules are deemed to be necessary to ensure that environmental dam-age is restored and costs arising from clean-up work associated with site contamination do not

---

[243] Lugano Convention of the Council of Europe on Civil Liability for Damage caused by Activities Dangerous to the Environment (21 June 1993). This convention provides a regime of strict liability that covers all types of damage (both tra-ditional damage and degradation of the environment) when caused by dangerous activities. Dangerous activities are defined with a reference to hazardous substances, biotechnologies and waste. However, the scope of application remains open and may be extended to activities other than those explicitly mentioned. The convention was signed by 9 States, six of which are Community Members (Greece, Italy, Luxembourg, the Netherlands, Portugal and Finland) None of which has yet ratified it. To look up the text of the Convention, see: http://conventions.coe.int/treaty/en/Reports/Html/150.htm

[244] The different options and instruments taken into consideration in the Commission's White Paper for Community action were: adhesion of the Community to the Convention of Lugano, a regime that covers exclusively trans-boundary damage, a recommendation by the Community to guide the actions of individual Member States, a Community horizontal direc-tive covering all potentially dangerous activities, and finally a Community directive that imposes a liability regime in spe-cific sectors (e.g. biotechnology).

[245] Official Journal – L143/56 – 30 April 2004).

end up being paid for with public money. As a **market based instrument**, liability for environmental damage is built on the assumption that the risk of liability for potential polluters would result in more careful and preventive behaviour. Making businesses that damage the environment legally and financially accountable for that damage is a way of **internalising external environmental costs** and urging industry to seek the most effective strategy for avoiding environmental damage[246]. Moreover, the Directive seeks to avoid environmental dumping, e.g. companies exploiting difference in liability rules in different countries, and to move towards a **level playing field**.[247]

V.3.3.3.2   Scope: Timing and Type of damage

Non-retroactivity of the system: Liability only applies to future cases of pollution - i.e. only those cases in which it can be proven that the pollution occurred after the entry into force of the directive - and it leaves the management of preceding or historical pollution cases up to Member States.[248] The Directive also provides for a long-stop period of 30 years after which polluters are no longer liable for the damage they cause.

"Environmental damage" covered by the liability regime is defined with reference to:

▶ species and habitats protected by Community legislation (*EC 1979 and EC 1992*) and national legislation

▶ waters covered by the Water Framework Directive (*EC 2000*)

▶ the dangers for human health deriving from contamination of soil and underground

Admittedly, the particular value of this Directive is that it imposes liability for damage to biodiversity. Regrettably, though, the definition of "biodiversity" is quite limited and differs from the authoritative definition contained in the 1992 Convention on Biological Diversity[249]. The focus is on species and habitats listed in certain Annexes of the Birds and Habitats Directives which are considered to be of Community interest and whose conservation status requires the designation of special protection areas or a system of strict protection.

Ultimately, the regime represents a further instrument in implementing Europe's nature conservation policy, under the Wild Birds and Habitat Directives and the Europe-wide ecological network "NATURA 2000"[250] (*see chapter IV.2*). Habitats and species protected under international con-

---

[246] Principle 16 of the Rio Declaration on Environment and Development calls on national authorities to endeavour to promote the internalisation of environmental costs and the use of economic instruments, taking into account the approach that the polluter should, in principle, bear the cost of pollution, with due regard to the public interest and without distorting international trade and investment.

[247] Not all EU countries have adopted rules governing liability for polluted sites. Portugal and Greece have no specific legislation on the matter. Moreover, environmental liability rules and clean up laws that already exist at Member State level differ from Member State to Member State and also from the recently adopted Directive.

[248] The non-retroactivity of liability constitutes one of the biggest differences between the European Directive and the American legislation in which this proposal for a directive was largely inspired (*CERCLA 1980*)

[249] Article 2 of the CBD defines biodiversity as the number, variety, and variability of all species of plants, animals and microorganisms as well as the ecosystem of which they are part.

[250] NATURA 2000 network is made up of "Special Protection Areas"(SPA) designated pursuant to Directive Wild Birds and "Special Areas of Conservation" (SACs) designated pursuant to the Habitat Directive. The Natura 2000 netwrok contitutes the central piece of the EC policy on nature conservation.

ventions to which the EU or Member States are signatories, as well as the areas protected by national and regional legislation that fall outside such a network, will not be automatically covered, unless a Member State expressly determines otherwise.

Under the Directive's regime, damage to waters is only covered if the damage "significantly adversely affects the ecological, chemical, and/or quantitative status and/or ecological potential of the waters concerned" as defined by the Water Framework Directive – e.g. breach existing chemical quality standards or change from high to good or good to moderate ecological status.

Land contamination acquires relevance for the purposes of the Directive insofar as it creates a significant risk for human health. The Directive does not cover damage to air. This omission is justified by the consideration that air is mainly a carrier of harmful substances, the deposition of which damages different media. Moreover, air pollution is caused in most cases by diffuse pollution, also excluded from the scope of the Proposal[251].

V.3.3.3.3   Scope: Type of activities

As for the scope of the Directive, the latter foresees a regime of strict liability for a limited number of activities considered intrinsically dangerous, and listed in its Annex IV[252]. Activities listed in Annex III of the Directive are those governed by EC legislation on the protection of the environment. The Directive thus creates a link to the established regulatory framework at EU level. For example, Annex III refers to the operation of the installations subject to permitting requirements under the Directive on Integrated Pollution and Prevention Control (EC 1996). This covers certain energy industries, such as mineral, oil and gas refineries, certain installations for the production and processing of metals, some types of mineral industry and chemical industry, waste management installations and also some agricultural activities exceeding certain threshold levels of production. The scope of the Directive is also extended beyond the operation of installations subject to permits to the handling of dangerous substances and compounds. It explicitly covers "any contained use, including transport" and "any deliberate release into the environment" of GMOs covered by the relevant directives (EC 1990 and EC 2001).

For those occupational activities not listed in Annex III, polluters will be held responsible only for damage to biodiversity if fault or negligence is found. Consequently, if an operator of any activity not listed in Annex III causes biodiversity damage without being at fault, or water or land damage, regardless of whether he is at fault or not, he will not be held liable for the damage he has caused.

V.3.3.3.4   Exemptions

The Directive also explicitly excludes from the scope of its regime, a wide range of industrial activities, such as the carriage of dangerous goods by sea, land or rail, and the nuclear industry.

---

[251] Article 4, paragraph 5 of the Proposal exclude from the scope of the Directive environmental damage or the imminent threat of such damage caused by pollution of a widespread, diffuse character, where it is impossible to establish a casual link between the damage and the activities of certain individual operators.

[252] Reference is made to legislation that regulates limits of discharge or emission of hazardous substances into water or air, legislation concerning dangerous substances and compounds also with the objective of protecting the environment, legislation aimed at preventing and controlling risks of incidents and pollution, legislation on management of hazardous waste and other waste; legislation concerning biotechnologies, legislation on transport of hazardous substances.

Liability arising from these activities is regulated by a number of International Conventions listed in Annexes V and V of the Directive and by the Euratom Treaty respectively. In addition, the Council of Ministers introduced a new paragraph taking into account two international instruments on liability for maritime and inland navigation[253], to allow ship owners to limit their liability under national legislation.

In practice, under the Directive's regime, primary liability for the prevention and remedying of environmental damage would lie with the operator that has caused the risk of damage or the damage itself. Member States, and public authorities appointed by them, should act as trustees of the system and "may" decide to take preventive and remedial action themselves if the author of the damage cannot be found (in cases of diffuse pollution), if he is otherwise exempted from liability under the Directive or also in the not-so-rare cases where the responsible operator does not possess the necessary financial resources to prevent or remedy the damage. What in the original Commission's proposal was a requirement for competent authorities to take preventive or remedial measures in cases where the responsible operator could not be found, or was exempted from obligation (subsidiary state liability), has been subsequently transformed by the Council of Ministers into a mere faculty[254].

Moreover, a variety of standard exemptions is available to polluters (armed conflict, *force majeure*, action by third parties, compliance with a compulsory order or instruction) under the Directive. In addition, the Directive further allows Member States to exempt operators from clean-up costs, if they can demonstrate that the damage has been caused by an authorised activity ("compliance with a permit", e.g. permits under IPPC Directive see chapter V.3.2) or by an activity not considered risky according to the technical and scientific knowledge at the moment of release of the emissions, or the development of the activity ("state of the art") and that they were not at fault or negligent. The said hybrid clause has been included by the Council of Ministers as a compromise solution between those Member States which were in favour of keeping "compliance with a permit" and "state of the art" as plain exemptions to liability, and those who favoured a "mitigating factors" clause.

V.3.3.3.5   Financial security

Another area of forceful controversy, both within the European Parliament and the Council of Ministers, was the opportunity to require operators of risky activities to be covered by some form of financial security, be it through insurance, reserves, bonds, dedicated funds or other mechanisms. Yet another poor compromise was agreed between the two Institutions during the conciliation procedure. As it stands, the Directive does not impose a compulsory financial guarantees system. It only requires Member States to encourage the development of financial security instruments and markets, and demands that the Commission report on the availability at reasonable costs of insurance and other types of financial security for the activities covered by Annex III, by 2010. The report shall also consider a gradual approach to mandatory financial security as well as a ceiling for liability, and exclusion of low-risk activities from a mandatory guarantees system.

---

[253] 1976 Convention on Limitation of Liability for Maritime Claims (LLMC) and 1978 Strasbourg Convention on Limitation of Liability in Inland Navigation (CLNI)

[254] For more information on the Council of Ministers and documents relating to see: http://register.consilium.eu.int/utfregister/frames/introfsEN.htm

In light of the report and of an extended impact assessment, the Commission shall, if appropriate, put forward proposals for a harmonised system of mandatory financial security.

Finally, precisely in consideration of the fact that environmental goods (water and biodiversity) are often not subject to property law and, consequently both the impulse to protect them and the control over the correct functioning of the regime can be lacking, it is foreseen that qualified entities or individuals with a sufficient interest may ask the authority to undertake the appropriate measures against potential or actual polluters and may challenge the legality of the action, or the lack of it, of the competent authority[255]. Member States are given the discretion to take away those rights in the case of imminent threat of damage - which is precisely when those participatory rights are more useful for protecting natural resources not associated with property rights.

V.3.3.4   STRENGTHS AND WEAKNESSES OF THE NEW LIABILITY REGIME

The Environmental Liability Directive has a two-fold aim: the prevention and remediation of environmental damage. Unlike what was initially envisaged in the White Paper, the Commission's Proposal left traditional damage (to persons and goods) out of the scope of the prospective regime. In fact, the Commission considered that Community action was strictly necessary to effectively address widespread site contamination and the alarming loss of biodiversity in its territory. How effectively the new Directive addresses these problems is arguable, and will ultimately depend on how Member States transpose key provisions.

As mentioned above, the Directive does not apply retroactively to past damage - damage which occurred before its entry into force. This could give polluters a defence that can produce lengthy legal controversies determining the time at which the damaging activity had occurred. In addition, the Directive does not apply to pollution of a diffuse character, where it is not possible to establish a causal link between the damage and the activities of individual operators. In those instances, there is a risk that environmental damage will be borne by the public and its restoration, where provided, will be paid for out of taxpayers' pockets, in contradiction with the Polluter Pays Principle.

The Directive covers damage to water, land and biodiversity. The innovative aspect and the particular value of this Directive is that it protects biodiversity. This is an area where the Directive goes beyond existing national regimes and could make a real difference. This is why it is particularly important that the Directive covers as much wildlife as possible. Regrettably, the Directive grants protection only to species and habitats listed in certain Annexes of the Birds and Habitats Directives, while Member States are free to include habitats and species designated pursuant to national laws, where they are not covered by EC legislation. ECOs have also expressed the concern that Member States may be reluctant to increase the number of conservation areas given

---

[255] Article 12 of the proposed Directive provides that persons adversely affected or likely to be adversely affected by environmental damage and qualified entities shall be entitled to submit observations relating to instances of environmental damage of which they are aware and to request the competent authority action. In addition, pursuant to Article 13, any person or qualified entity who had lodged a request for action with the competent authority shall have access to court or other independent body competent to review the legality of the decisions, acts and failure to act of the competent authority.

the liability risk it creates for economic actors on their territory and ultimately for themselves. Finally, damage to biodiversity is only relevant if a certain threshold is met. The damage needs to be such that it "has significant adverse effects on reaching or maintaining the favourable conservation status of such habitats and species"[256].

The Directive also covers waters protected by the Water Framework Directive, which includes basically all surface waters, groundwater and coastal waters. By including these waters the geographical boundaries of the liability regime are expanded considerably. However, damage to waters is only relevant for the purposes of the Directive if it "significantly adversely affects the ecological, chemical and/or quantitative status and/or ecological potential" of the concerned waters[257]. The meaning of this is yet unclear in reality as Member States still have to define the ecological status or potential and set key biological and chemicals standards. Nevertheless failing to meet existing chemical quality standards (under the Dangerous Substances Directive (76/464/EEC)) means falling short of good chemical status and thus represents a "significant adverse effect" (see chapter IV.5).

The focus of the definition of land damage shifts away from the main purpose of the Directive, namely the protection of natural resources in their own right towards the protection of public health. It refers to "any land contamination that creates a significant risk of human health being adversely affected as a result of the direct or indirect introduction, in, on or under land of substances, organisms or micro-organisms"[258]. The damage threshold to be met for land damage to fall under the scope of the Directive could pose problems. The criteria included in the Directive for determining whether land contamination is of such an extent as to exceed the damage threshold are very imprecise and open ended.

The Directive provides for a strict liability regime for the activities listed in its Annex III. As regards non-listed activities, operators could be held liable under the regime only for damage to biodiversity and when found to be at fault or if there is proof of negligence. This has the practical effect of completely excluding operators of non-listed activities from liability for damages to waters and land as defined in the Directive. This is unfortunate as it considerably weakens some of the positive impacts that the Directive might have had, notably on the protection of waters and biodiversity.

Damage caused by maritime and nuclear accidents is left outside the regime's scope. In this area, the Commission is required to report to the European Parliament and the Council in relation to the exclusion of pollution covered by the international instruments listed in Annexes V and V from the scope of this Directive in view of possible amendments. Since such conventions are much narrower in scope than the Directive in the way the environment is protected through their liability rules, the said exclusion will have the effect of a preferential treatment of these activities.

One of the most serious dangers for the effectiveness of the Directive as to its capability to make polluters pay for the environmental damage they cause –and thus to create strong incentives for potential polluters to operate as safely as possible- is the introduction of "compliance with a permit" and "state of the art" considerations as reasons to exempt polluters from all clean up costs.

---

[256] Article 2, Paragraph 1(a).

[257] Article 2, paragraph 1 (b).

[258] Article 2, Paragraph 1 (c).

As a matter of fact, this provision leaves Member States a great deal of discretion in the extent to which operators could be absolved from paying the costs of remedial action when they have complied with a permit or operated according to "state of the art" considerations. Such mitigating factors entail the risk of waiving the principle of strict liability for potentially dangerous activities, by re-introducing the necessity to prove the fault of the polluter. They are normally not allowed by existing national environmental liability regimes of the EU Member States. The practical consequence of their introduction would be that the Directive would only cover unauthorised, illegal, or negligent acts performed by an operator. Another practical outcome: activities with respect to Genetically Modified Organisms (GMOs), although formally covered by the regime, could fall outside its scope via the operation of "compliance with a permit" and "state of the art" considerations.

Another major risk for the effective functioning of the liability regime is that the insolvency of the operators can pose an obstacle for authorities which want to recover the costs of environmental remediation in line with the "Polluter-pays" principle. A compulsory system of financial security for risky activities is crucial to creating incentives for operating in an environmentally friendly fashion and ensuring that, once environmental damage has occurred, there are the financial resources to pay for its remediation. This is especially important in the absence of a requirement for Member States to remedy environmental damage where the polluter does not or cannot do so (subsidiary state liability). There is a danger that a regime of strict liability for environmental damage could induce large corporations to transfer their riskier activities to smaller enterprises, so that when the latter causes damage they do not possess the necessary financial resources to remedy it, and thus elude liability.

The availability of insurance or other financial instruments would impede such an eventuality. Not only would it render companies less able to elude their responsibility, but it would also it would induce them to adopt systems of efficient risk management to improve environmental performance so as to lower the cost of the insurance premiums. Notwithstanding the promising considerations contained in the White Paper (*CEC 2000*) in this respect[259], the Directive only requires Member States to encourage the development of these services. Along with the overcautious approach on the phase-in of mandatory financial guarantees, the Directive also leaves the door open to further weakening via the introduction of a limit to the financial liability of polluters.

Provisions on access to justice for citizens' organisations were deemed to be crucial for furthering the aims of the Directive, especially in those areas, such as biodiversity damage, where no property rights are associated with the natural resources damaged or in danger of being damaged. The White Paper specifically envisaged the rights for ECOs to bring claims directly against the polluters, if emergency action is required (injunctive relief), or in order to recover the costs incurred in taking urgent measures to avoid damage or further damage in line with the 1998 Arhus Convention (*UN/ECE 1998*). Regrettably, the Directive only provides for a two-tiered approach and gives qualified interest groups the right to request the competent authority to take action against polluters and bring judicial review proceedings against authorities' decisions. Moreover, Member States can waive those rights in the case of imminent threat of damage. These changes are unfortunate because ECOs can play an important role in enforcing liability rules and could make the regime more effective.

---

[259] White Paper on environmental liability, page 23. See note no. 7.

V.3.3.5   CONCLUSIONS AND ECO ACTION (ENVIRONMENTAL CITIZENS' ORGANISATION)

The newly adopted Directive is the result of 20 years of heated discussions within the European Union, and in the wider forum of the Council of Europe. It took over 10 years to produce a legislative text – a Directive – that is now in the hands of Member States which are required to transpose the new rules in their legal system by 2007.

The end result is indeed a compromise text which had to take into account the sensibilities of industrial groups and Member States and scale down from initial ambitions. The final text is very weak for the reasons enumerated above and by no means ideal from an environmental perspective. It is indeed a complicated piece of legislation with a limited scope and a great degree of freedom for Member States in transposing its provisions.

Member States are now left with a difficult task: transposing the terms of the Directive into national law. Environmental liability rules and clean-up laws that already exist at national level differ from Member State to Member State. The Directive will have significant overlaps with existing clean-up regimes and aspects of civil law in Member States. The latter will have to assess how much of the new Directive is already reflected in national laws and improve on national regimes when the latter are weaker and leave more stringent national rules unchanged or introduce more stringent national rules when both national laws and the new Directive are too weak. After all, the Directive explicitly mentions the right of Member States, under the EC Treaty (Art. 176), to make or keep their environmental liability rules more stringent than the provisions of the Directive[260].

Whether this Directive will improve the status quo vis-à-vis the remediation and prevention of environmental damage in Member States will ultimately depend on how the latter implements the Directive into national laws and, looking far ahead in the future, on how the Commission will use its review obligations.

ECOs have a crucial role to play in the national implementation process to prevent the latter becoming race to the bottom due to industry lobbying and competition arguments. There is a great danger in a number of Member States that not only will national legislators not use the most pro-conservation options in implementing the provisions of the Directive, but that they could also choose to weaken existing national laws where they are more stringent than the Directive's provisions.

In particular, ECOs should lobby their governments and support a sensible implementation of the critical provisions of the Directive in national law along the following lines:

▶ The biodiversity covered by the liability rules should be extended beyond the species and habitats listed in the relevant Annexes of the Birds and Habitats Directives to include those contemplated in International Conventions to which Member States are signatories and in national and regional laws.

---

[260] Article 16 provides that the Directive shall not prevent Member States from maintaining and adopting more stringent provisions in relation to the prevention and remedying of environmental damage.

▶ The thresholds of significance that trigger the liability of polluters for the damage to waters, land and biodiversity need to be lowered. In particular, land contamination should to refer to serious harm to the environment and not to human health.

▶ The system of strict liability needs to be extended to all occupational activities. Liability for damages to the environment should ultimately depend on the actual damage and not on the arbitrary nature of the activity causing the damage or the danger. This will also make the system more easily enforceable by public authorities and judicial bodies. Oil and nuclear accidents should be caught by the liability rules of the Directive where the relevant international conventions contain no satisfactory provisions in some respect (especially relevant for biodiversity damage) or have not been ratified by the concerned Member State.

▶ The option to introduce "compliance with a permit" and "state of the art" considerations as reasons to exonerate polluters from paying clean-up costs should in principle not be used. Where Member States implement this clause, permits and authorisations should only be granted under full public participation and strict environmental performance and reporting conditions. Sensible solutions should be sought that allow a mitigation and a reduction of appropriate parts of the clean-up costs for the polluters taking into account the level of fault and negligence but not the exoneration from any cost in every case.

▶ The use of financial security cover for dangerous activities should be made compulsory even if only in a phase-in approach. Especially with weak financial security provisions, subsidiary state liability should be reinstated as a safety net for environmental restoration if a polluter cannot be identified, cannot pay for clean-up costs or is otherwise exempted, in order to avoid instances where environmental damage simply does not get remedied.

▶ Finally, it is recognised that ECOs could play a vital role in making sure that the Directive's provisions are enforced. This is especially true when the national authorities whose primary responsibility it is to enforce the system are over-burdened or have a conflict of interests (e.g. when the same authority issues the permits.) Member States shall not be allowed to waive the additional participatory rights granted to ECOs by the Directive in relation to imminent threat of environmental damage. In addition, participatory rights for qualified entities (access to information and public participation in decision-making) need to be stepped up with clearer timeframes as well as allowing qualified ECOs to bring direct actions in court against operators when there is an imminent threat of damage (interim relief). In those cases, specifically, and in light of the often irreversible nature of environmental damage, the two-tiered approach for ECOs access to justice could be an impediment to an effective prevention of environmental damage.

▶ It will be then up to ECOs to use the rights they are granted by the Directive for the enforcement of its provisions to their full extent, especially in relation to damage to un-owned natural resources and vulnerable wildlife.

BIBLIOGRAPHY AND FURTHER READING

BRANS, Edward H.P. and UILHOORN, Mark (1997) Liability for Ecological Damage and Assessment of Ecological Damage, Erasmus University, Rotterdam 1997 Available at: http://europa.eu.int/comm/environment/liability/background.htm

CEC (2000) European Commission White Paper COM (2000) 66 final on Environmental Liability, 9 February 2000

CERCLA (1980)Comprehensive Environmental Response, Compensation, and Liability Act, of December 11, 1980 (CERCLA), more widely known as the Superfund. More about the Superfund at: http://www.epa.gov/superfund/action/law/cercla.htm

CMS Cameron McKenna (1995) Study of Civil Liability Systems for Remedying Environmental Damage, FINAL REPORT, 31st December 1995, http://europa.eu.int/comm/environment/liability/civiliability_finalreport.pdf

DELODDERE, Sophie and RYCKBOST, Donatienne (1997)Liability for contaminated Sites, University of Ghent 1997. Available at: http://europa.eu.int/comm/environment/liability/background.htm

EC (1979) Directive 79/409/EEC of the Council, April 2, 1979, concerning the conservation of wild birds [Official Journal L 103, 25.04.1979];

EC (1992) Directive 92/43 of the Council of 21 May 1992 on the conservation of natural habitats and of wild fauna and flora [Official Journal L 206, 22.07.1992].

EC (1996) Council Directive 96/61/EC of 24 September 1996 concerning integrated pollution prevention and control (OJ L 257, 10/10/1996).

EC (1990 and 2001) Council Directive 90/219/EEC of 23 April 1990 on the contained use of genetically modified micro-organisms (OJ L 117, 8.5.1990)and Directive 2001/18/EC of the European Parliament and of the Council on deliberate release into the environment and placing on the market of genetically modified organisms (OJ L106, 17.4.2001).

EC (2000) Directive 2000/60/EC of the European Parliament and of the Council of 23 October 2000 establishing a framework for Community action in the field of water policy [Official Journal L 327, 22.12.2000].

EEA (2000) Management of contaminated sites in Western Europe, EEA, June 2000. To access the text of the document, visit: http://reports.eea.eu.int/Topic_report_No_131999/en

KRAMER, L (1997), "Polluter-Pays Principle in Community Law. The Interpretation of Article 130R of the EEC Treaty" in Focus on European Law, Graham & Trotman, London 1997

WENNERAS, P. and PERMI DEDFENCE (2004) *Environmental Liability Regimes – Subsidizing Environmental Damage in the EC?*, The Yearbook of European Environmental Law, Volume 4, December.

V.3.4   Greenhouse Gas Emissions Trading Directive

By Matthias Duwe[261]

V.3.4.1   INTRODUCTION: CLIMATE CHANGE POLICY IN THE EU

There is growing consensus within the global scientific community that human activity is inter-fering with our climate. The enormous growth in the release of heat-trapping gases since the start of industrialisation, mainly caused by the combustion of fossil fuels such as coal, oil and gas, has already led to an increase in the average global temperature of 0.7 degrees centigrade. Projections suggest that this manmade climate change could lead to warming of 1.4 to 5.8 degrees centigrade by the end of the century (*IPCC 2001*), and recent studies indicate that it could be even higher[262]. Change on such a scale would have devastating effects for people, animals and plant life.

Growing scientific evidence, signs of impacts and public pressure to act on the problem have led European leaders to put climate change high on the agenda of environmental challenges. Furthermore, it has been recognised within the international negotiations on the issue, that rich industrialised countries are historically responsible for climate change and therefore have to take the lead in reducing them to counter the problem[263]. Despite the fact that the United States of America, the largest emitter worldwide, decided to withdraw from the Kyoto Protocol on climate change[264] - a treaty it had negotiated for years - the members of the European Union and 100 other countries stayed committed to the agreement and ratified the Protocol. It finally entered into force on the 16th of February 2005. This UN treaty stipulates that industrialised countries should reduce their greenhouse gas emissions globally by 5%, compared to 1990 levels, in the five-year period from 2008 to 2012. The EU15 had agreed to an 8% emissions reductions goal. This commitment was, however, redistributed within the EU to account for differences in histor-ical responsibility and reduction potential. It ranges from a −28% cut for Luxembourg to a +27% increase in Portugal[265]. Still, the EU as an international body remains responsible for the achieve-ment of the -8% overall target.[266] The Member States from Central and Eastern Europe that joined the EU in 2004 have taken on similar targets, but require almost no effort to achieve them, due to economic changes in the 1990s.[267]

In order to achieve the Kyoto obligations, EU countries agreed that in addition to national action, the Community as a whole would implement common and coordinated policies and measures to

---

[261] Policy Officer at the Climate Action Network (CAN) Europe asbl

[262] The International Climate Change Taskforce (January 2005) Meeting the Climate Challenge – www.ippr.org.uk

[263] See United Nations Framework Convention on Climate Change (UNFCCC) of 1992 - www.unfccc.int

[264] Kyoto Protocol to the United Nations Framework Convention on Climate Change (1997) Full text of the treaty at http://unfccc.int/essential_background/kyoto_protocol/items/1678.php

[265] See http://www.climnet.org/resources/euburden.htm for the full table

[266] See Article 4 of the Kyoto Protocol

[267] The enlargement of the EU does not affect the -8% target by the EU15 and their burden-sharing agreement. Latest data on emissions trends for the whole EU25 is contained in the EEA report No 5/ 2004 "Greenhouse gas emission trends and projections in Europe 2004" available from http://reports.eea.eu.int/eea_report_2004_5/en

lower distortion of competition within the EU and minimize overall compliance cost. The European Commission, therefore, initiated the European Climate Change Programme (ECCP) in the year 2000 - a process that was meant to identify the best policy options - in close consultation with Member States and a variety of other stakeholders, including environmental organisations and industry associations. A selection of measures was identified that addressed all sectors of the economy, and that could potentially deliver the reductions required by the Kyoto Protocol and more, at a cost of less than 20 Euros per tonne of CO_2[268].

The energy-intensive sectors of industry are major contributors to climate change. Cutting emissions from these sources is imperative for the feasibility of long-term climate goals. To facilitate such changes, businesses need the right price signals to reward technological innovation and channel investment into cleaner production. From the early 1990s, the introduction of a tax on CO_2 and energy use in industry was discussed in the EU as the instrument that was going to deliver emission reductions in these sectors. However, the debate was drawn out, and deadlocked by controversy and could not be realised at an EU-wide level, since decisions on harmonised taxes require unanimous agreement among EU Member States. A different measure was needed, and the focus shifted in the late 1990s to market-based measures such as the trading of emissions allowances, which had to that date not been used in Europe for environmental purposes. The concept had already been discussed under the negotiations on the Kyoto Protocol, which allows trading of climate credits between participating countries.

In 2000, the European Commission initiated discussions on the introduction of an EU trading system for greenhouse gas emission permits[269] with the release of a Green paper[270]. As part of the stakeholder consultation process established with the kick-off of the ECCP, a separate working group on Flexible Mechanisms was set up. that discussed the potential design of such a trading scheme among Commission officials, Member States, industry associations and ECO representatives.[271] The trading scheme emerged as one of the central pillars of the ECCP. In autumn 2001, the Commission then published a proposal[272] on the implementation of a European wide cap-and-trade system for large industrial points sources of carbon dioxide (CO_2), widely known as the Emissions Trading System (ETS). The Directive establishing the ETS was adopted in summer 2003 after controversial discussions within and between the European Council and the European Parliament (EC 2003). The ETS officially started operating on the 1st of January 2005. Its success is vital to achieving the EU's near-term Kyoto targets and plays an important role in enabling the necessary long-term transformation of our fossil fuel intensive to a low-carbon economy. However, the environmental effectiveness of the ETS is not guaranteed: it all depends on ambitious implementation and strict target setting.

---

[268] For more documents and more information, please consult the European Commission's DG Environment's webpage at http://www.europa.eu.int/comm/environment/climat/eccp.htm

[269] For more documents and more information, please consult the European Commission's DG Environment's webpage at http://europa.eu.int/comm/environment/climat/emission.htm

[270] COM(2000)87

[271] Documents from this working group are available at http://europa.eu.int/comm/environment/climat/flexiblemechanisms_ firstphase.htm

[272] COM(2001)581

V.3.4.2   EMISSIONS TRADING: THEORY AND PRACTICE

The EU Emissions Trading System (ETS) is the first of its scope and coverage world-wide. It is also a new instrument for European environmental policy. Previously, trading schemes had only been used as a climate policy tool in Denmark and the United Kingdom, and in other countries the trading of certificates had been employed to support renewable electricity. The biggest example of such a system in practice is the US trading system to reduce emissions of sulphur dioxide, which started operating in 1995.[273] The main theoretical benefits of a cap-and-trade approach are the guaranteed achievement of a given environmental objective at the least possible cost for the operators of pollution sources.

V.3.4.2.1   The concept: cap and trade

The idea behind a cap-and-trade system as an environmental policy tool is simple and straight-forward[274]. An absolute limit is put on the pollution that causes the problem (= the cap). This guarantees the environmental goal that is to be achieved (e.g. minus 10% of emissions within 5 years) and if left at that would work like standard environmental regulations. However, the pollution sources covered by this environmental goal are then allowed to exchange amongst each other the burden of fulfilment of this obligation (= the trading).

The global objective stays the same, but a market decides where the changes are being made. One pollution source may reduce its emissions more than required by the target because it can do so at a lower cost than another polluter, who will then pay the former to "acquire" the reductions made elsewhere. By limiting the total volume of pollution and allowing trading between sources a price is put on every unit of the emissions concerned. This means that environmental pollution - that used to be an external effect of its operation and came "for free" as far as emitters were concerned - now comes at a cost and will have to be taken into account in choices about future behaviour, just like any other production factor.

V.3.4.2.1.1   How it works in theory

The following example tries to explain in a simplified manner how the system should work in theory. It takes the design of the EU ETS as a model, to stay close to the system at hand. A cap-and-trade system could also be designed differently. In this example we look at only two emissions sources. Installations A and B both emit 100,000 tonnes of CO_2 per year. The government gives each of them 95,000 emission allowances (equivalent to a 5% reduction target), with one allowance representing the right to emit 1 tonne of CO_2. Both installations A and B are, therefore, 5,000 allowances short of covering their annual CO_2 output. The plant owners face the same choice: either reduce their emissions by 5,000 tonnes, or purchase 5,000 allowances on the market. In order to decide which option to pursue, they will compare the costs of reducing their emissions by 5,000 tonnes with the projected market price for allowances.

---

[273] For more information on the US trading system, please consult the EPA website at **http://www.epa.gov/airmarkets/arp/index.html**

[274] The description provided here is necessarily a simplification and limited to the case of emission permit trading.

We assume the market price for allowances is predicted to be 10 euro per tonne of CO_2[275]. Installation A's engineers advise the manager that the cheapest reduction possibility is one single investment (switching fuel input or optimising process efficiency) at a total cost of 60,000 euro which will cut emissions by 10,000 t CO_2. A's reduction costs are, therefore, 6 euro per tonne. The plant owner will (in economic theory) make the investment, because it presents the cheaper option of complying with the cap. So instead of having to BUY 5,000 shares for 50,000 euro, the company can SELL as many shares due to its reduced emissions. So the net costs for his response are 60,000 minus 50,000 euro - ie.10,000 euro - which is much less than the 50,000 that he would have paid for excess emissions if he had not acted. The assumption here is that this gain is enough to convince company A to make the investment.

Installation B, on the other hand faces reduction costs of 15 euro per tonne for an investment, allowing him to reduce emission by 5,000 tonnes, totalling 75,000 euro - which is higher than the expected market price. The owner will prefer to buy allowances for 50,000 euro, instead of reducing emissions at the respective installation, and thus saves 25,000 euro.

The market has helped identify where emission can be cut most cheaply, thus reducing overall compliance costs by over half in this example (60,000 euro instead of 135,000 euro - see graph) and the environmental goal of a 5% reduction is met nevertheless.

Figure 1: Effects on compliance cost with or without trading

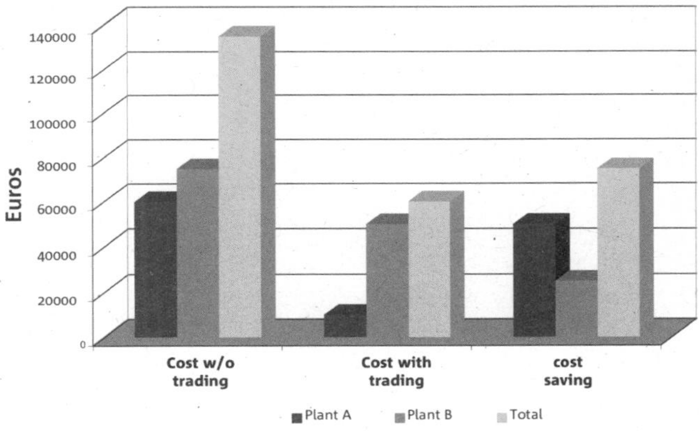

V.3.4.2.1.2 Environmental effectiveness

It is important to note, however, that in case there had been no trading in the example above, the environmental benefit would in fact have been 50% greater. This kind of unintended consequence often leads to criticism against the concept of trading. This criticism is to some extent misdirected, insofar as the target being set at the political level should correspond to the necessary urgency and extent of the environmental problem that is being tackled. In an ideal world, reductions beyond such a target should not be necessary. But the real problem lies with the political will and capability in setting the target which is environmentally sound and unbiased with respect to individual economic actors .or constituencies. The environmental effectiveness depends therefore inherently on the target-setting – the "cap" - as much as on the design of the system. In a non-trading scenario, the company that faces a higher burden than its competitors because of a design flaw in the policy (requirement to meet an objective resulting in more costly investments going beyond the target level) would feel hard done by - and probably rightly so.

---

[275] The market price is, of course, a function of supply and demand. This example is highly simplified to visualise the principle workings of such a system.

The fact that trading reduces the cost of compliance to basically all participants enhances the political feasibility of its introduction compared to other economic instruments.

There are other serious environmental issues to be considered when judging the suitability of a trading system. Climate change caused by greenhouse gases is independent of the geographical location of GHG emissions. But with respect to damage to human health and the environment from toxic pollutants, the geographical location and dilution is of crucial importance. An instrument that allows one source of pollution not to reduce its impact because this is being compensated for elsewhere has to deal with the problem of such locations with higher local pollution, so-called "hotspots", in which the emissions continue unabated. In fact, this is also an issue when a trading scheme is introduced for a greenhouse gas such as carbon dioxide, because other harmful gases are being produced alongside the CO_2. Continued CO_2 emissions in one plant must, therefore, not lead to a worsening of local air quality. This problem can, in theory, be overcome by ensuring adherence to separate air quality legislation to limit these other pollutants. In turn, such legislation could imply an indirect restriction on CO_2 emissions. The potential conflict between two such instruments serving different issues must be resolved without compromising either one of the environmental objectives (see also chapter V.3.2.4.4).

Emission trading designs other than the one described in the example above are possible. The distribution of allowances can also be carried out by auction, which has the benefit of creating revenue for the state, just like taxes, and is seen in economic literature as the more effective means of distributing permits. Environmentally it has the benefit that all emissions are paid for - not only those beyond the initial target - which leads to a stronger financial incentive for participants to reduce pollution. The revenue could be recycled back to society by lowering the cost of labour, which should lead to more employment. This way, a so-called "double dividend" could be reaped, a reduction in environmental pollution and a positive social effect (Jørgensen, C.E. 2003).

V.3.4.2.2   The practice: the EU Emissions Trading System (ETS)

In January 2005, the European Union Greenhouse Gas Emission Trading established by Directive 2003/87/EC, commenced operation. To start with, it covers only carbon dioxide emissions and concentrates on large point sources from key industry sectors. It is the first international trading system for emissions in the world, and covers around 12,000 installations (owned by ca. 5,000 companies) spread across Europe. The emission sources covered include combustion plants for electricity generation (capacities greater than 20MW), oil refineries, coke ovens, iron and steel plants, and factories making cement, glass, lime, brick, ceramics, pulp and paper. The operators of each of these installations now require a greenhouse gas permit issued by the relevant national authorities as part of the Trading System to continue running their plant. Member States grant each installation covered by the trading system a specific number of emission allowances, each worth 1 tonne of CO_2 (mainly for free), which make up the official currency of the system. At the end of each year, companies must have an allowance for every tonne they have emitted. If they have not received a sufficient number of free allowances from their government in the beginning, they will have to decide on a strategy for bringing allowances and emissions in line. They can either reduce their emissions or buy allowances from the market, effectively paying somebody else to make the reductions for them.

V.3.4.2.2.1   Design features of the EU ETS

The environmental targets for the ETS are being set for periods of years, in order to avoid the problem of specific circumstances in one year making achievement of the targets too easy or too difficult. The first trading period of the EU ETS runs from the beginning of 2005 to the end of 2007 (three years). From the 1st of January 2008 onwards, it will be operating in five-year periods. This design was chosen to allow compatibility with the Kyoto Protocol, the targets of which are for the period 2008 to 2012. Choosing the same time-line underlines the link with the Kyoto targets and facilitates taking these into account when setting ambition levels for the ETS. It also makes it easier to judge to what extent the industry sector is being asked to contribute to achieving the Kyoto targets.

While the ETS directive, negotiated at EU level, laid out the architecture of the system, the target-setting was left to the Member States. They have to decide on the total number of allowances (= the maximum CO_2 emissions from the sectors covered) and how many allowances each installation receives. These choices and other information have to be communicated to the European Commission and the public in so-called National Allocation Plans (NAPs) which Member States have to prepare for each trading period. The European Commission assesses the NAPs to determine whether they comply with the criteria set out in the directive that established the EU ETS. These criteria contained in Annex III of the directive include consistency with the Kyoto targets and projected and assessed progress towards them, as well as taking into account other relevant national and EU energy and climate policies. It should also consider economic and technological reduction potential, and not punish those that have already made improvements (so-called "early action") or operate a cleaner technology (such as power plants that produce electricity and heat at the same time).

For the first trading period, each Member State had to prepare and publish its first National Allocation Plan in early 2004. For future trading periods, every NAP needs to be submitted to the Commission 1 1/2 years in advance of the start of the trading period, so for 2008-12 the deadline is 30th of June 2006. The European Commission has three months to complete its assessment. In the first round of NAPs, the Commission sent letters to each Member State with technical questions that required clarification. On the basis of these answers, evaluation proceeded. The Commission can either approve or reject plans, with the possibility of conditionally approving them during the course of proceedings. No plans were rejected in the first round, although some were approved on condition only. Most MS amended their NAPs prior to the Commission decision to receive approval.

The distribution of allowances, referred to as "allocations" in the context of the EU ETS, is carried out mainly based on historical emissions. This approach, which uses past pollution levels, is called "grandfathering". Allocation decisions can be combined with future needs estimates (projections) or based on performance (using benchmarks) or on willingness to pay (via auctioning) and combinations of these. The ETS directive specifies that most of the allowances will be handed out for free. Auctioning is possible for up to 5% for the period 2005-7 and a maximum of 10% for 2008-12. Some MS that had experience with using benchmarks have used these for allocation (the Netherlands and Belgium), most others have used future growth projections and applied correction factors. Member States are supposed to hand out or auction allowances by the end of February of the year in progress.

The NAPs are required to give a lot of detailed information, which amounts to an outline of a Member State's Kyoto target implementation strategy with specifics for all sectors. For the trading sector there must be an explanation of the methodology used for allocating the overall cap on industry emissions and how they have been assigned to sub-sectors and the installations in them. Apart from making domestic efforts, Member States may also purchase emission credits from projects implemented in other countries in accordance with the mechanisms of the Kyoto Protocol, such as the Clean Development Mechanism (CDM) and Joint Implementation (JI)[276]. These projects generate officially certified credits, which enable governments to count the achieved reductions against their own Kyoto targets. NAPs must spell out in detail what volume of such credits a country is planning to purchase, what national authority will be in charge and where the funds will come from.

With the reception of their greenhouse gas permit that allows the participating installations to continue operating from the start of the EU ETS, operators are forced to monitor their CO_2 emissions according to the monitoring and reporting guidelines developed specifically for the EU ETS. This information will have to be evaluated by independent third parties, so-called verifiers, which will have to establish that the CO_2 monitoring has been done thoroughly and the resulting CO_2 emission figures are sufficiently accurate. Installations have to make the verified data available at the end of March every year for the emissions of the year before. One month later, at the end of April, they then have to show they have sufficient numbers of allowances equivalent to their emissions. These will then be "cancelled", meaning they are taken out of the system and are no longer available for trading. Information on the data and the allowances will be available in so-called registries, which are online systems with accounts for each participant that can be used similarly to an Internet-based banking system[277]. It is here that operators receive allowances for their installations and that they have to cancel them to comply with legislation.

V.3.4.2.2.2 Why the success of the ETS is important

The EU Emissions Trading System has been officially labelled a cornerstone of the European strategy to meet its Kyoto targets. It is the main instrument to starting the decarbonisation of the European economy that is necessary to achieve long-term emission reduction objectives. It is the first of its kind in terms of size and coverage, and its success or failure will therefore potentially constitute a precedent.

For the EU, showing that emissions can be reduced and at low cost, is an issue of environmental integrity as much as one of international credibility. The EU has declared itself a world leader on climate change and now it has to show it is true to its word. The outcome will influence how other countries judge the Kyoto Protocol and their relation to it - most importantly the United States, which has opted to step away from the treaty it helped to negotiate. Also current industrialised partners like Japan and Canada and developing nations who will have to start limiting their emissions in the future, will be watching to see if the EU can deliver. Showing that domestic emission cuts are possible and presenting ways to achieve them could bring the USA back on

---

[276] For more information on how citizens' organisations view these mechanisms, please consult www.cdmwatch.org or www.sinkswatch.org

[277] The European Commission provides access to all national registries via this webpage http://europa.eu.int/comm/environment/ets/

board and ease negotiations for future targets. Moreover, a functioning EU ETS could be linked with similar systems around the world, including regional or federal level systems in the USA or Australia, even if the country was to remain outside international agreements for the immediate future. Such a direct connection would build up the international climate mitigation system from the ground and facilitate the US's return to the fold. Moreover, the developing world is viewing the progress made by industrialised countries in reducing their emissions as a sign of how serious they are about taking action. Without positive signals here, developing countries are unlikely to agree to take up targets of their own any time soon.

On an intra-EU level, the consequences are manifold. If the EU ETS does not deliver on its promises, Member States will have little choice but to make up for the shortfalls towards their Kyoto targets by resorting to the purchase of external carbon credits from the Kyoto Protocol's flexible mechanisms.[278] This would most likely not only involve JI and CDM, but could go as far as buying excess emissions from countries like Russia and the Ukraine, that have plenty of unused CO_2 permits due to lax 1997 Kyoto targets (stabilisation at 1990 levels) and major reductions in their emissions after the economic decline in the 1990s. These excess credits, often dubbed "hot air" by critics because their existence was known when these targets were set, cannot be seen as equivalent to reductions made within the EU, as far as the benefit to the climate is concerned, since they allow an increase in emissions elsewhere[279]. In addition, the development and deployment of low carbon technology will suffer if there is no strong price signal from the ETS that requires innovation. Furthermore, if no critical mass can be achieved for the sound implementation of this instrument, successes in other areas of climate policy will become less likely. This would be disastrous for everyone affected by the future impacts of climate change.

How will ECOs measure if the EU ETS is a success or not? First of all, it needs to deliver emissions reductions over current levels that help the EU as a whole in order to comply with its Kyoto targets, mainly by cutting greenhouse gases at home. Secondly, its rules need to create a strong signal to reward clean technology and lower emission levels. This long-term effect is not to be underestimated - and it can be achieved, using proper rules together with targets. Thirdly, expansions to the system must not weaken its impact. Any changes should be for the better and help to stabilise the system.

V.3.4.2.2.3   The Review

In 2005, the Commission started the official review of the EU Emissions Trading System, as demanded by the directive in Article 30. This includes an evaluation of the allocation, reassessing certain provisions of the link with external credits and considers the expansion of the system to include other sectors and gases. A consortium of consultants has been awarded the task of assisting with an evaluation of the impacts of the system and the consequences of potential

---

[278] The Kyoto Protocol establishes three mechanisms to provide additional carbon credits that are eligible for official compliance with targets: projects in developing countries (Clean Development Mechanism (CDM)), projects in industrialised countries (Joint Implementation JI)) and trading of credits assigned to countries with targets for free (similar to the EU's allowances) between this group of countries.

[279] Generous targets for Russia and other Eastern European countries were part of the horse-trading in the final hours of Kyoto due to concessions to these countries' poor economic performance in the 1990s. The criticism against this "hot air" is similar to that described in subsection 2.1.2 and directed more at the targets rather than the trading mechanism as such. It does, however, nullify the actual environmental benefit of these credits.

changes. The review report by the Commission, accompanied by proposals to amend the directive if this is considered necessary, is due by mid-2006. Since such amendments will have to go through codecision procedures, meaning they require a compromise between the Environment Ministers in their Council and the European Parliament, they will most likely not be approved before the start of the next trading round in 2008. Major changes to the directive are, therefore, most likely to be applied from the period 2013-17 onwards.

The timing of the review means that little experience with the ETS will have been gained, but major decisions for the medium-term future need to be taken. The Commission's DG Environment has indicated that it would like to keep the system mainly as it is at the moment, rather than adding complexity through significant adjustments. Many actors agree, however, that a certain degree of harmonisation of the rules is desirable. Important discussions will also take place regarding the potential inclusion of other sectors and gases, for example, to link the aviation industry to the ETS, or to include methane emissions or fluorinated gases in trading.

V.3.4.3   ASSESSMENT OF THE EU EMISSIONS TRADING SYSTEM

The EU ETS as set up by Directive 2003/87/EC has a number of stronger and weaker elements from an ECO point of view. More or less all of them are potentially subject to change, on the structural side through amendments to the directive and procedurally through experience gains, capacity building and political will. The assessment presented here is therefore based on the information currently available.

V.3.4.3.1   Strong points

V.3.4.3.1.1   General

The public debate around the development of the EU ETS has really raised awareness of the implications associated with serious climate change policy and created a lot of interest in the media, and with many private and public actors. The European Commission made a conscious effort of involving a wide range of stakeholders, including environmental groups. An innumerable number of seminars and conferences on the subject has taken and is still taking place. Since the start of the system in January 2005, companies are officially obliged to account for their CO_2 emissions. They need to know how much they emit over the year and to ensure that they have a sufficient number of allowances to cover them. Even where the targets are low and the carbon constraint is small or practically non-existent, the psychological effect should not be underestimated: *an era of carbon constraint has begun for EU industry*.

Another inherent advantage of the system is that it transparently translates the respective national targets under the Kyoto Protocol (and potentially also future obligations of a similar kind) into maximum allowable emission limits for industry, basically assigning them a share of the burden vis-à-vis other sectors of the economy. In addition, similar emission shares have to be defined for non-trading sectors and policies spelt out on how to achieve them. It forces Member States to come up with a strategy, where there may not have been a clear plan to start with. This also gives the Commission important leverage in assessing MS compliance with Kyoto. Moreover, the system does provide MS with the tool to ensure that the necessary cuts in the industry sector are made where they are least costly. Whether they make use of the tool they have been given is another issue.

V.3.4.3.1.2 Coverage

The EU ETS covers around half of the EU's total CO_2 emissions and still only affects around 12,000 sources (which is around 1 for every 45,000 citizens in the EU25). This is a good start in terms of potential environmental control and it also limits the complexity of the system and therefore the administrative burden. The 12,000 installations are owned by ca. 5,000 companies, which is a manageable number of actors.

In addition, the industry sector that is covered by the ETS has in most cases not been subject to energy taxation or similar measures. The ETS allows opt-outs only for the period 2005-7; afterwards all installations covered must participate. This was a major issue of controversy, and it is an absolutely crucial element of the system - there are no exemptions and participation is mandatory. In many countries, the climate impact of the sectors covered by the ETS is being controlled for the first time.

In the future, it may be desirable to include more industry sectors and other greenhouse gases, provided that these can be monitored and verified with sufficient accuracy. However, such an expansion of the system can only take place once a proper functioning of the present set-up is guaranteed. Should the ETS turn out not to deliver on its promises, other instruments, like taxation, charges and product policies will need much greater political attention.

V.3.4.3.1.3 Compliance system

The question of how a policy such as the ETS is being enforced is obviously crucial to its environmental effectiveness. At the heart of this is the level at which penalties for non-compliance are being set, so that there is a strong disincentive to simply ignoring the legislation. The ETS directive puts a price tag on every tonne of CO_2 for which an installation cannot show an EU allowance. For the period 2005-7, this is a penalty of 40€; after this, it rises to and 100€ per tonne. Until the end of 2012, this is expected to be higher than the price of allowances on the market, so any rational actor will choose the market over the fine. In fact, in addition to the payment, any emission not covered by an allowance has be made up in the next trading period, meaning the operator will have to bring forward the missing allowances then. The level at which penalties are set may have to be increased in the longer term to account for higher carbon prices on the market, but this is unlikely to be necessary in the near future. This point will be addressed during the official review of the directive (see chapter V.3.4.4.3).

Rules for monitoring and reporting under the EU ETS have been developed by the Commission and were open for comments in the first half of 2005. It is important to have EU-wide rules rather than leaving these to MS, in order to ensure compatibility of effort and integrity of cross-border exchange of allowances. The fact that third party verifiers will assess the accuracy of company reporting is also an important provision to guarantee that no emissions remain unaccounted for.

V.3.4.3.1.4 Public participation and transparency

While there are still serious issues regarding public participation and transparency in some parts of the EU ETS, it is worth noting the positive elements. According to the directive, MS are obliged to consult the public while developing their NAPs and show how comments have been incorporated. This has not been a very inclusive process in some countries, where only industry representatives were heard, but in many Member States draft NAPs have been made available for public consultation. Furthermore, during the Commission evaluation of the NAPs, public comments are also officially to be taken into account. This provided an opportunity for ECOs and industry alike to communicate their perspectives to the Commission.

Important information on the implementation of the system will be made available via the registries. Information on how many free allowances a plant receives, and how much it actually emits will be provided for everyone to see. In cases where the initial allocation is much higher than the emissions, ECOs could try to verify the reason for what looks like a reduction. This way, cases of undue free allocations could potentially be identified and exposed to the public. It will not be possible to see how many allowances any given installation holds at any specific time or to check who trades with whom (this information will only become publicly available five years later, for reasons of commercial interests and business confidentiality). This information is, however, not as important as sound implementation of the system and the role of ECOs as watchdogs.

Another intriguing possibility is that anyone who wants can open a registry account and start trading EU allowances. This means it is possible for ECOs to organise campaigns among members or other players, even with companies, to buy allowances off the market (see chapter V.3.4.4.1).

V.3.4.3.2 Weak points

Unfortunately, there are a number of weak points in the system, which cast doubts over its environmental effectiveness. The extent to which these are a problem cannot be fully established at this point, so the assessment of the status quo can only serve as an indication.

V.3.4.3.2.1 Setting national targets and rules

The lax targets assigned to the industry sectors for the period 2005-7 present one of the greatest shortfalls of the ETS so far.

From an environmental point of view, the fact that a 'grandfathering' approach (see chapter V.3.4.2.2.1) was adopted rather than giving a clear priority to auctioning presents a significant weakening of the potential impact of the directive. Only the gap between the allowances that are being given out for free and actual emissions will have a price tag, diminishing the financial incentive to reduce emissions. Industry was very vocal in its opposition to auctioning, which it regarded as a different means of establishing a tax system. This opposition was reflected in the behaviour of many MS, despite the prospects of revenue from auctioning and advice from economists that this was the more efficient procedure. This means that no so-called "double dividend" is inherent in the ETS at the moment, which can be reaped with revenue-based policies. In the future, the possibility of introducing more auctioning will arise, when this issue is being discussed under the official review of the directive.

The fact that target-setting is being carried out by Member States and the rules that guide it left considerably vague is at present the weakest point of the EU ETS. The lack of common rules led to a situation where every country was developing its own methodology. A guidance document produced by the Commission[280] was available late in the process of preparing the first NAPs and was only informative and not binding. To some degree, this set-up was unavoidable as MS were unlikely to give away control over vital sectors of their economies. There is still a great deal to be done at national level to improve the implementation of the scheme.

Early assessments of the NAPs for the first trading period show that governments have been cautious in demanding ambitious reductions from their industry sectors and generous in their initial allocations, considering that many EU15 MS are considerably far from their Kyoto targets. This development threatens the achievement of the EU's Kyoto targets and thereby questions its international credibility, because it has invested a lot of political capital in the UN negotiations on climate change. Failure to deliver on the EU's promises to implement domestic policies will have serious repercussions on the feasibility of engaging other international partners to continue with the Kyoto framework of absolute reduction targets.

Furthermore, the financial incentives to change current practice and move to less-CO_2 intensive processes and plants depends on the level at which the targets are being set, because the degree to which companies lack available allowances determines the carbon market price. This signal is extremely important to help steer investment in the right direction. In the European power market for example, around a third of generation capacity needs to be replaced over the coming 10-15 years. These plants and their emissions will be around for many decades and it would be disastrous for the climate if these plants were running on coal or lignite that produces immense volumes of CO_2. These long-term effects also play a role in the kind of technology that will be developed to improve future processes. Strong targets are needed to send a clear signal that CO_2 emissions need to be avoided.

The rules that decide how many allowances a plant receives also have an important steering function. They signal that more pollution means higher cost. Factories of the same kind (e.g. cement production) but different levels of CO_2 output should, for example, be treated according to the same pollution or efficiency benchmark, so that there is an incentive for the worse plant to become cleaner. This is especially important in the treatment of new plants that cannot be judged on the basis of past emission levels. If investors know that the rules will penalise a plant with higher CO_2 emissions, this may play a role in their choice of technology. More details on how to improve these rules will be available from the CAN-Europe assessment of the NAPs for 2005-7, set for release in June 2005.

V.3.4.3.2.2 *Link to external credits from JI and CDM*

Representatives of the industry sectors covered by the EU ETS were looking for every possible way of lowering the cost of possessing an allowance for every tonne of CO_2. From the start, their demands therefore, included the possibility of using credits from the project mechanisms of the Kyoto Protocol (JI and CDM) as equivalent to EU allowances to meet their obligations under the

---

[280] The Guidance document is available from **http://europa.eu.int/comm/environment/climat/emission_plans.htm**

EU ETS.[281] The European Commission was also in favour of allowing these credits into the system, as a means of winning win industry acceptance as well as demonstrating to the outside world that the EU was interested in seeing these mechanisms being used.

ECOs had been opposed to allowing these credits to be used in the ETS for two main reasons: 1. Domestic emission reductions have inherent benefits and 2. At the low carbon prices demanded from JI and CDM these were prone to producing low quality credits not equivalent to emission reductions in reality.

1. The EU ETS was designed as an intra-EU policy to reduce emissions from the industry sector and foster cleaner technology to be developed and deployed. This signal for innovation and change in the EU would be diminished or lost if external credits were used to provide the bulk of credits, which companies use to comply with the EU ETS. Furthermore, international climate treaties recognise that industrialised countries, as historical climate polluters, have to take the lead in reducing emissions. Based on both the innovation argument and the insight that developed countries need to cut emissions first, achieving the Kyoto targets primarily through reductions at home became a negotiation position of the EU when the Protocol was being agreed. Showing that reductions in greenhouse gas emissions are possible in industrialised societies is an important signal to international partners in developed and developing countries. External credits should therefore be only a limited option, but already many EU Member States have declared they want to use them to comply with Kyoto. However, these purchases at a government level are different from also allowing them to be used in the ETS.

2. The early examples of what kind of projects would be generating credits were of low environmental quality or outright destructive[282]. They included large hydro-electricity dam projects that had already been under construction for years and would flood massive land areas and force people to leave their homes. The Kyoto Protocol also allows projects that involve the storage of carbon in plants and forests to count as emission reduction credits. ECOs feared that this could lead to large-scale monoculture plantations, potentially with genetically modified plants bred for maximum carbon uptake, which would have serious negative effects not only on biodiversity, but also on water and soil protection. In addition, any such so-called "sinks projects" contain the inherent problems of calculating and verifying the carbon uptake and guaranteeing that it will be permanent (what if the forest burns down?). In short, ECOs saw a system in its infancy that was focusing on low-quality projects at the lowest price. Allowing these credits into the ETS would effectively raise the cap by allowing additional emissions to take place, with doubts over whether real reductions had taken place elsewhere. The logic behind the exchangeability of emission credits of a different kind - that one tonne of CO_2 was the same wherever emitted - would be broken by project credits that were not equivalent to one tonne of CO_2 emitted from European chimneys.

With all non-ECO actors in favour of allowing JI and CDM credits into the system, the Commission drafted an amendment to the EU ETS directive, known as the "linking directive",

---

[281] The early actors in the JI and CDM market such as the World Bank's Prototype Carbon Fund (PCF) and the Dutch governments project tenders CERUPT and ERUPT were working with carbon credit prices of usually below 5€ per ton of CO_2.

[282] For more information on how citizens' organisations view these mechanisms, please consult **www.cdmwatch.org** or **www.sinkswatch.org**

which was adopted by Council and Parliament in 2004.[283]. It contains only a few vague restrictions on the use of JI/CDM credits in the EU ETS. A quantitative limit was introduced, which requires every Member State from the trading period 2008-12 to impose a cap of their choice on the use of JI/CDM credits per percentage of allowances initially given to an installation to be set by MS. In theory, this could still be set at zero, thus not allowing any credits, but most Member States are likely to settle for a few percent of the total number of allowances given out for free, which does not provide a proper limitation at current allowance levels. The directive also contains qualitative criteria for external credit use. It does rule out credits from projects involving nuclear power technology until 2012, but such projects are currently not allowed under the Kyoto Protocol anyway. Furthermore, it also prevents the use of credits from sinks credits for the first trading period 2005-7, but leaves this exclusion subject to a review for future periods. In addition, there is some non-binding language regarding hydro-electricity projects, referring to criteria developed by the World Commission on Dams and general sustainability concerns, but this does not constitute an acceptable safeguard.

The fact that JI/CDM credits are allowed into the system without significant restrictions on quantity and quality is a major flaw of the ETS from an ECO point of view, especially in terms of the quality of credits to be expected in the near future. Member States can set the installation level cap (which can be changed with every NAP) very low to remedy this problem to some extent.

V.3.4.3.2.3   Impact on other policies and sectors

The introduction of the EU ETS presents a significant change in the climate policy mix for many Member States: only Denmark and the UK had previously set up national trading systems. In a number of countries, the ETS is the first instrument that directly deals with the climate impact of the energy and manufacturing sector and is noteworthy as such. However, where such policies were already in place, the ETS has mostly replaced them, in full or in part, and it depends on each case whether this implies a strengthening or weakening of environmental legislation. At the same time, climate policies for other sectors not covered by the trading sector could be affected by the introduction of the ETS, and a weak ETS may mean that more stringent measures are needed in those other sectors.

In Germany, for example, a voluntary agreement had been in place between the government and industry that was meant to cut emissions from sectors now covered by the ETS. Most German industry associations were opposed to this change in instrument, but in terms of the environmental target they seem to have benefited from it. The German government has decided to give out a lot more CO_2 allowances than originally foreseen under the negotiated agreement. This is an indication of a low level of ambition in implementing the ETS in Member States.

In Finland, the peat industry has been lobbying the government to drop other taxation now that the ETS has been put in place, to alleviate the cost from the combined measures. Similarly, the Danish government is considering dropping taxation on CO_2 for companies covered by the ETS. However, Danish companies had previously been able to be almost exempt from the tax if they agreed to a voluntary commitment to improve energy efficiency instead. In Flanders, Belgium, a

---

[283] Directive 2004/101/EC (Official Journal 24.11.2004)

voluntary energy efficiency covenant received overwhelming industry support once the regional government announced this would be used as a basis for allocating allowances to companies. However, there are indications that the introduction of the ETS is being used as a justification for preventing any additional climate legislation for the sectors concerned. The Netherlands, on the other hand, used the arrival of the ETS to consider an energy tax increase on industry outside the trading system, an improvement for the climate.

It is important to highlight that taxation and emissions trading should be designed in such a way that they act as complementary instruments for covering the totality of emissions and should not be traded off against each other (*Sijm and van Dril 2003*)[284].

A potential impact of a CO_2 trading system for large stationary sources on air quality was already mentioned in the theoretical discussion of the concept of ET. In the EU, an existing legislation directly affected by the introduction is the Integrated Pollution Prevention and Control (IPPC) directive from 1996 (*see also chapter* V.3.2.4.4). The IPPC directive requires all installations covered by it to obtain a permit for operation, based on so-called best available techniques (BAT) standards which they need to implement, however in many cases with long lead times. The ETS directive directly brings changes to the IPPC resulting from the control on CO_2 that comes with it for those kinds of installations that are now covered by both the ETS and the IPPC. The ETS directive stipulates that the operating permits required by the IPPC should not contain emission limit values for gases covered by the ETS, "unless it is necessary to ensure that no significant local pollution is caused." (Article 26 Directive 2003/87/EC) This implies that greenhouse gas emissions potentially covered by the ETS in the future (at present only CO_2) could indeed be limited by Member State authorities if local pollution can be expected, regardless of the level of CO_2 allowances which the same installation would hold.

The amendment of the IPPC Directive allows Member States to decide not to impose energy efficiency standards when issuing permits to individual installations. The European Commission did, overall, not see negative impacts arising from these changes to the IPPC.[285] In fact the coverage of the ETS is wider in some categories, where for example combustion plants from a threshold of 20 MW are included (the IPPC sets a limit of at least 50 MW output capacity). Some environmental campaigners are concerned that local hotspots of increased pollution could be created, if Member States do not enforce the application of the IPPC directive properly.

To summarise, the introduction of a major new policy instrument inherently implies that changes to existing measures are likely. In many countries, these present an improvement over the status quo. This should not serve as an excuse for the weakening of other legislation, like relaxing energy taxation.

---

[284] The Interactions between EU Emissions Trading Scheme and Energy Policy Instruments in the Netherlands. ECN-C—03-06. See also "Environmental Fiscal Reform (EFR) and EU emissions trading scheme (EU-ETS): the link with National Allocation Plans (NAPs)", EEB Position Paper, February 2004

[285] The Commission's view is express in a non-paper on the issue available from **http://europa.eu.int/comm/environment/climat/emission/history_en.htm**

V.3.4.4   OUTLOOK AND PROSPECTS FOR ECO ACTION (ENVIRONMEN-
TAL CITIZENS' ORGANISATION)

ECOs have an important role to play in the implementation and further development of the EU Emissions Trading System. This subchapter will describe how environmental groups at all levels can engage in improving environmental effectiveness of the system or stop it from getting worse. There are three major areas for ECO action: 1) Monitoring implementation on the ground; 2) Influencing the targets and rules for 2008-12; 3) Participation in the review process.

Climate Action Network-Europe, which is a federation of European ECOs working on Climate Change, has been following the development of the EU Emission Trading System (ETS) since the first public discussions in 1999. ECOs cautiously welcomed the adoption of the ETS directive in 2003, which laid out the architecture of the system. The environmental impact, however, depends just as critically on the targets that fill the system with life and which are contained in the national allocation plans. Lessons need to be drawn from the NAPs process and from the methodologies employed that can be used to improve the allocation for the second trading period 2008-12.

At the same time, the architecture of the ETS is being reviewed and could face significant changes in the way of expansion. ECOs need to make their own assessment about the possibilities of including other sectors and gases into the scheme, which will become the focus of the discussion in 2005 and 2006. Early participation with an informed and, therefore, strong position will strengthen the voice of ECOs considerably and help bring about an environmentally more meaningful outcome. Climate Action Network Europe (**www.climnet.org**) is trying to coordinate these efforts.

From April 2006, ECOs will also be able to directly monitor the implementation of the system at company level. The information on how much CO_2 any participating installation emitted in a given year will be made public. This can be compared with the level of allowances received by the government, and ECOs can question instances in which there are significant differences between these free emission allowances and the actual pollution, to ensure that companies have not unduly received more credits than they needed to cover their emissions.

ECOs must also engage with the progressive actors at all levels, in international organisations, Ministries and businesses, to help ensure the environmental effectiveness of the EU ETS. In many areas, there is support for more stringent rules, but this support is often thwarted by pressure from other agencies or companies and it needs greater encouragement. A wider informal coalition of progressive actors is essential.

V.3.4.4.1   Monitoring implementation

Checking on how the ETS is being implemented at national and local level is an important field for ECO activity. This can be done by national groups or networks as well as initiatives that want to challenge local plants that are big CO_2 emission sources or that increase local air pollution. There are two main ways of engaging with the ETS and its participants as well as other actors: (1) via emission data that will be made public every year and (2) through being involved individually as a buyer on the emissions market .

1. At the end of March of every year (starting in 2006) all participating installations will have to publish their emissions data for the previous year. This will become public on online registries[286]. Equipped with this knowledge, ECOs can check what their local plants received for free and what they actually emitted. A potential campaign idea for pressuring the polluters could be to identify where significant gaps exist between what companies said they needed and what turned out to be the "emission reality". Especially in the first years, where ECOs suspect that in many cases installations might not need as much as they have been given, major reductions may not be genuine but present cases of them having been successful in convincing authorities that they needed more. This could be communicated to press and to plant owners, demanding clarifications for major changes in projected and actual emissions. The same is true for major increases, since these might entail a rise in local pollution as well, and also a massive additional climate impact. Such issues may well generate interest in the local media and receive some public attention.

   Going further into the details of the system, it is in fact also possible to identify where every single allowance comes from, in which country it was issued and to whom, etc. via unique serial numbers. This is especially interesting for ECOs that want to monitor the use of JI or CDM credits. These credits, that stem from the Kyoto system, also have their own serial numbers, to denote their origins. When they enter the EU ETS, JI/CDM credits are converted into an EU allowance, and all details necessary for project identification are retained. It should, therefore, be possible to trace these credits and to tell when an installation cancels a credit from a bad project to comply with the ETS. Publicising such cases will be important to raising awareness and will oblige companies to think twice about the kind of allowance they buy.

2. Another important potential option for ECOs is for them to become market actors themselves. The ETS allows anyone to open an account in a registry and to start buying and selling allowances. This provision, which was mainly intended for brokers and other financial service providers, can be exploited by ECOs at all levels, either directly or via a broker. The ETS provides ECOs with new opportunities to design schemes for carbon offsets, meaning that in this case allowances would be bought from the ETS market and cancelled, to make up for emissions caused by an ECO office or event (especially travel emissions from cars or planes). If we think about this option on a larger scale, it could involve engaging members of a network in raising funds to buy allowances. Companies could also participate in such a system.

   This kind of offset system has an interesting effect:. When non-industry actors start buying and cancelling allowances, this implies that actual CO_2 emissions from industry are avoided, since the allowances can no longer be used in the system. The participating companies have fewer allowances at hand to cover their emissions, a shorter market means prices rise, and the financial signal to go clean increases. A major buyer could significantly alter the market. On the down-side, this would mean that the money goes to industrial companies, a thought most likely not to the liking of every environmental group. However, they would be paid to reduce their climate emissions. Then again, in contrast to other high-quality offset schemes such as

---

[286] The European Commission provides access to all national registries via this webpage **http://europa.eu.int/comm/environment/ets/**

Atmosfair or myclimate[287], which use credits from projects of Gold Standard quality, a label designed by CDM experts to ensure high-quality credits with real benefits for both the global climate and local sustainability, there is no additional benefit for local communities from a system based on EU allowances. The pros and cons need to be thoroughly evaluated, but in particular from 2008 onwards, when the maximum limit on CO_2 allowances is expected to be less generous, this opportunity will be worthwhile exploring for ECO purposes.

225

V.3.4.4.2   Targets and rules for 2008-12

The lax targets assigned to the industry sectors for the period 2005-7 present one of the greatest shortfalls of the ETS so far. It is of crucial importance for the environmental effectiveness of the system that these targets are more ambitious in round 2, when the cuts count for the Kyoto targets. Lobbying on targets will mainly have to be done at a national level, and it will be important to communicate the importance to national media to increase pressure on policy-makers.

The level of targets will depend on a case-by-case analysis of the national situation regarding the current state of achievement of emission targets, and no general recommendation can be given other than that long-term climate goals demand a steady decrease: emissions need to go down. This is not always easy to argue. Most of the new EU Member States, for example, that have seen significant emission reductions due to economic changes in the 1990s, see no need to demand meaningful cuts from their industry sectors on the basis of complying with Kyoto targets. An important point to make in all EU countries is that wherever past emission levels are being used to decide on future allocations to installations, the year or period of years used in this calculation should not be changed to future years. This could create the perverse incentive that factories might try to inflate their emissions to higher than usual right now, in order to receive more free credits in the future.

Problems with target-setting based on the Kyoto Protocol are another reason why the allocation rules carry a particular weight, since most new Member States do not need to make reductions to meet their targets. Especially in those cases, the decisions on what kind of installations receive what quantities of allowances will influence the environmental impact (see chapter V.3.4.3.2.1). ECOs should lobby for the use of auctioning up to 10% of allowances for 2008-12, and demand product-specific benchmarks to ensure that dirtier technology is being penalised. The rules for new entrants, plants that are yet to be built, are especially important. Again, these discussions and decisions take place at national level.

Member States need to submit their NAPs for 2008-12 to the European Commission by the end of June 2006. This means that the second half of 2005 and early 2006 will be the most critical times for ECO lobbying. Finding out early what governments are planning can help influence the outcome. Member States have been meeting informally since early 2005 to discuss how these rules - all of which were developed by the Member States themselves for 2005-7 - could be harmonised across the EU for the period 2008-12. No major changes to the system can be expected for this second trading period. Smaller changes to criteria for developing NAPs are

---

[287] Atmosfair website http://www.atmosfair.de/ - myclimate website http://www.myclimate.org/

possible through the Climate Change Committee, a body made up by Commission officials and Member State representatives.[288] Member States also have the opportunity to request the unilateral inclusion of additional industry sectors for the period 2008-12. This would require approval by the Committee and the development of guidelines on how to record emissions from these new sectors.

CAN-Europe is aiming to facilitate the lobbying of national groups on the targets and rules for the second period of the ETS. A concrete example of communicating ECO experiences from different Member States will be the production of an ECO guide on NAPs. This report which contains an outline of an ideal NAP should help groups across the EU to engage with government officials. CAN-Europe maintains a network of experts on the ETS from ECOs around Europe and welcomes newcomers to its group.

V.3.4.4.3   The review process

The option to include the aviation sector in the EU ETS has received a great deal of attention in the first half of 2005. The UK had already announced in 2004 that it would like to see this option realised and would champion an amendment to this effect during its EU Presidency in the second half of 2005. Aviation is becoming more and more important for the climate, with its greenhouse gas emissions growing at 3-4% per year in the EU alone. Currently, there is no legislation to reduce or limit the climate impact of aviation, which stems from a combination of its CO_2 and NOx emissions as well as the vapour trails and the formation of cirrus clouds caused by airplanes. There are also other policy options to consider, such as fuel taxes or en-route emission charges. The ECO coalition T&E (European Federation for Transport & Environment) is coordinating an ECO discussion on these issues together with CAN-Europe.

It is difficult to predict how important the review will be for the future of the ETS. It presents a process that requires increased coordination and communication between ECOs in the Member States and representations based in Brussels. CAN-Europe coordinates ECO input and exchanges intelligence on the review with members of its network.

BIBLIOGRAPHY AND FURTHER READING

CAN-Europe, forthcoming (2005) An NGO assessment of the NAPs for 2005-7 — lessons for the future — **www.climnet.org**

Centre for European Policy Studies (CEPS) series of reports on the EU ETS   - **www.ceps.be**

EC (2003) Council Directive of the European Parliament and of the Council of 13 October 2003 establishing a scheme for greenhouse gas emission allowance trading within the Community, 2003/87/EC.

---

[288] This process is well-established in EU policy-making and is known as "commitology". An explanation provided by the European Parliament is provided here **http://www.europarl.eu.int/igc1996/fiches/fiche21_en.htm**

EEB Position Paper (2004): Environmental Fiscal Reform (EFR) and EU emissions trading scheme (EU-ETS): the link with National Allocation Plans (NAPs)

IETA (2005) Carbon Market 2004 – **www.ieta.org**

IPCC (2001) Third Assessment Report – www.ipcc.ch

JORGENSEN, C.E. (2003): Environmental Fiscal Reform: Perspectives for Progress in the European Union, EEB, June 2003

SIJM J.P.M. and VAN DRIL, A.W.N. (2003): The Interactions between EU Emissions Trading Scheme and Energy Policy Instruments in the Netherlands. ECN-C—03-06

UNICE (1998) *UNICE's Implementation of IPPC Working Group Opinion on the Implementation of the IPPC Directive*, 26 January 998

V.3.5   Environmental Impact Assessment

V.3.5.1   THE EIA DIRECTIVE

By Thisvi Ekmetzoglou-Newson[289]

V.3.5.1.1   Introduction and key elements

In 1985, the Council of European Communities adopted a Directive on environmental impact assessment for private and public projects ("EIA Directive") (*EC 1985*). This Directive was subsequently amended in 1997 (*EC 1997*) and in 2003 (*EC 2003*). The EIA Directive is based on the principle that the best environmental policy consists of preventing the creation of pollution at the source, rather than trying to minimise or mitigate its effects later (principle of precaution). The EIA Directive has been the pioneer in EU legislation as regards access to information and public participation and inspired subsequent legislation both at the EU and international level (Aarhus Convention – *see chapter V.2* and relevant community legislation). The operation of its "democratic" procedures has helped to improve the acceptance by society for certain projects.

Environmental impact assessment (EIA) is a tool to implement a procedure to evaluate environmental effects of certain public and private projects.[290] The term "environmental impact assessment" describes a procedure to be followed during the authorisation process of projects likely to have significant effects on the environment, by virtue, inter alia, of their nature, size and location. The procedure is a means of drawing together, in a systematic way, an assessment of a project's likely significant environmental effects and identifies the direct and indirect environmental effects on the activities of the following factors: human beings, fauna, flora, soil, water, air, climate, landscape, material assets, cultural heritage and the interaction between those factors. This helps to ensure that the importance of the predicted effects, and the scope for reducing them, are properly understood by the public and the relevant competent authority before the latter makes its decision. Projects subject to EIA are enumerated in Annex I and Annex II of the Directive. These annexes include projects from the industrial sector but refer to infrastructure projects and agricultural projects, too.

The first EIA Directive had to be transposed into national law by 1988. Almost all Member States were late in transposing the Directive and some provisions of the Directive were still open to discussion. The 5 year report on the implementation of the Directive 85/337/EEC *(CEC 1993)*[291] triggered a discussion on new issues and problems revealed. An amended Directive *(97/11/EC)* has been adopted by the Council setting a new deadline for transposition into national legislation: 14th March 1999.

---

[289] Legal and Policy Advisor to a Member of the European Parliament

[290] Article 1 of the Directive defines projects as
- the execution of construction works or of other installations or schemes,
- other interventions in the natural surroundings and landscape including those involving the extraction of mineral resources.

[291] The Directive provided for a report on its application and effectiveness five years after notification. In 1997 an update of this five year report has been prepared with additional information on EIA legislation in the new Member States Austria, Finland and Sweden (European Commission: Concise revision of the report from the Commission of the implementation of Directive 85/337/EEC on the assessment of the effects of certain public and private projects on the environment - Update 1995/96. Wagner, Dieter. Cologne, July 1997.

The amended EIA Directive enlarged the scope of application of environmental assessment and clarified the structure of Annex I (number of projects increased, with 14 new project types and extension of 4 others) and Annex II (the number of projects increased by 8 and others were extended, one project type was deleted). Projects listed in Annex I must always undergo an environmental impact assessment. For other projects listed in Annex II the Member States determine - on the basis of a process described in the Directive - whether an environmental impact assessment must be carried out because of their likely significant effects on the environment. This evaluation or pre-selection, called "screening" is done either by using thresholds or by a case by case examination or a combination of the two, while respecting the criteria found in Annex III. Some Member States define these Annex II-projects within their legislation, while other Member States choose a screening procedure to determine these projects and some provide for inclusion, exclusion or guiding thresholds. Very few members employ a case by case approach for all project types. Modifications or extensions to development projects included in Annex I are subject to an EIA, too. The screening procedure has been introduced in the Member States gradually but has gained importance within their authorisation procedures. The Commission has provided support for this approach by publishing guidance on screening (CEC 1996 and 2001). In addition, the European Court of Justice has clarified some elements of screening.[292]

After determining, whether a project needs an environmental impact assessment, the scope of the study has to be determined (CEC 1996 and 2001). The scoping procedure is non-obligatory. Today seven Member States (from the old 15) have a mandatory scoping procedure in place. There is also recognition in some Member States that public involvement at the scoping stage identifies the issues that are 'significant' to the people who will have to live with the project and not just the 'experts'.

Overall, the EIA Directive 97/11/EC introduced provisions that clarified, supplemented and improved the rules on the assessment procedure, in order to ensure that the Directive is applied in an increasingly harmonised and efficient manner.

After the screening and scoping processes, where applicable, the developer supplies the necessary information on the project. This comprises a description of the project, a description of measures to mitigate adverse effects, an outline of the main alternatives studied, data to identify and assess the main effects of the project on the environment and a non-technical summary. This information may be compiled in a report, sometimes called environmental impact study (EIS). Authorities likely to be concerned by the project by reason of their specific environmental responsibilities have to be consulted and the public has to be given a chance to comment on the EIS. If there are likely to be significant effects on the environment of another Member State, there has to be transboundary consultation. According to the amended Directive and the Espoo-Convention (UNECE 1991), transboundary consultation includes the affected public in the neighbouring countries, too.

---

[292] Commission v Belgium and the Dutch Dykes cases, Member States could not use thresholds to exclude whole classes of projects, but thresholds could be used to exclude very small or minor projects. In C-392/96 Commission v Ireland the ECJ ruled that thresholds could not be based on the size or other physical characteristics of a project alone and that they had to be based upon locational and other environmental factors.

The competent authority has to consider all the information gathered during the public consultation procedures and take it into account in the development consent procedure. Whether the results of the EIA are binding for the development consent depends on the various national systems.

The Directive can be transposed by integrating EIA into existing procedures for consent within the Member States. Almost all Member States choose to do so.

Finally, the amended Directive 2003/35/EC, that was adopted in order to align with the provision of the UNECE Aarhus Convention introduces more detailed provisions as regards the public consultation process as well as the notion of the "public concerned" where ECOs are explicitly mentioned. In addition, there is a new article that deals with the issue of access to justice.

V.3.5.1.2 Positive effects for the environment

There have been several studies on the effects of the environmental impact assessment procedure and the need for further research (*CEC 1996a, 1996b, 1996c, 1997*). Additionally, within the updated five-year-reports on Directives 85/337/EEC and 97/11/EC Member States commented on certain aspects of the EIA procedure. According to the comments received and the research results, projects have been modified and decisions been influenced by EIA, because the decision making authorities had to assess the results of the EIA and take them into consideration. It appears that the EIA process is having a notable effect on the number of project modifications. Improvements are not always as obvious because elements of a specific project are changed even before they are presented for public consultation. At least the planning process has been improved.

The principal benefits of an EIA are:

 ▶ the identification of key environmental issues,

 ▶ improvement of project design,

 ▶ higher standards of mitigation,

 ▶ better decision-making.

Additionally there have been cases where a well structured and managed EIA process led to savings in cost and time. One example where savings on costs were achieved is the extension of the Billund airport, in Denmark. The EIA contributed to extending the capacity of the airport and at the same time reducing the number of homes exposed to noise from 1,290 to 328, while at the same time concluding that an extension of the airport with a new runway was not necessary. It saved 300 million Kroner (€40.4 million) — the cost of the planned new runway - and approximately 350 hectares of farm land, as well as preserving an ancient Danish forest. This example shows that although even today some developers regard EIA as a bureaucratic obstacle, used constructively it can be an efficient way of planning which often turns out to be more cost effective.

It could also be claimed today that one additional success of EIA is its contribution to safeguarding the money of the European tax-payer. Due to the constant monitoring of the application of EIA there have been cases where funding has been blocked due to bad application or incomplete transposition of the EIA Directive. This has led the beneficiary Member States to take the necessary measures to comply with the EIA Directive. Special attention was also given to the pre-accession funds provided during enlargement, thereby contributing to integrating EIA into the decision making process of accession countries.

Finally, the operation of EIAs has gradually formed a new generation of decision-makers. Both national authorities and developers are better able to integrate environmental consideration into their everyday planning.

V.3.5.1.3   Weaknesses and constraints

The EIA Directive is only about procedures, it does not set new environmental standards or provides for new and more stringent material requirements. Therefore the results of the EIA procedure are very much dependent on the transposition into national law and the already existing national standards and environmental requirements but also on the implementation by the relevant authorities. EIA did not - as sometimes was wrongly expected - change environmental policy and it is certainly not a substitute for it. The integration of EIA into already existing procedures makes it sometimes very difficult for outsiders to gain insight into the process and find out the essential steps, where pressure can be applied. Sometimes the increase in cost and time is criticised. One has to consider that delays can occur for many reasons not related to the EIA process itself, such as modifications proposed by the promoter of the project or technical or economic reasons. The main reasons for delay are usually a lack of proper scoping and a failure on the part of the developer or the consultant to undertake a systematic study and resulting from this the need for supplementary information.

The five year report on the application of EIA Directive 97/11/EC (*CEC 2003*) revealed some important deficiencies in the implementation of the Directive by Member States. It appeared that the main problem lies with the application and implementation of the Directive and not, for the most part, with the transposition of the legal requirements of the Directive[293].

---

[293] The five year Report comprises a list of actions that the Commission undertook to carry out: Firstly continue the on- going enforcement of the Directive. Secondly to initiate research as to the way that thresholds are set out as well as the operation of screening. Thirdly to prepare an interpretative and practical guidance with the participation of Member States, the new Member States and other stakeholders such as local and regional authorities, ECOs and industry representatives. More guidance might help to overcome weaknesses. Lastly but not least to consider the initiation of capacity building comprising training programmes tailored made for each particular Member State.

Some of those deficiencies are:

1) **Unsystematic screening of Annex II projects, inadequate setting of thresholds and criteria and of their application**[294]. The report revealed that while many Member States have set thresholds for the same project types, there are very large differences in the levels at which thresholds have been set[295]. Some Member States have made EIA mandatory for some project types regardless of size. This means that a project will be subject to EIA on a mandatory basis in one Member State while the same project type, of the same size, will require EIA only after case-by-case screening in another Member State.

2) **Insufficient consideration of the cumulative effects of projects and use of "salami slicing"**[296]. There seems to be growing awareness of the issues raised by the requirement to assess the cumulation of impacts, and measures have been put in place in many Member States to address this. The Commission has also published guidelines for the assessment of indirect and cumulative impacts (*CEC 1999*). The issue of possible salami-slicing is recognised by the Member States and some States have established measures to reveal and prevent such practice, including setting low thresholds or calling for assessment of "the whole programme" where this is appropriate. However, in day to day application, those two issues need special attention.

3) **Alternatives.** One of the main criticisms of the EIA Directive 85/337/EEC was that it did not require, in a more formalised way, developers or competent authorities to examine alternatives, where they exist, from the outset of the development of EIA. The element of taking into account the alternatives and the obligation on the developer to supply to the competent authority an outline of the main alternatives "studied" was inserted in the amended EIA Directive 97/11/EC and is in line with the development of Community environmental legislation[297] In some Member States, the consideration of alternatives is a central focus of the EIA process; elsewhere the consideration of alternatives appears to be less complete than it might be. Most Members States require assessment of the zero alternative and other project alternatives, which may include options for location, process, design, etc. A variety of institutions, and sometimes the public, may contribute to the selection of alternatives for assessment, and these may include the most "environmentally friendly" alternative. However, since the Directive applies to projects only, in most cases the discussion of alternatives - either a differ-

---

[294] See also the jurisprudence of the ECJ. Case C-392/96 Commission versus Ireland: Member States are not allowed to exempt whole classes of projects from the requirement for EIA, nor to avoid taking into account when establishing thresholds only of the size of projects but also their nature and location (Annex II 1.b projects for the use of uncultivated land or semi-natural areas for intensive agricultural purposes).It has also ruled that the setting of thresholds or criteria for Annex II projects could not be set at such a high level that the objectives of the Directive would be circumvented by the splitting of projects into smaller units and that the cumulative effects of such an approach would need to be assessed. Case C-435/97 ITA/WWF (Bolzano/Italy): No general exemptions of entire class of Annex II projects are permitted. Member States are not allowed to se thresholds/criteria that practically exclude classes of projects from EIA.

[295] Examples include thresholds used for wind farms: some Member States have used a basis the capacity of turbines, other the number of turbines and others monetary thresholds (cost of the project).

[296] Salami-slicing includes the practice of dividing projects up into two or more separate entities so that each individual element does not require an EIA and therefore the project as a whole is not assessed. It also refers to the perceived practice of a developer obtaining permission for a project that is below a threshold, and therefore not subject to EIA, and at a later date extending that project or its capacity above the threshold limits.

[297] In particular with the Habitats 92/43/EEC and Birds 79/409/EEC Directives.

ent location or a different production method - comes too late at this stage. The important decisions are made at the policy, plan or programme level. The new SEA Directive that deals with the environmental assessment of plans and programmes will be valuable in this direction (*see chapter. V.3.5.2*).

4) **Annex I and II project types.** While most Member States have experienced difficulties with the interpretation of Annex II 10 (a) (industrial estate development projects) and Annex II 10 (b) (urban development projects), the approach taken by the Member States on how to tackle this issue varies. It seems that most prefer that such definition be left to Member States, so that they can take their own economic and social aspects to take into consideration. Other types of projects that can create problems with interpretation include animal carcass disposal sites and means of disposal.

5) **Poor quality control systems for the EIA process (voluntary).** Directive 97/11/EC introduced new minimum requirements for the information to be supplied by the developer. Failure to provide adequate information constitutes grounds for refusal of development consent in the majority of countries, under a variety of arrangements. Some Member States have formalised a review procedure to ensure that the environmental information supplied to the competent authority is in compliance with the Directive. In some cases an independent specialist review commission or panel of experts carries out the review of the EIS. However, since there is no explicit obligation in the Directive to provide for such a review as well as a general quality control provision[298] there is no harmonised approach to the matter. It is worth noticing that research conducted in about half of all Member States on the quality of information and on the overall quality of the assessments showed that up to 50% of EIS did not fully meet the requirements of the Directive. Nevertheless, checking the quality of EIA process is provided via national judicial review processes. Also judicial review is further strengthened in the amended Directive 2003/35/EC.

6) **The inadequate incorporation of EIA results in development decisions and monitoring.** Without formal monitoring of the outcomes of the EIA process and more detailed research, it is difficult to assess the effectiveness of the EIA Directive on decision making. Nevertheless, it is apparent from the five year report on the application of the EIA Directive 97/11/EC that the environmental considerations raised by the EIA process are balanced against other societal and economic considerations in decision making. The lack of central monitoring of the key stages of EIA make it difficult for Member States to ensure that their EIA systems are consistently and correctly applied.

V.3.5.1.4   Opportunities for action

Environmental impact assessment has not changed environmental policy, but it is a very valuable, internationally accepted tool, for integrating environmental concerns into decision making. The EIA Directive enables decision makers and developers to protect the environment by finding alternative solutions, minimising impacts and taking appropriate mitigation measures. It should therefore be used and strengthened and ECOs should familiarise themselves with the basic

233

---

[298] The EIA Directive does not prescribe the manner in which competent authorities carry out their screening duties, how the scoping process should operate, how the assessments should be completed, at what level of detail or how their outcome should be reported.

requirements and national procedures. This is particularly true for ECOs in new Member States. The intervention of ECOs and citizens has been growing throughout the years of EIA implementation and the high number of complaints submitted to the European Commission would indicate that civil society shows great interest in seeing that the Directive is effectively applied in practice.

At the European level the adoption of the Strategic Environmental Assessment (SEA) Directive could solve some of the deficiencies in the implementation of the current EIA Directive, particularly in the case of alternatives and monitoring of plans and programmes.

At national level, ECOs should concentrate not only on correct transposition but also most importantly on effective implementation of the 97/11/EC, as well as the newly adopted 2003/35/EC, into national law. Screening procedures have to be further enhanced, in particular the issues of 'salami slicing' and cumulation with other projects. Special attention should be given to thresholds and project selection. Rather than preaching to the converted, ECOs should form coalitions with other groups - including progressive groups from the business sector - to strengthen EIA. The advantages of EIA as a management tool for consent procedures should be emphasised.

At the project level, ECOs should use their rights within the scoping and participation procedures. Sometimes these are rather complicated and difficult issues and therefore early training is of importance. National EIA centres that collect and document environmental impact studies can play an important role. The amended Directive 2003/34/EC, which provides for access to justice, will further enhance the right to challenge such decisions on development projects. The last resort will always be to go to court or to file a formal complaint with the European Commission (see chapter VI). Because the Directive only establishes the procedure to be followed and is less concerned with substance, the success of complaints to the Commission is rather limited. Legal developments in this field can be followed by checking the annual report on monitoring the application of Community law.[299] The responsibility for EIA within the European Commission rests with the Directorate General for Environment (Directorate D - Water and Environmental Programmes, D.3 Cohesion Policy and Environmental Impact Assessments). The Commission has a homepage especially for EIA, including links to the national EIA centres (http://europa.eu.int/comm/environment/eia/home.htm).

---

[299] Official Journal of the European Communities - Information and Notices.

Figure 2. Environmental Impact Assessment (Flow Chart)

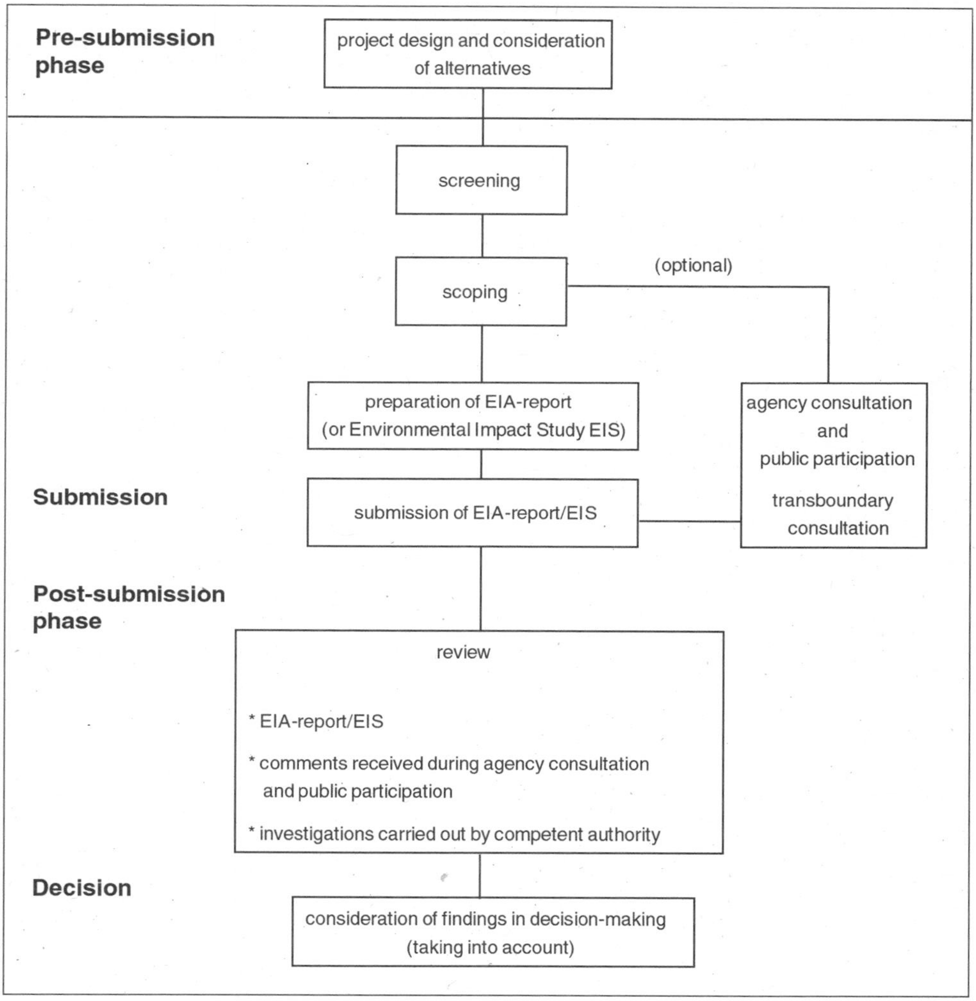

BIBLIOGRAPHY AND FURTHER READING

CEC (1993) Report from the Commission of the Implementation of Directive 85/337/EEC on the assessment of the effects of certain public and private projects on the environment. COM (93) 28 final. Brussels, April 1993.

CEC (1996 and 2001) Environmental Impact Assessment - Guidance on Screening, Guidance on Scoping, Guidance on EIA Review. European Commission, Directorate General for Environment, Nuclear Safety and Civil Protection, Brussels, 1996 and 2001.

CEC (1996a) European Commission: Evaluation of the Performance of EIA process. Final Report. Volume 1: Main Report. Wood Ch., Manchester October 1996.

CEC (1996b) European Commission: Environmental Impact Assessment in Europe. A Study on Costs and Benefits. December 1996. Land Use Consultants in association with Eureco, Luxembourg, and Enviplan, Athens.

CEC (1996c) European Commission: An Analysis of Environmental Impact Studies of Installations for the Treatment and Disposal of Toxic and Dangerous Waste in the EU (Ispra Study on Projects under Directive 85/337/EEC, Annex I.9). Brussels, Luxembourg, August 1996. Colombo, A. G. et. al. Institute for Systems, Informatics and Safety.

CEC (1997) European Commission: A Study to Develop and Implement an Overall Strategy for EIA/SEA Research in the EU. April 1997. Colombo, G.; Haq, G.; Melaki, I. European Commission - Joint Research Centre Institute for Systems, Informatics and Safety.

CEC (1999) European Commission: Guidelines for the Assessment of Indirect and Cumulative Impacts as well as Impact Interactions., May 1999, Hyder.

CEC (2003) Report from the Commission to the European Parliament and the Council On the Application and Effectiveness of the EIA Directive (Directive 85/337/EEC as amended by Directive 97/11/EC) - How successful are the Member States in implementing the EIA Directive/* COM/2003/0334 final */

EC (1985) Council Directive of 27 June 1985 on the assessment of the effects of certain public and private projects on the environment (85/337/EEC). Official Journal of the European Communities, No L 175/40, 05.07.1985.

EC (1997) Council Directive 97/11/EC of 3 March 1997 amending Directive 85/337/EEC on the assessment of the effects of certain public and private projects on the environment. Official Journal No. L 073. 14/03/1997. p. 0005.

EC (2003) Directive 2003/35/EC of the European Parliament and of the Council of 26 May 2003 providing for public participation in respect of the drawing up of certain plans and programmes relating to the environment and amending with regard to public participation and access to justice Council Directives 85/337/EEC and 96/61/EC. L 156/17, article 3. Deadline for transposition was 25th June 2005.

UNECE (1991) Convention on Environmental Impact Assessment in a Transboundary Context, Espoo, 1991

V.3.5.2   THE SEA DIRECTIVE

By Frederik Hoedeman[300]

V.3.5.2.1   Introduction

237

The Strategic Environmental Impact Assessment Directive 2001/42/EC (*SEA Directive*) from the year 2001 is a powerful tool for greening policy areas like spatial planning, urban development policies, transport policies and industrial policies. As such, Strategic Environmental Impact Assessment Directive is the extension of the previous Environmental Impact Assessment (EIA) Directive from 1985, from the individual project to the plans and programmes level.

Like the EIA Directive, one important objective of the SEA Directive is to create the basis for better decision-making and integration of environmental considerations. However, the big difference is that SEAs only apply to public programmes and programmes and consequently are much more directly linked to political decision-making. Hence, SEAs can make political decision makers more directly accountable for decisions which have a negative effect on the environment. The political discussions of whether politicians "dare" to approve a plan with significant environmental consequences are thus much more qualified.

The overall aim of the SEA Directive is to "provide for a high level of protection of the environment", by integrating a wide range of environmental considerations into spatial planning. . The scope of the SEA Directive is potentially very broad as it will require SEAs of all plans and programmes produced in eight different sectors and land use planning. Member States must also pay special attention to environmental problems in relation to other EU environmental laws including Air, Nature Protection, Waste and Water. Most other plans and programmes with significant environmental effects will also require a strategic environmental impact assessment.

A SEA is a process requirement which should enable different government authorities to better coordinate their planning and citizens to have a say in this, on the basis of better environmental information and a systematic assessment of alternative options. Whereas EIAs of projects were not effective in assessing cumulative effects like urban sprawl, the effects of ongoing intensification of agriculture or fragmentation of nature sites, SEAs of much larger plans or programmes have the potential to capture a clearer picture of their cumulative effects.

However, in the end, changes to the plans or programmes to take account of environmental impacts is left to political decision makers. This means that in practice and in legal terms, the SEA only starts "biting" when brought together with the wealth of existing EU environmental objectives and requirements. The Environmental Report provides an excellent access point for checking and ensuring such cross-compliance.

In order for these benefits to occur in reality, it will be important that ECOs check the quality of the national transposition of the SEA Directive to ensure that no blanket exemptions of plans or programmes takes place, all assessment levels are included and the public receives a clear framework and support for its participation. Furthermore, ECOs must be aware of the specific protection objectives and requirements in existing EU environmental legislation, including air, nature,

---

[300] Nature Protection Officer, Danish Society for Nature Conservation.

waste and water legislation (*see chapter IV*). A SEA can provide them with important access and a control point to ask for full compliance with those objectives and requirements.

So far there has been little experience in applying SEAs. In most EU Member States, the Directive has just been transposed, but only a few countries have started to apply the national SEA legislation in their daily administration of plans and programmes.

This overview of Strategic Environmental Impact Assessment intends to give an idea of what SEA is about and how SEA can specifically contribute to biodiversity protection and assessing climate change.

V.3.5.2.2 Objective and Tools

The main objective of the directive is enhancement of sustainable development, by integrating environmental protection issues into spatial planning and thus moving towards achieving the integration requirement of Article 6 of the EU Treaty.

The SEA requires assessment of significant environmental effects of a plan or programme, and addressing those during its preparation. Furthermore, the SEA requests the identification of strategic alternatives and their effects before final decisions or choices are made. It also gives the public quite substantial possibilities for participating in the SEA and in the preparation of the plan or programme. One important requirement is that the significant actual effects during the implementation of the plan or programme need to be monitored.

The scope of the SEA Directive (article 3.1) is very broad as SEAs must be carried out for most land use plans and programmes, including the following eight different sectors : forestry, fisheries, energy, industry, transport, waste management, water management, telecommunications, tourism, town and country planning or land use. In the case of a plan or programme having likely effects on Natura 2000 sites an SEA is also required. Other plans and programmes with significant environmental effects also need an SEA (*see Figure 3*).

Figure 3: Plans and Programmes covered by the SEA Directive

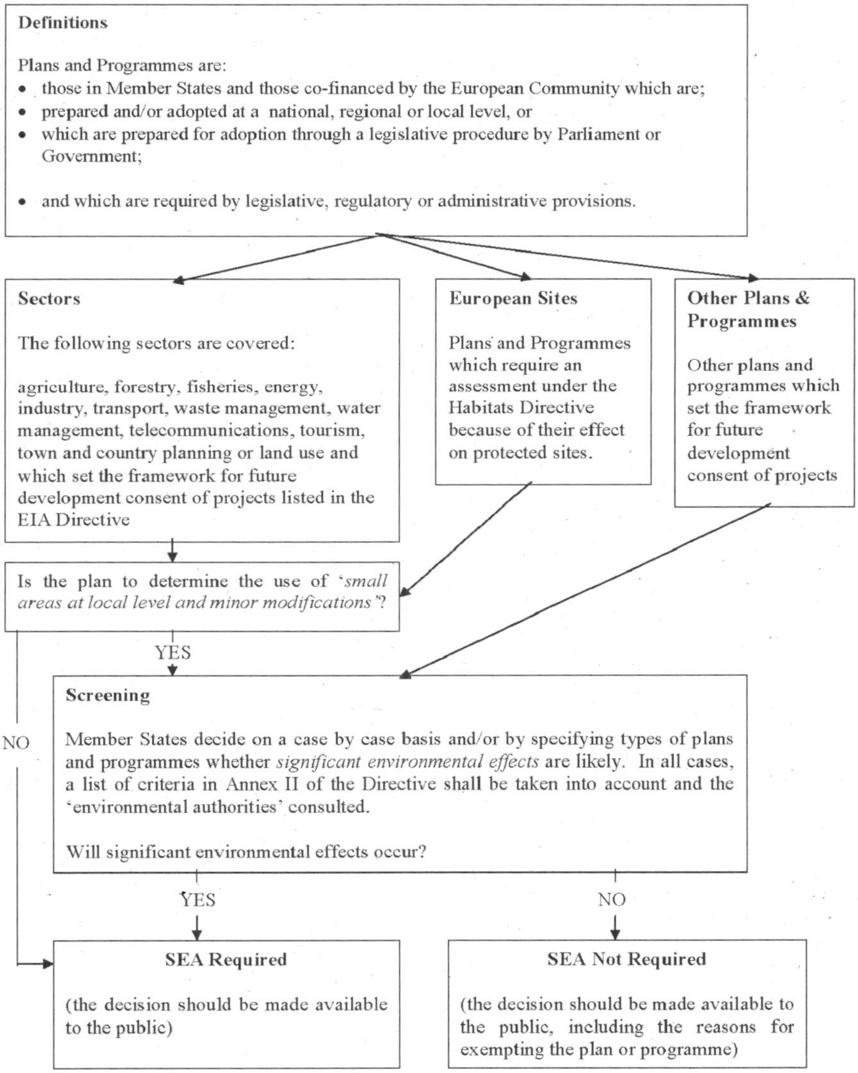

Definitions

Plans and Programmes are:
- those in Member States and those co-financed by the European Community which are;
- prepared and/or adopted at a national, regional or local level, or
- which are prepared for adoption through a legislative procedure by Parliament or Government;

- and which are required by legislative, regulatory or administrative provisions.

Sectors

The following sectors are covered:

agriculture, forestry, fisheries, energy, industry, transport, waste management, water management, telecommunications, tourism, town and country planning or land use and which set the framework for future development consent of projects listed in the EIA Directive

European Sites

Plans and Programmes which require an assessment under the Habitats Directive because of their effect on protected sites.

Other Plans & Programmes

Other plans and programmes which set the framework for future development consent of projects

---

Is the plan to determine the use of '*small areas at local level and minor modifications*'?

YES

NO

Screening

Member States decide on a case by case basis and/or by specifying types of plans and programmes whether *significant environmental effects* are likely. In all cases, a list of criteria in Annex II of the Directive shall be taken into account and the 'environmental authorities' consulted.

Will significant environmental effects occur?

YES

NO

SEA Required

(the decision should be made available to the public)

SEA Not Required

(the decision should be made available to the public, including the reasons for exempting the plan or programme)

The SEA applies to most plans and programmes that started after 21st July 2004 or before this date, but to be completed after 21st July 2006.

240

V.3.5.2.2.1   Public consultation, the environmental report and monitoring

A SEA requires consultation at several stages and can give ECOs new opportunities for greening state plans and programmes. In most EU Member States, the decision not to undertake a SEA of a plan or a programme in the screening phase must be made public, and can be challenged in a administrative court or in the civil court system. When preparing the Environmental Report, the plan-making authority must consult other authorities, which, by reason of their environmental responsibilities, are likely to be concerned by the environmental effects. Apart from that, the planning authority must "consult the public affected or likely to be affected or having an interest", and the authorities with environmental responsibilities, in the "draft plan" and the Environmental Report. Other Member States must also be consulted if their environment is likely to be significantly affected by the plan, but this is likely to be relatively rare for plans covered.

When a plan or a programme is adopted, the Member State must ensure that the following information is made available to the public:

▶ The adopted plan;

▶ How the environmental considerations were integrated;

▶ How the environmental report and the results of consultation were taken into account;

▶ The reasons for choosing the plan or the programme in the light of other reasonable alternatives; and

▶ The arrangements for monitoring environmental effects.

The SEA Directive leaves it up to Member States to define what the "public" is. It is possible to do this case by case or by developing national guidelines. The Aarhus Convention (see chapter V.2) will also be a relevant tool for the public in gaining access to information, participation and justice, since it broadens the requirements of most Member States in matters of public consultation.

The Environmental Report

In cases where an environmental assessment is required, an environmental report shall be prepared, which describes the likely significant effects on the environment of implementing the plan or programme as well as reasonable alternatives. Annex 2 of the Directive sets out criteria for determining the likely significance of the effects, which include the relevance of the plan or programme for existing EU environmental policies and objectives, i.e. Air, Nature, Waste and Water legislation – see chapter IV. Reasonable alternatives taking into account the geographical scope of the plan or programme must be identified, described and evaluated. A long list of information is required in the Environmental Report, and the term "environment" has to be understood in a very broad sense:

a) an outline of the contents, main objectives of the plan or programme and relationship with other relevant plans and programmes

b) the relevant aspects of the current state of the environment and the likely evolution thereof without the implementation of the plan or programme

c) the environmental characteristics of areas likely to be significantly affected

d) any existing environmental problems which are relevant to the plan or programme including in particular any areas of particular environmental importance, such as areas protected pursuant to the Habitats Directive and Birds Directive

e) the environmental protection objectives, established at international, Community or Member State level, which are relevant to the plan or programme and the way those objectives and any environmental considerations have been taken into account during its preparation.

f) The likely significant effects on the environment, including issues such as biodiversity, population, human health, fauna, flora, soil, water, air, climatic factors, material assets, cultural heritage including architectural and archaeological heritage, landscape and the interrelationship between the above factors.

g) The measures envisaged to prevent, reduce and as fully as possible offset any significant adverse effects on the environment of implementing the plan or programme

h) An outline of the reasons for selecting the alternatives dealt with and the description of how the assessment was undertaken including the difficulties encountered in compiling the required information.

i) A description of the measures envisaged concerning monitoring

j) A non-technical summary of the information provided

The Environmental Report must be written simultaneously with developing the plan or the project to ensure that the results of the report are also used in the planning process. SEA should start from the first planning ideas, not with an already drafted plan or project. This integration, into one common process, fosters the integration of environmental impacts into the final project or plan.

Monitoring

The monitoring of environmental effects is an important and interesting strength of the SEA Directive. According to Article 10 of the SEA Directive, "Member States shall monitor the significant environmental effects of the implementation of plans and programmes in order, inter alia, to identify at an early stage unforeseen adverse effects, and to be able to undertake appropriate remedial action". The SEA directive does not, however, specify what the term "monitoring" means.

A key objective of monitoring is to check whether the described significant environmental effects in the Environmental Report also occur over time, in order to identify unforeseen environmental impacts. Furthermore, the monitoring should enable the planning authorities to initiate mitigation in cases where monitoring identifies unexpected environmental impacts. Nevertheless, such action is not specifically required in the Directive.

The amount of detail of environmental information needed to set up the monitoring programme depends also on the likely significant environmental effects of the plan or the programme. In case a plan or a project is changed as a result of an unexpected impact on the environment, the new plan may require a new SEA.

242

V.3.5.2.2.2   The SEA Directive and Biodiversity

The SEA Directive is intended to achieve a high level of environmental protection and is identified in key international agreements as an important tool for promoting the conservation and sustainable use of biodiversity. There are two key principles for biodiversity protection, which should be considered:

▶ The precautionary principle implies a presumption in favour of biodiversity protection where the knowledge required ensuring effective mitigation or compensation for a significant adverse impact is lacking. The principle guides decision-making in order to be able to respond and prevent situations of potentially irreversible or long-term negative effects. Therefore the availability of alternatives plays a prominent role, whereby alternatives must be understood in a very wide sense of providing a service rather than technological options.

▶ The "no net loss principle"[301] requires a status quo to be maintained in terms of quantitative and qualitative aspects of biodiversity. Member states have agreed that further losses of biodiversity must be halted by 2010.

SEA can become a good tool for protecting and enhancing biodiversity because it may build biodiversity objectives into the development of plans and programmes. SEA is also a suitable method for considering the full range of threats to biodiversity in an area, enabling assessments of the cumulative effects. Furthermore, the SEA Directive implies the suggestion of effective mitigation strategies to ensure no net-loss of biodiversity as a course of the development and implementation of plans. Finally, monitoring programmes can provide and coordinate biodiversity data and enable remedial measures to be taken.

In particular, SEA should follow the "positive planning" approach, e.g. avoiding biodiversity loss and damage in the first place and mitigating only where impacts cannot be avoided and there are no alternative solutions. Biodiversity should be enhanced where possible. This includes consolidation of existing designated sites, enhanced connectivity between biodiversity hotspots and the compensation of unavoidable biodiversity loss..

The SEA directive is therefore a potentially significant tool to support the achievement of the Natura 2000 network and Water Framework Directive objectives. All plans and projects, which may have a significant environmental effect on Natura 2000 sites and other areas of environmental importance, including wetlands, rivers, makes and coastal areas require an environmental assessment.

V.3.5.2.2.3   Strategic Environmental Assessment and Climate Change

The SEA Directive requires authorities to assess the likely significant effects of their plans and programmes on "the environment, including on issues such as biodiversity...climatic factors...and the interrelationship between the above factors". To deal efficiently with climate change, greenhouse gas emissions need to be reduced sharply and EU Member States need to start adapting to climate change. Assessing climate change in SEA is very different to assessing other effects. Climate change is one of the most significant and complex cumulative effects. It is due to the accumula-

---

[301] As stated in: "Strategic Environmental Assessment and Biodiversity: Guidance for Practitioners" by the Countryside Council for Wales, English Nature, Environment Agency and the Royal Society for the Protection of Birds, UK, June 2004

tion of many actions, each of which has only a limited impact but all of which together cause serious effects. Examples of adaptation measures to reduce the impact of climate change are

▶ planning for land use change

▶ providing wildlife corridors

▶ type and location of infrastructure

▶ type and level of flood defences

▶ designing buildings and urban areas to cope with new climate extremes.

Existing climatic data and models have highlighted general changes to the global climate. However, uncertainties remain over the rate and severity of change at the regional and local level as well as the interrelationship with natural climatic factors. The challenge facing landscape planning and nature protection planning in the future is therefore less about dealing with specific impacts and more about developing strategies to manage uncertainties created by climate change. Above all, the management of climate change involves adaptive management that focuses on transparency and mutual learning between sectors.

Used effectively, the SEA Directive can initiate such processes towards both reducing greenhouse gasses and adapting to climate change.

V.3.5.2.3   Transposition and Implementation

All EU governments should have transposed the directive by 21 July 2004. Nevertheless, The European Commission found that by February 2005 only 11[302] out of the 25 Countries had transposed the SEA into national law, compared with nine by July 2004. Application of the SEA requirements and progress on the ground is very difficult to measure, but appears to be running ahead of legal transposition.

Likewise, the number of SEAs being carried out is gradually increasing, though reliable figures are non-existent. Quite a number of SEAs are being performed in countries like Belgium and Malta that have yet to transpose the directive. In the Netherlands - another country still to transpose the directive - the first SEA was announced January 2005, covering the province of Gelderland's 2005 regional plan. In the UK, there was already some experience of SEA a number of years prior to the directive. Five SEAs were carried out for oil and gas exploration and more recently wind farms were also included. Since the Directive has been implemented, much of the SEA work has centred on development plans.

In Portugal, the transposition proposal excludes urbanisation plans and 'detail' plans. This exclusion seems not to be in compliance with the directive. In fact, art. 3.2 refers specifically to urban planning. The exclusion of urban planning is jeopardising the SEA's objectives, as inappropriate land management and excessive urbanisation are the cause of several of Portugal's environmental problems[303].

---

[302] Czech Republic, Denmark, Germany, Estonia, Ireland, Latvia, Lithuania, Hungary, Poland, Slovenia and the UK

[303] The problem in Portugal is that, in fact, Portugal has no real urban planning, so it is even more difficult to control construction, especially outside protected areas.

The Scottish Executive has decided that it wants to be "the world leader in SEA". It has drafted a SEA Bill and if passed in its present form it would mean that all public sector strategies, plans and programmes would need SEA. Scotland will also go far in ensuring quality control. The intention is to set up a body, probably within the Executive, to advise responsible authorities on meeting SEA requirements and co-ordinate the administrative requirements.

In Denmark, one of the first Member States to transpose the directive, the level of ambition, however, seems to be quite limited. The Danish Ministry of the Environment has not so far published any guidelines for municipalities and counties on how to administrate the new legislation. Since the national legislation on SEA came into practice in July 2004, there have been no decisions about non-requirement of SEA for particular plans or programmes and only one Environmental Report has been sent out to a public hearing. This report did not fulfil the minimum requirements of the Danish SEA legislation or the SEA Directive and the Danish Society for the Conservation of Nature has complained about this to the Higher Nature Complaint Board.

V.3.5.2.4   Strength and Weaknesses of the SEA Directive

The SEA Directive is a procedural Directive which requires changes in administrative behaviour and working style. While it systematically requires the integrative assessment of environmental effects of plans and programmes involving different authorities and requires public hearings, it is still a political decision as to whether or not the assessment leads to changes. Even in cases where the monitoring programme identifies unexpected environmental impacts, the authorities are not obliged to initiate mitigation. Nevertheless, a well carried out SEA provides a good tool for challenging the political decision makers in the light of documented negative impacts on the environment. When such impacts are checked against existing EU environmental objectives and requirements the SEA provides an excellent opportunity to ensure cross-compliance and enforcement of existing environmental legislation.

Article 3.3 of the SEA directive could reduce the scope of applying an SEA, since it says that *Plans and Programmes referred to in 3.2 determine the use of small areas on the local level and minor modifications to plans and programmes referred to in paragraph 2 shall require an environmental assessment only when Member States determine that they are likely to have significant environmental effects.* The terms *small* and *local* and *minor modifications* leave a lot of possibilities for authorities to "screen out" a plan or programme from SEA requirement. In any case, however, the authorities must document that such a small, local and minor plan has no significant environmental effects. As the discussion of article 3.3 illustrates, it is indeed possible that the classic EIA problem of "salami-slicing", as it happened for example with the Spanish National Hydrological Plan[304] or the Ribe County in Denmark[305] - splitting plans into smaller ones in order to hide cumu-

---

[304] Spain adopted on 5 July 2001 a water management plan involving 863 individual works and over 100 dams with total costs of 20,050 million EURO, seeking with 2/3 funding from EU funds. Central part of the plan was a transfer of 1050 million m? water per year from the Ebro river in the south east to regions in the sout west. The government sponsored environmental impact assessment only covered individual parts of the plan in the clear attempt to find support for EU funding for «positively » assessed parts of the plan while financing harmful parts itself. The plan was stopped after a change in government in the year 2004.

[305] Each of the 351 extension projects of intensive livestock farms in the Ribe County in Denmark in the period 2000-2004 was assessed separately under the EIA requirements escaping the assessment of the overall massive environmental impact.

lative effects and disaggregating unfavourable findings from favourable ones, to escape the obligation of carrying out an evaluation of the impacts - will continue when implementing the SEA. Linked to this problem is the issue of authorities not assessing environmentally unsound or politically unpopular details of plans stating that the details cannot be assessed at all.

It is also likely that many authorities in Member States will have a tendency to understand SEA as a larger EIA of several projects, in spite of the fact that SEA requires the consideration and assessment of a much wider range of alternative options and issues to be examined.

The problems of mobilising the public to participate in the EIA process are also likely to be evident in the implementation of SEA. It may be difficult for people to become involved as it is difficult for them to understand what the effects of a plan will be in 20 years time.. Linked to this problem is the lack of capacity and expertise among local communities or ECOs to respond to consultations and public hearings. As a consequence, a great number of SEAs are likely to receive only a sporadic response from the public.

The exclusion from the directive of plans and programmes serving solely national defence or civil emergency is also a problem since such plans and programmes often have considerable environmental effects.

Quality control is a major problem with project level EIA, and it will be important to learn from the implementation experience and establish member state based external instruments of quality control. The EU Commission has to prepare a report on the application and effectiveness of the SEA by July 2006 and propose amendments with a view to further integrating environmental protection requirements according to article 6 of the EU Treaty.

V.3.5.2.5   Outlook and action

ECO participation in SEA public processes is extremely important if the SEA directive is to work as a way of challenging environmentally unsound plans and enforcing EU environmental targets. But it has to be kept in mind that the Directive is a process rather than an end in itself. It can force in-depth scrutiny, leaving well-trodden paths and engaging in developing alternatives, thereby increasing the possibility of putting pressure on political decision makers tofind solutions which are less damaging to the environment. Furthermore, during the carrying out of a SEA, identification of an incorrect application and failure to achieve environmental objectives as laid down in EU Air, Nature Protection, Waste and Water legislation (see chapter IV) can lead to conditions and requirements of those laws being applied in a more rigorous manner.

But first of all the SEA Directive has to be correctly transposed into national law, and implemented. This involves the scrutiny and control of transposition legislation in each Member State by ECOs and the European Commission. The focus of such quality control should include the correct scope of the Directive and the right of the public to participate in the SEAs and to complain to civil or administrative courts (see also chapter V.2).

A further important task of ECOs is to build capacity at local and regional level to engage in the process, explore the opportunities for their specific work areas within SEAs and get their views expressed and taken into account during the SEA process. The SEAs Environmental report should help ECOs as well as the European Commission to insist more firmly on the correct application of existing EU environmental laws.

ECOs will also have a difficult task in checking whether decisions over which Plans and Programmes should be subject to SEA are correct and making a quality control of SEA reports and raising issues. ECOs should also make complaints at the national level or through the European Court of Justice to ensure the early implementation of SEA and that it leads to positive case law that backs a strong interpretation of the Directive.

One of the best ways to help members of the public and ECOs to understand the SEA process is to develop simple checklists, which explain what to do. One example of this is the "Strategic Environmental Assessment and Biodiversity: Guidance for Practioners" by RSPB and English and Welsh Environmental Agencies

Making the SEA Directive work strengthens the call for thematic coalitions including governments, ECOs, business, local communities and research institutions on issues like climate change, protection of landscapes and biodiversity. These coalitions should aim to bring about change in planning practice, reaching a high protection status of the environment. This directive in particular, allows ECOs to work simultaneously, directly and hopefully together with the national authorities, giving them strength to green the proposed plans and programmes.

BIBLIOGRAPHY AND FURTHER READING:

EU Commission's guidance on implementation of the Directive 2001/42/EC on the assessments of the effects of certain plans and programmes of the environment
http://europe.eu.int/ comm/environment/eia

"Strategic Environmental Assessment and Biodiversity: Guidance for Practitioners" by the Countryside Council for Wales, English Nature, Environment Agency and the Royal Society for the Protection of Birds, June 2004.
http://www.rspb.org.uk/Images/SEA_and_biodiversity_tcm5-56786.pdf

"The Strategic Environmental Assessment Directive: Guidance for Planning Authorities", Office of the Prime Minister, London, October 2003.
http://www.odpm.gov.uk/stellent/groups/odpm_planning/documents/page/odpm_plan_025198.pdf

"Strategic Environmental Assessment and Climate Change: Guidance for Practitioners", produced in cooperation between the UK Climate Impact Programme, the Environmental Institute, English Nature, Environment Agency, Countryside Council for Wales, CAG consultants and Levet-Therivel sustainability consultants, May 2004.
http://www.sea-info.net/SEA%20climate%20change%20final.pdf

EEB Seminar Report: Strategic Environmental Assessment Making a Difference, Brussels 5. September 2003

V.3.6 Environmental Management and Audit Scheme

By Dr. Karola Taschner[306]

V.3.6.1 INTRODUCTION

The Fifth Environmental Programme of the EU - "Towards Sustainability" - put forward the idea that the range of environmental policy instruments should be broadened and that EU environmental policy should no longer rely exclusively on command and control measures. One policy measure which has been revised, and which will be revised again since the current revision of ISO 14001 was finished in 2004, is the Regulation on Environmental Management and Audit Schemes - the EMAS.

Environmental Management Systems (EMS) can be an effective tool for improving environmental protection inside companies in a systematic manner. Organisations that have an EMS in place can reduce their risk and cope more easily with legal and performance requirements .

Given the fact that inspection by regulatory authorities is insufficient most of the time, such a system can give support to government control of companies without necessarily replacing it.

The goal of the EMAS Regulation (*CEC 2001*) is that EMAS registered companies have an EMS in place that includes, among other requirements, legal compliance. An EMS serves as a tool to help companies to improve their environmental performance, i.e. to continually reduce their significant environmental impacts. The process requires regular checking by internal and external auditors.

The Community Regulation on EMAS sets up a system which consists of several contiguous, inter-linked parts which mutually reinforce each other. EMAS applies to all types of enterprise, including service industries like banks. Such an inclusion could lead to an interesting development, if the lending policies of banks are also deemed to be subject to the requirements of EMAS. Products and services are covered as well.

The Regulation specifies the requirements involved in dealing with environmental aspects: defining an environmental policy, making an initial environmental review, determining the significant environmental aspects, setting up an environmental programme and environmental management system and the audit criteria to be put forward in the public environmental statement.

An initial review results in the definition of the significant environmental impacts of the company's activities, products and services. For that purpose, the company has to develop comprehensive, reproducible criteria, which can be checked independently and which should be made publicly available. Significant environmental aspects determine the next step in the process.

The company must then formulate an environmental policy in the light of the initial review. This is important because it demonstrates the commitment of the top of the company hierarchy to environmental goals and its willingness to invest in activities which might not deliver immediate economic return.

---

[306] Scientific advisor to the European Environmental Bureau

248

The company should then set up an environmental programme setting reduction targets and time tables, with the environmental management system organised accordingly. The environmental performance of the company is then checked every year in a well-defined audit cycle by an internal (or external) auditor. Third parties, the "verifiers", deliver private inspection services to companies which check the auditors' work i.e. the documentary evidence submitted by the auditor on its validity. Finally, the company has to issue a public environmental statement that is validated by the verifier and requires a yearly update. Only then is the company entitled to register with the national competent body.

The public statement can be shaped according to the different groups at which it is targeted. The Commission has issued a guidance document(CEC 2003) to implement the Regulation. Mild forms of benchmarking are envisaged: "Companies may select indicators relevant to their business (e.g. energy use per tonne of product)"

Environmental aspects, policy, management programmes, monitoring and auditing are determined by Annex I of the Regulation and are identical with the international standard ISO 14001:1996[307] Since 14001 has been revised and a new version published in 2004 (*ISO 14001:2004*), as a consequence the EMAS Regulation also has to be revised.

EMAS requirements for the initial environmental review, legal compliance, independent verification, accreditation system for verifiers, validated public environmental statements and employee involvement go beyond the requirements of the ISO standard and are laid down in detailed annexes.

Member States have set up two competent bodies to

1. establish a system for the accreditation of independent environmental verifiers and the supervision of their activities

2. register those companies which have fulfilled EMAS Regulation requirements

The implementation of the Regulation was discussed in meetings of the EMAS experts from Member States in a Committee envisaged under Article 14 of the Directive. Representatives from industrial federations and the EEB were admitted.

Recently the EU Commission reacted to two developments

1. the necessity that small and medium enterprises should also have the possibility of becoming EMAS registered,

2. a growing number of private initiatives to apply some kind of EMAS "light".

Based on the BSI standard 8555 ([308]) the Commission issued the Mandate 356 ([309]) to CEN and CENELEC for the development of a new standard for the "staged implementation" of EMS. The European standard bodies have meanwhile asked ISO to develop the standard.

---

[307] ISO 14001:1996 on Environmental Management Systems – Specification with guidance for use

[308] British Standard BS 8555:2003
Environmental management systems —Guide to the phased implementation of an environmental management system including the use of environmental performance evaluation

[309] M/356 Standardisation mandate to CEN and CENELEC for the development of an en guidance standard concerning a staged implementation of environmental management systems (EMS)

V.3.6.2   POTENTIAL OF EMAS

EMAS offers many advantages. The management of a company may decide to commit itself, in a top-down approach, to continually improving its environmental performance. Since the regulation also requires the involvement of employees, it provides for a bottom-up approach as well. This can have an integrating effect on the corporate climate.

Companies that have gone through the process of EMAS registration, often report that the installation of an EMS has ultimately been a cost saving measure. Investment for lower resource use mean less waste, and better control leads to reduced emissions – both of which result in financial savings. A study conducted by the Austrian Economic Chamber came to the conclusion that investments into EMAS had been written off after less than fourteen months on average. Often changes were small: "a valve here, a switch there...". End-of-pipe solutions for waste to be "placed" somewhere were no longer regarded as an optimal solution and, instead, recycling of some waste suddenly appeared to be cost-effective.

Overall, a company which has an EMAS in place, reduces its environmental risk and makes an ongoing commitment to do so- the idea being that once a company has been registered it then needs to go on making improvements in order to avoid losing face were its registration to be subsequently cancelled.

V.3.6.3   THIRD PARTY AUDIT

As verification is a key element of EMAS, the credibility of EMAS depends to a large extent on the credibility of the verifiers, i.e. their competence, independence and objectivity. Member States have to guarantee an efficient and reliable accreditation system and supervision for its verifiers.

Verifiers can sign up as individuals or as a team with a responsible team leader. All Member States have set up accreditation boards. Verifiers are granted accreditation only for the industrial sectors in which they have experience and require to have specialised knowledge on the environmental impacts caused by these sectors. Furthermore they must demonstrate their competence in conducting an audit. Different Member States have constructed very different accreditation boards. Some allow environmentalists to be part of them, others do not. In Germany, for instance, environmentalists are entitled to contribute to defining the conditions for the accreditation of verifiers.. . A very effective method of checking an applicant's abilities is to have him conduct "witnessed assessments", i.e. in order to become a verifier, the applicant must be examined while working "on the job".

The quality of the accreditation boards is assessed by "peer review", i.e. accreditation boards of the different countries evaluate each other.

Apart from their competence, it is also very difficult to guarantee the objectivity and independence of a verifier. Some guidelines have been set up by the Commission. A verifier should not have been involved in the consultancy process prior to EMAS verification, nor be financially dependent on that company, e.g. because he has no other clients.

Nevertheless, in practice, verifiers seem to have very limited powers. One expert commented that if verifiers were to examine cases too closely "they would soon be out of business".

V.3.6.4   EMAS AND ENVIRONMENTAL MANAGEMENT STANDARDS

The EMAS Regulation's requirements on EMS are copied from ISO 14001:1996. Industry had asked for the standard to facilitate EMAS registration. Therefore the Commission issued a mandate to develop an EMS standard which was eventually conducted by ISO. The requirements of the ISO standard can be certified by accredited certifiers and by carrying out the additional requirements of legal compliance, independent verification and validated public environmental statement, a company can be verified and registered under EMAS. A company that has entered EMAS via the ISO standard is not obliged to conduct an initial review.

Site operators undergoing EMAS have commented that there is little difference between the two schemes, which may mean that EMAS registration is too easily obtained when starting from the ISO standard. In Germany, certifiers and verifiers are identical, so that verification is obtainable almost without any extra effort.

Countries excelling in EMAS registrations are Austria, Germany, Denmark and Sweden. Recently the number of registrations became stagnant with the exception of Spain where, because of Government incentives, EMAS registrations and certifications for ISO 14001 are booming. Other countries also have registrations but only in a limited number. EMAS has not been a major success at EU level.

European industry and the Commission both have reasons to be reluctant about accepting EMAS without linking it to global markets. Neither of them want to set up trade barriers and so their preferred option is an international standard i.e. the standard developed by the International Standardisation Organisation (ISO). It would be in contravention of WTO to grant the products and services of EMAS registered companies a competitive advantage in the EU e.g. in public procurement. An international standard on environmental management systems gives companies the same access world-wide. This is the reason why EMAS is perceived as an ISO 14001 certificate with some add-ons – such as the public statement.

The ISO standard is negotiated among representatives from standard institutes from all over the world. US representatives, who have been extremely in negotiations, are not interested in a standard that sets environmental requirements which are too precise and instead prefer vague wording. This can be explained by the fact that the US has very strict legislation on environmental liability and that Americans do not want to sign up to any requirements which could later be used against them in cases of litigation.

A critical view on the International Standard on Environmental Management Systems, ISO 14001:1996 concluded (*Krut 1998*)

▶ The goal of ISO 14001:1996 is conformance with the organisation's environmental policy. The environmental policy of companies "includes a commitment to comply with relevant environmental legislation and regulations and with other requirements to which the organization subscribes". Quite a number of Member States found this wording insufficient for making use of ISO 14001 certificates for surveillance purposes, since a "commitment to legal compliance" does not necessarily mean that legal requirements are really met.

▶ ISO 14001 does not audit the actual environmental performance but the performance of the environmental management system hoping (without stating it) that with the improvement of the system the environmental performance would also be improved.

251

▶ The goal is not defined as "continual improvement of environmental performance" but rather vaguely as "prevention of pollution"

▶ The audit cycle is left open and

▶ Past activities are not considered either

Also the ISO standard does not require the company to request its suppliers to have an EMS in place

The revised text of ISO 14001 was issued at the end of 2004. One of the conditions of the revision has been that it should not add on any new requirements and align to ISO 9001 to facilitate organisations to apply both standards in one go.

According to the EU-Commission

▶ The new standard ISO 14001:2004 clarifies the assessment of legal compliance and how organisations treat legal compliance. This renders it easier to move from ISO 14001 to EMAS, as the gap between them regarding non-compliance with legal requirement will be closer. Full compliance with legal legislation as requested in EMAS is still not an ISO 14001 requirement.

▶ On documented procedures: the new regulation leaves it up to the companies and makes it simpler what document need to be covered by documentation. Not every procedure needs to be documented.

▶ The new standard explains on which parts of the organisation the environmental management system applies.

▶ A definition for the auditor has been introduced to ensure compatibility with ISO 9001, while the definition for "auditor" is not included in the EMAS regulation.

▶ A check list for minimum requirements of the management review has been included.

The importance of communicating environmental issues has been emphasized.

The main problem is that a text which resulted from negotiations conducted in a democratically non-legitimised body like the ISO (standardisation institutes from all over the world) has entered EU legislative text.

While EMAS registrations have stagnated and are even declining, ISO 14001 certifications are still rising. The main reason seems to be that companies are reluctant to make information public. As any advantage granted to EMAS will immediately be demanded for ISO 14001 as well, there is little incentive for companies to make the extra effort in applying for EMAS registration.

From an environmental point of view, EMAS is without any doubt the superior of the two systems.

252

V.3.6.4 SHORTCOMINGS OF EMAS AND ISO 14001:1996

In view of the revision, the Commission is considering extending the field of application of EMAS. The question arises whether it will be appropriate to use EMAS as an intrument to accomodate the multiple uses envisaged for it:

▶ as implementation tool as envisaged in the draft directive for setting eco-design requirements for energy-using products([310])

▶ as label for services, like tourism

▶ as instrument for suppliers to demonstrate their compliance with the legal requirements with respect to producers' responsibility

▶ as a framework for integrating life-cycle thinking into the Commission Communication on Integrated Product Policy (IPP) in order to improve the exploitation of the product dimension of EMAS([311])

▶ as a method for European standards to implement the directive on packaging and packaging waste 94/62/EC ([312])

EEB and ANEC([313]) were very alarmed that the EMS might be used as an instrument to delegate controls to industries themselves. The two NGOs have taken the opportunity to make an assessment of EMAS and ISO 14001 to demonstrate that both are not fit to serve surveillance let alone self surveillance purposes.([314]) They point to studies which provide evidence that neither the application of EMAS nor EMS standards have clearly resulted in superior environmental performance.([315]) ([316])

The Regulation requires the "continual improvement of environmental performance". It is, however, entirely at the discretion of the company as to how much they want to improve and in what time frame. No indication as to when the verifier should withhold his signature because he has found that the improvement made was not worthwhile.

---

[310] COM(2003) 453 final

[311] COM(2003) 302 final

[312] Official Journal L 365 , 31/12/1994 p 10 -23

[313] European Association for the Co-ordination of Consumer Representation in Standardisation

[314] ANEC/EEB Position Paper on Environmental Management System Standards, 2003-02-21

[315] ANEC/EEB Position Paper on Environmental Management System Standards, 2003-02-21

[316] The state of EMAS in the EU, Eco-management as a tool for sustainable development, Literature study, Jens Clausen, Michael Keril and Martin Jungwirth, IÖW and Ecologic, Berlin 2002

Both a company with good environmental performance and a comparatively "dirty" one can become registered. This is only acceptable because the "dirty" company will also have committed itself to the improvement of environmental performance and legal compliance. If they do not meet the demands immediately, they are required to sign up to do so in future.

The initial environmental review leaves considerable leeway for the company in choosing criteria for determining their significant environmental aspects. The only pressure on them is that these criteria have to be disclosed. When defining targets and objectives, companies have to "consider" "environmental aspects that have a significant environmental impact, a wording that does not establish any priority or obligation.

It is of utmost importance that EMAS provides guarantees that EMAS registered organisations meet legal environmental requirements. Environmental policy includes "compliance with all relevant regulatory requirements regarding the environment" (Article 2(a). The competent registration body has to consult the enforcement authorities concerning the organisation's regulatory compliance. In some Member States regulatory authorities have reacted by inspecting the company in question which was just what the company had wanted to avoid. In other Member States authorities have reacted more to the organisations' taste by attributing relief to EMAS registered companies. There have been one or two cases in the EU where an EMAS registration has been suspended or withdrawn due to evident lack of compliance.

The public environmental statement asks for only very general information. Benchmarking is purely voluntary and the Commission implementation guide is not mandatory. What is worse, EMAS registered sites have even not respected the very general information requirements of the Regulation. In 1996, the Swedish EMAS Council assessed 58 environmental statements and came to the conclusion that almost none of the industrial sites had met them (*Swedish EMAS Council 1996*).

The nature of the registration bodies varies greatly throughout the EU. In some countries, environmental agencies serve as registration points. In others, this function is fulfilled by Departments of Industry, and in Germany the Chambers of Commerce of the different 'Länder'. Unfortunately, the registration point is generally viewed not as a control but only as a formal registration point.

One shortcoming of the system which can never be overcome is that verifiers are paid by the very people they are verifying. There are, in fact, too many offering their services, which in the past has led to the situation where a verifier has refused to give his accord, only for another one to do so 24 hours later. From an optimistic viewpoint, this can be regarded as merely a 'teething trouble' and one which is symptomatic of any new system still in the early stages of development.

One solution to this problem would be if verifiers making a rejection then informed the registration body of their decision. The latter could then refuse to register such an audit until and unless considerable improvements were made.

An additional problem arises from the fact that companies that are ISO 14001 certified can become EMAS-registered without repeating their audit. They only require to demonstrate to the verifier that they have been fulfilling the additional requirements of EMAS.

The ANEC/EEB position paper provides evidence from studies showing that it is not even certain that EMS has improved the environmental performance of companies.

V.3.6.8   PROMOTION OF EMAS

The question remains as to what credit can be given to companies which have been making an effort to have their environmental management systems checked by third parties and to issue a validated environmental statement to the public. Such a commitment to future improvement in environmental performance is extremely valuable if it can be monitored. Firms should be encouraged to subscribe to EMAS and gain credibility. Only if enterprises see a worthwhile return will they get involved in the exercise. This is certainly one of the reasons for current stagnation, and even a roll back in EMAS registrations (*Austrian Economic Chamber 1996*). After the initial benefits have been reaped companies then tend to ask themselves questions about the further value of EMAS to them.

The first ones to appreciate the effects are certainly people living in the neighbourhood of the company, even though the public at large may not be the primary target. Banks in their lending policies and insurance companies in fixing their premiums could consider the fact that EMAS registered companies have better control over their risks and liabilities. However, they usually find the securities delivered by EMAS insufficient to merit a lower premium.

Some enterprises demand that their suppliers have a certified EMS in place. Public authorities could make EMAS registration a condition in their purchasing policy. Public procurement in which some credit is given to EMAS could be a vigorous incentive for encouraging it. However, encouraging EMAS means encouraging ISO 14001 at the same time, without providing the same environmental gains.

The strongest element of EMAS is in theory its public statement. Many companies have started to issue environmental reports. These used to be of limited value especially as they were not validated and not comparable. In principle the EMAS public statement could provide a framework for environmental reporting. For this purpose the requirements should be more demanding and be based on environmental indicators and benchmarks. The introduction of producers' responsibility in the Directives on end-of-life vehicles, waste from electric and electronic equipment and reduction of hazardous substances (*see chapter IV.4*) have opened up a new field of interest in EMAS. Some producers think EMAS could serve as a basis for supply chain control in that suppliers could introduce environmental requirements into the scope of their environmental policy programmes and EMS.

EMAS can also help companies to reduce risks and could help to prevent civil liability cases under the Environmental Liability Directive (see chapter V.3.3). The US act on environmental liability caused severe problems to companies when it was introduced because insurance companies were not willing to provide insurance cover. They found the risk too high. Only when companies established full-blown environmental management systems and procedures to help ensure that they remained in compliance with rules, regulations and laws, did insurers gain confidence.

The main advantage of EMAS as far as companies are concerned is that it can supposedly allevi-ate the controls imposed upon them by regulatory authorities, perhaps even concerning their permits. Here care has to be taken that EMAS does not become an easy way out. EMAS asks for "continual improvement of environmental performance", which is in no way conducive to being lax on permits. One could imagine, however, that reporting obligations under regulatory require-ments and EMAS should not be duplicated. Some Governments argue that since EMAS represents for them a guarantee of environmental conformity with EMAS objectives, this allows them to con-centrate their scarce inspection resources on more problematic companies. Here it has to be made clear that EMAS companies have no right to obtain preferential treatment from authorities unless EMAS can guarantee full compliance with environmental regulations.

255

V.3.6.9 OPPORTUNITIES FOR ENVIRONMENTAL CITIZENS' ORGANISA-TIONS AND DEMANDS FOR EMAS REVISION

The EEB has deprioritised its work on EMAS after the last revision, but EMAS offers a number of opportunities to make use of the information that must be made publicly available:

▶ environmental policy (Annex I-A.2.f),

▶ criteria for determining significant environmental impact (Annex 6.4),

▶ environmental statement that has to be validated annually. (Annex 3.1 and especially 3.3 set-ting reporting criteria)

EMAS will have to be revised following the publication of the revised ISO 14001 and ECOs should be involved. ECOS ([317]) - of which EEB is a member - provides access to the International Standardisation Organisation (ISO) but the standard was almost finalised when ECOS was able to join. So it could only provide limited input.

ANEC/EEB have put forward the following demands:

▶ EMS standards must contain a clear obligation to comply with legal provisions (a commitment to comply as in ISO 14001:1996 is insufficient)

▶ continual improvement of Environmental performance means a measurable reduction of environmental burdens and resource consumption rather than an enhancement of the man-agement system and its measurable output

▶ EMS standards shall require compliance with international environmental agreements irre-spective of whether or not the home country of the organisation has ratified the agreement. In addition, time scales for the incorporation into EMS standards of such provisions are short-er than the legal implementation. Both agreements and time scales shall be indicated in the relevant standards

▶ organisations shall be required to apply the same environmental criteria in all countries in which they operate

---

[317] European Environmental Citizens Organisation for Standardisation (ECOS)

▶ a set of key environmental indicators measuring environmental performance and not the performance of the EMS, including appropriate scales, shall be developed

▶ the key environmental indicators shall allow comparisons between different organisations

▶ EMS standards shall contain detailed standardised reporting requirements including key performance indicators and data, which allow a comparison between legal provisions or BAT and the actual performance

▶ minimum performance levels based on the state-of-the-art shall be defined

> The Commission argues that the Directive on Integrated Pollution Prevention and Control (IPPC) demands application of best available technology by law and therefore it is not necessary to mention it in a regulation which is voluntary. This is not quite true. The EU reference documents for what is deemed to be "best available technology" is only to be taken into account by the permitting authorities, which may also consider local and geographical conditions in issuing a permit under the IPPC Directive (*see chapter V.3.2*).

> Only very large plants are regulated under the IPPC Directive. Many are not - e.g. small and medium enterprises - and for them it would be important to apply economically viable application of BAT because they are often not regulated otherwise.

▶ organisations which have obtained an EMS registration shall be subject to evaluation studies. The performance achieved shall be compared with the performance of companies which have not implemented EMAS or ISO 14001. Methods for performance ranking shall be developed.

▶ companies shall not be awarded for the mere implementation of an EMS. Instead, incentives (e.g. more lenient enforcement control) should be linked to compliance with environmental laws. It is also not adequate that EMS firms are given preference in the context of public procurement or enforcement unless their excellence has been proven using comparable data.

▶ under no circumstances shall the EMS approach be applied as a substitute for product specifications.

The recently revised ISO 14001 has not taken these demands into account.

The current EMAS is still far from being a "full blown environmental management system which helps companies to be in compliance". However, we should try to give the future EMAS ambition so that it becomes a rigorous and convincing instrument.

ECOs will be confronted with the development of the new mandated standard for the "staged approach to EMAS". As with ISO 14001, ISO will conduct the task and so heavily influence the legislative framework of the EU's EMAS. ECOs should follow this issue closely in order to make sure that certificates are not issued before the enterprise conforms fully to the requirements any other organisation has to meet under ISO 14001. Any earlier certification with only "key elements" or some relevant environmental aspects dealt with, would undermine the credibility of all other EMS certificate.

BIBLIOGRAPHY AND FURTHER READING

Austrian Economic Chamber (1996) Ecomanagement Pays Off. Evaluation of a Subsidised Programme for Introduction of Environmental Management and Audit Systems (EMAS) in Austria, commissioned by the Federal Ministry of Science, Transport and Arts, and the Federal Ministry of the Environment, Youth and Family, Land Government of Upper Austria, Austrian Economic Chamber, Vienna, June 1996.

Council Regulation (EC) No 761/2001 of 19 March 2001; allowing voluntary participation by organisations in a Community eco-management and audit scheme (EMAS); OJ L 114 , 24.4.2001, p.1

Commission Recommendation of 10 July 2003 on guidance for the implementation of Regulation (EC) No 761/2001; notified under number C(2003)2253; OJ L 184, 23.7.2003, p 19

ISO 14001:1996 on Environmental Management Systems – Specification with guidance for use

KRUT, R (1998) ISO 14001 – A missed opportunity for sustainable global industrial development, Riva Krut and Harris Gleckman, Earthscan, 1998

TASCHNER, Karola, Environmental Management Systems: The European Regulation in New Instruments for Environmental Policy in the EU, edited by Jonathan Golub, Routledge Ltd., London and New York, 1998.

ANEC/EEB Position Paper on Environmental Management System Standards; Brussels, 2003-02-21

British Standard BS 8555:2003 Environmental management systems —Guide to the phased implementation of an environmental management system including the use of environmental performance evaluation

M/356: Standardisation Mandate to CEN and CENELEC for the development of an EN guidance standard concerning a staged implementation of environmental management systems (EMS)

V.4   Product Policy

258

V.4	**Product Policy**	**258**
V.4.1	Introduction	261
V.4.2	The EU Ecolabel – the 'Flower'	263
	V.4.2.1 The purpose of the Flower	263
	V.4.2.2 Deliverables of the Flower 1992 - 2003	264
	V.4.2.3 EEBs evaluation of the Flower - 2004	266
	V.4.2.4 Second Regulation revision on the horizon	268
	V.4.2.5 Links with other EU and national labels	269
	V.4.2.6 Ecolabels and cross-links with other legislation	270
	V.4.2.6.1 Public Procurement	270
	V.4.2.6.2 Integrated Product Policy (IPP)	271
	V.4.2.7 Tips for ECO Action (environmental citizens' organisation)	271
	V.4.2.7.1 National ECO support/use of EU Ecolabel resources	271
	V.4.2.7.2 The EU Ecolabel and EU Directives on Public Procurement	272
V.4.3	Integrated Product Policy	276
	V.4.3.1 The purpose of Integrated Product Policy (IPP)	276
	V.4.3.2 Brief history of EU IPP Policy	276
	V.4.3.2.1 The white paper- The IPP Communication	276
	V.4.3.2.2 The EEB proposals	277
	V.4.3.3 Implementation of IPP to date	279
	V.4.3.4 Links with ISO Standards on Environmental Product Declarations	280
	V.4.3.5 Tips for ECO Action (environmental citizens' organisation)	280
V.4.4	The Ecodesign framework for energy using products (EuP) Directive	282
	V.4.4.1 The purpose of EuP	282
	V.4.4.2 Strengths and Weaknesses of the final text	283
	V.4.4.3 Future Implementation	285
	V.4.4.4 Links with other legislation / policy tools	286
	V.4.4.5 Tips for ECO Action (environmental citizens' organisation)	288
V.4.5	New Approach and Standardisation	289
	V.4.5.1 Introduction	289
	V.4.5.2 Community action	290

V.4.5.2.1   Notification and standstill obligation..290

V.4.5.2.2   "New Approach" 290

V.4.5.3   Role of Standards in EU legislation from "business to business" to "business to public"? 292

V.4.5.3.1   Non-Mandated Standard 292

V.4.5.3.2   Public Participation...293

V.4.5.4   Who develops standards and what are the environmental problems.........294

V.4.5.4.1   Standard bodies and how they work 294

V.4.5.4.2   Structural problems linked to the use of standards in legislation – lack of accountability 295

V.4.5.4.2.1   Lack of legislation mandating integration of environmental aspects. 295

V.4.5.4.2.2   Accountability 296

V.4.5.4.2.3   Accessibility 296

V.4.5.4.2.4   Distinguishing political and technical competencies 296

V.4.5.4.3   Examples of Environmental requirements in mandated standards ... 296

V.4.5.4.3.1   Packaging and packaging waste375 - The first and only example of the New Approach in the field of the environment 297

V.4.5.4.3.2   Waste of electric and electronic equipment376..................298

V.4.5.4.3.3   Heating appliances standards...298

V.4.5.4.3.4   Construction Product Directive 298

V.4.5.4.3.5   Refrigerants...298

V.4.5.4.3.6   Environmental Management Systems 298

V.4.5.4.3.7   Standard for the harmonised UNICA bottle. 299

V.4.5.4.4   Greening standardisation: Attempts to improve environmental accountability 299

V.4.5.4.4.1   Commission...299

V.4.5.4.4.2   Member States..300

V.4.5.4.4.3   Standard bodies 300

V.4.5.4.4.4   European Environmental Citizens Organisation for Standardisation (ECOS)...301

V.4.5.4.5   Outlook and Action for environmental citizens' organisations 302

V.4.6   Chemicals and Pesticides Regulations..306

V.4.6.1   Introduction 306

V.4.6.2   Chemical intensification of Europe's economy...............................307

V.4.6.3   From large scale chemical contamination to applying precaution 307

V.4.6.4   Industrial Chemicals 309

V.4.6.4.1 Hazard characterisation, classification, labelling and safety data sheets...309

V.4.6.4.2 Risk assessment at the core of EU chemicals policies....................309

V.4.6.4.3 Restrictions on marketing and use..310

V.4.6.4.4 Related policy fields: worker and consumer protection 311

V.4.6.4.5 Weaknesses in current EU chemicals legislation............................311

V.4.6.4.6 Towards a new European chemicals policy (REACH).......................312

V.4.6.5 Pesticides 314

V.4.6.5.1 The Authorisation Directive 314

V.4.6.5.1.1   Introduction 314

V.4.6.5.1.2   EU positive list and national authorisation 314

V.4.6.5.1.3   Transition period - All pesticides to be safety checked by 2003? 315

V.4.6.5.1.4   Assessment 315

V.4.6.5.2 Towards a Thematic Strategy on Sustainable Use of Pesticides 316

V.4.6.5.2.1   Revision of the Pesticides Authorisation Directive 317

V.4.6.5.2.2   Sustainable use or use reduction strategy?.........................317

V.4.6.5.2.3   Assessment 318

V.4.6.6 Outlook and ECO Action (environmental citizens' organisation)................318

V.4.6.6.1 Substituting "hazardous"390 chemicals and ending emissions by 2020 319

V.4.6.6.2 Core elements of the REACH proposal which should be defended... 319

V.4.6.6.3 Pesticides Authorisation and Use Reduction 320

V.4.7 Motor fuels and environmentally relevant legislation...323

V.4.7.1 Introduction 323

V.4.7.2 Fuel components and their environmental problems 323

V.4.7.2.1 A problem of the past: lead...323

V.4.7.2.2 Oxygenates as substitutes for lead in petrol 323

V.4.7.2.3 Sulphur content 324

V.4.7.2.4 Olefins 324

V.4.7.2.5 Aromatics and poly-aromatics 324

V.4.7.3 Different aspects of regulating fuel composition...325

V.4.7.3.1 Objectives..325

V.4.7.3.2 Fuel consumption versus emission reduction 325

V.4.7.3.3 Alternative fuels..325

V.4.7.3.4 Testing 326

V.4.7.4 Current legislation...326

V.4.7.4.1 Lead free petrol...326

V.4.7.4.2 Quality of petrol and diesel fuels (Directive 98/70/EC) 326

V.4.7.4.3 "Mini-Directive" on sulphur-free (EC 2003a) fuels. 327

V.4.7.4.4 Alternative fuels..328

V.4.7.4.5 Taxation 328

V.4.7.5 Future legislation 329

V.4.7.5.1 The world-wide fuel charter 329

V.4.7.5.2 Ultra-fine particles 329

V.4.7.5.3 Alternative fuels..329

V.4.7.6 Outlook and ECO action (environmental citizens' organisation)................329

V.4.1   Introduction

Up to now most EU environmental policy responses have been based on end-of-pipe measures (such as emission limits) or the establishment of environmental quality objectives (such as air quality values) - see chapter IV. Environmental standards have only been introduced at EU level for a limited number of products - such as energy labels (*see chapter V.4.2*), restriction of hazardous substances in electrical equipment (*see chapter IV.4*), market ban for carcinogenic substances (*see chapter V.4.6.4.3*), or authorisation schemes for open uses of chemicals like pesticides and biocides (*see chapter V.4.6.5.1*).

While the first have lead to important environmental improvements - for example in cleaning up polluting emissions to water from industrial installations - the effects of those such as energy efficiency, vehicle emissions and exposure to carcinogenic chemicals, have been overwhelmed by increasing consumption patterns or new chemical threats, like interference with hormone systems. At the same time, the increasing complexity in sourcing materials and components in manufacture supply chains has made product information and estimation of their lifecycle impacts almost impossible to calculate and actual recycling equally complicated. For this reason the addition of a structured product lifecycle focus to European environmental policy can supplement existing measures as well as enabling a more coherent approach to addressing the environmental challenges set down in both the European Sustainable Development Strategy and the Sixth Environmental Action Programme.

Environmental product policy focusing on articles – usually understood as consumer articles, but not exclusively - is commonly understood as a policy approach aimed at addressing environmental impacts `at source' and moving thus from addressing point source (installation based) pollution and impact sources to diffuse (product associated) impact sources.

In the past, EU environmental product policy (EPP) was based on the approach that there should be free circulation of goods and that questions of protecting health, safety and the environment, should in principle be left to Member States. Only national measures were taken to prepare EU-wide provisions. Subsequently an EU wide policy was developed for product safety taking into account human health – but leaving out environmental considerations.

Previously most initiatives on EPP came from international forums –for example the Montreal Protocol phasing out substances that deplete the ozone layer, POPs under the Stockholm Convention etc

At present the EU policy on products is still fragmented, lacking data and mostly of a voluntary nature (consisting principally of the European Ecolabel scheme). The recent political agreement (April 2005) on the Ecodesign framework for Energy Using Products is an important step towards establishing legislative product policy. However, the umbrella framework for EU product policy – a role purportedly to be played by the Commission's Integrated Product Policy Strategy – has so far failed to materialise in any concrete form.

In terms of the specific focus of reductions in impacts from hazardous substances used in products there is an EU chemicals policy in place. However, this policy addresses only a small percentage of all 100,000 or so chemicals and leaves us with insufficient information to effectively prevent damage (see chapter V.4.6.4). The Commission has proposed a much needed replacement of this flawed system with a new regulation – REACH , the Registration, Evaluation and Authorisation of Chemicals - which applies to chemicals produced in quantities of over one tonne per year and used in consumer articles. This regulation would, for the first time, collect and make available safety data for 30,000 chemicals by making chemical producers responsible for the safe management of chemicals, and implement a precautionary system for unacceptable risks by encouraging the substitution of chemicals with potential long-term or irreversible impacts. REACH is currently being debated in the European Parliament and Council, were it has met with strong opposition from chemical producers.

V.4.2   The EU Ecolabel – the 'Flower'

By Dr. Karola Taschner[318] and Melissa Shinn[319]

V.4.2.1   THE PURPOSE OF THE FLOWER

The idea behind the Flower as expressed in the current Regulation on the Community eco-label award scheme (*EC 2000*) revised in 2000[320] is *to promote products which have the potential to reduce negative environmental impacts, as compared with the other products in the same product group, thus contributing to the efficient use of resources and a high level of environmental protection.*

This objective shall be pursued through *the provision of guidance and accurate, non-deceptive and scientifically based information to consumers on such products.* The environmental impacts are *identified on the basis of examination of the interactions of products with the environment, including the use of energy and natural resources, during the life cycle of the product.* .

The Flower scheme covers goods and services, and the 'consumers' addressed include professional purchasers. The eco-label may be awarded to *a product possessing characteristics which enable it to contribute significantly to improvements in relation to key environmental aspects,* these environmental aspects (namely Air quality, Water quality, Soil protection, Waste reduction, Energy savings, Natural resource management, Global warming prevention, Ozone layer protection, Environmental safety, Noise and Biodiversity) are identified in the light of an indicative matrix (laid down in the annex of the Regulation). The matrix guides the assessment of the environmental impact over the life cycle of products (covering preproduction/ raw materials, production, distribution (including packaging), use, reuse / recycling / disposal and waste management) to determine *where the most significant contribution for improvement exists, from a life cycle perspective.*[321]

The Flower provides for an environmentally motivated product standard, which is developed through a transparent stakeholder process, and thus represents an important counterbalance to the purely business and economic driven standardisation process (*see chapter V.4.5*).

On the other hand the scheme is still far from delivering satisfactory results and suffers from a number of deficiencies. Firstly, non business stakeholder participation is somewhat limited and results sometimes in rather mediocre criteria rather than being on the cutting-edge of environmental performance. Secondly, market penetration and uptake by producers is still low in many countries. Some of the criteria set for environmental product performance are relatively weak in some product groups. From the EEBs experience this is also partly due to the fact that there is no level of ambition defined in the regulation (such as for example the top 10% of products). The

---

[318] Scientific advisor to the European Environmental Bureau

[319] Senior Policy Officer for Product and Waste Policies, European Environmental Bureau

[320] For more information on the previous regulation an account can be found in the EEBs previous handbook – EEB industry handbook –A critical evaluation of available European Legislation on Industry and the Environment, December 1998.

[321] For more information on the functioning and set up of the scheme an account can be found in the EEBs previous handbook – EEB industry handbook –A critical evaluation of available European Legislation on Industry and the Environment.

regulation leaves it open as to which products and criteria are selected, and gives no guidance on priority issues. The wording of the revised Regulation (from 2000) is slightly more explicit in setting the objectives[322], and the regulation envisages that products of the same product group will be compared with each ither. Information for consumers should not only be better, but accurate and non-deceptive. However, the regulation text makes it quite clear that the ambitions of the Community Eco-label schemes were never high. The intention was more to provide a modest marketing instrument for producers who wanted to make an effort. The EEB argues that the one of the reasons the Community Eco-label has not been accepted by manufacturers is because it has not been signalling excellence.

Furthermore, there has been no requirement in the regulations to respect, throughout the criteria setting process, the original selection of, and focus on, the most relevant criteria (established in the life-cycle considerations study). In the EEB's experience, the final criteria frequently drift from the most important environmental impacts.

V.4.2.2   DELIVERABLES OF THE FLOWER 1992 - 2003

To date, the performance of the Flower can be illustrated partially by the following statistics.

In 1999, there were 15 product categories for which Flower criteria existed. Forty manufacturers and two importers held a total of 55 Flower licences. Together, they manufactured 240 different product items that carried the Flower on their label. The licences all belonged to eight product categories: paints and varnishes, textiles, washing machines, laundry detergents, refrigerators, tissue paper products, personal computers and soil improvers. By March 2005, there are 7 more product categories (total 22) with Flower criteria, 6 more under development, and a total of 208 licences held by manufacturers. These licences cover 17 product categories and over 500 different individual products.[323]

Between 2000 and 2001, the number of product items (available on shelves) rose from 17 million to 54 million, and ex-factory sales volume from €38 million to €119 million. The most successful countries so far are Denmark, France and Italy, followed by Spain, Greece and Sweden. The most successful product groups are textiles and paints and varnishes, followed by soil improvers and dishwashing detergents.

What the Flower actually delivers in terms of reduction in environmental impacts and overall ecological burden is difficult to calculate. It delivers these effects through a number of mechanisms, some of which are direct, such as a shift in production processes and product composition or design. At the same time good promotion can increase the market share for the more ecological products and deliver on potential impact reductions.

---

[322] Regulation 880/92 (revised by 1980/2000 in 2000) stipulated that the Eco-label award 'is intended to promote the design, production, marketing and use of products which have a reduced environmental impact during their entire life cycle, and provide consumers with better information on the environmental impact of products.' Note that 'a reduced environmental impact' can mean any kind of impact.

[323] See http://europa.eu.int/comm/environment/ecolabel/product/index_en.htm

Other mechanisms are more indirect, such as the creation of a product benchmark that puts pressure on non-licensed manufacturers to evolve (mimicking all or some of the Eco-label criteria), or simply guides them as to what is expected of them, even though they may not apply for the Flower. For example, in the case of washing machines, the Eco-label has certainly resulted in the creation of a standard, although it is difficult to establish how much this is due to the Eco-label and how much it is also thanks to the EU energy label. In the case of light bulbs, it is remarkable, given the original resistance of the industry, that there has been an applicant in this sector.

A study on direct and indirect benefits was recently carried out by the UK for the EU Eco-label[324].The study concluded with estimates that the current potential for reduced environmental impacts in terms of Indirect Environmental Benefits (assuming nine types of indirect benefits and a market penetration of 5%) of the Ecolabel within the EU25 (per year) was the potential saving/avoidance of:

▶ 43 TWh energy,

▶ 27 million tonnesCO_2,

▶ 35 million m? water,

▶ 39 thousand tonnes of hazardous substances; and

▶ 1.5 million tonnes in materials.

In the absence of quantifiable statistics, trust in the Flower's ability to deliver reduced environmental impacts is heavily dependent on two pre-conditions:

1) It represents the environmental leader – i.e. the best available in terms of technology, choice of raw materials, design for use, etc - and it maintains this leadership position through regular revision.

2) It can achieve a significant market uptake - through the number of products that have Flower criteria, and the quantity of labelled products on the market or sold (the current work plan target is a 25% increase per year).

Essentially, therefore, whether the Flower is delivering can also be qualitatively judged on its respect for, and achievement of, these conditions.

Within the European Union Eco-labelling Board (EUEB) and among the Member States, however, the debate is often over whether it would be better to have an Eco-label with strict criteria that have no, or only a few, applications and little market penetration, or to have less strict Eco-label criteria that would have more ease of application and more likelihood of market penetration. Such debate has never been resolved. It makes the assumption, however, that stricter criteria mean market uptake is automatically lower (as less producers would automatically be able to meet them). It is the EEB's opinion however that this assumption is not correct. The EEB maintains that an EU Ecolabel that makes a name for itself as a label of excellence would have far greater ability to attract producers (through the very nature of its reputation) than a mediocre label that has a luke-warm reputation, and subsequently lower promotion attraction to ECOs.

---

[324] Study available at http://europa.eu.int/comm/environment/ecolabel/pdf/market_study/benefitsfinalreport_1104.pdf

Based on the trends observed (most recently in the EEB EU Ecolabel evaluation study[325]), it can be concluded that there is a tendency to contemplate less ambitious criteria as one of the prerequisites to facilitating market uptake (by manufacturers). This is especially evident in the revision of pre-existing criteria where there are current Flower licence holders.

266

Example: In the recent revision (revision of criteria - 2002) of the criteria for paints and varnishes it was evident that concern over maintaining the current holders was clearly dominant over the necessity of strengthening the criteria, e.g. in the content of Volatile Organic Compounds (VOCs). This is, however, a short-sighted approach. While it may indeed facilitate maintaining the current licence holders and encourage the entry of new ones, it doesn't not actually achieve the environmental leadership the Flower should represent. In the case of paints and varnishes, it will probably not engender environmental and occupational safety credibility with the truly progressive paint manufacturers who should be the target audience.

In the case of (compact fluorescent) light bulbs (revision of criteria - 2002), while there was not sufficient will to demand lower levels of mercury, there was an indirect recognition of the emergence of zero mercury bulbs[326] in the reduction of the lifetime of the new criteria from five to three years. The light bulbs case also illustrated however that the life-cycle assessment approach can potentially be used detrimentally if applied in too simplistic a form. There was a proposal to remove the demand for the use of recycled paper packaging as it didn't represent a 'major environmental impact in the total life cycle of the product'. Fortunately, a majority of competent bodies did not agree with this perspective, maintaining that packaging had an important visual role in terms of providing consistency with the green image of the product, and the criterion was maintained.

This did however highlight that the fact that the Flower needs to be thought about in a larger context than just the immediate and direct impact of each product, i.e. the use of recycled packaging is part of the 'market' of another more ecological product – recycled fibres.

Packaging is one of the horizontal issues that were identified as needing discussion in the revision of the last EU Eco-label work plan in 2001. (Flame-retardants is another.) These issues, along with the Flower's approach to establishing preferences for renewable over non-renewable resources, need to be addressed urgently and have not yet been timetabled for discussion. (The EEB attempted to advance the discussion in 2003, and presented a proposal for a working group on horizontal packaging criteria, but so far the proposal has not been given working group time.)

V.4.2.3   EEBS EVALUATION OF THE FLOWER - 2004

In 2003/2004 the EEB performed its own evaluation of the Flower Scheme (in the context of the 2004 EU Flower Week promotion campaign and) in anticipation of the upcoming revision of the regulation. The aim of the evaluation was to collect and summarise the EEB's experience over the last 10 years of the existence of the Flower both in terms of governance (balance of interests),

---

[325] See EEB evaluation of the European Ecolabel and Scheme – What we wanted, what we got, July 2004 – http://www.eeb.org/activities/product_policy/EEB-Ecolabel-evaluation-What-we-wanted-what-we-got-July2004.pdf

[326] These bulbs will be the new benchmark, due on the market within two years, according to the Swedish CB.

quality (ambition level of the criteria) and impact on the market (is the label making a difference, delivering environmental improvements and changing market dynamics and changes in manufacturing?).

The conclusions of the evaluation were the following:[327]

1) As concerns **governance** the evaluation concluded that there was a satisfactory uptake of the EEB recommendations in some product areas, but this is not true for the majority of products (only 42.5% of the cases showed satisfactory uptake results). Furthermore, there is still too frequent evidence (at least 25%) of industry participation 'imbalances' (ie too many industry representatives compared to non-industry participants), with no guidelines available to clarify this. It would be advisable therefore to establish some clearer guidelines on stakeholder balance and guidelines on pro-active steps that should be taken to consult and integrate recommendations from product eco-design frontrunners.

2) As concerns the **quality of the criteria** and level of ambition (using the EEBs original demands as a benchmark) the evaluation concluded that the global quality (the level of environmental ambition) is still on average mediocre. Only three products (footwear, soil improvers and hand-dishwashing detergents) were evaluated as showing a good level of ambition, and two products (mattresses and dishwashers) were evaluated as showing a mixed good/mediocre level of ambition. A majority,75%, evaluations were classified as mediocre[328]. This was often felt to be a result of some participant Member State representatives (often S. European and the UK) fearing that too ambitious criteria would hamper the number of applications. However, given the relatively low level of some criteria, and the continued low level of applicants in some product categories, EEB's conclusion is that the applicant deficit is more likely to be due to active producer boycott (such as in the case of electrical goods and copy paper) and a lack of leverage from retailer and consumer demand due to a lack of, and grave deficits in and lack of continuity in, national promotion activities.

During the EEB´s eco-label working group meetings that accompanied this study it was recognised that whilst the ambition level of the criteria are the relevant aspects for evaluation of the scheme's performance, other aspects related to **the bigger Product Policy picture** were also relevant. These include the fact that the ecolabel scheme is the only concrete and productive EU Ecological Product Policy tool that is setting any requirements and establishing an EU network of national competent bodies - the only existing formal network of eco-design competence at the moment. Another important factor is that there continues to exist the necessity for a credible alternative to producer self-claims and producer own eco-labels and a more suitable consumer communication tool than Environmental Product Declarations[329].

---

[327] For more full conclusions and details on methodology etc see EEB evaluation of the European Ecolabel and Scheme – What we wanted, what we got, July 2004 – http://www.eeb.org/activities/product_policy/EEB-Ecolabel-evaluation-What-we-wanted-what-we-got-July2004.pdf

[328] NOTE: The evaluation of ambition level was admittedly particularly tricky, given that there is a normally a process of compromise involved. The benchmarking done here was against the EEBs original and most demanding ambitions for each criteria process – ie our demands at the beginning of the process which may not have been the same by the end of the discussions (due to either political 'fine tuning' or adjustments due to new information)

[329] Environmental Product Declarations are a declaration of information about products and services and what type of environmental problems they cause in a life cycle perspective. These may be certified or self-declared, but very little standardisation of the content exists till now.

The EU Ecolabel also can, and does, deliver several other indirect benefits. These include:

a) possibility of use of eco-label criteria in greening public procurement (*see chapter V.4.2.6.1*),

b) informal use of ecolabel criteria for eco-design benchmarking for individual companies,

c) creation of an information database on the best available technologies, substitution feasibility and a network of expertise and contacts with frontrunners on different product areas.

3) Concerning **impact on the market** (is the label making a difference and delivering environmental improvements and market dynamics (changes in manufacturing /consumer demand)?) the evaluation concluded that this was an almost impossible question to answer. The proxy of number of applicants as an estimate for environmental improvements and impact reduction is far from satisfactory and contributed heavily to the negative evaluation of these aspects[330]. Direct environmental improvements (or reduced environmental impacts) was judged to be poor to mediocre in more than 3/4 of the cases and there was no clear evidence available of market dynamics created (or at least no data is available). It seems fair to say that if the data does not exist the scheme has not been sufficiently active in attempting to create some transparency about its own benefits, which for a consumer communication tool is not a strong recommendation.

The study concluded overall that the EEB's involvement in the scheme may have resulted in criteria with a higher level of ambition than if the EEB had not been involved, even though the results have a low market dynamics impact eventually due to the problem of lack of promotion of the label. The EEB Eco-label working group and eco-label experts made it clear at the time that expectations were that it was not sufficient that the EEB simply settle for the passive role of scheme 'watchdog', but that the scheme must start to deliver direct and indirect benefits and that these should be documented.

The evaluation provided **several operational and policy recommendations** – including the necessity for seeking a more active engagement of EEB member organisations (only 7 actively contribute to the eco-label working group at the moment). Ideally the EEB's Ecolabel resources should be increased to allow it to play a more pro-active role.

V.4.2.4   SECOND REGULATION REVISION ON THE HORIZON

The revision for the EU Ecolabel Regulation was due to be presented by the Commission before October 2005 (but the process has been delayed). In preparation for this and in an attempt to find some consensus on issues to be included in the revision the Commission made quite intensive use of the existing Policy Management group, an informal parallel structure to the EU Ecolabel Board. Whilst no formal consensus was reached many useful ideas were generated and the ideas were summarised and documented in a series of reports[331] and experiences shared. This

---

[330] Note that lack of data was used as a negative classification.

[331] These reports and presentations are available at the Ecolabel website http://europa.eu.int/comm/environment/ecolabel/whats_eco/pm_managementgroups_en.htm

has undoubtedly increased the understanding of the different National Competent Bodies and hopefully will lead to better political compromises when the political phase of revision comes to Council and Parliament. At the end of 2004, the Commission decided to perform an additional own study of its own – on the revision of the Ecolabel and EMAS schemes. The results of this study are still pending so it is not clear yet what the Commission intends for the revision.

269

V.4.2.5   LINKS WITH OTHER EU AND NATIONAL LABELS

The implementation of the Scheme must *comply with the provisions of the Treaties, including the precautionary principle* and the Community environmental policy, as specified in the Community Programme of Policy and Action in relation to the Environment and Sustainable Development (the Fifth Environmental Action Programme), and shall be *coordinated with other labelling or quality certification arrangements as well as ... the Community Energy Labelling Scheme established by Directive 92/75/EEC and the Organic Agriculture Scheme established by Regulation (EEC) No 2092/91.*

The hierarchy of the Flower with reference to the national labelling schemes is an uneasy one as historically one of the objectives of the scheme, from the point of view of the Commission, was to harmonise EU ecolabelling by eventually replacing the national schemes[332]. Member States however actively resist this, preferring to maintain their freedom to initiate criteria based on their own ambition levels and on products relevant to their own markets and consumers. Partly thanks to the collaboration through the Cooperation and Coordination ad-hoc working group set up by the 2002-2004 Flower work plan (a more detailed implementation timetable of work items) there has been a growing realisation that the co-existence of the two levels of labelling has significant benefits – not only in terms of resources and experience gained (it is unlikely that the EU label would succeed in commanding similar resources via the Regulation) but also in terms of keeping the pressure on EU label in terms of ambition level, especially as the existing labelling schemes are mostly from Northern European and Nordic countries who traditionally have higher environmental ambitions.

The Ecolabel regulation further states that *participation in the Scheme shall be without prejudice to environmental or other regulatory requirements of Community or national law applicable to the various life stages of goods, and where appropriate to a service* – in other words if a product is awarded the label it does not avoid any other Community obligations.

Indeed the Ecolabel criteria frequently take up legal obligations (such as the phase-out of hazardous substances in electrical and electronic equipment) and introduce them earlier (eg in the 2004 revision of the criteria on portable and personal computers), or the use of the energy efficiency labelling levels (eg specifying level A+ energy efficiency as a criteria in refrigerators – criteria revised 2004).

The labels criteria can now be used in other policy tools – namely public procurement criteria setting (in conformity with the EU directives on public procurement – more below) and the setting of and conformity to Energy Using Product Ecodesign Implementing Measures.

---

[332] Examples of national schemes are – the German Blau Engel, The Nordic Swan (covering Denmark, Sweden, Finland, Iceland and Norway), the Dutch MileuKeur and private labels such as the Good Environmental Choice run by the EEBs member The Swedish Nature Conservation Association. For more information on ecolabels exsiting in the EU

V.4.2.6 ECOLABELS AND CROSS-LINKS WITH OTHER LEGISLATION

V.4.2.6.1 Public Procurement

The link with **Public Procurement** legislation is perhaps the most significant, with the greatest potential for expanding the impact of the EU Ecolabel scheme at this point in time. In 2004, the European Union adopted two directives revising European rules for Public Procurement[333]. Public procurement accounts for over 16% of the European Union's GDP equivalent to 1500 billion Euros. Moreover, it has a determining effect on certain sectors such as construction and public works, energy, telecommunications and heavy industry.

In addition to their economic importance, such rules have a direct impact on the daily lives of European citizens. It is through public procurement that decisions are made on public transport, road building,, making our towns and cities better places to live in, building and cleaning our schools and hospitals, as well as running their canteens, our water supplies, waste treatment, equipment of public offices, schools, hospitals etc.

The revised Directives offer scope for contracting authorities to consider social, employment, ethical and environmental issues. The revision process has helped to clarify and strengthen the scope for considering such objectives. In the past, this scope was vague, and discouraged many public authorities from pursuing such objectives. For several of the brave, it lead to a generation of court cases (many of them helpful to our concerns[334]) to clarify the rules, which have also helped to shape the revised Directives.

There are several references to environmental considerations in the new directives. Following the requirements of Article 6 of The Treaty, Recital 5 provides that "contracting authorities may contribute to the **protection of the environment and the promotion of sustainable development**" and Recital 6 states that the Directive should **not prevent the protection of animal and plant life**.

These principles are then spelt out in the provisions covering different phases of the public procurement procedure. The use of environmental characteristics in technical specifications is identified in article 23, and annex VI of the directives. Recital 29 and article 23 provide further explanation whereby technical specifications **may include environmental characteristics and may use for thatpurpose requirements that are laid down in eco-labels**. These eco-labels may be European, multi-national or national. However the requirements must have been adopted through a procedure involving all relevant stakeholders, including environmental organisations,

---

[333] Directive 2004/18/EC of the European parliament and of the Council on the coordination of procedures for the award of public work contracts, public supply contracts and public service contracts, (O.J. L 134, 30.4.2004) and Directive 2004/17/EC of the European Parliament and of the Council of 31 March 2004 coordinating the procurement procedures of entities operating in the water, energy, transport and postal services sectors (O.J. L 134 30.4.2004). Public procurement rules relate to the procedures for the award of public supply, public service, public works and utilities contracts all over the European Union and the European Economic Area.

[334] One famous and influential example is the Concordia Bus Finland Oy Ab v Helsingin kaupunki and HKL-Bussiliikenne - In its judgment of 17 September 2002 the ECJ ruled that *"where, in the context of a public contract for the provision of urban bus transport services, the contracting authority decides to award a contract to the tenderer who submits the economically most advantageous tender, it may take into consideration ecological criteria such as the level of nitrogen oxide emissions or the noise level of the buses, provided that they are linked to the subject-matter of the contract, do not confer an unrestricted freedom of choice on the authority, are expressly mentioned in the contract documents or the tender notice, and comply with all the fundamental principles of Community law, in particular the principle of non-discrimination."*

and be based on scientific information. Technical specifications are further defined in Annex VI and shall include **environmental performance levels, production processes and methods** (Annex VI).

The very important aspect of using Ecolabel criteria for public procurement is the simplicity with which local authorities, who do not have the capacities to define their environmental criteria, can simply "copy and paste" the Ecolabel criteria into their call for tender, as suggested by the Commission handbook on applying Public Procurement[335]. Energy Using Product (EuP) Ecodesign Framework

As far as the Energy Using Product (EuP) Ecodesign Framework and implementing measures[336] are concerned (*see chapter V.4.4*), both can be used as sources of information for setting criteria for the EuP implementing measures and for the conformity assessment process. EU Ecolabeled products *shall be presumed to comply with the ecodesign requirements of the applicable implementing measure insofar as those requirements are met by the eco-label.* In addition to which, for national ecolabel schemes for the same purpose the Commission *may decide that other eco-labels fulfil equivalent conditions to the Community eco-label pursuant to Regulation (EC) No 1980/2000).*

V.4.2.6.2   Integrated Product Policy (IPP)

The Ecolabel working plan identifies coordination with ongoing activities under the EU IPP Strategy(CEC 2003b) – in particular within the planned IPP working group on information instruments. As the IPP has not yet started this work and because the outcome of the IPP working group on product information is not clear, no direct results are foreseen.. Possible synergies could be the use of Ecolabel information in common IPP tools/Ecolabel product performance database for subsequent use in future setting of Ecolabel criteria and any minimum binding product requirements that the Energy Using Products Ecodesign framework will set and that the IPP may one day see fit to set for other product sectors.

V.4.2.7   TIPS FOR ECO ACTION (ENVIRONMENTAL CITIZENS' ORGANI-SATION)

V.4.2.7.1   National ECO support/use of EU Ecolabel resources

(for more information on the criteria, background studies, technical information, products labelled, where they are sold, your national EU Ecolabel competent bodies etc see http://europa.eu.int/comm/environment/ecolabel/whats_eco/index_en.htm and the Flower online catalogue - http://www.eco-label.com/)

1) **Improvement in National and EU level ECO engagement** - The identified shortcomings in **promotion of the scheme** and the associated dilution of the impact of importance of the scheme to manufacturers as well as active producer boycott, could also be addressed by a more active uptake of the Eco-label in ECO campaigns.

---

[335] Commission handbook on applying Public Procurement - Commission Staff Working Document - **Buying green!** A handbook on environmental public procurement, SEC(2004) 1050 of August 2004

[336] According to the second reading political agreement adopted by the Council COREPER April 2005 and due for EP plenary adoption shortly afterwards.

However, this would in turn depend on higher ECO confidence in the ambition level of the criteria, either generically, or on specific symbolic thematic issues (ie those particularly relevant to the ECO campaigns – e.g. PVC free, Chlorine free paper, fully and credibly certified wood fibre sources etc) which itself depends on ECOS seeking more resources for better communication and coordination between the ECO EU Eco-label coordinator (at this moment carried out by the EEB) and other ECOs and their relevant campaigns. This should be done both at the National level – direct ECO involvement in National EU Ecolabel Competent Body decision making and at EU coordination level (for example through increased and targeted funding of dedicated ECO EU Ecolabel coordination).

2) **Improvement in National ECO use of the EU Ecolabel as a reference benchmark**. Apart from direct promotion of the Flower in terms of `greening consumption´ campaigns (which can be done alone or in conjunction with relevant national labels) the EU Ecolabel can serve a multiplicity of other ends. For example:

a) use of the criteria (or background technical information and studies) for political demands in various thematic campaigns – eg the criteria on energy efficiency, emissions, hazardous substances have been scrutinised for manufacturing feasibility at an EU level, which means that in principle they should represent no or very low barriers to access to markets in any EU MS.This can be a strong (and pragmatic) argument when demanding product based solutions for thematic campaigns on climate change, toxic substance releases, global emission reductions, or can simply be a useful benchmark to use in reference to the demands being made (are they stronger or weaker than on EU labels). The fact that the EU ambition level is compromised by political consensus should/ can always be recalled.

b) the information and criteria levels can provide a good basis for further work on establishing (and even going beyond) ECO's own ecolabels , on single or multiple issues. Similarly the Ecolabel working group could potentially be a rich network of information sharing as some of the EEB MOs participating in the EEB Ecolabel working group already have their own labels.

V.4.2.7.2   The EU Ecolabel and EU Directives on Public Procurement

For more detailed information and recommendations see the joint guide developed by the coalition of European level non-governmental organisations in co-operation with European trade unions – ***Making the most of public money*** – *a practical guide to implementing and contracting under the revised EU Public Procurement Directives, November 2004*[337], see also the Commission handbook on applying Public Procurement - Commission Staff Working Document - **Buying green!** A handbook on environmental public procurement, SEC(2004) 1050 of August 2004 (and other useful links - available on EEB website on Public Procurement) (*CEC 2004*).

ECO activities concerning Ecolabels and greening public procurement can focus on 2 phases:

1) **Transposition phase** - *in the national level implementation phase of applying the revised Public Procurement Directives (normally with a deadline of 31 January 2006)* it is crucial that National and Regional Governments with the responsibility for implementing these provisions into

---

[337] See http://www.eeb.org/activities/public_procurement/Practical-Guide-Public-Procurement-FINAL-NOVEMBER2004.pdf

their legal frameworks maximise the scope to include these considerations, and that public authorities develop their contracting procedures to ensure their use in practice.

Several aspects of the EU Directives offer Member States or public authorities' choices in applying certain provisions, and decision-makers should be urged to allow a maximum scope for applying social, employment, ethical and environmental considerations. A contracting authority may include award criteria that are not of a purely economic nature, but they must be linked to the subject matter of the contract. This has to be made quite clear in national law.

273

2) **Implementation phase** – ECOs can attempt to alert authorities to the opportunities presented by the Public Procurement Directives. Environmental characteristics should take account of the entire life cycle of a product. Often the most important environmental impact occurs in the production phase. Apart from allowing public authorities to consider this, the Directives allow also for the use of eco-labels as selection criteria.

National legislation should make differences between the different types of eco-labels. Full use should be made of the leeway given by the Directives in favour of Type I (independently verified) eco-labels. Type I eco-labels like European Flower, German Blue Angel, Nordic Swan and others have the advantage of providing quantitative criteria developed in multi-stakeholder processes and they put the producer under the obligation to either give detailed evidence that they meet the criteria or have their products certified.

Note: Self-claims (Type II eco-labels) should not be recognised.

Environmental declarations Type III (ie supposedly verified) should only be used if they either

a) fulfil the stricter elements of the options envisaged in Environmental declarations International Standard currently under development - ISO 14025 i.e. they should be based on:

▶ a complete LCA

▶ product category rules (PCR) of a representative PCR programme[338]

▶ both LCA and rules should be reviewed by a panel of interested parties.

▶ the environmental product declaration at the end should be certified by an external independent third party.

or

b) **alternatively they fulfil an eventual (possibly to be developed) EU equivalent standard for Environmental Product Declarations (EPD),**

Overall the requirements on the data should ensure that the information conveyed in the product declaration is:

▶ **comparable:** meaning all necessary conventions (system borders) and calculation (e.g. data bases) rules have to be equivalent within a product category

---

[338] Product requirements can only be defined and therefore hold for products that have a common functional unit and are comparable on the basis of this unit as the requirements (on the data) are specific of products (within one product category).

274

▶ **relevant:** meaning it covers relevant environmental aspects.

▶ **transparent:** meaning the user of the information can assess for himself the preconditions by which the information was gathered and processed.

▶ **credible**: meaning the user of the information can be sure it is plausible and collected in a scientifically sound way as well as according to publicly accessible rules. This also ensures that the data is of appropriate quality.

▶ **consistent:** meaning the data was collected to produce a consistent characterisation of the environmental performance of a product in terms of the geographical, temporal, and technological borders of the product system.

BIBLIOGRAPHY AND FURTHER READING

Austrian Economic Chamber (1996) *Ecomanagement Pays ·Off. Evaluation of a Subsidised Programme for Introduction of Environmental Management and Audit Systems (EMAS) in Austria*, commissioned by the Federal Ministry of Science, Transport and Arts, and the Federal Ministry of the Environment, Youth and Family, Land Government of Upper Austria, Austrian Economic Chamber, Vienna, June.

CEC (2003) Commission Recommendation of 10 July 2003 on guidance for the implementation of Regulation (EC) No 761/2001; notified under number C(2003)2253; OJ L 184, 23.7.2003, p 19.

CEC (2003b) *Integrated Product Policy (IPP) – Building on environmental Lifecycle Thinking*, COM (2003) 302 final Brussels, 18th of June.

CEC (2004): *Buying green! A handbook on environmental public procurement.* SEC(2004) 1050, Commission Staff Working Document

EC (2000) Council Regulation on a revised Community eco-label award scheme of 17 July 2000, No 1980/2000, OJ L 237 p.1

EC (2001) Council Regulation allowing voluntary participation by organisations in a Community eco-management and audit scheme (EMAS), No 761/2001 of 19 March 2001; OJ L 114 , 24.4.2001, p.1

EC (2001b) Directive 2001/95/EC of the European Parliament and of the Council of 3 December 2001on general Product Safety.

EEB (2004): *Evaluation of the European Ecolabel and Scheme – What we wanted, what we got.* July 2004

EUROCITIES, CECOP, EDF, EEB, EPSU, ETUC, GMB, The Social Platform (2004): *A practical guide to implementing and contracting under the revised EU public procurement directives: Making the most out of public money*. Published by a large coalition of environmental and social groups, Unions and Employers as well as cities organisations.

KRUT, R. and GLECKMAN, Harris (1998) *ISO 14001 – A missed opportunity for sustainable global industrial development*, Earthscan

KVISTGAARD Consult (2005): *Evaluation of the EU Flower Week Campaign 2004*. Final Report April 2005

Swedish EMAS Council (1996) *EMAS Environmental Statement Survey*, Svenska Miljöstyrningsradet AB, April.

TASCHNER, K. (1998) *Environmental Management Systems: The European Regulation in New Instruments for Environmental Policy in the EU*, edited by Jonathan Golub, Routledge Ltd., London and New York.

275

V.4.3 Integrated Product Policy

By *Melissa Shinn*[339]

V.4.3.1 THE PURPOSE OF INTEGRATED PRODUCT POLICY (IPP)

A potentially far-reaching way to integrate a product focus into mainstream manufacturing is through an Integrated Product Policy (IPP) approach. This approach is based on Life-Cycle Thinking – it takes the product's life-cycle and aims for a reduction of the environmental impacts of products from the cradle to the grave. In so doing it aims to give comprehensive coverage of all life cycle impacts and all environmental aspects. This can also be used to prevent individual parts/impacts of the supply-chain from being addressed in a way that results in the environmental burden simply being shifted to another part or to another environmental impact type. This implies that eventually for each stage of manufacture (from extraction to end of life) environmental impacts are recorded or estimated in such a way that total lifecycle impacts can eventually be calculated.

V.4.3.2 BRIEF HISTORY OF EU IPP POLICY

After a first discussion document (Green Paper) was produced in 2001, the Commission then came up with a more detailed document (White Paper) in June 2003, in which it presented intentions and plans for a European Integrated Product Policy

V.4.3.2.1 The white paper- The IPP Communication

In June 2003 the Commission communicated its further developed intentions and plans for a European Integrated Product Policy (*CEC 2003*). However, the content was disappointing as it failed entirely to give the European Community what it needed - a strong policy tool capable of driving significant change in product design and manufacture. The shortcomings explained above were not overcome and only minor steps in a leadership direction have been made – namely putting some Commission resources behind studying a product selection methodology and some form of future product focus (a future discussion document on product design obligations). However, previous references to four product groups of particular environmental relevance - namely cars, buildings, furniture and cleaning products - were no longer in evidence. It also proposed setting up Commission-run working groups and possibly some steering (like) group. It made a committment to developing IPP indicators to assess environmental improvements, requiring member states to report on IPP implementation, and calling on member states to draw up action plans on greening Public Procurement.

However, it lacked the most important policy component of them all: active leadership through a legislative platform from which to make effective use of these actions and any results thereof and to lay down a common vision on what concrete environmental objectives were trying to be achieved, and by when.

---

[339] Senior Policy Officer for Product and Waste Policies, European Environmental Bureau

Additionally the 2003 IPP communication failed to provide:

▶ Concrete support of tools such as reduced VAT on EU Ecolabels (indeed this was specifically dismissed), this is especially regrettable in view of pending negotiations on the amendment of the 6th VAT directive

▶ Concrete activities on consumption - such as a working group towards a policy to deal with this - despite recognition of the problem of increasing consumption

▶ A clear map on the development of Product Information requirements or Product Data Files on environmental performance

▶ Mention of the potential of Producer Responsibility for Environmentally sound products - such as a driver for 'mass (widespread) integration' of life-cycle thinking and environmental considerations into manufacturing and the standardisation process

▶ Recognition of the necessity to create efficient, tailor-made resources such as the IPP or Benchmarking Institute, in the style of the IPPC Seville Bureau and setting up of important 'consumer information tools' such as EU Eco-test networks

▶ Recognition of the importance of extended producer responsibility in general and of the potential of individual producer responsibility

▶ Recognition of the need to address the short comings of the New Approach and the lack of integration of environmental interests into the EU's privatised standardisation process

The main activities generating from the Communication consisted of 3 studies (on product prioritisation, life-cycle information and interpretative tools, development of indicators for IPP) and voluntary product pilot projects to start 'learning by doing'. It however remains to be seen what the Commission can do with the results of such 'learning by doing' if there is no legal mandate to use the results.

V.4.3.2.2   The EEB proposals

In February 2003 the EEB published its vision and demands for an appropriate EU integrated product policy framework[340]. The vision paper does not necessarily foresee the immediate launch of product-specific legislation, but first, a general legislative platform in the form of a general IPP Framework Directive. This would create the appropriate setting for the Commission and Member States to turn the IPP into a 'live' policy and gives the Commission the mandate to invest resources in its implementation.

The vision includes laying down clear environmental objectives based on existing policies - especially the 6th EAP - and using existing targets and commitments in the 6EAP and other existing policy ambitions such as Kyoto, Ospar (marine pollution) and Persistent Organic Pollutant (POPs) Conventions, the marine strategy, the air quality objectives, the waste prevention strategy etc...

---

[340] See An NGO vision paper towards a European Integrated Product Policy , EEB, March 2003 at www.eeb.org/activities/ product_policy/main.htm – under Position Papers

A second key element is creating a general legal provision for environmentally-sound products, to deny market access to those product designs that have the greatest negative impacts and to drive environmental considerations and the lifecycle methodology fully into product manufacture and consequently the European Standardisation process.

Obliging producers to supply product lifecycle information is another crucial step, and setting up the necessary resources such as a benchmarking institution, to be able to provide producers and the standardisation process with important tools such as product benchmarks and lifecycle data is essential for enabling further product innovation, setting product performance criteria and providing information for future policy. It foresees a Commission 'working framework' to establish working groups on specific IPP tools, such as economic incentives.

Simulation of a directive on environmentally sound products

The EEB subsequently developed further the idea of building an IPP legal framework in the form of **a generic environmentally sound product responsibility directive**, building on the approach of the Product Safety Directive.

Overview of EEB proposal for an EU IPP Strategy and Policy framework

EU IPP STRATEGY:
Should –
Define Timeframe / scope / ambition and environmental objectives / indicators/ reporting
Set–up the following implementing mechanisms:
▶ **EU level Steering Group**
   With IPP tools Working groups
      - Economic tools
      - Procurement
      - Product data
      - Producer responsibility on environmentally sound products
      - Product Prioritisation
      - Consumption Policy...
▶ **Framework Directive on Producer responsibility on environmentally sound products**
   • General producer responsibilities on environmentally sound products
   • Obligatory product lifecycle information
▶ Operationalising **Daughter Regulations** – on specific products
   • Providing Politically set minimum requirements
▶ **Resources (€100 Million)**
      - Research Bureau (or use JRC?)
      - European Benchmarking Institute .
      - Eco-test network

The main objectives of an **environmentally sound product responsibility directive** would be to:[341]

1. Lay down the principal that all consumer products should be *environmentally sound* and enforce this by restricting market access for products that are environmentally not sound.

2. Make the integrated methodology[342] and lifecycle approach to environmental soundness a general requirement of product design and manufacture.

3. Establish the general requirements for the environmentally sound products in the absence of specific requirements.(Provide an instrument that gives the Benchmarking Institute/ Commission the information to set cut-off criteria and to deny market access to products which do not meet the requirements and whose producers do not disclose information on the significant environmental impacts of their products)

4. Foresee the establishment of specific minimum criteria for environmentally sound products

5. Require producers to supply life-cycle data on their products and foresee standardisation of the reporting of this data. The data should refer to sector specific baseline criteria and be third party certified.

V.4.3.3   IMPLEMENTATION OF IPP TO DATE

The Commission began implementation of the EU IPP in 2004 establishing twice yearly 'Regular' meetings with stakeholders and Member State competent authorities. The studies to develop indicators for IPP (end June 2005) to make life-cycle information and interpretative tools available (end November 2004) and identify the products with the greatest potential for environmental improvement are all underway and due in 2005. The Commission made use of the internet for stakeholder consultation on working documents.

Working groups on product information and one on Member State reporting have been established and are due to start work soon (an EEB proposal for a working group on an environmentally sound products directive was rejected). Two voluntary pilot projects were launched in 2004 involving analysing a mobile phone and a teak garden chair and their environmental impacts during the life-cycle. The Commission had asked all stakeholders, through its Internet site, to supply the Commission with information on the environmental impacts of the products in question.

Meanwhile the Commission has restated its commitment to taking a decision on Environmental Product Declarations (whether EU level guidance is necessary or the work done in ISO is sufficient) in 2005.

---

[341] For more details see EEB Explanatory memorandum on Sound Products June2004, at http://www.eeb.org/activities/product_policy/EEB-working-document-Explanatory-memorandum-on-Sound-Products-June2004.pdf

[342] integrating in this case means integrating all the environmental aspects (water, air, materials, substances etc)

V.4.3.4 LINKS WITH ISO STANDARDS ON ENVIRONMENTAL PRODUCT DECLARATIONS

The International Standards Organisation (ISO) is currently in the process of revising ISO 14025 on Environmental product declarations. The standard has meanwhile changed its name to "Environmental declarations". However the ISO standard will unfortunately not address important elements such as comparable data. Furthermore it will allow companies to issue self-audits. ECOs could still try to improve the ISO standard via their national standard institutes. If they do not succeed they have to insist that the Commission, when intending to use this standard for EU purposes, add stricter requirements on comparability and credibility of the data. To achieve comparability, criteria and parameters have to be fixed and measuring methods standardised. Credibility can be guaranteed by verification through independent third party verification or – as the second best solution – through an independent external review panel[343].

V.4.3.5 TIPS FOR ECO ACTION (ENVIRONMENTAL CITIZENS' ORGANISATION)

Given the lack of concrete outcomes of the IPP process at EU level (legislation etc..) there is not much that can be used at the National campaigning level other than to **push national governments to develop their own IPP Strategies** and possibly incorporate some of the EEB recommendations on activities. As product policy is seen more and more as an EU domain (following the objectives of a harmonised internal market and free movement of goods) it becomes more difficult for Member States to find the political will at National level to act in a Strategic way on environmental product policy, pending EU action. This makes the `stand still` at EU level doubly problematic as the `eventual` promise of an EU product policy holds off action at National level but does not deliver at EU level.

As far as pushing for National Strategies is concerned, ECOs can **make use of the IPP Regular meeting minutes** for inspiration. These reports identify interesting studies, reports and **policy activities in other EU Member States**. These are publicly available at the European Commission website - http://europa.eu.int/comm/environment/ipp/meetings.htm For example in Denmark two studies are ongoing: on product prioritisation; and on how IPP can be integrated into other EU policy areas. The national stakeholder panel on professional green procurement has now adopted an action plan. In Sweden, the Swedish EPA was working with UNEP and Lund University on preparing an eco-design handbook that was due to be completed in 2004. The EPA has also produced some new information for public purchasers and is also in the process of putting together a plan on how to develop information flow along the product chain. In Austria a ministerial order was drawn up in 2004 that will develop some guidelines containing obligatory environmental considerations etc.

Unfortunately the Commission has not allowed one very IPP active country, Switzerland, to join the EU's regular IPP meetings (even on an informal basis - due to their non member status) the EEB does.. Interesting activities that Switzerland is carrying out include establishing a Swiss

---

[343] For more information please request ECOS expert reports– see http://www.ecostandard.org

Centre for LifeCycle inventories, the already existing Ecological Purchasing Service of the Swiss Agency of Environment, forests and Landscapes, providing advice to the Competence Centre for Federal Public Procurement, including a new information tool for public purchasers (*GIMAP*)[344].

The **link with Public Procurement** legislation is maybe the most significant and that which has the biggest potential of working on IPP in practice at this point in time. Public procurement accounts for over 16% of the European Union's GDP equivalent to 1500 billion Euros. The European Union adopted two directives in 2004 revising European rules Public Procurement (*EC 2004a and EC 2004b*)[345] (*see also chapter V.4.2.7.2*).

BIBLIOGRAPHY AND FURTHER READING

CEC (2001): Green Paper on Integrated Product Policy. (2001) 68 FINAL

CEC (2003) *Integrated Product Policy (IPP) – Building on environmental Lifecycle Thinking*, COM (2003) 302 final Brussels, 18 June.

EC (2004a) Directive 2004/17/EC of the European Parliament and of the Council of 31 March 2004 coordinating the procurement procedures of entities operating in the water, energy, transport and postal services sectors, 2004/17/EC, O.J. L 134 30.4.2004.

EC (2004b) Directive of the European parliament and of the Council on the coordination of procedures for the award of public work contracts, public supply contracts and public service contracts, 2004/18/EC, O.J. L 134, 30.4.2004

EEB (2001): *EEB response to the Commission Green Paper on Integrated Product Policy*

EEB (2003): *An Environmental NGO vision paper towards a European Integrated Product Policy*

EEB (2004): *Explanatory memorandum on Sound Products Directive*

---

[344] For more information see **www.ecoinvent.ch** and http://www.environment-switzerland.ch/buw—-al/eng/fachgebiete/fg_produkte/strategie/index.html.

[345] Public procurement rules relate to the procedures for the award of public supply, public service, public works and utilities contracts all over the European Union and the European Economic Area.

V.4.4 The Ecodesign framework for energy using products (EuP) Directive

----

By *Melissa Shinn*[346]

In August 2003 the Commission published its proposal for a directive establishing a framework for the setting of Eco-design requirements for Energy- Using Products (EuP) *(EC 2003)*. Its aim was *to ensure free movement of products; it contributes to sustainable development through increased security of energy supply and high environmental protection.*

The EuP was a result of the merger of a planned Energy using Products energy efficiency directive and the Ecodesign of Electrical and Electronic Equipment proposal (previously part of a triple package on Electrical and Electronic Equipment addressing Waste EEE, Restriction of Hazardous substances in EEE and Ecodesign).

At the time of writing the EuP Directive has just been agreed between the European Parliament and Council.

V.4.4.1 THE PURPOSE OF EUP

The explanatory memorandum states that with *the wide dissemination and considerable environmental impacts of energy-using products, growing public awareness of environmental issues will result in the environmental aspects and performance of such products being increasingly targeted. It is important to avoid market fragmentation through diverging national requirements as regards the environmental aspects of these products. It consequently appears necessary to create a coherent harmonised Community framework in which to address these "eco-design" requirements.* It reminds us that it is estimated that over 80% of all product-related environmental impacts are determined during the product design phase[347]. **Integrating environmental considerations as early as possible into the product development process is therefore the most effective way of introducing changes and improvements to products.**

The EuP foresees the setting up of `daughter´ implementing measures using a Commission steered Regulatory Committee (preparing and taking Commission decisions by consulting and receiving approval from Member States experts). Whilst the implementing measures foreseen are legally binding instruments, making full use of EU harmonised standards, the EuP proposal gives considerable priority to voluntary tools (such as voluntary agreements) as alternative avenues of action - where such actions are likely to deliver the policy objectives faster or in a less costly manner than mandatory requirements. *Legislative measures may be needed where market forces fail to evolve in the right direction or at an acceptable speed* (recitals).

----

[346] Senior Policy Officer for Product and Waste Policies, European Environmental Bureau

[347] "How to do EcoDesign?", a guide for environmentally and economically sound design edited by the German federal Environmental Agency, Verlag form, 2000

Concerning the environmental focus, the EuP has a very strong emphasis on energy efficiency over other environmental aspects (due to its merger with the Energy Efficiency proposal and the high political attention given to climate change and the demand side efficiency agenda at the time of its adoption).

V.4.4.2   STRENGTHS AND WEAKNESSES OF THE FINAL TEXT

In April 2005 a political agreement was reached between the European Parliament, Council and Commission on the EuP directive text. Analysis of this agreement reveals that the European Parliament's efforts to improve the governance and environmental ambition of the EuP proposal have only been partially successful, with the exception of the aspects concerning action on energy efficiency where some significant changes were made.

The most significant positive element of the EuP framework is that is does foresee the setting of EU legally binding 'minimum' cut-off criteria through the setting of 'specific requirements (annex II) for all products placed on the EU market including importers. It foresees (theoretically) that these should be used for all environmental impact categories (with a detailed clarifying list in annex I). This has the potential to get rid of some of the worst products on the market.

The functioning of the directive can be divided into two fundamental aspects – those concerning the environmental ambition level of the criteria setting and those concerning the governance foreseen for the implementation of the directive.

The main critical elements concerning the environmental ambition of the directive are:

▶ the article basis – the directive is based on article 95 of the treaty ("free movement") which means Member States may only go beyond the prescribed EU level of protection under restricted and predefined conditions.

▶ the references (in recitals and main implementing article – article 12) to environmental policies and objectives thereof. Unfortunately there is only fleeting reference to the 6EAP and the European Climate Change Programme (ECCP), lacking specific references to e.g. waste and chemical policy objectives.

▶ the use of Commission defined benchmarking foreseen in the directive for setting requirements with no levels (so called generic requirements – Annex I) is a significant element as it ensures there is some level of accountability and comparability of manufacturer progress on Ecodesign for these requirements. High ambition levels for requirements where levels ARE set – so called specific requirements – Annex II). The directive requires that *impacts should be* minimised, instead of just reduced, and international benchmarking should be included in analyses – especially concerning energy efficiency in use.

▶ The directive anticipates a number of elements specifically on energy efficiency – including the use of the Least Lifecycle Costing methodology, a obligation for the Commission to have a scheduled working plan (including an obligation to address as a matter of priority the products identified by the European Climate Change Programme (ECCP) as having the greatest potential for demand side energy efficiency).

Key elements of the directive as concerns governance are the following:

▶ a parallel stakeholder 'consultative' committee to oversee the Commission's Regulatory Committee work (to be further defined by its own regulation),

▶ foresees (but does not oblige funding) of ECO participation in the consultative committee and in associated scrutiny of standardisation activities

The final negotiation on the directive unfortunately lead to prioritising elements of national enforcement measures and special resources for SMES - articles 6 and 7 and 11 respectively) rather than many important environmental ambition and governance aspects. In order to give priority to environment over free movement, article 175 would have been preferable to article 95 and would have allowed Member States the freedom to go beyond the EU levels set (which based on article 95 they may only do so under certain restrictive predefined conditions – new scientific evidence etc).

Furthermore, stronger links to other existing policies, objectives and targets (such as waste prevention, hazardous substances etc) would have created clearer obligations on the Commission when implementing. At the same time the prioritisation of the status of voluntary agreements and criteria on protection of competitivity will potentially weaken efforts to maintain the upper hand for binding, environmentally focused instruments.

Most of the specific elements on energy efficiency were taken up by the agreement - Article 13 and Annex II. An exception to this was one of the elements with the greatest potential, namely a specific requirement to create a standalone measure on energy consumption in standby, now relegated to a recital.

Other significant elements that were added/strengthened include the setting up of the stakeholder consultative forum (originally an EEB proposal) – Article 14 - and requirements for the setting of a common Commission defined benchmark when setting generic requirements (i.e. product requirements where setting a specific cut-off level is not foreseen) – Annex I paragraph 3. There was a significant mitigation of the criteria protecting competitivity applied to the selection procedure (Article 12) and the use of voluntary agreements were made conditional to multiple criteria assessment on their potential performance (but are regrettably still seen as priority measures).

Crucially, however, the agreement failed to introduce many changes on important governance aspects, many were requested by the Parliament in the first reading. In particular, the following did not make it through to the final directive:

▶ a Parliamentary call back mechanism on the Implementing Measures adopted by the Commission (so that stakeholders can call the alarm if implementing measures are not acceptable)

▶ a feedback mechanism between the process of development of EU harmonised standards and the consultative forum to allow an external `quality control´ of standards whilst they are being developed

▶ independent eco-design and lifecycle impacts expertise (independent from industry interests) to reinforce the Commission's capacity to perform balanced analyses towards the implementing measures

▶ stronger possibilities to challenge the (automatic) use of industry self-declaration for conformity as a default, as opposed to stronger models of conformity assessment (for example models whereby the producer has to involve third parties to verify their claims)

Last, and very crucially, including more detailed requirements on producers to provide information on the performance of their products as regards their lifecycle impacts (data ideally to be standardised and collected centrally). Parliamentary amendments foresaw specific requirements on manufacture provision of standardised product information. These were not taken up in the agreement so all that remains are the general provisions in Annex I - Part 2 on *Requirements relating to the supply of information* and article 11 on *Administrative cooperation and exchange of information* that unfortunately foresee only information for consumer, supply chain or end of life purposes or enforcement monitoring information between authorities, Member States and the Commission.

Some additional negative elements – mainly introduced by the Council were also taken up – in particular the provision that so called `negative´ requirements can be set (*implementing measures may also provide that no ecodesign requirement is necessary for certain specified ecodesign parameters …*). This is the first time we see such a measure in EU policy whereby a harmonisation measure can define that `no requirement´ is necessary, and this could be dangerous taking into account the fact that the framework is based on Article 95 of the EU treaty (ie national measures may not, without difficulty, be more ambitious than EU measures.). It remains to be seen if these negative requirements can be interpreted as 'addressing' an issue. If that is the case this could effectively cap national activities despite a continued absence of EU requirements.

V.4.4.3   FUTURE IMPLEMENTATION

Given the obligation for the Commission to establish a work plan of products to be addressed (foreseen to be published by 2007) any implementing measures on specific products are not to be expected until 2008/2009. However the mandate (article 13.2) for the Commission to adopt implementing measures during the transitional period on those products which have been identified by the ECCP means that the Commission may actually move forward with some implementing measure proposals before then. Methodological studies are already underway under the management of the Directorate General for Enterprise.

However, given the lack of obligatory funding for civil society organisations involvement in the stakeholder consultative forum it will depend very much on the Commissions desire to engage environmental advocates and the guidelines set in the subsequent Commission regulation that will set up the forum (still to be defined). The achievement of the other governance and environ-mental ambition objectives missing in the regulation itself will in its turn depend on the resources dedicated to the involvement of ECOs in the forum and the Commissions receptiveness to compensating for some of the missing elements in the regulation – such as the focus on issues other than energy efficiency (for example waste prevention and recyclability) and product infor-mation requirements and the use of independent expertise for analyses and studies (the latter being an issue the European Parliament is keen to follow up on).

V.4.4.4   LINKS WITH OTHER LEGISLATION / POLICY TOOLS

The EuP Directive has direct links to other legislation – in particular it is supposed to be 'comple-mentary' to legislation on labelling and standard product information of the consumption of energy and other resources by household appliances and the energy efficiency labelling pro-gramme for office equipment   (Council Directive 92/75/EEC of 22 September 1992[348] and Regulation (EC) No 1980/2000 [349] respectively). The same applies to Directive 2002/96/EC of the European Parliament and of the Council of 27 January 2003 on waste electrical and electronic equipment (WEEE) [350], Directive 2002/95/EC of the European Parliament and of the Council of 27 January 2003 on the restriction of the use of certain hazardous substances in electrical and elec-tronic equipment[351] and Council Directive 76/769/EEC of 27 July 1976 relating to restrictions on the marketing and use of certain dangerous substances and preparations[352]. *Synergies between this Directive and the existing Community instruments should contribute to increasing their respec-tive impacts and building coherent requirements for manufacturers to apply* (recitals).

There are also complementary links to Council Directive 90/396/EEC of 29 June 1990 on the approximation of the laws of the Member States relating to appliances burning gaseous fuels[353] and Directive 2002/91/EC of the European Parliament and of the Council of 16 December 2002 on the energy performance of buildings[354] especially as concerns performance of heat generators for space heating and the production of hot water and the insulation of heat and domestic hot-water distribution in new non-industrial buildings.

The EuP Directive also integrates some existing legislation that already contained provisions for the energy efficiency requirements (and foresaw future revisions themselves) giving them the new status of implementing measures of the EuP framework – namely the Directive 92/42/EEC of 21 May 1992 on efficiency requirements for new hotwater boilers fired with liquid or gaseous

[348] OJ L 297, 13.10.1992, p. 16. Directive as amended by Regulation (EC) No 1882/2003 of the European Parliament and of the Council (OJ L 284, 31.10.2003, p. 1).

[349] OJ L 332, 15.12.2001, p. 1.

[350] OJ L 37, 13.2.2003, p. 24. Directive as last amended by Directive 2003/108/EC (OJ L 345, 31.12.2003, p. 106).

[351] OJ L 37, 13.2.2003, p. 19.

[352] OJ L 262, 27.9.1976, p. 201. Directive as last amended by Commission Directive 2004/98/EC (OJ L 305, 1.10.2004, p. 63).

[353] OJ L 196, 26.7.1990, p. 15. Directive as amended by Directive 93/68/EEC (OJ L 220, 30.8.1993, p. 1).

[354] OJ L 1, 4.1.2003, p. 65.

fuels[355], Directive 96/57/EC of 3 September 1996 on energy efficiency requirements for household electric refrigerators, freezers and combinations thereof[356] and Directive 2000/55/EC of 18 September 2000 on energy efficiency requirements for ballasts for fluorescent lighting[357].

The EuP foresees that the Directive 92/42/EEC providing for a star rating system intended to ascertain the energy performance of boilers be amended *to open the way for more effective schemes....since Member States and the industry agree that the star rating system has proved not to deliver the expected result.*

287

The EuP framework also absorbs (and so effectively repeals) Council Directive 86/594/EEC of 1 December 1986 on airborne noise emitted by household appliances[358] which lays down the conditions under which publication of information on the noise emitted by such appliances may be required by Member States, and defines a procedure to determine the level of noise (although *Member States may continue to apply existing national measures adopted under Directive 86/594/EEC until such time as implementing measures for the products concerned are adopted under this Directive*). These integrations and repeals will have to be watched to ensure there is no regression as concerns environmental performance/information requirements.

Concerning the link to IPP the EuP's explanatory memorandum states (echoed in the directives recitals) that the implementation of the present Directive will contribute to the integration of life-cycle thinking, one of the basic principles of the Integrated Product Policy (IPP), into product design. The experiences with its implementation *will contribute to judging the appropriateness of establishing similar parallel framework Directives for other products*, or general obligations on producers to undertake eco-design, activities that will be pursued in the follow-up to the Communication on IPP.

As mentioned above (see above under *The EU Ecolabel – The Flower*) the EuP makes use of both the EU and National Ecolabels (for both content and proving conformity) and foresees the use of EU harmonised standards to ease implementation. Concerning standards there are significant issues of lack of democracy in setting such standards (for more details see Chapter V.4.5 on Standardisation and New Approach and the example of the packaging directive given in Chapter IV.4.4 on Waste).

---

[355] OJ L 167, 22.6.1992, p. 17. Directive as last amended by Directive 2004/8/EC of the European Parliament and of the Council (OJ L 52, 21.2.2004, p. 50).

[356] OJ L 236, 18.9.1996, p. 36.

[357] OJ L 279, 1.11.2000, p. 33.

[358] OJ L 344, 6.12.1986, p. 24. Directive as amended by Council Regulation (EC) No 807/2003 (OJ L 122, 16.5.2003, p. 36).

V.4.4.5 TIPS FOR ECO ACTION (ENVIRONMENTAL CITIZENS' ORGANI-SATION)

Concerning transposition, many of the governance elements listed in *chapter V.4.4.2* concerning the final EuP directive's weaknesses can be reinforced at National level. In particular:

288

a) funding of national ECO participation in national level consultative committees and in asso-ciated scrutiny of standardisation activities;

b) stronger pressure/responsibilities on the national authorities to asses the effectiveness of using industry self-declaration for conformity as a default as opposed to stronger models of conformity assessment (for example models whereby the producer has to involve third parties to verify their claims);

c) dedicated national/ regional product Ecodesign resources (national expert body or bureau) that can be responsible for following the implementation and/or enforcement of the eco-design requirements placed on producers. This body can also collect product performance information and make it a) publicly available, b) comparable for wider use (in research, labelling, business to business product declarations and future ecodesign requirements).

Perhaps most important however is the addition to the national transposition of the Ecodesign directive of more detailed requirements on producers to provide information on the performance of their products as concerns their lifecycle impacts for expert usage — ie not just consumers, their own manufacture chain and end-of-life-purposes, but also for researchers, ECOs, authori-ties etc (data ideally to be standardised and collected centrally at a national level).

BIBLIOGRAPHY AND FURTHER READING

EC (2003) Proposal for a directive On establishing a framework for the setting of Eco-design requirements for Energy- Using Products and amending Council Directive 92/42/EEC, COM(2003) 453 final - Brussels 1 August.

EEB & WWF (2004): *Recommendations on the proposed Eco-design Framework for Energy Using Products (COM 2003 (453).*

UBA - German federal Environment Agency (2000): *How to do EcoDesign?, a guide for environmen-tally and economically sound design.* Verlag form, 2000

V.4.5   New Approach and Standardisation

By Karola Taschner[359]

V.4.5.1   INTRODUCTION   289

One of the four freedoms of the EU is the free circulation of goods. To achieve this goal, it turned out that it was even more important to remove technical barriers to trade than customs and tariffs. Apart from national legislation and technical regulations referring to standards, standards themselves have been revealed as an important source of market distortion and potential trade barriers and have so become the target of Community action.

Standards have been developed over the past 100 years or so, and are one of the backbones of industrial societies. Standardisation is a method of setting up technical specifications in order to facilitate the technical compatibility of products, parts of products and production equipment etc. Nuts and bolts are a striking example of standardisation at work. Standardisation institutes are private and standards used to be developed by industry for industry although anybody with an interest in them is invited to comment.

Standard writing is a highly formalised process and not easy to influence. Technical committees and their attached working groups are rather autonomous and defend the right to produce standards that serve the needs of the participating industry. Standard writing experts perceive the purpose of a standard to be creating compatibility and not to serve the protection of the environment. This stated, the fact is that standards can have considerable effects on the environment.

Technical standards are recommendations from the European standardisation organisations[360] to their members. There is therefore no obligation to apply European standards. However, overwhelming reasons of practicability (creating common understanding, for monitoring etc and facilitating trade) underline the drive for the use of so called (EU) harmonized standards. Often the legislation uses different techniques to favour the application of standards. In the most direct form, the legislation itself contains a reference to technical standards, e.g. for measuring methods, tests for certain substances, development of certain procedures and procedures and management systems. Other methods consist of establishing a presumption that a product complies with legislative requirements, where the product conforms with the standards. Furthermore, the legislative framework for the market access of a range of 20 product groups follows the so called "new approach"

---

[359] Scientific advisor to the European Environmental Bureau

[360] There are at present three European standardisation organisations: CEN (Comité Européen de Normalisation), CENELEC (Comité Européen de Normalisation Electrotechnique) and ETSI (European Telecommunications Standards Institute); the national standardisation organisations are members, in part also individual companies. Membership extends beyond the area of the European Union.

V.4.5.2   COMMUNITY ACTION

In order to prevent the uncontrolled publication of ever more technical standards and regulations, the Community developed rules on how to deal with technical regulations and standards
in two different ways.

V.4.5.2.1   Notification and standstill obligation

In 1983 the Council adopted a Directive (*EC 1983*) which has been amended several times since
and consolidated in 1998 (*EC 1998*).

Member States must inform the Commission and other MSs of draft technical regulations[361], the
adoption of which is delayed in order to give the Commission and other Member States the
opportunity to make comments or to decide whether there is a technical trade barrier and thus
a need for the Commission to intervene and make its own proposals. During this time MSs have
to put the adoption of any such technical regulations on hold.

The Directive furthermore requires the Commission and national standard bodies (NSBs) to be
informed at least every four months of the new items the NSBs have added to their standards programme. The Commission and other NSBs either comment or the Commission issues a request
to European standard bodies[362] (ESBs) to develop a European standard. During this time NSBs are
obliged to follow the stand-still rule, i.e. they are not to adopt the standard and Member States
are not to create legislative or other references to it. The membership of CEN, CENELEC and ETSI
consists of the national standard bodies of EU and EFTA[363].

This directive has given standard bodies a prominent role in internal market policy.

For implementing the Directive, the Commission is accompanied in its action by the standing
98/34 Committee whose members are appointed by Member States.

V.4.5.2.2   "New Approach"

Rather often the explanation for deviating requirements in MSs lies in different levels of protection with regard to the safety of consumers, workers at their workplace or the protection of the
environment. (Thus, to a certain extent much of Europe's consumer and environmental policy has
been a by-product of single market policy.) In the past, the Community made attempts to make
proposals and to regulate technical issues but soon realised that neither the Commission, nor
the Council or the Parliament were able to cope with the complexity of the technical details.

In order to find a systematic solution the Commission presented the "New Approach" policy
which was eventually adopted as a Council Resolution on 7 May 1985.[364]

---

[361] technical specification or other requirement in laws, regulations or administrative provisions of a MS the observance of
which is compulsory

[362] CEN (European Committee for Standardisation), CENELEC (European Committee for Electrotechnical Standardisation) and
ETSI (European Telecommunications Standards Institute)

[363] European Free Trade Association

[364] Council Resolution of 7 May 1985 on a "new approach" to technical harmonisation and standards, OJ N°C136 of 4 June
1985, p.1.

It represents an innovative program of technical harmonisation and standardisation. It is thus named as it significantly differs from the way in which European Directives were drafted in the past. The "old Directives" were in fact characterised by three main shortcomings:

▶ as technical harmonisation Directives, they showed a high level of technical specification, making them very complex

291

▶ their adoption, requiring unanimity within the Council, took a long time to be finalised

▶ as a consequence, they were often out-dated by technological developments before they were accepted as binding norms.

The "new approach" is instead characterised by a clear separation of tasks and responsibilities between the European legislator and the European standards bodies (European Committee for Standardisation-CEN, European Committee for Electrotechnical Standardisation-CENELEC and European Telecommunication Standards Institute-ETSI):

▶ The New Approach Directives define the "essential requirements" that goods must meet when they are placed on the Community market

▶ the EU Commission issues a Mandate according to which European standards bodies draw up the corresponding technical specifications meeting the essential requirements of the Directives, compliance with which will provide a presumption of conformity with the essential requirements themselves. These specifications are known as "harmonised standards".

In terms of achieving environmental ambitious standardisation Europe has failed so far. In the only case of a New Approach for the environment - the Directive on packaging and packaging waste - this has become very visible, where lack of clear guidance in the Directive and the Commission mandate, lack of balanced participation of stakeholders and lack of control of the quality of the end product resulted in inadequate standards to support EU environmental obligations and objectives (*EEB 2005*) (see **http://europa.eu.int/comm/enterprise/newapproach/ index.htm** for a list of Directives following the New Approach).

The Commission also gives Mandates for standards outside the framework of the New Approach that deal with the access of products to the market. There are Commission Mandates for products, services, installations, management systems, pollution concentrations or emissions etc. The crucial difference to legislation is simply that standards are drafted and agreed not by the legislative institutions (European Parliament and Council) in (relatively) accessible and democratic codecisions, but by the industry and/or the standardisation associations, dominated by industry interests.

If a Directive requires assistance from standard bodies for its implementation – be it "New Approach" or otherwise – the EU Commission issues a mandate as negotiated with a European standard body (ESB). The Commission pays ESBs to develop standards according to the mandate. The Technical Committee in charge is none the less autonomous in completing the mandate. If they do not comply with legislation or a mandate, the only sanction consists of the Commission not publishing the reference of the standard.

The standardisation procedure follows the internal rules of standard bodies. After having received and accepted the Mandate, the respective European standardisation body sets up a technical committee to develop the appropriate "technical specifications" and a business plan for defining the scope of the "harmonized standards" needed. This is an important phase where participation is extremely important.

When the text of the standard has finally been agreed upon in the technical committee and the Commission is also satisfied, the latter publishes the reference of the standard in the Official Journal of the EU (OJ). Such standards enjoy a quasi legal status, since the "national authorities are obliged to recognise that products manufactured in conformity with requirements established by the Directive (This signifies that the producer has the choice of not manufacturing in conformity with the standards but that in this event he has an obligation to prove that his products conform to the essential requirements of the Directive)", i.e. applying the standard demonstrates the assumption of conformity.

After having issued the mandate, the Commission is given little room to intervene. The Commission can comment while the standard is under development or at the enquiry phase. A Commission official coming to the working group meeting would have no more rights than any other member of the group. Also after assessment of the adopted standard, the Commission has no right whatsoever to change or amend the standard text unilaterally. (Here it should be mentioned that the screening of the standard is by no means common practice.) There only exists one sanction for the Commission in case the standard does not meet the essential requirements of the directive or the terms of reference of the mandate: not to publish the standard reference in the OJ.

V.4.5.3 ROLE OF STANDARDS IN EU LEGISLATION FROM "BUSINESS TO BUSINESS" TO "BUSINESS TO PUBLIC"?

Considering the growing importance of standards for products within the EU and the fact that no standard is just the result of technical considerations alone, but rather the result of economic interests, the environmental impact of standards is only taken into account if there is an obligation to do so either by legislation or a Commission mandate.

V.4.5.3.1 Non-Mandated Standard

Apart from the standards under the **New Approach** or other mandates of the EU Commission, other voluntary non-mandated technical standards have been entering into EU environmental legislation as reference methods etc. and – as mentioned above - fulfil quasi legal functions.

On one side the Commission issues mandates for test procedures for monitoring purposes: measuring and sampling (e.g., air pollutants), for environmental product performance (e.g., motoring fuels, solid recovered fuel).

On the other side European and international standard bodies develop technical standards spontaneously without a mandate which they offer to the Commission when it is defining a new policy area (e.g., water framework directive, soil strategy) or are requested to do so when the Commission sets up a programming mandate e.g. under the Directive on energy using products (*see Chapter V.4.4*).

Many legislative acts of the EU refer to technical standards that have been developed by European or international standard bodies[365]. Standard references on measurement, testing, sample taking, calibration methods etc. are an integral part of annexes and are thus compulsory. Sometimes more standards for the same purpose are added without their compatibility having been tested. Such references to standards can be found in the directives concerning air and water quality, motor vehicle exhaust gases. They are also very common in the decisions the Commission issues under the EU eco-label Regulation. The Directives on Public Procurement (*EC 2004a and 2004b*) propose standards as a reference for defining technical specifications.

The entire text of a standard has even been taken up in an EU legislative text: Annex I of the EMAS-Regulation (*EC 2001b*) consists of Section 4 of EN ISO 14001:1996.

V.4.5.3.2 Public Participation

As outlined above the participants in standard groups are usually industry experts defending certain interests, i.e. the patents of the companies paying them.

It was obvious that at least under the New Approach mandates this industry domination could not continue. Industry could not take over the task of developing the standards on its own for quasi legal purposes. Instead, other stakeholders needed to be involved. To guarantee at least some participation of stakeholders, consumers (ANEC), trade unions (TUTB) and later also small and medium enterprises (NORMAPME) were given support from the Commission to organise themselves and to participate in the process. While environmental organisations initially were denied funding for setting up their own technical bureau, in 2002 the Commission supported the start of ECOS (European Environmental Citizens Organisation for Standardisation).

The Commission issued two Communications in 2004 that also highlight the need for public participation.[366,367]

---

[365] ISO (International Standardisation Organisation), IEC (International Electrotechnical Committee)

[366] The role of European standardisation in the framework of European policies and legislation; Commission Communication COM(2004) 674 final of 18.10.2004

[367] Commission Communication (2004) on Integration of Environmental Aspects into European Standardization, (COM (2004)130 final)

V.4.5.4 WHO DEVELOPS STANDARDS AND WHAT ARE THE ENVIRON-MENTAL PROBLEMS

Standardisers have to respect safety requirements not only according to the new approach directives but also due to Community legislation on product liability[368] and product safety (*EC 2001a*). This is different for the environment where no such legislation exists.

294

When developing non-mandated standards as described above, standard writers have no obligation whatsoever to take the environment into account. Standards bodies have made several attempts to cope with this problem by incorporating respect for environmental aspects while drafting standards. However, the success of these initiatives has remained limited as these internal rules have no sanctions with which to enforce them.

Finally, in 2004, the European Commission reacted by presenting its Communication on the "Integration of Environmental Aspects into European Standardisation". While this Communication puts the emphasis on the use of standardisation for the purpose of environmental protection, ECOs give priority to binding legislation or at least insist on the necessity that essential requirements have to be clearly defined in the legislative text and mandates, before standards can be regarded as an effective and accountable way forward for the environment.

For the time being, the problem of how non-mandated standards take up environmental aspects is left to free negotiation inside the standards groups. The voice of the environment, however, is usually not strong enough as economic interests prevail. The Communication envisages training of standardisers. This may be useful in some cases although environmental experts have often observed that their colleagues from industry are very well aware of the environmental problems.

V.4.5.4.1 Standard bodies and how they work

In the European Union, the standardisation system consists of the three above mentioned organisations CEN[369], CENELEC[370] and ETSI[371]. These organisations are responsible for the establishment of European Standards in their respective fields of activities. They achieve this goal by managing Technical Committees, consisting of national delegations established by the national standardisation institutes (or bodies). Standard bodies are composed of a great number of technical committees working autonomously at their own cost and for their self determined goals.

The latter should ensure that interested parties are informed about the text of European standards being discussed and be given opportunity to comment on the text. For that purpose they create and convene national committees mirroring European technical committees, organise public enquiries and transfer national positions. The national member of ESB's, i.e. the National Standard Bodies (NSBs) vote on the final texts to approve European standards, according to a weighted-vote procedure similar to the voting system in the Council of Ministers. The participants

---

[368] Directive 85/37/EEC as amended by Council Directive 1999/34/EC on the approximation of laws, regulations and administrative provisions of the Member States concerning liability of defective products; OJ L 210, 7.8.1985, p.29; OJ L 141, 4.6.1999, p. 20

[369] Europen Committee for Standardization

[370] European Committee for Electrotechnical Standardization

[371] European Telecommunications Standards Institute

of these groups are industry dominated in spite of the efforts of some countries and NSBs to obtain more balanced participation. NSBs have the obligation to transpose all European standards into national standards and to withdraw any conflicting national standards. While a European standard is under development, NSBs have to stop all work on standards projects dealing with the same issue. There are 10 000 European standards.

ISO[372], IEC[373] and IUT[374] work on a similar basis to their counterparts in Europe. However, participation in an international standardisation item does not prevent national standards institutes from publishing national standards on the same subject. This is the difference with EU harmonized standards: international standards do not have to be accepted as national standards.

Agreements exist between ISO and CEN on the one hand, and IEC and CENELEC on the other hand, to promote synergies between international and European standardisation. Some 30 % of the work programme of CEN is indeed developed in collaboration with ISO (either within ISO Technical Committees or in CEN Technical Committees), whilst the proportion reaches 80 % in CENELEC/IEC.

The co-operation between ISO and European standard bodies is problematic. ISO standards transformed into European standards make it compulsory for NSBs to withdraw their national standards, whereas ISO and national standards can coexist in non-EU countries. It is very difficult to come to an agreement on the level of requirement in ISO work as it has to cover the needs of countries which have very diverse degrees of technical and economic achievement.

Recently, in cases where unique international solutions have not been possible because of legitimate market differences, ISO has introduced the concept of 'options' to facilitate global harmonization. This will mean that ISO standards of different levels of environmental stringency have to compete on the world market. The idea is that any products that meetthe ISO standard, although at different levels, have access to any market and can compete there to the advantage of those who have the lowest demands to fulfil. Where margins are slim this will ask savings to be made and EU producers will ask for relief from regulatory requirements.

V.4.5.4.2   Structural problems linked to the use of standards in legislation – lack of accountability

V.4.5.4.2.1   Lack of legislation mandating integration of environmental aspects.

As there does not exist any horizontal legislation for the protection of the environment,(corresponding to the product safety and product liability directives for consumers), the only possible reference remains article 6 of the Treaty, requiring the integration of environmental concerns into all policies of the European Union. This requirement should be implemented by EU legislation making the integration of environmental aspects mandatory for any standard.

---

[372] International Organisation for Standardisation

[373] International Electrotechnical Commission

[374] International Telecommunication Union

V.4.5.4.2.2   Accountability

According to their own description, however, ESBs are purely technical in nature. Their activity is not meant, therefore, to extend to legal issues. This probably also explains the composition of their decision-making boards, which are made up exclusively of representatives of national standard bodies, whose brief is determined by national stakeholders (in practice this has been mainly industry) having an interest in standards on economic grounds, and not for the benefit of society at large. All standardisation activities fulfilling quasi legislative functions are characterised by a democratic and information deficit.

V.4.5.4.2.3   Accessibility

In principle, the standardisation process is open to all interests. In reality, its accessibility to nonprofit interests is only limited. If the EU delegates the task of defining technical standards to private standard bodies, the concerned industry will be well equipped with human and financial resources. Under the present circumstances, organisations defending public interests like the protection of the environment cannot yet meet that challenge and are still in a minority position.

The Commission gives support to consumers groups, trade unions and small, medium enterprises and, since 2002, ECOs as well, so that they might participate as observers in European standardisation.

V.4.5.4.2.4   Distinguishing political and technical competencies

The Council Resolution of 7 May 1985 itself restricts the scope of "mandated" standardisation. The "new approach" and other directives envisaging mandates define "essential requirements" to be harmonised and laid down by the Directives in accordance with article 95 of the Treaty. The "general reference to standards" can operate only in areas in which these "essential requirements" and "manufacturing specifications" can be clearly distinguished. It will often be difficult to make such a distinction. When legislation and mandates are not clear enough it means that policy decisions are frequently referred to standard bodies and sometimes set the agenda for the formulation and application of European environmental law.

The crucial problem of "mandated" standardisation lies in the fact that, where the distinction between legislative and "merely" standardisation activity is unclear, the standards institutions with their close links to industry can take important decisions affecting the efficient application of European environmental regulations.

Since, however, all "technical standardisation " also involves a political decision there is a danger that responsibility for EU environmental policy decisions will be shifted from the political to the diverse business sectors.

V.4.5.4.3   Examples of Environmental requirements in mandated standards

Only a few studies have looked into the question, if mandated standards – under the "new approach" or others - are meeting environmental requirements. It is rather difficult, even for experts, to assess standards if they have not participated in the process. That means that there is not much evidence on standards meeting or not meeting the environmental requirements apart from cases where experts have participated in the writing of the standard. The following examples do not concern mandates with exclusively environmental objectives. None of the examples of standards have taken environmental aspects sufficiently into account. While assessing standards in which they have expertise, ECOs have not been able to find positive examples.

V.4.5.4.3.1 Packaging and packaging waste[375] - The first and only example of the New Approach in the field of the environment

The Directive on packaging and packaging waste stipulates "A standard is necessary to set criteria for assessment and measurement of source reduction". Instead of the expected thresholds and quantitative criteria the standards developed by CEN eventually opted for a management system approach that did not allow the determination of compliance with essential requirements.

Denmark, Belgium, and the Commission, as well as NGOs (EEB and ANEC) came to the conclusion that the requirements had not been met.

In 2002 the Commission issued a new mandate for the same packaging standards. This differed only slightly to the original one and the standards have only marginally improved. Participating NGOs concluded that the standards do neither satisfy the essential requirements of the Directive not the provisions of the Mandate and identified some major shortcomings, in particular [ANEC/ECOS (2005)]:

▶ The European standard EN13428:2004 (Requirements specific to manufacturing and composition – prevention by source reduction) offers no measurable criteria nor incentive to companies to reduce excessive packaging. In addition, the standard allows for the use of substances that are harmful for the environment.

▶ The European standard EN 13429:2004 (Packaging – Reuse) does neither specify a minimum number of trips or rotations nor does it contain a test method for the verification of such a requirement. ANEC and ECOS are also very concerned about the inclusion of hybrid systems in the standard.

▶ The European standard EN 13430:2004 (Requirements for packaging recovering by material recycling) does not define material specific requirements to facilitate recycling nor a minimum percentage of the package that has to be recyclable.

▶ The European standard EN 13231:2004 (Requirements for packaging recoverable in the form of energy recovery) contains a minimum caloric value, which is, however, much too low. Further, substances or materials that can adversely affect energy recovery have been insufficiently covered.

Despite these serious shortcomings, the Commission published the standards in February 2005 in the Official Journal, arguing that they could not be better than the directive.

---

[375] Council Directive 94/62/EC on packaging and packaging waste

V.4.5.4.3.2   Waste of electric and electronic equipment[376]

The Directive envisages standardised marking on electric and electronic equipment to facilitate its separate collection. The mandated standard developed by CENELEC will not be published since it does not comply with the requirements of the Directive. The marking is envisaged for 13th August 2005 and no harmonized standard is available.

V.4.5.4.3.3   Heating appliances standards

There are standards of heating appliances (boilers)[377] that allow for higher emissions of air pollutants than the legal limit values would allow in certain countries.

V.4.5.4.3.4   Construction Product Directive

Some standards mandated under the construction product directive did not meet Mandate and Community legislation although the Commission had demanded explicitly that EC legislation on dangerous substances be respected. This should have been an obligation even without the mandate.

V.4.5.4.3.5   Refrigerants

The producer of an environmentally benign hydrocarbon refrigerant with a global warming potential no higher than that of CO_2 found that he was no longer able to compete because of exaggerated fire protection considerations imposed by the producers of fluorigenated hydrocarbons. Those chemicals have a global warming potential of 100 to 1000 times higher than CO_2. The dimensions of the refrigerant containing compartments had to be reduced to such an extent that the hydrocabon refrigerant was no longer efficient.

V.4.5.4.3.6   Environmental Management Systems

In 1995, the CEN Programming Committee on the Environment was to be entrusted with the Commission mandate for a standard on environmental management systems envisaged under the EMAS Regulation. The business representatives, however, insisted that such a standard should not be developed at European but at international level. This, in the end, was accepted by the Commission. It happened to be a somewhat awkward construction in that the Americans in particular saw no reason to accommodate European legislation. It raised endless controversy in the so-called "Bridging Committee" (between the international and the European level) over how much of the standard 14001 was in conformity with the Regulation. Only afterwards did it turn out that four of the delegates representing national standard bodies had come from EXXON. Finally only parts of ISO 14001 where declared to be in conformity with the regulation and these were published in the Official Journal. The subsequent revision of the Regulation deleted the detailed original annexes and replaced them by the vaguely and often ambiguously worded text of ISO 14001.

If NGOs had not participated in discussing the mentioned standards, these cases would never have surfaced.

---

[376] Directive 2002/96/EC of 27 January 2003 on waste electrical and electronic equipment (WEEE)O.J. 13.2.2003 L 37/24

[377] NOx emission limits from gas heating appliance standards EN 297/EN483 and EN 676, CO emissions of gas heating appliance standards EN 303-3, NOx emissions of oil heating appliance standards EN 267/EN 303-2

V.4.5.4.3.7 *Standard for the harmonised UNICA bottle.*

ECOs are in principle in favour of standards and recognize their advantages. The German ECO BUND, and some bottle producers and retailers agreed on having the same type of reusable milk bottle to avoid unnecessary transport of empty bottles. They defined the specifications on the bottle's environmental performance, e.g. the number of trips the bottle was to achieve. The system has been in place for ten years.

V.4.5.4.4 Greening standardisation: Attempts to improve environmental accountability

Standardisation structures and processes have been the subject of many public discussions. This has facilitated improved understanding of what standardisation can serve and what its shortcomings are. For many years, environmental organisations have called for legislative measures that would make it mandatory for standardisation to take environmental aspects into account and in conjunction with such measures they asked for the means to·be able to participate in standardisation at all levels.

The arguments and complaints from ECOs have not remained unheard. EU institutions became aware, some Member States reacted and also the standard bodies themselves have been trying to improve their accountability. ECOs demonstrate their willingness to co-operate.

V.4.5.4.4.1 *Commission*

DG Environment employs a desk officer on standardisation whose task it is to follow the standardisation policy of the Commission and of the standard bodies, looking at their repercussions on the environment. DG Industry has desk officers to deal with environmental issues and tries actively to improve working conditions of ECOs in ESBs.

In 2004 the Commission tabled the Communication on "Integration of Environmental Aspects into European Standardisation"[378] . It does not envisage any binding measures and instead emphasises extending the use of standardisation for purposes of environmental protection by envisaging more funding for stakeholder participation. The Council confirmed this in its Resolution of 23 July 2004.

In 2002 DG Environment granted a three-year service contract to the newly founded European Environmental Citizens Organisation (ECOS) It will expire October 2005 and the Commission has promised continuity.

ECOS has welcomed the Commission's attempt to make standardisation accountable also for environmental protection. However, ECOS demands binding legal requirements.

As in general the EU wants to broaden the scope for the role of standardisation, the revision of the New Approach has long been on its agenda. ECOS has taken the position that the New Approach should not be opened to environmental issues unless it can be ensured that integrating environmental aspects into product policy is made a general requirement. (*ECOS 2004*).

---

[378] COM(2004)130 final

The Council and the European Parliament have time and again pronounced themselves in favour of extending the use of standards. At the same time they stress the need for the participation of all interested parties in the standard setting process.

Environmental citizens' organisations like ECOS depend on Commission funding. If funding became unavailable they could no longer participate in the work and would have to oppose further application of the New Approach on the basis that one of the main preconditions i.e. public participation, was no longer being met.

V.4.5.4.4.2 Member States

Only a few Member States have set themselves an environmental agenda for standardisation. One way for governments to achieve this is by giving financial support to ECO participation in the standardisation process.. Their involvement has to be triggered by financial support as ECOs have neither the financial means nor the expertise at hand to follow the cumbersome work involved in technical committees.

Already in 1992 the German Government granted environmental citizens' organisations financing and negotiated a contract with DIN, the German standards body, giving these organisations access to standard committees. Furthermore, the Government pays DIN to screen its standards for environmental aspects.

Three years ago the Danish Government attributed a contract to Denmarks Naturfredningsforening to follow some standardisation projects. DS, the Danish standard body has shown itself to be more open than others to environmental concerns.

The French ECO, France Nature Environnement was able to participate in the horizontal Committees of AFNOR, the French standard body. French ECOs receive restricted financial government support for certain projects like the participation in the Plenary of ISO TC 207 on Environmental Management.

In Belgium ECOs can and do participate in selected standard committees, e.g. on packaging.

The Swedish organisation SNCC can participate via the consumer ticket and if they pay in standard committees. They themselves do not receive public funding for their participation in standardisation.

V.4.5.4.4.3 Standard bodies

Since it has become evident that standardisation has enormous political implications, a process of reflection has started.

ESBs have learned to cope with the issues of health and safety. It has to be regarded as a sign of trust that consumer organisations are starting to request technical standards and are proposing their own items. Co-operation between environmental citizens' organisations and ESBs is, however, still in its early stages.

Since 1992 CEN has an environmental committee. The Strategic Advisory Body for the Environment, SABE. SABE is concerned with the problems of environmental aspects in standard-isation. SABE is encouraging industrial sectors to develop environmental guidelines, invites chair-persons dealing with TCs on air, water, soil related standards to report, has several subgroups, monitors the activity of the Environmental Helpdesk etc

301

Apart from SABE, CEN has set up the Environmental Helpdesk (EHD) which depends on SABE and whose task it is to screen all standards for their environmental aspects and to send comments to the technical committees in question. ECOs have been disappointed by the results achieved and the low acceptance that the EHD has received from standard committees (ANEC / EEB 2002). Following well-founded criticism the terms of reference of the EHD have been reformulated in order to improve its transparency of working .

SABE's role should be strengthened towards greater implication in the CEN's overall work pro-gramme (ECOS 2004). Recommendations could be made and monitored, with respect to the envi-ronmental dimension of the work of product- or process-related technical committees. CENELEC, TC 111X (Environment), a hybrid body with both an advisory and a standard-writing role, is in charge of all environmental issues. An ad-hoc working group (WG) has been set up to report to and carry out the work for the TC after signature of the Commission Mandate on the standardis-ation programme under the Directive on energy using products. DIN will convene this WG. The results of this preparatory work will be endorsed and developed, if necessary, by TC 111X.

CENELEC has recently decided to make it an obligation for their technical committees to report on their environmental achievements in a data bank that will be accessible to a wide public.

Involvement in ISO is by definition more costly than work in CEN. The costs incurred in holding and attending meetings outside Europe and lengthy discussions amongst more than 170 NSBs may lead to discouragement and a search for more cost-effective solutions. Great care should therefore be taken when deciding to transfer standardisation work from the European level to ISO. The main interest of ECOs is with ISO/TC 207 "Environmental Management" which has shown a willingness to attract ECO representatives. To make sure that this participation is reflect-ed in the final text of standards, ISO shall make sure that recommendations regarding the partic-ipation of ECOs, as proposed by the ECO Task Force Group within ISO/TC 207, are implemented. These would include, for example, a right of appeal for ECOs and procedures that make it pos-sible to sustain minority positions.

V.4.5.4.4.4 European Environmental Citizens Organisation for Standardisation (ECOS)

The EEB, after having participated as an observer for seven years in CEN environmental commit-tees, suspended its observer status when it realised that it was in danger of fulfilling nothing but alibi functions. It was especially disappointed that the CEN Environmental Helpdesk was not given enough power to be efficient and was even declined the right to publish its comments internally.

In order to have more weight in the standardisation process, environmental organisations joined forces. EEB, WWF-EPO, Friends of the Earth Europe and BirdLife International and some national environmental organisations have founded a new environmental federation, the European

Environmental Citizens Organisation for Standardisation (ECOS). This organisation applied for a three year service contract to "green" standardisation -which it was eventually awarded.

The members of ECOS decided on the work programme according to their own interests. Only items on which ECOS members are working and employ staff are included in the organisation's work programme.

ECOS is meanwhile represented by technical experts in about 30 working groups and the technical committees of standard bodies. ECOS is an associate member of CEN and allowed to comment on any standard, is a Liaison A member of ISO TC 207 and is in the process of acquiring a similar status in CENELEC. ECOS wants to become a reliable partner and to be given the chance to influence programmes of work and the strategic decisions of standardisation bodies to the benefit of the environment. It hopes to achieve this through its observer status in administrative and technical boards. In addition, ECOS will play a similar role in technical committees. Its experts, involved in technical discussions, will be able to make contributions regarding the final texts of standards. Their progressive involvement will help to increase the environmental dimensions of European standards and thus their reputation in the eyes of the public inside and outside the Community and EFTA.

In addition, ECOS may be able to promote new standardisation works with a full environmental purpose. In such cases, alliances will need to be sought with front-runner industries as well as with consumers or workers representatives.

Standardisation remains largely based on national participation. Votes on final standards texts are an exclusive competence of the national standardisation bodies. Therefore, ECOS sees it as one of its main tasks to contribute to the establishment of close relations between national ECOs and national standards bodies. It is of utmost importance that the action of ECOS at European level can be complemented by the involvement of its members. This, however, will require the availability of sufficient resources.

V.4.5.4.5   Outlook and Action for environmental citizens' organisations

European environmental citizens' organisations have realised the implications of standardisation and are concerned that the repercussions of standards on the environment are still not sufficiently taken into account. They fear that the imbalance in the composition of the standardisation groups gives rise to standards that do not reflect societal needs. ECOs should continue to ask for both binding legislation on environmental aspects to be considered and the means to further participate in standardisation. They should attempt to obtain government funding for this from a budget line that does not drain the already scarce resources of ECO support for other projects.

European environmental citizens' organisations can become members of ECOS.[379]

They should comment on business plans and standard texts under preparation at national standard institutes and could base their comments on advice provided by ECOS.

---

[379] For more information: www. Ecostandard.org

Demands

▶ ECOs should be aware that only binding requirements guarantee the consideration of the environment in standardisation and should demand this. Only a generalised requirement to respect the environment will eventually make standardisation as equally satisfactory for the environment as it has become for health and safety. The revision of the New Approach should take care of this. Furthermore, an equivalent to the product safety directive should only allow market access to environmentally sound products.

▶ In the past mandated standards have taken political decisions as legislation and mandates were never clear enough and did not draw the division line between both. This should be avoided in future by using clear and unambiguous language in legislation and mandates.

▶ If ECOs want to have an impact on standardisation, it will not be enough for them to participate in standardisation. They would also need some minority rights and to be able to take recourse to an appeal structure in the standard bodies.

▶ Furthermore, all standards organisations should work towards a more balanced representation of all interested parties in their technical groups. Access to technical committees should not be hindered by a lack of resources within important stakeholders representatives.

▶ In order to give ECOS a permanent status it needs continuous financing. The Budget of the EU must envisage financing for ECOS in the same way as it does for ANEC, TUTB and NORMAPME. Public participation and thus public funding is the justification for the application of the New Approach.

▶ ECOs should be aware of the repercussions of the new policy of ISO on the environment to allow for options in ISO standards.

BIBLIOGRAPHY AND FURTHER READING

ANEC / EEB (2002) Position Paper on CEN Environmental Helpdesk (EHD), October 2002, Brussels

ANEC / ECOS (2005) ANEC-ECOS Position Paper on the revised Packaging Standards prepared under the second Standardisation Mandate M317, January 2005, Brussels

CEC (2004) The role of European standardisation in the framework of European policies and legislation; Commission Communication COM(2004) 674 final of 18.10.2004

CEC (2004) Commission Communication (2004) on Integration of Environmental Aspects into European Standardisation, (SEC(2004)206)

CEN (European Committee for Standardisation)

CENELEC (European Committee for Electrotechnical Standardisation)

EC (1983) Council Directive 83/189/EEC of 28 March 1983 laying down a procedure for the provision of information in the field of technical standards and regulations O.J. N° L109 of 26 April 1983, p.8

EC (1985) Council Resolution of 7 May 1985 on a "new approach" to technical harmonisation and standards, OJ N°C136 of 4 June 1985, p.1.

EC (1985) Directive 85/37/EEC as amended by Council Directive 1999/34/EC on the approximation of laws, regulations and administrative provisions of the Member States concerning liability of defective products; OJ L 210, 7.8.1985, p.29; OJ L 141, 4.6.1999, p. 20

EC (1998) Council Directive 98/34/EC of 22 June 1998, laying down a procedure for the provision of information in the field of technical standards and regulations O.J. N° L204 of 21 July 1998, p.37

EC (2001a) Council Directive 2001/95/EC of 3 December 2001 on general product safety, OJ L 11, 15.1.2002, p.4

EC (2001b) Council Regulation allowing voluntary participation by organisations in a Community eco-management and audit scheme (EMAS), No 761/2001 of 19 March 2001, OJ L114, 24.4.2001, p.1

EC (2004a) Directive 2004/18/EC of the European Parliament and of the Council on the coordination of procedures for the award of public works contracts, public supply contracts and public service contracts;

EC (2004b) Directive 2004/17/EC of the European Parliament and of the Council coordinating the procurement procedures of entities operating in the water, energy, transport and postal services sectors

ECOS (2004) ECOS contribution to the review of CEN SABE Strategy, January 2004, Brussels

ECOS (2004) ECOS comments on the European Commission Working Paper "The role of European standardisation in the framework of European legislation and policies", January 2004, Brussels

ECOS (2004) Position on Commission Communication on Integration of environmental Aspects into European Standardisation

EEB (2005) PRESS RELEASE Commission encourages increases in packaging waste!, 22 February 2005, Brussels

ETSI (European Telecommunications Standards Institute)

IEC (International Electrotechnical Committee)

ISO (International Standardisation Organisation)

TASCHNER, Karola (1999), Standardisation's role in caring for society, Conference reports of "Standardization for the 21st Century", hosted by EUCommission and EFTA, 15-17 March 1999

TASCHNER, Karola (1995), Standardisation européenne et politique de protection de l'environnement - l'exemple de l'eco-audit -, Cahiers des étudiants allemands en droit à Genève 15 ; 31 octobre 1995

TASCHNER, Karola (2004), CEN Annual Meeting, Presentation at Conference on standardisation and sustainable development, Lyon, 30th September, 2004

V.4.6 Chemicals and Pesticides Regulations

By Andreas Ahrens[380], Hans Muilermann[381], Stefan Scheuer[382] and Catherine Wattiez[383]

V.4.6.1 INTRODUCTION

Of all environmental problems, that of environmental degradation caused by chemicals was probably one of the first to be recognised by environmentalists and governments. However, at that time in the 1970s a regulatory framework on classification and labelling of chemicals was already in place, mainly driven by the need to harmonise classification of health hazards related to workers safety. The first EU-legislation on chemicals dates back to 1967, in the days of the European Economic Community.

Since then a respectable body of legislation has been constructed, including:

▶ classification, labelling and packaging of dangerous chemicals (since 1967);

▶ restrictions on marketing and use of chemicals (since 1976);

▶ introduction of new substances into the market (since 1981);

▶ market authorisation of pesticides (since 1991);

▶ risk assessment framework for new chemicals (since 1993);

▶ risk assessment framework for chemicals "having existed" on the market in 1981 (since 1994);

▶ for the integrated prevention of pollution of air, water, soil and sediments (since 1996);

▶ market authorisation for biocides (since 1998); and

▶ protocol for the export of chemicals to countries outside the EU-region (Rotterdam Convention since 2004).

In this chapter the core of EU legislation on chemicals and pesticides will be briefly discussed with regard to the environment. A cursory glance will also be given to occupational safety and health and consumer safety policies, where chemicals are involved. The chapter also includes the current Commission's proposal for a chemicals policy reform in Europe (REACH) and the latest plans towards a European Strategy for the Sustainable Use of Pesticides.

---

[380] Senior Advisor, Institute for Environmental Strategies, Ökopol, Hamburg

[381] Policy Advisor, Stichting Natuur en Milieu

[382] EU Policy Director, European Environmental Bureau

[383] Pesticides Action Network, Europe

V.4.6.2   CHEMICAL INTENSIFICATION OF EUROPE'S ECONOMY

Chemicals play an ever-increasing role in modern society. They are added to almost all consumer products in a growing variety of mixtures, which increasingly determines their function and design. They also play a greater role in manufacturing and farming processes than ever before. Chemical production is increasing far above GDP growth (*EEA 2003*) starting with 1 million tonnes per year in 1930, reaching 400 million tonnes in 2000. 100,196 chemicals are listed by industry as having been on the market before 1981. Unfortunately no statistics are available to analyse this "chemical intensification" of our societies in more detail. In particular, no information is accessible on how many chemicals are dangerous to the environment or human health.

Since 1981 more than 3,700 chemicals have been registered with around 300 chemicals per year registered since the mid-nineties (*Nordbeck and Faust 2002*). Only for those "few" and mostly low volume chemicals is safety information available.

Annual pesticides sales have increased in EU countries by around 10% (*Eurostat 2002*) despite the implementation of a number of national pesticide use reduction programmes (in The Netherlands, Denmark and Sweden) and at the same time pesticide ingredients are now more "effective" at lower doses. These figures and developments strongly suggest that farming is increasingly dependent on pesticide use (*PAN 2002*).

V.4.6.3   FROM LARGE SCALE CHEMICAL CONTAMINATION TO APPLY-ING PRECAUTION

Concurrent with the chemical intensification of Europe's economy, chemicals are increasingly found in places for which they are not designed: industrial chemicals are accumulating in human bodies and in wildlife (*EEA 2003 and WWF 2003*). As far as the scarce monitoring data reveals, pesticide levels in Groundwater are above legal standards (*EEA 2000*) in many parts of Europe and pesticides are commonly found in food (*PAN 2002*). Many of those chemicals are known for (or often found later to have) the potential to cause serious harm, including allergies, cancers, hormone disruption and neurodevelopmental disorders. Up to now the regulatory framework has not been effective in preventing such problems.

This is due to a systematic failure of the current system

- to provide incentives for manufacturers of chemicals to assess the safety of their products;
- to place the burdens of proof on those who have interest in the market and use of substances; and
- to handle the scientific uncertainties around the prediction of exposure and harmful effects of chemicals in an efficient way.

These failures are illustrated by the risk assessment programme for EU's "existing" chemicals (see chapter V.4.6.4.2). Before evidence of a risk becomes available or the EU risk assessment concludes that there is concern over a risk the chemical manufacturer can assume that his product is safe. This means that there is no effective incentive for the chemical manufacturer to make data

available on toxicity and exposure. On the contrary, manufacturers making data available, which would lead to a classification of a chemical as hazardous, would be "punished" with a negative label as untested chemicals do not carry any label. Thus the burden of proof rests with the authorities and the general public to prove chemical risks.

308

In the case of chemicals put on the market after 1981 or active substances in pesticide and biocide products, the manufactures have to provide a risk assessment before marketing, or by a certain deadline in the case of the latter. Thus some burden of proof lies with the manufacturers, although enforcement is hampered due to lack of public capacity to evaluate industry data and enforce the deadlines (see chapter V.4.6.5.1.3).

However, both systems suffer from the limitation of the risk assessment approach and its failure to communicate uncertainties. Firstly, exposure data is rarely available due to lack of monitoring or uncertainty in the case of predicting concentrations in the environment. Secondly toxicity data, for each chemical considered individually only relate to some specific species in a specific environment, making conclusions for other species contaminated by a cocktail of additional chemicals highly uncertain (DEPA 2003) (Zeliger 2003). Moreover progress in toxicology is not adequately considered and potential toxicological endpoints such as neurotoxicity, immunotoxicity, hormone disruption, and developmental toxicology are not systematically investigated, even for pesticides, which are currently the most intensively tested chemicals (Seralini2003) (Howard 2003) (Meyers 2002) (WHO 2002).

Helping to address this problem an important principle is available to Community policy via Article 6 of the Maastricht Treaty – the precautionary principle (de Sadeleer 2003). This principle should help to overcome the stalemate surrounding decision taking in cases of high scientific uncertainties or lack of data. But it has never been defined with great clarity as to how it should be implemented and most significant progress has been made mainly through Case Law[384] and secondary laws.[385] With respect to chemical risks the discussion has moved towards establishing criteria for identifying so-called "hazardous" chemicals[386] - those which have high persistency in the environment, accumulation in human tissue and/or irreversible effects including cancer or infertility, for which the risk assessment procedure delivers the highest uncertainties. Such sub-

---

[384] European Court of Justice rulings on BSE and Hormones in the Cases C-180/96 (United Kingdom versus Commission 1996) and C-331/88 (Fedesa 1990)

[385] Examples for the application of the precautionary principle, include:
- Directive 2001/18/EC on the deliberate release into the environment of genetically modified organisms: "Member States shall, in accordance with the precautionary principle, ensure that all appropriate measures are taken to avoid adverse effects on human health and the environment which might arise from the deliberate release or the placing on the market of GMOs.";
- Water Framework Directive 2000/60/EC (article 4 and 16) "Cessation of emission of vPvB and PBT substances";
- Directive 2002/95/EC on restriction on hazardous substances in electrical and electronic equipment (ROHS) – banning common brominated flame retardants and heavy metals; and
- Council Directive 76/769 - market and use restrictions for all CMRs.

[386] The term "hazardous" is used here as in the context of the Marine Conventions and the Water Framework Directive 2000/60/EC. It addresses substances of very high concern for the environment (and human health via the environment) due to their persistence, liability to bioaccumulate and liability to cause irreversible effects an low or unknown dose.

stances should then primarily be regulated on the basis of availability of safer alternatives and socio-economic considerations (substitution principle[387]).

Additionally the proposed chemicals policy reform, REACH, foresees shifting the burden of providing data and proving the safety/danger of chemicals from public authorities to industry and banning chemicals in the case of the highest possible uncertainty – a complete lack of data by a given deadline.

309

V.4.6.4 INDUSTRIAL CHEMICALS

V.4.6.4.1 Hazard characterisation, classification, labelling and safety data sheets

Hazard characterisation and classification and labelling of chemicals is regulated by the Council Directive 67/548/EEC, including its key amendments of 1979 and 1992, as well as 29 adaptations on technical progress. According to this Directive all chemical substances and preparations (mixtures of substances, e.g. glues, solvents, reagents) available to the general consumer or to professional users must be assessed for hazardous characteristics. For classification and labelling of preparations a separate Directive (1999/45/EC) applies. Also a material safety data sheet must be available for professional users (see Directive 2001/58/EEC).

A substance or preparation is characterised as hazardous when it meets the criteria for one or more hazard characteristics. This is usually determined on the basis of test data. Chemicals can be flammable or explosive, irritating or corrosive, harmful or toxic, carcinogenic, mutagenic or teratogenic, etc. The hazards are further characterised by a system of standardised risk phrases used on the labels of hazardous chemicals.

In principal, the implementation of these requirements is the responsibility of the producers and importers of the chemicals. However for existing substances, the generation of new test data if information is lacking is not required. In addition, there exists a programme run out by Member States, in collaboration with the European Commission, to harmonise classification and labelling data for dangerous substances based on existing information. About 3,000 dangerous substances and groups of substances with a harmonised classification are listed in annex 1 of Directive 67/548.

V.4.6.4.2 Risk assessment at the core of EU chemicals policies

At the core of EU chemicals policy there is a harmonised risk assessment method. On the basis of a standardised set of data for a chemical, a no-effect-level is established and exposure to man or the environment is estimated. Thus a Predicted No Effect Concentration (PNEC) and a Predicted Environmental Concentration (PEC) can be established. On the basis of the PEC/PNEC ratio it is decided whether risk management is needed.

[387] "Safer Chemicals within Reach. Using the Substitution Principle to drive Green Chemistry", REACH Report prepared for the Greenpeace Environmental Trust By Clean Production Action, October 2003 http://eu.greenpeace.org/downloads/chem/SaferChemsWithinReach.pdf

In this risk assessment procedure the risks for consumers, workers and the environment are assessed using an integrated methodology. The procedure is applied for newly marketed substances on a regular basis within the notification procedure. For these substances there is an obligation to generate test data on human and environmental hazards before a substance may be marketed.

The procedure is also used in the EU programme on assessment of "existing" substances under Council Regulation (EEC) 793/93. This programme applies to 100.196 substances having been registered by industry 1981 as being on the market (EINECS[388] Substances). The programme started in 1994, 13 years after the requirements for new substances entered into force. At that time and until now, new substances still had a marginal share in the market and hence action on existing substances became necessary. Based on the registration of all substances > 10 t/a between 1994 and 1998, the member states selected about 140 substances to carry out a community risk assessment. The selection was based on an agreed computerised priority setting methodology. Today, for about 80 substances the assessment is finalised at technical level. In addition, in 2003 and 2004 about 120 high production volume substances have been screened in order to determine whether they possibly meet the criteria of being persistent, bioaccumulative and toxic. At this time, the assessment is still going on.

The outcome of the programme seems quite poor compared to the number of substances still to be assessed. However, it should be taken into account, that the 10 years of risk assessment work at EU level has created a common understanding on the risk assessment approach and a full package of methodology is readily available. Hence even though the performance of the process seems low, it has increased the capacity to carry out future risk assessments more effectively.

Nevertheless the risk assessment does not deliver what was expected. The comprehensive nature of the current risk assessment approach, the burdens of assessment placed on authorities, and in particular the lack of knowledge on uses and exposure of substances make it difficult to speed up the process based on the current legislation (see chapter V.4.6.3).

V.4.6.4.3 Restrictions on marketing and use

The final policy goal of these selection and assessment procedures is to decide whether restriction on the marketing and use of the chemicals under scrutiny is necessary to protect human health and/or the environment. The EU has the possibility of restricting the marketing and use of hazardous chemicals under Council Directive 76/769. Many restrictions have already been set up even without a comprehensive risk assessment (like e.g. the ban of CMRs in consumer products).

This first EU regulation for the restriction of marketing and use of chemicals contains restrictions for polychlorinated biphenyls, polychlorinated terphenyls and vinyl chloride monomer. In subsequent amendments of Council Regulation 76/769 a substantial list gives substances which were banned or severely restricted in the EU. The list includes: benzene; asbestos; 8 chlorinated hydrocarbon solvents; several compounds and products containing heavy metals such as lead, mercury, cadmium and nickel; substances that are considered to be carcinogenic, mutagenic or toxic to reproduction for humans; a list of benzidine dyestuffs; pentachlorophenol and its salts and esters; organic compounds.

---

[388] European Inventory of Existing Chemical Substances

At the basis of these restrictions to marketing or use lies a risk assessment, which shows a risk to human health or the environment of a chemical substance in a particular application. This may be a risk to workers, to consumers, to the public at large or to the environment.

Also a cost/benefit analysis is performed for proposed restrictions on marketing and use. The authorities must demonstrate that the selected measures are proportionate and that the restriction leads to alternatives posing a smaller or no risk.

ECOs campaigning for the ban of a chemical product or substance will be confronted with this burden of proof one way or the other. When they call upon their national government to ban a substance, this government will have to prove to the European Commission the necessity of such a ban. When the ECO decides to lobby directly at EU level , the European Commission will have to fit the substance into a risk assessment procedure and perform a cost/benefit analysis.

The alternative for ECOs is a campaign against the use of a substance via public awareness and concern. This means flagging products or trademarks in case they contain hazardous chemicals. However, in this case, it is the market which decides, regardless of comparative risk considerations on alternatives, costs or benefits.

V.4.6.4.4   Related policy fields: worker and consumer protection

As stated earlier, risks to the health of workers and consumers is assessed in the same procedure and using the same methods of analysis as environmental risks. This opens up the possibility of forming coalitions with trade unions or consumers unions in campaigns against the use of specific substances or groups of substances. In the case of asbestos the coalition of worker and consumer groups was very strong, leading to bans on asbestos in most European countries.

In the Netherlands a coalition of environmental, consumer and worker organisations was successful in introducing legally binding measures at source against the indoor use of solvent rich products in the work place. These measures are also beneficial for the environment because reducing the use of solvent rich products also leads to a reduction in emissions of volatile organic compounds in the air.

V.4.6.4.5   Weaknesses in current EU chemicals legislation

There are five major weaknesses in current EU chemicals legislation. Firstly, not all chemicals on the market are treated equally. Substances placed on the market after 1981 must be notified based on an obligatory and costly set of information. For substances which have already been on the market,, no such requirement has existed until now. As a consequence more than 95% of substances on the market in amounts > 1 t/a fall into the category of these "old" substances and the barriers to innovation are quite high.

Secondly, as a consequence of the above, there is a lack of hazard classification for many chemicals. Even for the better known substances there are gaps in knowledge about their hazards or environmental effects. For many commercially used chemicals, data about their hazards to human health or the environment are inexistent, not documented or not publicly accessible. There is no obligation in any EU legislation to perform the basic set of tests upon a chemical substance before it is used in production. This means that without changes in the policy, ignorance about the risks of substances will continue to exist. Moreover, the current system requires substances with known dangerous properties to be labelled as such but does not ask the same of substances which have simply not yet been tested.. Such a system does not provide any incentives to generate more knowledge on substance properties since doing so may be "sanctioned" with a hazard label.

Thirdly, in practice it is not the producer of the existing substances who is required prove the safety of his products, but governments or environmental groups who have to make the case against a substance. With the limited capacity for assessing risks, both in Member States, in ECOs and in the European Commission, only a few chemicals per year can be assessed.. The task is simply too onerous and moreover rests on the wrong shoulders. Under the current system, the majority of commercially applied chemicals is and will continue to be non-assessed.

A fourth problem in chemicals policy is the fact that measures must be based on quantitative risk assessment. This is based on the assumption that all chemicals have a threshold concentration below which exposure causes no harm. This supposition does not hold true for many carcinogens and mutagens, where there is no safe level of exposure. The supposition does not hold true either for persistent and bio-accumulating substances which may lead to life time exposure of biota and humans, starting already in pre-natal phase. Again, a safe level of exposure can hardly be determined based on scientific methodology.

Last but not least, current legislation fails to provide a mechanism based on which the responsibility for risk assessment, risk communication and risk management can be located along supply chains in a transparent and effective way. The duty of manufacturers and importers to provide available information on substance properties and corresponding risk management advice and the duty of employers to assess the risk caused by exposure to chemicals in the work place are not sufficiently adjusted to each other in practice. As the present situation stands, they are not required to provide an assessment relating to consumer and environment safety.

V.4.6.4.6 Towards a new European chemicals policy (REACH)

The shortcomings mentioned above have triggered an intensive debate on chemical policy reform during the last 6 years. Initiated by an informal meeting of the Environment ministers of several EU countries in Chester, UK in 1998, the EU Commission started a systematic analysis of the performance of the four major elements of EU chemicals legislation. Based on this, a White Paper on a new Chemicals Strategy was published in 2001. Finally in October 2003 the Commission issued a proposal and drafted a regulation on the Registration, Evaluation, Authorisation and Restriction of Chemicals COM 2003/644, the so called REACH System. This draft regulation integrates about 40 single pieces of chemicals legislation into one consistent framework for all chemicals produced in or imported into Europe. The marketing and use of biocides, plant protection products, pharmaceuticals and cosmetics will be authorised under separate legislation. Apart from consolidating existing legislation, the REACH systems introduces 6 rather new elements into the legislation:

1. Further marketing of existing substances is only allowed based on a registration including an obligatory set of information, a safety assessment covering the whole life cycle of the substance in the market, and appropriate risk management advice for intended uses. Hence the producers and importers are not only obliged to carry out sufficient testing, they are also requested to decide on the type and conditions of use they consider to be safe. The justification of this decision must be documented in a chemicals safety report. This report will cover workers' health, consumer safety and the environment.

313

2. The information will be transferred to the user by means of a slightly modified safety data sheet. It will be the task of the formulator of preparations (like glues, paints, lubricants, plastic additives, ..) to check whether his process and the type of intended use of the preparation are covered in the suppliers safety assessment. If this is not the case, the formulator of the preparation will have three possibilities: i) adjusting his practice to comply with the risk management advice of the manufacturer, ii) asking the manufacturer to extend his safety assessment or iii) to carry out his own safety assessment and to notify this to the authorities. These mechanisms also apply to the next step of the supply chain, the use of preparations in industrial manufacturing or professional services. In theory, this mechanism is expected to avoid gaps of responsibility in future.

3. For substances of particular concern, like CMRs, PBTs or vPvBs[389], an authorisation requirement can be set up. In such cases, the only uses permitted will be those for which the manufacturer or user is able to demonstrate that risks can be prevented or that no safer alternatives are available. Quantitative risk assessment will not usually be applied in the authorisation procedure.

4. The information in the registration dossiers will be stored and processed in a central EU data base managed by the new European Chemicals Agency. Data related to the hazardous properties of substances, classification, labelling and safety data sheet information will be publicly accessible via the internet.

5. For the first time substances in articles are fully covered in chemicals legislation and it will therefore become easier to reduce the risk related to diffuse losses of dangerous substances from articles like building material, textiles, vehicles, electronic equipment, etc..

6. In future, authorities will focus on four types of work : (this does not mean however that their administrative capacity can be reduced):

   i. spot-checking the registration dossiers for quality and supervising the implementation of the down stream user obligations at local level,

   ii. evaluation of manufacturers or importers proposals for vertebrate testing

   iii. substance evaluation if i) the risk management measures advised by the manufacturer are not sufficient as such or ii) if they are not sufficiently implemented in the market or iii) if cumulative risk occurs which cannot be controlled based on the strategies proposed by the single manufacturer.

---

[389] Carcinogens, mutagens, substances toxic to preproduction; persistent, bioaccumulative and toxic substances (PBT); very persistent and very bioaccumulative substances (vPvB)

iv. Setting up an authorisation requirement for very hazardous substances or taking measures for restrictions on marketing and use.

The whole system aims at putting producers and users responsibility into practice, achieving greater transparency of substance properties and types of use, and improving the effectiveness of regulatory interventions by the authorities.

The phase-in-process of about 30,000 existing substances into the new system is expected to last about 11 years, starting with high production volume chemicals and CMRs in 2007 and finishing with all chemicals between 1 and 10 tonnes (about 17000 substances) by 2017. For substances < 10 t/a, no safety assessment is required.

V.4.6.5   PESTICIDES

V.4.6.5.1   The Authorisation Directive

V.4.6.5.1.1   Introduction

The Pesticides Authorisation Directive from 1991 (91/414/EC, OJ L230) is one of Europe's central legislations for controlling pesticides (another Authorisation Directive deals with Biocides (98/8/EC, OJ L123). It concerns the placing on the market of "plant protection products" with the intention of harmonizing national pesticide registration schemes, and the introduction of a Community-wide market approval scheme. One of its key objectives is: "...when plant protection products are authorised, to make sure that, when properly applied for the purposes intended, they are sufficiently effective and have no unacceptable effect on plants or plant products, no unacceptable influence on the environment in general, and in particular, no harmful effect on human or animal health or on groundwater".

V.4.6.5.1.2   EU positive list and national authorisation

The Directive requests that Member States ensure that only pesticides which are authorised according to the Directives provisions and contain ingredients approved at EU level, are marketed and used.

The Directive provides so-called Uniform Principles in Annex VI that set common criteria for evaluating pesticide formulated products to be authorised at national level. One of the important authorisation conditions is that the concentration in ground and surface water of the pesticides active ingredient, its breakdown products or metabolites should not exceed established standards under the proposed conditions of use. For groundwater this includes a maximal concentration of 0.1 microgram/litre - the drinking water standard introduced in 1980 as being equivalent to the detection level and thus as a proxy for zero.

For the EU level approval of pesticide active ingredients the Directive establishes an EU wide positive list (Annex I of the Directive), which means that all unlisted active substances may not be marketed or used in pesticides products. While this is a simple mechanism the process employed to establish the list is complex:

i) A company applies to a Member State for an active ingredient to be accepted in Annex I.

ii) This Member State ensures that the company provides other Member States and the Commission with data according to Annex II (related to the active substance) and Annex III on at least one preparation containing that active substance

iii) The commission refers the dossier to a Member States' expert committee (the Standing Committee on Plant Health) for examination; and finally

iv) The Commission prepares a decision for inclusion or non-inclusion, which can be accepted or rejected by the Standing Committee on Plant Health.

It has to be noted that for the decision to include a pesticide ingredient in the positive list the Directive does not provide criteria as the Uniform Principles (annex VI of the Directive) for the authorisation process of pesticide products at Member State level.

Based on those EU authorised active ingredients Member States can than authorise pesticide products uses at national level according to the Directive assessment procedures and data requirements. Once a pesticide product is accepted in one Member State, other Member States have to accept this product on their market, according to the mutual recognition clause, unless they can prove harm according to specific conditions in their countries.

V.4.6.5.1.3   Transition period - All pesticides to be safety checked by 2003?

In order to phase in the system the Directive allowed Member States until July 2003 to authorise pesticides products containing active ingredients, which are not yet on the EU positive list (Annex I). Furthermore, the Commission was mandated to engage a process to gradually examine active ingredients for inclusion on the positive list by July 2003.

Progress turned out to be too slow to assess all known (ca. 800) pesticide ingredients in time. The Commission decided to apply the so-called ' one-use' principle, which means that if a pesticide use is authorised in at least one place in Europe, those ingredients are placed on the EU positive list. This shifted a lot of work and responsibility back onto Member States. Furthermore, the Commission decided in 2001 to extend the 2003 deadline by five more years.

Until July 2003 only 28 "old" and some 40 "new" substances were included and 13 excluded from Annex I. A further 377 substances have been excluded from the market as there was no notifying company. Because some Member States claimed they needed excluded active ingredients, the Commission created a loophole through establishing a so-called " essential use" exemption, which allows the further use of around 50 active ingredients in specified products and Member States until 31 December 2007.

V.4.6.5.1.4   Assessment

The Pesticide Authorisation Directive presents a step forward in regulating chemical risks, com-pared to how non-pesticide chemicals are regulated. At the same time the assessment procedure has proven to be cumbersome and the decision making process incapable of dealing with unac-ceptable risks from hazardous pesticides.

1. The Directive places too much burden on Member State Authorities, which do not have the capacity to deal with the procedures required and which are under huge pressure from pesticide producers and farmers. Moreover, they cannot rely on a level-playing field in Europe.

2. The Directive does not provide effective sanctions for industry delaying the safety assessment by very slow data delivery.

3. Pesticides producers have ample room for manoeuvre, e.g. by promoting and funding 'scientific' meetings ending in a proposal for 'guidelines' with the aim of being accepted by the Commission. Several of these industry-sponsored guidelines are available (on so-called non-relevant metabolites, on statistical methods of risk assessment called HARAP, on microcosm tests called CLASSIC replacing regular tests with the most sensitive organism, etc.)

4. The decision about inclusion of pesticide active ingredients in Annex I uses neither clear cut-off criteria which are necessary to implement the precautionary principle in dealing with highly persistent and bioaccumulative substances or substances with irreversible effects, like cancer or hormone disruption, nor the substitution principle.

5. Finally decisions about including pesticide ingredients in Annex I take place behind the closed doors of a Member States Committee, were the Commission makes a proposal and has to find support from Member States. It is clear that in such a process political trade-offs are made.

For more information: http://www.pan-europe.info

V.4.6.5.2   Towards a Thematic Strategy on Sustainable Use of Pesticides

Europe's pesticide policy is weak, resting entirely on one Directive, the Authorisation Directive 91/414/EEC, which is only capable of addressing the most problematic pesticides through market exclusions or restrictions. This has not addressed pesticide use – where farmers are involved - and has allowed an increasing use of and dependency on pesticides in Europe. The contamination of water, soil and organisms continue at an alarming rate.

Already in its 5th Environmental Action Programme covering the period 1993 to 2000, the EU committed itself to achieving "a substantial reduction of pesticide use per unit of land under production". But nothing happened except a series of 7 studies in the mid –1990s and a 1998 stakeholder meeting which recognised the need for complementary pesticide policy.

The 6th Environmental Action Programme, covering the period 2001-2010, aims at a high level of protection of human health and the environment and at halting biodiversity loss by 2010. It calls for Community-level action to "reduce the impacts of pesticides on human health and the environment and more generally to achieve a more sustainable use of pesticides as well as a significant overall **reduction in risks and of the use of pesticides** consistent with the necessary crop protection" and for the elaboration of a *Thematic Strategy on the sustainable use of pesticides* by 2005. It is foreseen that the revision of the Pesticides Authorisation Directive will be part of that strategy.

316

V.4.6.5.2.1   Revision of the Pesticides Authorisation Directive

The 1991 Directive requests an evaluation after 10 years of operating and the Commission presented this report in July 2001 (DG SANCO 822/2001). Its main conclusion is that the harmonization cannot be finalised as intended in 12 years, but needs an extra 5 years to complete (up to 2008) . Furthermore, the Commission announced a revision of Directive 91/414. The Environmental Council responded with Conclusions in November 2001. The European Parliament decided on a Resolution in May 2002 accepting the delay on the condition that:

▶ exclusion criteria for pesticides being CMR (carcinogenic, mutagenic, reprotoxic) or endocrine disruptors or P or B or on priority lists in EU legislations or international EU ratified Treaties will be used for inclusion of active ingredients in Annex I,

▶ all metabolites of active ingredients will be considered relevant,

▶ the substitution principle and the comparative risk assessment will be introduced including only the least harmful active ingredient in Annex I and

▶ a Directive for pesticide use reduction will be proposed by Commission.

In July 2002 and January 2003 the Commission organised stakeholder meetings on the revision of the Authorisation Directive. One of the main discussions was the comparative risk assessment, which would help implement the substitution principle and promote less harmful pesticides. The majority wanted such comparative assessments to be carried out at Member State level; industry wanted it only at farm level and on a voluntary basis. The revision of the Authorisation Directive was expected mid 2004 but has been delayed.

V.4.6.5.2.2   Sustainable use or use reduction strategy?

In its communication "Towards a Thematic Strategy on the Sustainable Use of Pesticides" from 4 July 2002 the Commission outlined its views for a Strategy on Sustainable Use of Pesticides following the 6EAP. It also mentioned the need to review Directive 91/414/EEC to include the substitution principle of particularly active substances for which certain concerns persist, the need to address the problem of stockpiles of obsolete pesticides in New Member States and Accession countries, and some international aspects. However, it focuses on measures to be taken to reduce the risks to health and environment from pesticides use; in particular:

▶ establishment of national plans to reduce hazards, risks and dependence on pesticides,

▶ introduction of best practices in river basin management such as mandatory field margins,

▶ introduction of measures to increase protection of NATURA 2000 areas,

▶ banning of aerial spraying but with specific derogations,

▶ initiation of epidemiological research for plant protection product (PPP)users at risk and monitoring programmes on pesticides residue levels for consumers,

▶ better and coordinated data collection for poisoning cases,

▶ support of cost /benefits analyses of PPP and alternatives,

317

▶ research into and development of less hazardous methods of application and handling, definition of IPM techniques, combination effects of pesticides particularly on vulnerable groups and point source pollution.

A stakeholders meeting on the development of this Thematic Strategy was organised by the Commission on the 4th November 2002 (see http://www.europa.eu.int/comm/environment/ppps/home.htm). Council Conclusions on the Commission Communication were adopted 9 December 2002 as well as a European Parliament Resolution on 27 March 2003. An extended impact assessment was started on September 2003 for some measures proposed by the Commission for the Thematic Strategy.

V.4.6.5.2.3   Assessment

So far the Commission has failed to show a clear vision and objective for pesticide authorisation and use. The urgently needed criteria for the inclusion of active pesticide ingredients on the positive list of the Authorisation Directive in order to ensure a consequent substitution of hazardous substances, including CMR, endocrine disruptors, neurotoxics, persistent or bioaccumulative pesticides, as well as increased transparency are not high on the agenda. The de-coupling of agricultural production and pesticides use is not spelled out and the proposed patchwork of, mostly voluntary, measures will be completely insufficient for a responsible and long-term pesticides policy. The Commission's Communication addresses many elements crucial to reducing pesticides impact, but leaves it either to Member States' discretion or falls short of the legislative measures needed.

For more information: http://www.pan-europe.info

V.4.6.6   OUTLOOK AND ECO ACTION (ENVIRONMENTAL CITIZENS' ORGANISATION)

Europe has a massive backlog in assessing and controlling chemical risks. Risk ignorance is rewarded and policy responses are often too late to prevent serious and long-term environmental and health damage. Chemicals used as pesticides have received much higher attention and stricter control measures. But European pesticide regulation faces problems similar to those of industrial chemical control. It encourages chemical producers to invest huge amounts of resources in defending "problem" substances instead of encouraging the market to invest in safer alternatives including chemical free ones. At the same time public authorities and regulators remain in a weak position as they carry the burden of proof on their shoulders.

The new chemicals policy REACH, as proposed by the Commission, has been described by many stakeholders as a "once in a life time opportunity" to implement a transparent producer responsibility for chemical safety and a precautionary substitution of hazardous chemicals. But resistance from chemical producers is high and has already led to a dramatic watering down of the Commission proposal. Current developments in the political debate suggest that REACH will be adopted – but it is as yet unclear as to how strong the final proposal will be. In any case, REACH will always be a procedural framework, whose success depends on the attitude of all involved parties towards working together to improve the environmental and human health performance of modern products. Besides the obvious players - industry and public authorities – this also includes ECOs, who need to prepare for participation in a long implementation process. It is a process which is likely to exceed 10 years and ECOs should use increasingly available chemical data to point out problems and deficiencies.

V.4.6.6.1 Substituting "hazardous"[390] chemicals and ending emissions by 2020

In order to achieve Europe's commitment to ending emission of all "hazardous" chemicals, including chemicals with CMR, PBT or vPvB properties, by 2020[391], comply with the Water Framework Directive (*see chapter IV.5*) and achieve a safe management of all chemicals by 2020[[392]] the new chemicals policy under REACH and the revision of the Pesticides Authorisation Directive have to implement a clear programme to substitute hazardous chemicals with safer alternatives. This can be achieved by banning all uses of chemicals meeting the properties above, except certain uses which have been specifically authorised. For such a time-limited authorisation, industry should be obliged to prove that there are no safer alternatives and that there exists an overriding societal interest. Such a system would provide all market players with a clear direction for innovation and reward frontrunners. Finally it would ensure the gradual diminution of uses of "hazardous" chemicals.

V.4.6.6.2 Core elements of the REACH proposal which should be defended

Although REACH puts into practise very much what the Chemicals Industry promised for years under its responsible care commitment , industry has been lobbying heavily against the new system. There are several reasons behind this lobbying, which will not be analysed here. However ECO campaigning is needed to defend the core elements of the regulatory proposal against industry. This means in particular:

▶ Promoting the need for a new system by illustrating at which point the current systems fails to prevent damage to health, to the environment and to business. In particular the unintended occurrence of dangerous substances in consumer articles is always a strong argument for better supply chain management.

▶ Highlighting the need for good quality information on chemicals to be made accessible through a public data base. In addition better information on dangerous substances in products is needed, an element not yet sufficiently covered in the current proposal, in particular with regard to imported articles.

▶ Advocating the need for a standard set of basic information requirements for all substances, regardless of their intended use, in order keep to the burden of proof on industry.

▶ Addressing the need for a third party quality assurance mechanism in order to prevent the new system from processing information of poor quality.

---

[390] The term "hazardous" is used here as in the context of the Marine Conventions and the Water Framework Directive. It addresses substances of very high concern for the environment (and human health via the environment) due to their persistence, liability to bioaccumulate and liability to cause irreversible effects at low or unknown dose.

[391] 1998, Sintra, Ministerial statement under the OSPAR Convention for the Protection of the Marine Environment of the North-East Atlantic

[392] 2002, Johannesburg, United Nations World Summit on Sustainable Development

▶ Bringing forward cases of substances which require to be put under authorisation requirements due to inappropriate control of exposure and the existence of safer alternatives.

▶ Maintaining credibility as a more reliable information source for parliamentarians than chemical industry lobbyists.

▶ Motivating Member States to concentrate on the enforcement strategies for REACH, including
increasing capacities in competent authorities.

V.4.6.6.3 Pesticides Authorisation and Use Reduction

While awaiting the revision of the 1991 Pesticides Authorisation Directive, which urgently needs
to bring exclusion criteria for the EU positive list of pesticides ingredients, a great deal remains
to be done on the issues of national implementation and enforcement. If the Directive is correctly applied, Europe's groundwater should not be polluted from pesticides. Authorisation for the
use of pesticides, which are found in concentrations above 0.1 microgram/litre in groundwater,
should either be withdrawn or stricter controls, including bans, at farm level applied. Otherwise
citizens will have to continue to pay the bill for cleaning up these polluted waters for their drinking water supply.

In order to reverse the growing use of and dependency on pesticides Europe needs a mandatory use reduction programme and support measures. Therefore a Directive that sets a clear
timetable and indicators is urgently needed. Emphasis should be put on use reduction. A global
reduction of human and environmental exposure to a complex mixture of chemicals, including
pesticides, is necessary. In addition, a ban on the most hazardous chemicals and pesticides needs
to be achieved through a revised and improved Pesticides Authorisation Directive.

ECOs are proposing a 50% reduction in the frequency of pesticides application for each country
within 10 years. 'Frequency of application' is the average number of times per year agricultural
land can be treated with the prescribed dose of a pesticide, based on the quantities sold.

The Directive should contain tools to achieve this target, including:

▶ Improved controls on the use and distribution of pesticides

▶ Set standards for Integrated Crop Management for every crop and crop rotation system as
basic requirement for all non-organic crops

▶ Promotion of organic farming

▶ Levies on pesticides.

A text proposal for a Directive on pesticides use reduction was published in 2002 by Pesticides
Action Network Europe and co-signed by EEB and about 80 environment, public health, consumer and farmers groups including European federations of these groups as well as by trade
unions from 25 European countries (see http://www.pan-europe.net). Such a Pesticides Use
Reduction Directive should be part of the future Thematic Strategy. In particular, successful
national use reduction strategies should be held up as an example and lead they way by showing the feasibility of breaking the link between agricultural production and pesticides use.

Denmark, Sweden and the Netherlands have already shown that use reduction without unacceptable costs for farmers and society is possible.

Even if the EU will not issue mandatory targets and deadlines for pesticides use reduction, it is likely that countries will be requested to develop and implement national strategies. ECOs should participate in the planning and design process of such strategies and request implementation of pesticide use reduction at national level.

BIBLIOGRAPHY AND FURTHER READING

AHRENS, A. (1999) What is wrong with EU's Chemicals Policy? European Environmental Bureau, Brussels 1999

BEUC, EEB, Danish Consumer Council, Danish Society for the Conservation of Nature, Danish Ecological Council (2000) Copenhagen Chemicals Charter – Chemicals under the Spotlight. International Conference Copenhagen, 27-28 October 2000.

DEPA Danish Environmental Protection Agency (2003) Report on the health effects of selected pesticide co-formulants, Pesticide Research series n° 80. **http://www.mst.dk/udgiv/publications/2003/87-7614-057-1/html/default_eng.htm**

DE SADELEER, N (2003): Environmental Principles - From Political Slogans to Legal Rules. Jan 2003

EEA (2003) Third Assessment, European Environment Agency, Copenhagen 2003

EEA (2000) Environmental Signals 2000, European Environment Agency, Copenhagen 2003.

EEA (2002): Late lessons from early warnings: the precautionary principle 1896-2000, Environmental issue report No 22, European Environment Agency, Copenhagen, 2002

Eurostat (2002), New Cronos, January 2002

Greenpeace (2003): Safer Chemicals within Reach. Using the Substitution Principle to drive Green Chemistry. Report prepared for the Greenpeace Environmental Trust By Clean Production Action, October 2003

HOWARD, C.V. (2003): The inadequacies of the current licensing system for pesticides. Proceedings of the PAN Europe Policy Conference Nov 2003 in Denmark

KORZINEK, A., SCHEUER, S. and Dr WARHURST, A. M. (2003): A new chemicals policy in Europe – new opportunities for industry. A response to the claims made regarding the business impact of a new chemicals policy that is designed to protect the environment and human health. A discussion paper from WWF European Toxics Programme and the European Environmental Bureau, January 2003

MEYERS, J.P.(2002): From Silent Spring to Scientific Revolution. Part 1 and 2, Rachel's Environment and Health News n° 757, November 28, 2002 and n° 758 , December 12, 2002

NORDBECK, R. and FAUST, M. (2002): European chemicals regulation and its effect on innovation: an assessment of the EU's White Paper on the Strategy for a future Chemicals Policy. UFZ Discussion Papers, July 2002

PAN Pesticides Action network Europe (2002): Draft explanatory Memorandum for a suggested text for a Directive on pesticides use reduction in Europe

SERALINI, G.E. (2003): A new concept useful for pesticide assessment: ecogenetics. Proceedings of the PAN Europe Policy Conference Nov 2003 in Denmark.

WHO Regional Office for Europe and European Environment Agency (2002):,Children's Health and Environment : A review of evidence WWF (2003): Contamination - The results of WWF's biomonitoring survey. November 2003

WHO (2002) Children's Health and Environment : A review of evidence, WHO Regional Office for Europe and European Environment Agency. http://reports.eea.eu.int/environmental_ issue_report_2002_29/en/eip_29.pdf

WWF (2003) Contamination - The results of WWF's biomonitoring survey, November 2003.

ZELIGER, H. I. (2003) Toxic effects of chemical mixtures, Archives of Environmental Health, January. http://www.findarticles.com/cf_dls/m0907/1_58/101860467/print.jhtml

322

V.4.7 Motor fuels and environmentally relevant legislation

By *Karola Taschner*[393]

V.4.7.1 INTRODUCTION

The reduction of exhaust gas emissions is dependent on the interaction of engine technology, after-treatment devices and fuel quality. The optimisation of fuel composition in particular can help to exploit the technical potential of engine technology and after - and end-of-pipe treatment. From among the multitude of motor fuel components a few have been singled out below, as they have particularly beneficial or adverse effects.

V.4.7.2 FUEL COMPONENTS AND THEIR ENVIRONMENTAL PROBLEMS

V.4.7.2.1 A problem of the past: lead

In the past a lead compound was added to petrol as an anti-knock agent and octane booster or enhancer. Lead alkyl additives were used historically as inexpensive octane enhancers for gasoline.

Health effects of lead in petrol
Results from research studies into pupils' classroom behaviour revealed that there was a positive correlation between the levels of lead in milk teeth and behavioural disturbances among pupils. Furthermore, there was a direct correlation between blood lead level and lead pollution in the ambient air where children lived. Motor vehicle exhaust gases were identified as the source of lead in ambient air.

This evidence resulted in a consumer campaign in the UK to remove lead from petrol and to substitute it by ethyl-alcohol or ethyl- or methyl tertiary butyl ether (i.e. ETBE or MTBE).

Poisoning of catalytic converters through lead in petrol
Another strand of the campaign against lead in petrol has followed quite different objectives. Air pollution from vehicle exhaust gases started to become a serious problem in cities. Modern three-way catalysts promised to be the solution but they lost their efficiency rapidly due to poisoning from the lead in petrol.

V.4.7.2.2 Oxygenates as substitutes for lead in petrol
Oxygenates are alcohols or alcohol derivatives.
Oxygenates work as an antiknock agent and octane booster in petrol, making a tankful of petrol last longer. They have proven to be particularly efficient in reducing CO formation in the US. CO, however, has rarely been a European problem.

▶ **Ethanol** can have negative effects on driveability. Complaints about evaporative emissions of alcohol and its giving rise to ozone formation made it relatively unpopular as a petrol additive.

---

[393] Scientific advisor to the European Environmental Bureau

▶ **ETBE** was, in the end, the additive of choice as it was the best substitute for lead. Also US farmers were in favour, since cereals are used as raw material for bioethanol production, thus providing a boost to their income. **MTBE** is produced from natural gas and has given rise to some problems concerning water pollution.

There is a limit to the positive effects of oxygenates as they are offset by their ozone forming potential.

V.4.7.2.3   Sulphur content

Health effects

There has long been evidence that diesel particles increase the risk of lung cancer. The relationship between the emission of diesel particles and sulphur content of diesel fuel has also been known for many years. This has lead to a continuous reduction of sulphur in diesel fuel. In addition evidence has emerged over the last decade that ultra-fine particles can represent a serious risk for people suffering from heart and lung diseases.

Sulphur affecting catalytic converters

Emissions of NOx and hydrocarbon continue to decline significantly at ultra-low sulphur levels for advanced technology vehicles.

Stringent emission requirements, combined with long-life compliance, demand extremely efficient and durable, after-treatment systems. In addition, many manufacturers are developing and introducing lean-burn technology. This technology has the potential to reduce fuel consumption by up to 15 to 20%, but require NOx control technologies which can function under lean conditions.

The performance of new exhaust after-treatment devices such as catalytic converters is very sensitive to fuel sulphur and their effectiveness is dependent on sulphur free fuel. Sulphur free fuels also reduce the polluting emissions of older vehicles as they allow catalytic devices fitted in these vehicles to operate more effectively

V.4.7.2.4   Olefins

Olefins are unsaturated hydrocarbons and can boost octane numbers. If low molecular weight olefins are emitted unburnt or evaporated, however, they can give rise to ozone formation.

V.4.7.2.5   Aromatics and poly-aromatics

Aromatics are the most energy containing components of motoring fuels and so they are important in the provision of octane and – as a consequence - increased emissions of CO_2, but also toxic substance, like benzene.

A higher aromatic content in the fuel increases the flame temperature during combustion which results in increased NOx emissions.

Aromatics in petrol

Benzene causes cancer in humans and the reduction of aromatics reduces benzene emissions.

Poly-aromatics in diesel fuel

Poly-aromatics in diesel fuel contribute to Polyaromatic Hydrocarbons (PAH) in the exhaust. There is evidence that PAHs contribute to the formation of particular matter (PM – finest dust, see chapter IV.3) and exhaust PAH.

Total aromatics also have been shown to influence exhaust emissions, particularly NOx emissions from heavy-duty diesel engines.

V.4.7.3   DIFFERENT ASPECTS OF REGULATING FUEL COMPOSITION

V.4.7.3.1   Objectives

Regulating fuel quality is a complicated equation with many unknown factors, where the overall result has to be maximum energy efficiency and near zero exhaust emissions.

The most important objective of regulating fuel composition is to reduce, in conjunction with the most advanced vehicle and after-treatment technology, the classical pollutants from tail-pipe emissions: NOx, (HC), CO and diesel particles. All of them have proven to be air toxics and their emission needs to be prevented altogether. A balance has to be achieved in trade-offs especially between particle and NOx formation.

V.4.7.3.2   Fuel consumption versus emission reduction

CO and HC emissions can be dealt with effectively by two- and three-way catalytic converters. NOx removal through 3-way catalysts depends on stochiometric combustion conditions which are not optimal for fuel consumption. Part of the fuel leaves the engine unburnt and has to be oxidized in after-treatment. This is of course a waste of energy. Diesel engines combust fuel in excess oxygen - as do "lean-burn" petrol engines. However, the greater energy efficiency of diesel and lean-burn technology is accompanied by higher NOx emissions. Upcoming de-NOx catalyst technology needs sulphur free fuels.

Physical properties like density, volatility and the boiling point of fuels play a role in avoiding evaporation. On the other hand, the combustion of fuels has to result in maximum energy efficiency. Compounds of high energy density like aromatics raise the combustion temperature and thus give rise to excessive NOx formation. But aromatics are the "bread and butter" of fuels and can thus only be reduced to a certain extent.

Additives of many kinds, like oxygenates and stabilisers, have effects that have to be bridled as well.

V.4.7.3.3   Alternative fuels

A completely new approach will have to be applied with respect to alternative fuels - apart from natural gas and LPG which have already been regulated. The properties of biofuels, for example, are not stable and blending them or using them in a pure form would not guarantee the carefully fixed environmental performance of 'normal' motor fuels.

The EU has left the task of defining the specifications of biofuels to standardisation. This decision has had important consequences as it means that interested parties will decide among themselves.

V.4.7.3.4   Testing

Test results using reference fuels do not mirror real world conditions as they are of better environmental quality than market fuels. The use of reference fuels is part of the procedure of testing for type approval. Limit values that can be met with reference fuels cannot necessarily be met by market fuels. This factor should be kept in mind.

Regulating fuel quality is a complicated equation with many unknown factors where the objectives should be maximum energy efficiency and near zero exhaust emissions.

V.4.7.4   CURRENT LEGISLATION

There have been various changes in the regulations governing fuel composition in the Community. From the outset, motor fuel composition was standardised within certain tolerance limits. Only later was action taken for environmental reasons. In 1985 lead-free petrol (*EC 1985*) was allowed in the Community, and three years later, the Community adopted its ground breaking Directive 98/70/EC (*EC 1998*) which fixed market fuel quality by means of environmental specifications.

According to a Directive adopted in 2003, Member States have to introduce sulphur free motoring fuels by 2005 at the latest and make it a market access requirement by 2009.

Since 2000 natural gas and LPG fuelled vehicles have had to meet emission requirements (EC 1999). In 2003 the EU adopted legislation to promote the use of biofuels and other renewable fuels for transport (EC 2003b).

V.4.7.4.1   Lead free petrol

For reasons of environmental protection, the Community adopted legislation that made it mandatory for Member States to make unleaded petrol available to consumers.

Member States had to ensure the availability of unleaded petrol within their territories by 1989 and at the same time a balanced distribution of leaded petrol as the engine valves of some car engines depended on the lead as lubricant. The same Directive fixed the maximum benzene content at 5% and also the minimum motor octane number at 85, where it has stayed ever since.

Tax incentives have done what the legislator had not dared achieve – a wide distribution of lead-free petrol As Directive 85/210/EEC did not ban lead in petrol right away, this eventually happened only in 1998.

V.4.7.4.2   Quality of petrol and diesel fuels (Directive 98/70/EC)

In 1996 it had become clear that even the introduction of stricter emission limits over time had not succeeded in providing air quality in the EU that would meet WHO guidelines on the different air pollutants.

The Commission did two things :

a) it asked WHO to revise its recommendations and provided the funding for this task

b) it set up a partnership with motor vehicle manufacturers and oil companies to evaluate the technical potential of optimising fuel and motor technology with the objective of reducing exhaust emissions.

327

The decisive part of the Auto-Oil Programme was that it made use of refined modelling methods for air pollutants - CO, NOx, and hydrocarbons (HC) - developed under the UN-ECE Convention on Long Range Transport of Air Pollutants, and applied these methods to seven European test cities to show by how much these air pollutants had to be reduced. These air pollution models were the backbone of the process even if negotiations with the motor manufacturing and oil industries did not result in sufficiently stringent fuel specifications in the ultimate Commission proposal.

The European Parliament in the end completed the task supported by a broad coalition of political·parties and in agreement with the Council (EC 1998).

V.4.7.4.3 "Mini-Directive" on sulphur-free (EC 2003a) fuels.

The 2003 Directive introduces sulphur free motoring fuels by 2005 at the latest and ensures exclusive marketing of sulphur-free fuels by 2009.

As most Member States granted tax incentives and the costs of desulphurisation in refineries were less onerous than expected, sulphur free fuels rapidly became available on the market.

Table 10: The environmental specifications for market fuels are thus envisaged as follows:

PETROL (MAXIMUM CONTENT IN % V/V)

	before	2000	2005	2009
Sulphur content mg/kg	500	150	50	50010
Olefins	none	18	18	18
Aromatics	none	42	35	35
Benzene	5.0	1	1	1
Oxygen content	2.7	2,7	2,7	2,7

DIESEL (MAXIMUM CONTENT MG/KG)

	before	2000	2005	2009
Sulphur content	500	350	50	10
Polycyclic aromatic hydrocarbons	none	11	11	11

328

V.4.7.4.4   Alternative fuels

Natural gas and Liquefied petroleum gas (LPG)

Natural gas and LPG fuels have entered the market and demonstrate low exhaust gas emissions. Emissions for LPG and natural gas vehicles are meanwhile regulated and specifications for reference fuels for the purpose of testing have been introduced (EC 2002). Gas fuelled buses and heavy duty vehicles are likely to be the first to meet the very strict emission requirements of EEVs (Environmentally enhanced vehicles).

However, under specific driving conditions , natural gas fuelled vehicles can emit uncombusted methane which has a higher global warming potential than CO_2.. Therefore the Directive stipulates that emissions of uncombusted methane stay below a creation level. Natural gas is by no means a very homogeneous fuel, differing widely in composition depending on its origin.

LPG is mainly a by-product of the refining of petroleum and consists largely of propane and butane.

Biofuels

France has been allowed to apply a differentiated rate of excise duty to biofuels (EC 2003c).

Meanwhile the promotion of the use of biofuels or other renewable fuels for transport (EC 2003b) has been generalised and Member States are encouraged to make use of tax incentives. Member States should ensure that a minimum proportion of biofuels or other renewable fuels is placed on the market, indicatively 2% by 2005 and 5.75% by 2010. No legislation on the quality of biofuels is envisaged. Instead CEN standards are being developed. In particular, the production of bio diesel increased by 34 % in 2003. Germany's share of this is about half and France's one quarter[394].

V.4.7.4.5   Taxation

Member States were encouraged to apply new minimum tax rates for conventional motor fuels as well as for LPG and natural gas[395] in 2003. The latter are obviously privileged due to their positive effects on exhaust emissions but above all because of their low CO_2 emissions.

Table 11: Minimum taxation for 1000 l fuel in ?

FUEL TYPE	1.1.2004	1.1.2010
Leaded petrol	421	421
Unleaded petrol	359	359
Gas oil (diesel)	302	330
Kerosene	302	330
LPG	125	125
Natural gas	2.6	2.6

[394] Biofuel Barometer

[395] Council Directive 2003/96/EC of 27 October 2003 restructuring the Community framework for the taxation of energy products and electricity (Text with EEA relevance); Official Journal L 283 , 31/10/2003 P. 0051 – 0070

V.4.7.5   FUTURE LEGISLATION

V.4.7.5.1   The world-wide fuel charter

Past legislation was determined by the international alliance of motor manufacturers[396], whose members prepared for markets with further advanced requirements for emission control, especially to enable sophisticated NOx after-treatment technologies.

The demands of "The world-wide Fuel Charter" of 2002 have been met, with the exception of some concerning *Petrol*, where the demand was that Olefins be reduced to 10%.

Diesel: Demands continue to be made for high cetane, low aromatics divided into two classes: total aromatics 15% and poly-aromatics 2% and for low limits on biodiesel (5%) and a ban on any ethanol-diesel ("e-diesel") blends.

V.4.7.5.2   Ultra-fine particles

The issue most under discussion at this time is the problem of ultra-fine particles which are usually emitted by diesel vehicles but also – although to a lesser extent – by petrol vehicles.

Member States are encouraged to measure PM 10 in ambient air but evidence of health studies recommends focussing on PM 2.5. Fuel quality has an influence on PM concentrations but abatement technology is more directed to after-treatment.

V.4.7.5.3   Alternative fuels

Research and development on alternative fuels is to be encouraged as decreasing dependence on mineral oil will lead towards greater sustainability. However, only proper life cycle assessments will help in making the right choices. The EEB has come to the conclusion when assessing life cycle studies that biofuels are not the first choice given the environmental impact of intensive farming on biodiversity. Additionally, the effect of the growing use of biofuels as alternatives would be outweighed in only four years by the increase in overall fuel consumption. Driving smaller cars, using them less often and using advanced vehicle technology etc. would be more effective in reducing dependence on fossil fuels.

V.4.7.6   OUTLOOK AND ECO ACTION (ENVIRONMENTAL CITIZENS' ORGANISATION)

Fuel quality is critical to reducing vehicle emissions, and the following recommendations for action deserve careful attention:

1. monitor fuel quality, as marketed fuels often do not live up to requirements

2. follow the development of alternative fuels. A wide variety of alternative fuels is being developed and all their environmental advantages and shortcomings need to be discussed beyond a pure CO_2 balance

---

[396] www.autoalliance.org/fuel_charter.htm

330

3. assess the standards developed for specifications of biofuels by CEN

4. observe to what extent fuel quality can reduce or increase ultra-fine particle formation

5. get involved in the debate when it comes to evaluating the trade offs of enhanced fuel quality with positive effects on air quality and thus human health and, on the other hand, the energy consumption that may increase when producing these improved fuels.

BIBLIOGRAPHY AND FURTHER READING

EC (1985) Council Directive 85/210/EEC of 20 March 1985 on the approximation of the laws of the Member States concerning the lead content of petrol, *Official Journal L 096 , 03/04/1985, P. 0025 – 0029*

EC (1998) Directive 98/70/EC of the European Parliament and of the Council of 13 October 1998 relating to the quality of petrol and diesel fuels and amending Council Directive 93/12/EEC, *Official Journal L350, 28.12.1998, P. 0058 – 0068*

EC (1999) Directive 1999/96/EC of the European Parliament and of the Council of 13 December 1999 on the approximation of the laws of the Member States relating to measures to be taken against the emission of gaseous and particulate particulate pollutants from compression ignition engines for use in vehicles, and the emission of gaseous pollutants from positive ignition engines fuelled with natural gas or liquefied petroleum gas for use in vehicles and amending Council Directive 88/77/EEC *Official Journal L 40, 16.2.2000, P. 0001 – 0022*

EC (2002) Commission Directive 2002/80/EC of 3 October 2002 adapting to technical progress Council Directive 70/220/EEC relating to measures to be taken against air pollution by emissions from motor vehicles (Text with EEA relevance.), *Official Journal L 291 , 28/10/2002 P. 0020 – 0056*

EC (2003a) Directive 2003/17/EC of the European Parliament and of the Council of 3 March 2003 relating to the quality of petrol and diesel fuels and amending Council Directive 98/70/EC, *Official Journal L 76 22.3.2003 P. 0010 – 0019*

EC (2003b) Directive 2003/30/EC of the European Parliament and of the Council of 8 May 2003 on the promotion of the use of biofuels or other renewable fuels for transport, *Official Journal L 123 , 17/05/2003 P. 0042 – 0046*

EC (2003c) Commission Decision of 15 May 2002 on the aid scheme implemented by France applying a differentiated rate of excise duty to biofuels (notified under document number C(2002) 1866) (Text with EEA relevance), 2003/238/EC, *Official Journal L 094 , 10/04/2003 P. 0001 – 0042*

EC (2003d) Council Directive 2003/96/EC of 27 October 2003 restructuring the Community framework for the taxation of energy products and electricity (Text with EEA relevance), *Official Journal L 28 , 31/10/2003 P. 0051 – 0070*

JONK, Gerie (2002) European Environmental Bureau (EEB) background paper: On the use of biofuels for transport

EEB (2002) Biofuels are not green enough. Press Release May 21, 2002

www.autoalliance.org/fuel_charter.htm

VI.

Environmental Legislation – information on structure, implementation and enforcement

334 VI.1 EU environmental legislative structure

334 VI.2 Implementation and Enforcement

VI.1   EU environmental legislative structure

EU environmental policy is based on the **provisions and principles provided in the EC Treaty mainly in Articles 2, 95 and 174**, which cannot be changed by subsequent legislation.

334

Subsequent or secondary legislation in the field of the environment includes:

a) Regulations

Regulations are binding, directly applicable in all Member States and are **immediately enforceable** at Member State level before national courts. Regulations are normally adopted to provide legislation on issues requiring uniform provisions throughout the Community.

b) Directives

Directives are addressed to the Member States and impose upon them an obligation to achieve a specific result within a certain period of time. However, it is up to the Member States to decide how to achieve this result. Member States must **transpose directives' provisions into domestic legal orders**, but it is up to each Member State to decide what kind of legislative act is more appropriate to achieve the imposed result as long as the transposition is 'complete' and 'correct'.

c) Decisions

Decisions are not legislative instruments aimed at the general public, unlike the regulations, and are binding in their entirety, unlike the directives. Decisions are binding upon those to whom they are addressed. A recent and important example of an environmental Decision is the Decision of the European Parliament and Council for the Sixth Environmental Action Programme, which i.a. obliges the European Commission to present Thematic Strategies within a given time and including certain elements.

VI.2   Implementation and Enforcement

As many EU laws are adopted in the form of Directives the quality of transposition into national legislation and their application at national level are key aspects for successful environmental protection. The European Commission has a special obligation to enforce EU laws and thus must control the quality of transposition and application of provisions.

Transposition:

The Commission receives reports about the transposition of Directives into national laws and is thus in a position to check compliance – although due to a lack of resources the checks are rather limited. In case of delayed, incomplete or wrong transposition the Commission can launch an infringement procedure.

The room for manoeuvre to transpose Directives, especially so-called Framework Directives into national law is often substantial and can lead to large interpretation differences and subsequently in different environmental ambition levels.

Implementation:

The second hurdle for a proper application of environmental Directives is the actual implementation of the provisions, measures, monitoring etc. in practice. Most Directives foresee reporting of how those obligations are to be carried out – but often in a very general and summarised format, which is often insufficient to check compliance.

As there is no EU inspectorate for the environment to check what is happening in practice, the European Commission has only limited possibilities of ensuring proper enforcement. The Commission acknowledges that complaints sent by citizens, ECOs or others play a vital role in keeping the Commission informed of (non-)compliance with certain EU environmental legislation.

This complaints procedure is available to everybody, there is no need to be personally specifically concerned by the infringement and it does not involve any costs. It has, however, also a number of shortcomings. The process is very slow (worst cases remain pending for 10 years and more); there is no fixed time frame; the complainant often lacks information on the development of the case confidentiality being invoked with regard to Member States' replies; and cases cannot be reopened when a Member State does not keep its promise to remedy an offence.

Stages of the infringement procedure

1) Suspected Infringement

   - Complaints launched by citizens, ECOs, corporations
   - The own initaitive of the Commission
   - Petitions and Questions by the European Parliament
   - Non-communication of the transposition of Directives by Member States

2) Formal Letter of Notice (Art. 226)

3) Reasoned Opinion (Art. 226)

4) Referral to the European Court of Justice (ECJ) (Art. 226)

5) ECJ Judgement (Art. 226)

6) Proceedings, Financial Penalties (Art. 228)

FURTHER READING AND INFORMATION SOURCES

EEB (2002) EC complaints procedure: EEB's seven key recommendations for a change, Position Paper December 2002, Brussels

EEB (2004) Your Rights Under the Environmental Legislation of the EU, Special Report by the EEB, December 2004, Brussels

Legislation: can be found at most easy through the natural number year/number ate EUR-lex at http://europa.eu.int/eur-lex/lex/RECH_naturel.do

Transposition: a calendar for the transposition of directives can be found at the Website of the Secretariat-general of the Commission **http://europa.eu.int/comm/secretariat_general/ index_en.htm** under the heading "Application of Community law"

Infringements: The Commission's "Annual report on monitoring the application of Community law" provides a detailed overview of the application – or lack of application – of Community law per Member State and per issue. It can found at the Website of the Secretariat-general of the Commission **http://europa.eu.int/comm/secretariat_general/index_en.htm** under the heading "Infringements"

Complaints: A standard form for complaints to be submitted to the European Commission for failure by a Member State to comply with Community law can be found at the Website of the Secretariat-general of the Commission **http://europa.eu.int/comm/secretariat_general/ index_en.htm** under the heading "Infringements"

VII.

Authors

338

AUTHOR	CONTACT
AHRENS, Andreas 1956 *MSc Biology and Chemistry (University of Hamburg)* *Senior Advisor, Institute for Environmental Strategies (Ökopol, Hamburg)*	Nernstweg 32-34 22765 Hamburg Tel: +49 40 391 002 0 Fax: +49 40 391 000 233 ahrens@oekopol.de www.oekopol.de
DUWE, Matthias 1974 *MSc Development Studies (SOAS, London)* *Policy Officer at Climate Action Network (CAN) Europe asbl*	48 Rue de la Charité 1210 Brussels Tel: +32 2 229 52 20 Fax: +32 2 229 52 29 matthias@climnet.org www.climnet.org
EKMETZOGLOU-NEWSON, Thisvi 1967 *Diploma in Greek and European Law* *Legal and Policy Advisor to MEP* *Director of the European Environmental Policy and Law Institute (EEPALI)*	Tel: +32 2 534 16 63 +32 2 284 79 37 bs178011@skynet.be nvakalis-assistant2@europarl.eu.int
FALTER, Christine 1976 *MSc European Studies (from London School of Economics and Political Science)* *former policy officer for agriculture and biodiversity at the EEB (now working for the European Commission)*	christine.falter@cec.eu.int christine.falter@lse.ac.uk
HEY, Christian 1961 *Diploma in Public Administration, Ph.D. in Political Sciences* *Secretary General of German Advisory Council on the Environment*	christian.hey@uba.de

AUTHOR	CONTACT
HOEDEMAN, Frederik 1970 *Masters in Central European Studies, Nature Protection Officer, The Danish Society for Nature Conservation*	Masnedøgade 20, 2100 Copenhagen Ø Tel: +45 39174000 FH@DN.DK
MEYER, Kerstin 1975 *MA Political Science and Sociology* *EU Policy Officer at the European Environmental Bureau*	Bld. de Waterloo 34 B-1000 Brussels Tel: +32 2 289 10 90 Fax: +32 2 289 10 99 kerstin.meyer@eeb.org www.eeb.org
MICCICHE, Rosanna *followed the legislative process of the Environmental Liability Directive and worked actively throughout its key phases for the European Policy Unit of Greenpeace*	rosanna.micciche@tiscali.it
MUILERMAN, Hans *Degree in Biochemistry* *Policy Advisor, Stichting Natuur en Milieu*	Donkerstraat 17, 3511 KB Utrecht The Netherlands Tel: +31 30 234 82 93 Fax: +31 30 231 27 86 h.muilerman@snm.nl www.natuurenmilieu.nl
SCHEUER, Stefan 1972 *Diploma in Hydrology* *Policy Director at the European Environmental Bureau* *Chairperson of the European Environmental Citizens Organisation for Standardisation*	Bld. de Waterloo 34 B-1000 Brussels Tel: +32 2 289 10 90 Fax: +32 2 289 10 99 stefan.scheuer@eeb.org www.eeb.org

340

AUTHOR	CONTACT
SHINN, Melissa 1973 *BSc in Biology, MSc in Environmental Policy* *Senior Policy Officer at the European* *Environmental Bureau*	Bld. de Waterloo 34 B-1000 Brussels Tel: +32 2 289 10 90 Fax: +32 2 289 10 99 melissa.shinn@eeb.org www.eeb.org
TASCHNER, Karola *Doctor of Natural Sciences, Engineer in* *Environmental Management* *Scientific Advisor of the European Environmental* *Bureau*	Bld. de Waterloo 34 B-1000 Brussels Tel: +32 2 289 10 90 Fax: +32 2 289 10 99 karola.taschner@eeb.org www.eeb.org
TAYLOR, Mary *Campaigner* *Friends of the Earth (England, Wales & N Ireland)*	26-28 Underwood Street London N1 7JQ, UK Tel: +44 (0) 20 7490 1555 Fax: +44 (0) 20 7490 0881 maryt@foe.co.uk
WATTIEZ, Catherine 1948 *Doctor of Biological Sciences* *Position: Pesticides Action Network Europe and* *European Environmental Bureau Board Member*	70 avenue des Tilleuls 1640 Rhode-Saint-Genèse Belgium Tel/Fax: + 32 2 358 29 26 catherine.wattiez@skynet.be

Key words

Aarhus Convention *160-163, 165, 166-168, 171, 228, 240, 230*

access to justice *26 ,96, 114, 161, 163-164, 166, 171, 204, 206, 230, 234, 236*

accreditation (system for environmental verifiers) *248-249*

active substance *308, 314-315, 317*

Air quality daughter directives *26, 113,118*

Air quality framework directive *46-47, 49-51, 178*

Air quality limit values *50, 53, 66*

alternative fuels *325, 328-329*

associated emission values (BAT) *115, 176, 184, 194*

authorisation *15, 127, 135, 139, 140, 150, 158, 191, 228-229, 260-262 ,306, 314-320*

BAT (best available technique) *34-35, 94-95, 100, 110, 115, 146, **175-184**, 186, 188-189, 191-196, 222, 256*

benchmarking *32, 181, 248, 253, 267, 268, 277, 278, 279, 283*

better regulation *187*

biodiversity *9, 12-14, 27, 32-34, 36-38, 40, 44, 49, 55, 115, 126-128, 132, 147, 151, 158, 198, 199, 200, 202-206, 220, 238, 241-242, 246, 316, 329, 338*

biofuels *325-326, 328-331*

biogeographical regions *38*

BREF (BAT-Reference Document) *94-95, 115, 176, 179-180, 183-191, 193-194*

CAFE Programme *48-49, 55-56, 58-59, 62-64, 76*

cancer *188, 307-308, 316, 324*

cap-and-trade *209-210*

carrying capacity *32, 34, 47-48, 50, 127, 139*

catalytic converter *66, 323-324, 425*

CDM (Clean Development Mechanism) *214-215, 219, 220-221, 224-225*

CEN (European Committee for Standardisation) *16, 106, 248, 257, 289-291, 295, 297-298, 301-305, 328, 330*

CENELEC (European Committee for Electrotechnical Standardisation) *106, 248, 257, 289-291, 295, 298, 301-302, 304*

certification *250-251, 256, 269*

classification (of chemicals) *8, 259, 268, 306, 308, **309**, 311, 313*

classification (of waste) *79, 88 - 9, 116*

classification (of ecological water status) *132*

climate change *13, 15, 21, 27-28, 37, 39, 49, 56-57, 65, 75, 102, 144-145, 152, 173, 175, 208-209, 212, 214-216, 219, 223, 226, 238, 242-243, 246, 272, 283-284*

CMR (Carcinogens, mutagens, substances toxic to preproduction) *308, 310, 313-314, 317-319*

coastal water *33, 44, 113, 126, 128, 133, 203*

combined approach (pollution prevention) *34, 138; 146, 179, 186*

complaint *13, 36, 42-44, 112, 116, 130, 161, 166, 183, 234, 244, 246, 299, 323, 335-336*

composting *92-93, 101, 103, 106, 111, 117, 189*

conservation *13-15, 18, 33, 36-44, 55, 116, 126, 130, 141, 147, 154-155, 178, 199, 202, 203, 205, 207, 237, 242, 244, 269, 321, 339*

cross compliance *237, 244*

cumulative effects *232, 237, 242*

dangerous substances *134, 145-146, 148, 150, 181, 200, 203, 286, 298, 309, 313, 319*

decentralisation *23, 189*

demand side *283, 284*

ECCP (European Climate Change Programme) *209, 283, 284, 285*

Ecolabel *15, 91, 263-268, 270-274, 293*

ecological status *13, 14, 33, 42, 44, 127-128, 131-132, 138, 141, 153-154, 200, 204*

economic instruments ***21-22**, 107, 144, 199, 212*

ecosystem *13, 21, **32-34**, 36, 47-50, 52, 62, 108, 126-128, 131-133, 138-140, 143, 150, 152-153, 155, 199*

ELV (Emission Limit Value) *12, 13, 20, **33**, 34, 66-75, 95, 101-102, 110, 129, 139, 148, 149, 176-180, 182-183, 186-187, 189, 192, 194, 222*

emission control *12, 21, 26, 35, 70, 114, 127, 129, 133-134, 137-139, 149, 154, 158, 175, 186, 189, 191, 194, 329*

emission register *115, 167-168, 170, 176, 180-181, 183, 188*

emission trading *13-14, 25-26, 158, 174-175, 176, 194, 212, 223*

EMS (environmental management system) *247-252, 254-257*

endocrine disruption *317-318*

end-of-life cars *100, 118*

energy using products 6, 252, 258, 262, 271, 282, 288, 293, 301

Environmental Action Programme 10, 18, 22, 23-24, 30, 35, 37, 71, 79, 82, 84, 91, 129, 144, 148, 261, 269, 316, 334

environmental information 118, 161-163, 233, 237, 241

EPER (European Pollutant Emission Register) 71, 85, 115, 167-170, 180, 219, 271, 299, 302, 303

EQS (Environmental Quality Standard) 12-13, **33-34**, 54, 129-130, 136, 138-139, 146-148, 159, 172, 179, 186, 192-193

essential use (pesticides) 315

ETSI (European Telecommunications Standards Institute) 106, 289-291, 304

EU Ecolabel Board 268

EU ETS (EU Emmission Trading Scheme) 210, 213-224, 226, 227

EURO 5 67, 69

EURO VI 68, 70

European Environment Agency 32, 47, 62-63, 149-150

Eutrophication 33, 47, 62-63, 149-150

existing substance 309-310, 312, 313-314

extended producer responsibility 10, 277

external costs 26, 105, 178

fine particles 62-63, 324, 329

Fish 125, 126, 128-130, 146-148, 153

flexible mechanism 209, 215

flood 128, 141-142, 154, 220, 243

Good Status 126, 130-133, 140-142, 152, 178

Groundwater 80-81, 102, 113, 126, 128-132, 139-143, 145-146, 150, 155, 192, 203, 307, 314, 320

Habitat 12-13, 33, 36-44, 47, 116, 131, 142, 146, 151, 158, 178, 199, 202-203, 205, 207, 323, 241

hazardous waste 79, 81, 83, 85, 87-89, 94-95, 98-102, 118, 164, 200

hormone disruption 307-308, 316

hydropower 145, 154

Incineration 14, 26, 80-81, 86-87, 89-90, 92-96, 98-99, 101-102, 108, 110-111, 113-115, 117-118, 123, 183, 185

Indicator 32, 35-36, 44, 109, 116, 147, 151, 155, 175, 248, 256, 276-277, 279, 320

industrial activity 186

initial environmental review 247-248, 253

inspection (environmental) 188, 247-248, 255

integrated approach 18, 21, 27, 181, 192-193

Integrated Product Policy 15, 27, 252, 262, 271, 274, 276-277, 287

Integration (environmental) 8, 12, 13, 15, **20**, 22, **25**, 36, 43, 44, 126, 147, 150-151, 154, 158, 174-175, 231, 233, 237-238, 241, 244-245, 277, 282, 293-295, 299, 303, 304

Internalisation (of pollution, environmental costs) 14, 26, 105, 146, 174, 199

ISO (International Standardisation Organisation) 174, 247-248, 250-255, 257-258, 273, 275, 279-280, 293, 295, 298, 300-304

JI (Joint Implementation) 214-215, 219-220, 221, 224

Kyoto Protocol 22, 62, 158, 208-209, 213-216, 219-221, 225

labeling 8, 84, 265, 269, 286, 288, 306, 209, 313

lake 33, 44, 48-49, 126, 128-129, 133, 147, 149, 153

Large Combustion Plant 26, 47, 70-76, 183-185, 192

legal compliance 247-248, 250-251, 253

lifecycle approach 108, 279

market based instruments **14-15**, 158, 174-176, 190, 199, 209

minimum criteria (environmental for products) 279

National Emission Ceilings Directive 33

national standard bodies 290, 293-294, 296, 298

Natura 2000 13, 37-45, 199, 238, 242, 317

navigation 142, 154, 201

new substance 15, 306, 310, 315

Nitrate 33, 62, 126-131, 138, 146, 149, 150, 153-154, 164

NOx 33, 51, 52, 60 ,62, 64-75, 101, 113, 192, 226, 298, 324-325, 327, 329

Olefins 324, 327, 329

OSPAR 132, 134-135, 188, 191, 277, 319

Oxygenates 260, 323-325

Packaging 14, 16, 24, 81, 85, 86, 91-93, 103-104, 106, 109, 117, 119-120, 123, 164, 252, 259, 263, 266, 287, 297, 300, 303-304, 306

Permit 14-15, 75, 85, 87, 89, 94-95, 100, 114, 118, 176-177, 179-182, 187-191, 193, 201, 203-204, 206, 210, 212, 214, 22, 256

344

pesticide *21, 128, 134-136, 167, 307-308, 314-318, 320-322*

pesticide use reduction *307, 317, 321*

petrol *67, 69, 83, 88, 98, 114, 121, 323-330*

plant protection products *213, 414, 317*

Policy Approaches *8, 25, 27, 85*

Polluter Pays *44, 62, 81-83, 87, 114, 143-145, 158, 174, 190, 197-198, 202, 204, 207*

pollution reduction *51, 57, 181, 188*

poly-aromatics *324-325, 329*

Precaution *24, 36, 158, 228, 307*

Precautionary Principle *28, 82, 90, 127, 134, 138, 150-151, 154, 188, 190, 242, 269, 308, 316, 321*

Predicted Environmental Concentration (PEC) *309*

Predicted No Effect Concentration (PNEC) *309*

prevention at source *33, 34, 81, 82, 84, 85, 90, 121, 153, 158, 175, 261*

priority substances *125, 131-132, 134-136, 138-139, 148, 188, 194*

product eco-design *267*

product policy *91, 146, 157, 252, 261-262, 266-267, 271, 274, 276-277, 279-281, 287, 299*

product prioritisation *277, 278, 280*

proximity principle *82, 84-85, 87, 108, 110, 120*

PRTR (pollutant release and transfer register) *167-170*

public health *53-54, 59, 76, 119, 204, 320*

public participation *14-15, 77, 95-96, 113-114, 143-145, 152-153, 159, 161, 163-168, 170, 175, 180, 193, 206, 218, 228, 236, 293, 300, 303*

public procurement *15, 91, 250, 256, 268, 269, 270-272, 274-275*

risk assessment *81, 134-135, 138, 306-313, 316-317*

REACH *16, 34, 99, 127, 137, 147, 191, 262, 306, 309, 312, 318, 319, 320*

recycling *14, 27, 79-80, 82-85, 87, 89-93, 95, 97, 99-110, 114, 116, 117, 121-123, 189-190, 194, 249, 263, 297*

registration *85, 127, 139, 191, 249, 250-254, 256, 262, 310, 312-314*

reporting *32, 57, 61, 97, 98, 112, 118, 127, 130, 149, 151, 169-170, 206, 211, 217, 254-256, 278-279, 335*

re-use *84-85, 88-89, 95, 105,107*

river *127-128, 130, 134, 141-143, 145, 148, 152, 153, 244, 317*

river basin *127-128, 130, 134, 142-143, 148, 152, 317*

safety data sheet *309, 313*

safety factor *33*

significant environmental aspects *247, 253*

spatial planning *19, 187, 237, 238*

Special Areas of Conservation *37, 38, 199*

Special Protection Areas *37-38, 40, 199*

subsidiarity *24, 107, 109, 159, 180, 182*

substitution *16, 22, 35, 110, 137, 147, 192, 262, 286, 309, 316-318, 321*

substitution principle *309, 316-317, 321*

sulphur content *54, 324, 327*

supply chain *254, 261, 276, 285, 312-313, 319*

surface water *80, 101 ,113, 125-126, 128-133, 134, 138-141, 145, 147-148, 203, 314*

Sustainable Development Strategy *8, 37, 261*

Thematic Strategy on air pollution *61, 63*

toxicity *138, 189, 308*

uniform principles *314-315*

voluntary (measures) *318*

voluntary pilot projects *278*

vulnerable species *38*

waste electrical and electronic equipment *80, 83, 85, 104, 122, 286, 298*

waste management *14, 80, 82, 84-86, 97, 89-97, 100, 103, 104-105, 108-112, 115-120, 122-123, 158, 188-189, 200, 238, 263*

waste oils *86, 92-93, 103-104, 110, 111, 113-114, 121-122*

waste prevention *14, 23, 79, 89-91, 108, 116-117, 123-124, 272, 284, 281*

waste recycling *84, 87, 92, 116, 189-190*

waste shipment *14, 85, 99-100, 118, 189*

waste streams *14, 21, 26, 79-81, 84, 86, 92, 96-97, 103-104, 117, 123, 190*

waste water *87, 94, 101, 126-130, 145, 149, 154, 185*

water quality *33, 35, 113, 126, 131-132, 139, 153, 194, 253, 293*

wetlands *44, 128, 147, 242*

wild bird *33, 37-39, 44, 146, 199, 207*

ENGLISH SIMPLIFIED

Tenth Edition

Blanche Ellsworth
John A. Higgins

PEARSON
Longman

New York   San Francisco   Boston
London   Toronto   Sydney   Tokyo   Singapore   Madrid
Mexico City   Munich   Paris   Cape Town   Hong Kong   Montreal

Senior Acquisitions Editor: Steven Rigolosi
Senior Supplements Editor: Donna Campion
Electronic Page Makeup: Dianne Hall

Exercises for Ellsworth/Higgins *English Simplified,* Tenth Edition

ISBN: 0-321-10430-7

2345678910–ML–060504

CONTENTS

(ESL denotes that the exercise has an ESL component. C denotes choice-type items; W denotes items asking for original written responses.)

Exercise	Name	English Simplified Sections Tested	Type of Exercise	Page
DIAGNOSTIC TESTS				
1.	Grammar, Sentences, and Paragraphs	1–27	C,W	1
2.	Punctuation	30–56	C	5
3.	Mechanics, Spelling, and Word Choice	60–74	C	9
GRAMMAR AND SENTENCES				
4.	Parts of a Sentence	1–3	C,W	13
5.	Parts of Speech	4–9	C	15
(ESL) 6.	Parts of Speech	4–9	W	17
7.	Uses of Nouns	11	C,W	19
8.	Complements	11B	C,W	21
9.	Uses of Nouns	11	W	23
(ESL) 10.	Verb Tenses and Forms	14A, B	C	25
(ESL) 11.	Verbs—Kind, Tense, Voice, and Mood	13, 14, 15	C,W	27
12.	Verbals	14D	C,W	31
(ESL) 13.	Verbs	14, 15	W	33
(ESL) 14.	Using Verbs	14, 15	C	35
15.	Adjectives and Adverbs	16, 17	C	37
(ESL) 16.	Articles and Determiners	16E	C,W	39
17.	Pronoun Kind and Case	18, 19	C	41
18.	Pronoun Case	19	C	43
19.	Pronoun Reference	20	C,W	45
20.	Phrases	21, 25E	C	47
21.	Verbal Phrases	21B	C,W	49
22.	Review of Phrases	21 (25E)	C,W	51
23.	Recognizing Clauses	22	C	55
24.	Dependent Clauses	22B	C,W	57
25.	Noun and Adjective Clauses	22B (25D)	W	59
26.	Adverb Clauses	22B (25D)	W	63
27.	Kinds of Sentences	22C (25B–E)	C,W	67
28.	Subject-Verb Agreement	23	C	71
29.	Pronoun-Antecedent Agreement	24	C	75
30.	Agreement Review	23, 24	C	77
31.	Effective Sentences	25	C	79
32.	Effective Sentences	25B–E	W	83
33.	Parallel Structure	25F	C,W	87
34.	Fragments	26A	C,W	91
35.	Comma Splices and Fused Sentences	26B, 26C	C,W	93
36.	Fragments, Comma Splices, and Fused Sentences	26	W	97
37.	Placement of Sentence Parts	27A	C	103
38.	Dangling and Misplaced Modifiers	27A, B	W	105
39.	Effective Sentences Review	25–27	W	111
40.	Review	1–27	C	115
41.	Review	1–27	W	119
(ESL) 42.	Review for Non-Native English Speakers	7, 14, 16E	C	121
PUNCTUATION				
43.	The Comma	30–32	C	123
44.	The Comma	30–32	C	127
45.	The Comma	30–32	C	131
46.	The Comma	30–32	C,W	135
47.	The Period, Question Mark, and Exclamation Point	33–38	C	139
48.	The Semicolon	39	C	141
49.	The Semicolon and the Comma	30–32, 39	C	143

Exercise	Name	*ES* Sections Tested	Type of Exercise	Page
50.	The Semicolon and the Comma	30–32, 39	W	145
51.	The Apostrophe	40–42	C	147
52.	The Apostrophe	40–42	C	149
53.	The Apostrophe	40–42	C	151
54.	Italics	43	C	153
55.	Quotation Marks	44–48	C	155
56.	Quotation Marks	44–48	C	157
57.	Italics and Quotation Marks	43, 44–48	C	159
58.	Colon, Dash, Parentheses, and Brackets	49–54	C	161
59.	The Hyphen and the Slash	55, 56	C	163
60.	Review	30–56	C	165
61.	Review	30–56	C	167

MECHANICS AND SPELLING

Exercise	Name	*ES* Sections Tested	Type of Exercise	Page
62.	Capitals	60–62	C	169
63.	Capitals	60–62	C	171
64.	Numbers and Abbreviations	65–69	C	173
65.	Capitals, Numbers, and Abbreviations	60–68	C	175
66.	Spelling	69–73	C	177
67.	Spelling	69–73	W	179
68.	Spelling	69–73	C	181
69.	Review	60–73	C	183

WORD CHOICE

Exercise	Name	*ES* Sections Tested	Type of Exercise	Page
70.	Conciseness, Clarity, and Originality	80	W	185
71.	Standard, Appropriate American Usage	81	C	189
72.	Standard, Appropriate American Usage	81	W	191
73.	Nondiscriminatory Terms	82	W	195
74.	Words Often Confused	83	C	197
75.	Words Often Confused	83	C	199
76.	Review	80–83	W	201
77.	Review	80–83	W	205

PARAGRAPHS AND PAPERS

Exercise	Name	*ES* Sections Tested	Type of Exercise	Page
78.	Topic Sentences and Paragraph Unity	92A, D	C	207
79.	Paragraph Development	92B	W	209
80.	Paragraph Coherence	92C	C	211
81.	Paragraph Review, Netiquette	92C, 93	W	213
82.	The Thesis Sentence	94B	C,W	215
83.	Planning the Essay	94A–D	W	217
84.	The Essay Outline	94E	W	219
85.	The Essay Introduction and Conclusion	95 (91, 92)	C,W	221
86.	The Essay Body	95 (91, 92)	C,W	223
87.	Research Paper Topics and Theses	96A, B	C,W	225
88.	Locating and Evaluating Sources	96C, E; 98	W	229
89.	Taking Notes, Citing, Avoiding Plagiarism	96F; 97A, B	W	233
90.	The Works Cited/Reference List	98	W	241

REVIEWS AND ACHIEVEMENT TESTS

Exercise	Name	*ES* Sections Tested	Type of Exercise	Page
91.	Review: Proofreading	30–73	W	245
92.	Review: Editing and Proofreading	1–83	W	247
93.	Review: Revising, Editing, and Proofreading	1–92	W	249
94.	Achievement Test: Grammar, Sentences, and Paragraphs	1–27	C,W	251
95.	Achievement Test: Punctuation	30–56	C	255
96.	Achievement Test: Mechanics, Spelling, and Word Choice	60–73	C	259

LIST OF GRAMMATICAL TERMS   263

PREFACE

The tenth edition of *Exercises for English Simplified* provides instructors with two types of exercises for assessing and developing students' writing and researching skills. For instructors who want quick, easy scoring, there are nearly 3000 choice-type items (exercises including these are labeled C in the table of contents). For instructors who prefer original, open-ended responses, there are hundreds of items for which students compose their own answers—words, sentences, paragraphs, even a full essay (exercises including these are labeled W in the table of contents).

The tenth edition retains the extensive revisions of the ninth, with further improvements including—

- Updated research exercises reflecting latest MLA, APA, and COS form revisions, and including evaluation of sources

- An expanded exercise on note-taking, avoiding plagiarism, and citing

- A new exercise section on online writing

- Hundreds of new items, adjusted to reflect revisions in *English Simplified* 10

- A more legible answer-key format for easier scoring

Revisions retained from the ninth edition include—

- Expanded diagnostic and achievement tests

- ESL exercises

- Simplified student answering keys

- Added review exercises in most sections

- Optional collaborative instructions in many exercises

- Perforated, hole-punched pages for easy removal, scoring, and keeping of exercises

Longman Publishers wishes to acknowledge the expert advice offered by the following during the revision of *Exercises for English Simplified:* Jeff Andelora, Mesa Community College; Dr. Eileen Ariza, Florida Atlantic University; Christine Gray, Community College of Baltimore County; Rita Hamada-Kahn, Cal Poly Pomona; Kristen Holland, Franklin University; Dr. Carol A. Lenhart, Arizona Western College; David Siar, Winston Salem State University; and Stephen A. Smolen, Saddleback College.

1. DIAGNOSTIC TEST: Grammar, Sentences, and Paragraphs

Part 1: Sentences

In the blank after each sentence,

Write **S** if the boldfaced expression is **one complete, correct sentence**.
Write **F** if it is a **fragment** (incorrect: less than a complete sentence).
Write **R** if it is a **run-on** (incorrect: two or more sentences written as one—also known as a **comma splice** or **fused sentence**).

Example: The climbers suffered from hypothermia. **Having neglected to bring warm clothing.** _____F_____

1. State inspectors have found doctors-in-training at State Hospital working excessive hours. **However, the hospital has promised to reduce their work load at once.** 1._____

2. The bank's computers have broken down again. **All transactions halted, and hundreds of customers angry.** 2._____

3. **In Russia, pork is sold for 465 rubles a pound that amount is equivalent to the average monthly salary.** 3._____

4. The Yankees made two big trades after the season had begun. **First for a shortstop and then for a center fielder.** 4._____

5. The African American Society put Martina Jones in charge of the Multicultural Festival. **A responsibility that appealed to her.** 5._____

6. **The boys are learning traditional Irish dancing, they really seem to enjoy their dance class.** 6._____

7. The President eventually seemed happy to retire from politics. **His family looking forward to spending more time with him.** 7._____

8. **Although American society may seem uncaring, more people are volunteering to help with the homeless.** 8._____

9. **The reason for her shyness being that she knew no one at the party except her hostess.** 9._____

10. **The experiment to produce nuclear fusion was both controversial and exciting, scientists all over the world attempted to duplicate its results.** 10._____

11. **Scientists have learned that sick bison can infect livestock with a serious bacterial disease.** 11._____

12. She loved all styles of art. **She said she particularly loved the impressionists, she had studied them in Paris.** 12._____

13. We walked over to the lost-and-found office. **To see whether the bag had been turned in.** 13._____

14. **The shift lever must be in neutral only then will the car start.** 14._____

15. **Buenos Aires, Argentina, is a lively city, the streets are safe at all times.** Movie theaters stay open all night. 15._____

16. **If you want an unusual form of exercise, learn to play the bagpipes.** 16._____

In the blank,

Write **C** if the boldfaced expression is used **correctly**.
Write **X** if it is used **incorrectly**.

Example: There **was** dozens of dinosaur bones on the site.   X

1. We need to keep this a secret between you and **I**.   1. _____
2. Bill was fired from his new job, **which** made him despondent.   2. _____
3. Each member will be responsible for **their** own transportation.   3. _____
4. There **was** at least five computers in the office.   4. _____
5. Several of **us** newcomers needed a map to find our way around.   5. _____
6. Every administrator and faculty member **was** required to attend the orientation program.   6. _____
7. The graduate teaching assistant and **myself** met for a review session.   7. _____
8. Surprisingly enough, presidential candidate Joan Smith was leading **not only** in the cities **but also** in the rural areas.   8. _____
9. In each sack lunch **were** a cheese sandwich, an apple, and a soda.   9. _____
10. Leave the message with **whoever** answers the phone.   10. _____
11. **Having made no other plans for the evening,** Tony was glad to accept the invitation.   11. _____
12. Everyone in the Hispanic Society **was** urged to join the movement to bring more Hispanic faculty to campus.   12. _____
13. If I **were** driving to Pennsylvania this weekend, I would take along my sketch pad.   13. _____
14. I bought one of the printers that **were** on sale.   14. _____
15. There **were** five different Asian student organizations on campus.   15. _____
16. The prosecutor demanded that the witness tell her **when did she hear the shot**.   16. _____
17. The director, as well as the choir members, **has** agreed to appear on television.   17. _____
18. The supervisor is especially fond of arranging training programs, working on elaborate projects, and **to develop budgets.**   18. _____
19. A faux pas **is when you commit a social blunder**.   19. _____
20. **Who** do you think mailed the anonymous letter to the editor?   20. _____
21. Neither the students nor the instructor **knows** where the notice is to be posted.   21. _____
22. Are you sure that it was **him** that you saw last evening?   22. _____
23. Between you and **me**, her decision to transfer to another department was not well received by her current supervisor.   23. _____
24. If the dog **had been** on a leash, it would not have been hit by a car.   24. _____
25. He joined the Big Brothers Organization and coached in Little League. **It** was expected of him by his law firm.   25. _____
26. Given the candidates, it's painfully clear that **us** voters didn't have much of a choice.   26_____
27. Customers should check the fruit carefully before paying; otherwise, **you** may end up with rotten or spoiled fruit.   27. _____

28. Anyone who forgets his book will not be able to take **their** report home   28. _____

29. While carrying my books to the library, **a squirrel darted across my path.**   29. _____

30. Norma **only** had one issue left to raise before she could rest her case.   30. _____

31. I had no idea that **my** giving a report would create such turmoil at the meeting.   31. _____

32. We didn't think that many of **us** substitutes would get into the game.   32. _____

33. Dean Robert Patterson gave Karen and **I** permission to establish a volunteer organization to tutor students from city schools.   33. _____

34. Although he often spoke harshly to others, his voice sounded **pleasant** to us.   34. _____

35. Neither the librarian nor the students in the reference room **was** aware of the situation.   35. _____

36. Professor Rogers looks very **differently** since he dyed his beard and moustache.   36. _____

37. There is no question that it was **she** under the table.   37. _____

38. The audience comprised **not only juniors but also seniors.**   38. _____

39. Dr. Smith, together with thirty of his students, **are** working at a community service site.   39. _____

40. Each of three employees **were** given a set of business cards.   40. _____

41. The only kind of transportation running **are** buses.   41. _____

42. **Standing motionless on the windswept, dreary plain,** the rain pelted my face.   42. _____

43. I had agreed to **promptly and without delay** notify them of my decision.   43. _____

44. The dean agreed to award the scholarship to **whomever** the committee selected.   44. _____

45. **Knowing that I should study,** it seemed important to unplug the phone.   45. _____

46. **Who** were you looking for in the auditorium?   46. _____

47. The noise and the general chaos caused by the alarm **were** disturbing to the visitor.   47. _____

48. As hard as I try, I'll never be as thin as **her.**   48. _____

49. Only one of these stamps **is** of real value.   49. _____

50. Only my brother and **myself** were allowed in after visiting hours.   50. _____

Part 3: Paragraphs (not included in scoring)

On the back of this page, write a **paragraph** of six to eight sentences on **one** of the topics below (you may also use scrap paper):

I will never do *that* again

My room (or clothes, car, etc.) as a reflection of me

If I were mayor (or governor or president) for one day

The best (or worst) film I have seen in the past year

The most unfair law

2. DIAGNOSTIC TEST: Punctuation

In the blank after each sentence,

Write **C** if the punctuation in brackets is **correct**;
Write **X** if it is **incorrect**.

(Use only one letter in each blank.)

Example: Regular exercise[,] and sound nutrition are essential for good health. _____X_____

1. The legislature has voted to close the old County Nursing Home[;] a larger, more modern home will replace it. 1._____

2. "What hope is there that the war will end soon[?]" the ambassador asked. 2._____

3. "Why can't a woman be more like a man["?] the chauvinist asked. 3._____

4. I learned that the newly elected officers were Marzell Brown, president[;] Leroy Jones, vice president[;] Sandra Smith, treasurer[;] and James Chang, secretary. 4._____

5. The class expected low grades[. T]he test having been long and difficult. 5._____

6. It[']s hard to imagine life without a VCR, a personal computer, and a microwave oven. 6._____

7. Eventually, everybody comes to Rick's[;] the best saloon in Casablanca. 7._____

8. Recognizing that busing places stress on younger students[,] the state officials are restructuring the school transportation system. 8._____

9. Richard Hernandez was unhappy at his college[,] he missed hearing Spanish and enjoying his favorite foods. 9._____

10. That is not the Sullivans' boat; at least, I think that it isn't their[']s. 10._____

11. When it rains, I always think of the opening lines of Longfellow's poem "The Rainy Day": "The day is cold, and dark, and dreary [/] It rains, and the wind is never weary." 11._____

12. Inspector Trace asked, "Is that all you remember?[" "]Are you sure?" 12._____

13. "The report is ready," Chisholm said[,] "I'm sending it to the supervisor today." 13._____

14. Didn't I hear you say, "I especially like blueberry pie"[?] 14._____

15. Joe enrolled in a small college[;] although he had planned originally to join a rock band. 15._____

16. Stanley moved to Minneapolis[,] where he hoped to open a restaurant. 16._____

17. That was a bit too close for comfort[,] wasn't it? 17._____

18. The advertiser received more than two[-]hundred replies on the Internet. 18._____

19. Sarah is asking for a week[']s vacation to visit relatives in Canada. 19._____

20. On February 21, 2008[,] Robin and Sam are getting married. 20._____

21. The womens['] basketball team has reached the state finals. 21._____

22. Recently, researchers have discovered that rhesus monkeys have some hidden talents[;] such as the ability to do basic math. 22._____

23. She received twenty[-]three greeting cards on her nineteenth birthday. 23._____

24. He caught the pass[,] and dashed for the end zone.
24. _____

25. Many weeks before school was out[;] he had applied for a summer job.
25. _____

26. Dear Sir[;] Please accept this letter of application for the teaching position.
26. _____

27. Schweitzer summed up his ethics as "reverence for life[,]" a phrase that came to him during his early years in Africa.
27. _____

28. Our communications professor asked us if we understood the use of extended periods of silence often found in conversations among Native Americans[?]
28. _____

29. "As for who won the election[—]well, not all the votes have been counted," she said.
29. _____

30. ["]The Perils of Aerobic Dancing["] (This is the title at the head of a student's essay for an English class.)
30. _____

31. Any music[,] that is not jazz[,] does not appeal to him.
31. _____

32. "Election results are coming in quickly now," the newscaster announced[;] "and we should be able to predict the winner soon."
32. _____

33. More than 42 percent of all adults eighteen and over are single[,] however, more than 90 percent of these adults will marry at least once.
33. _____

34. The children went to the zoo[;] bought ice-cream cones[;] fed peanuts to the elephants[;] and watched the seals perform their tricks while being fed.
34. _____

35. ["]For He's a Jolly Good Fellow["] is my grandfather's favorite song to sing at birthday parties.
35. _____

36. In the early 1900s, department stores provided customers electric lighting, public telephones, and escalators[;] and these stores offered countless other services, such as post offices, branch libraries, roof gardens, and in-store radio stations.
36. _____

37. Watch out[,] Marlene, for icy patches on the sidewalk.
37. _____

38. The rival candidates for the Senate are waging an all-out campaign[,] until the polls open tomorrow.
38. _____

39. Because he stayed up to play computer games[,] he didn't make it to his early class.
39. _____

40. The weather[—]rain, rain, and more rain[—]has ruined our weekend plans for an entire month.
40. _____

41. The first modern drive-in was called[,] The Pig Stand, which was a barbecue pit along a highway between Dallas and Fort Worth.
41. _____

42. The scholarship award went to Julia Brown, the student[,] who had the highest grades.
42. _____

43. Some of the technologies developed after World War II were[:] television, synthetic fibers, and air travel.
43. _____

44. The Lincoln Highway[,] which was the first transcontinental highway[,] officially opened in 1923 and was advertised as America's Main Street.
44. _____

45. Esther Greenberg[,] who is my roommate[,] comes from a small town.
45. _____

46. I hav[']ent made up my mind whether I want a computer system with an attached video camera.
46. _____

47. The talk show host[,] irritated and impatient[,] cut off the caller who insisted he was calling from aboard a flying saucer.
47. _____

48. Author Mike Rose writes[:] "When a local public school is lost to incompetence, indifference, or despair, it should be an occasion for mourning. . . ."
48. _____

49. A note under the door read: "Sorry you weren't in. The Emerson[']s."
49. _____

50. Most of my friends are upgrading their computer systems[,] they want to use CD-ROM software at home.

50. _____

51. This spring we began a new family vacation tradition[:] we flew to Florida to watch the Indians' spring training.

51. _____

52. No matter how cute they look, squirrels[,] in my opinion[,] are very destructive rodents.

52. _____

53. We are planning a trip to Chicago[,] the children will enjoy the city's museums.

53. _____

54. By saving her money[,] Laura was able to build her cottage on the lake.

54. _____

55. To gain recognition as a speaker[;] he accepted all invitations to appear before civic groups.

55. _____

56. Charles Wright[,] who survived an avalanche in the Himalayas[,] thought he heard a flute just before the storm occurred.

56. _____

57. Any candidate[,] who wants to increase social spending[,] will probably be defeated during the upcoming elections.

57. _____

58. "Oh, well[!]" the officer yawned, "I guess I'll stop in for coffee and a bagel."

58. _____

59. "I cannot believe that you have not read my book!"[,] shouted the author to the critic.

59. _____

60. In his painting [*The Red Dog,*] the French artist Paul Gauguin painted people from Tahiti in a bold and bright style.

60. _____

61. A group of workers in Westerville, Ohio[,] won a multi-million-dollar lottery prize.

61. _____

62. According to my family's written records, my great-grandfather was born in 1870[,] and died in 1895.

62. _____

63. My hometown is a place[,] where older men still think white shoes and belts are high fashion.

63. _____

64. She spent her student teaching practicum in Johnstown[,] where she went to hockey games each week.

64. _____

65. Having learned that she was eligible for a scholarship[,] she turned in her application.

65. _____

66. Living in his car for three weeks[,] did not especially bother him.

66. _____

67. The novel ["Underworld"] uses a famous home-run baseball as both a symbol and a unifying device.

67. _____

68. Many Americans remember family celebrations from their childhood[,] moreover, they are seeking ways to incorporate some of these rituals into their busy lives.

68. _____

69. In 1888 a bank clerk named George Eastman created the first amateur camera, called the Detective Camera[;] this camera was a small black box with a button and a key for advancing the film.

69. _____

70. After the long, harsh winter, I needed a soak[-]in[-]the[-]sun vacation.

70. _____

71. Veterans of World War I[,] who were hit hard by the Great Depression[,] received a government bonus in the 1930s.

71. _____

72. The girls['] and boys['] locker rooms had no heat.

72. _____

73. The parking lot always is full[,] when there is a concert.

73. _____

74. Dan was proud that he received all *A*[']s.

74. _____

75. You can reach Delaware Avenue by turning left[,] and following Route 19.

75. _____

7

3. DIAGNOSTIC TEST: Mechanics, Spelling, and Word Choice

Part 1: Capitalization

In each blank, write **C** if the boldfaced word(s) **follow** the rules of capitalization.
Write **X** if the word(s) **do not follow** the rules.

Example: The Mormons settled in what is now Salt Lake **City**.   _C_

1. The *Andrea Doria* sank in the 1950s.   1. ____

2. My **college** days were stressful.   2. ____

3. He attends Taft **high school.**   3. ____

4. The **President** vetoed the bill.   4. ____

5. They drove **east** from Tucson.   5. ____

6. We presented **Mother** with a bouquet of roses.   6. ____

7. I finally passed **spanish**.   7. ____

8. She is in France; **He** is at home.   8. ____

9. "Are you working?" **she** asked.   9. ____

10. I love **Korean** food.   10. ____

11. We saluted the **american** flag.   11. ____

12. Last **Summer** I drove to California.   12. ____

13. My birthday was **Friday**.   13. ____

14. I am enrolled in courses in **philosophy** and Japanese.   14. ____

15. She went **North** for Christmas.   15. ____

16. Please, **Father**, lend me your car.   16. ____

17. "But he's my **Brother**," she wailed.   17. ____

18. "Stop!" **shouted** the officer.   18. ____

19. Jane refused to be **Chairperson** of the committee.   19. ____

20. "If possible," he said, "**Write** the report today."   20. ____

Part 2: Abbreviations and Numbers

Write **C** if the boldfaced abbreviation or number is used **correctly.**
Write **X** it is used **incorrectly.**

Example: They drove through **Tenn.**   _X_

1. **Six billion** people now inhabit the world.   1. ____

2. I participated in a **five-hour** workshop on interpersonal communications.   2. ____

3. The play begins at **7** p.m.   3. ____

4. Aaron was born on November **11th,** 1988.   4. ____

5. The rent is **$325** a month.   5. ____

6. The interest comes to **8** percent.   6. ____

7. **Sen.** Levy voted against the bill.   7. ____

8. There are **nineteen** women in the club.   8. ____

9. **1999** was another bad year for flooding.   9. ____

10. I wrote a note to **Dr.** Rhee.   10. ____

11. [Opening sentence of a news article] The **ACDYM** has filed for bankruptcy.   11. ____

12. She lives on Buchanan **Ave.**   12. ____

13. We consulted Ricardo Guitierrez, **Ph.D.**   13. ____

14. Our appointment is at **4** o'clock.   14. ____

15. I slept only **3** hours last night.   15. ____

Part 3: Spelling

In each sentence, one boldfaced word is **misspelled.** Write its number in the blank.

Example: (1)**Its** (2)**too** late (3)**to** go.   _____1_____

1. Jane's (1)**independent** attitude sometimes was a (2)**hindrence** to the (3)**committee.**   1._____

2. (1)**Approximatly** half of the class noticed the (2)**omission** of the last item on the (3)**questionnaire.**   2._____

3. The (1)**mischievous** child was (2)**usualy** (3)**courteous** to adults.   3._____

4. At the office Jack was described as an (1)**unusually** (2)**conscientous** and (3)**indispensable** staff member.   4._____

5. Even though Dave was (1)**competent** in his (2)**mathematics** class, he didn't have the (3)**disipline** required to work through the daily homework.   5._____

6. The sociologist's (1)**analysis** of the (2)**apparent** (3)**prejudise** that existed among the villagers was insightful.   6._____

7. She was (1)**particularly** (2)**sensable** about maintaining a study (3)**schedule.**   7._____

8. It was (1)**necesary** to curb Tad's (2)**tendency** to interrupt the staff discussion with (3)**irrelevant** comments.   8._____

9. (1)**Personaly,** it was no (2)**surprise** that (3)**curiosity** prompted the toddler to smear lipstick on the bathroom mirror.   9._____

10. Tim developed a (1)**procedure** for updating our (2)**bussiness** (3)**calendar.**   10._____

11. As a (1)**sophomore** Sue had the (2)**perseverence** and (3)**sacrifice** needed to work three part-time jobs and to raise her three sons.   11._____

12. Her (1)**opinion,** while (2)**fascinating,** revealed an indisputable (3)**hypocricy.**   12._____

13. Every day our (1)**secretery** meets a colleague from the (2)**Psychology** Department at their favorite campus (3)**restaurant.**   13._____

14. During (1)**adolescence** we often (2)**condemm** anyone who offers (3)**guidance.**   14._____

15. Based on Bill's (1)**description,** his dream vacation sounded (2)**irresistable** and guaranteed to (3)**fulfill** anyone's need to escape.   15._____

Part 4: Word Choice

To be correct, the boldfaced expression must be standard, formal English and must not be sexist or otherwise discriminatory.

Write **C** if the boldfaced word is used **correctly.**
Write **X** if it is used **incorrectly.**

Examples: The counsel's **advice** was misinterpreted.   __C__   3. The plane began its **descent** for Denver.   3.____

They **could of** made the plane except for the traffic.   __X__   4. Economic problems always **impact** our enrollment.   4.____

1. Her car is different **than** mine.   1.____   5. My glasses **lay** where I had put them.   5.____

2. He **hadn't hardly** any chance.   2.____   6. We didn't play **good** in the last quarter.   6.____

10

7. I selected a **nice** birthday card. 7. ____

8. The float **preceded** the band in the parade. 8. ____

9. No one predicted the **affects** of the bomb. 9. ____

10. My aunt always uses unusual **stationery**. 10. ____

11. I dislike **those kind** of cookies. 11. ____

12. We are going to **canvas** the school district for the scholarship fund. 12. ____

13. The computer **sits** on a small table. 13. ____

14. College men and **girls** are warned not to drink and drive. 14. ____

15. The **principal** spoke to the students. 15. ____

16. I **had ought** to learn to use that software. 16. ____

17. He made **less** mistakes than I did. 17. ____

18. The family **better** repair the furnace. 18. ____

19. The package had **burst** open. 19. ____

20. Mrs. Grundy **censured** so much of the play that it was unintelligible. 20. ____

21. We are taught to consider the feelings of our **fellow man**. 21. ____

22. **Irregardless** of the warning, I drove in the dense fog. 22. ____

23. The next **thing** in my argument concerns my opponent's honesty. 23. ____

24. The new carpet **complements** the living room furniture. 24. ____

25. A different **individual** will have to chair the service project. 25. ____

26. I **ought to of** made the flight arrangements. 26. ____

27. **Numerical statistical figures** show that an asteroid may collide with Earth. 27. ____

28. **Due to the fact that** it rained, the game was canceled. 28. ____

29. **That sort of** person is out of place in this salon. 29. ____

30. The parade float was **round in shape**. 30. ____

4. GRAMMAR AND SENTENCES: Parts of a Sentence

(Study 1–3, The Sentence and Its Parts)

Part 1

In the blank, write the number of the place where the **complete subject** ends and the **(complete) predicate** begins.

Example: Immigrants (1) to the United States (2) have helped greatly (3) in building the country.   ____**2**____

1. A new convention center (1) will be constructed (2) on North Main Street (3) within five years.   1. _____

2. Many of the abandoned railroad stations (1) of America and Canada (2) have been restored (3) for other uses.   2. _____

3. The junction (1) of the Allegheny (2) and Monongahela rivers (3) creates (4) the Ohio River.   3. _____

4. The editor (1) wrote a kind note (2) after the long list of changes (3) to be made before final printing.   4. _____

5. The United States, (1) Mexico, (2) and Canada (3) now have (4) a free-trade agreement.   5. _____

6. I (1) recently completed (2) a twenty-page research paper (3) on the new common currency for European countries.   6. _____

7. Which (1) of the three word-processing software packages (2) has (3) the best thesaurus?   7. _____

8. Rarely would she drive her car after the earthquakes.
[This inverted-word-order sentence, rewritten in subject-predicate order, becomes:
She (1) would rarely (2) drive (3) her car (4) after the earthquakes.]   8. _____

9. None (1) of the polls (2) shows Stanton (3) winning.   9. _____

10. When did the committee select the candidate for comptroller?
[Rewritten in subject-predicate order:
The committee (1) did select (2) the candidate (3) for comptroller when?]   10. _____

Part 2

Write **S** if the boldfaced word is a **subject** (or part of a compound subject).
Write **V** if it is a **verb**.
Write **C** if it is a **complement** (or part of a compound complement).

Examples: Wendell played a superb game.   ____S____

Wendell **played** a superb game.   ____V____

Wendell played a superb **game**.   ____C____

1. **Many** paid their taxes late.   1. _____

2. Many paid their **taxes** late.   2. _____

3. Champion athletes **spend** much time training and competing.   3. _____

4. Champion athletes spend much **time** training and competing. 4. _____

5. **Sue** and Janet enjoy gardening. 5. _____

6. Sue and Janet enjoy **gardening**. 6. _____

7. Many **athletes** worry about life after the pros. 7. _____

8. Many athletes **worry** about life after the pros. 8. _____

9. The populist **theme** from the last election may survive until the next election. 9. _____

10. The populist theme from the last election **may survive** until the next election. 10. _____

11. The **neighborhood** worked hard to clean up the local playground. 11. _____

12. The neighborhood **worked** hard to clean up the local playground. 12. _____

13. The clustered lights far below the plane were **cities**. 13. _____

Part 3

In each sentence, fill in the blank with a word of your own that makes sense. Then, in the blank at the right, tell whether it is a **subject** (write **S**), **verb** (write **V**), or **complement** (write **C**).

Example: The builders needed a _ladder_ for the new job. _____C_____

1. Beautiful _____ grow in our garden. 1. _____

2. We grow beautiful _____ in our garden. 2. _____

3. Cabbages, onions, and _____ grow in our garden. 3. _____

4. Three students in English 101 _____ their final examination. 4. _____

5. The instructor was a _____. 5. _____

6. With the ball on the ten-yard line, the crowd _____. 6. _____

7. Shaw wrote a _____ about a speech professor and an uneducated young woman. 7. _____

8. The crisp, clear _____ refreshed us. 8. _____

9. Too many people in this country _____ unconcerned about their health. 9. _____

10. Materials necessary for this course include a(n) _____ and a calculator. 10. _____

5. GRAMMAR AND SENTENCES: Parts of Speech

(Study 4–9, The Parts of Speech)

Write the **part of speech** of each boldfaced word (use the abbreviations in parentheses):

noun	adjective (**adj**)	preposition (**prep**)
pronoun (**pro**)	adverb (**adv**)	conjunction (**conj**)
verb		interjection (**inter**)

Example: Shaw wrote many **plays**.   _noun_

1. This device is a **modem**.   1. ____
2. Applicants must **complete** this form.   2. ____
3. **She** anticipated the vote.   3. ____
4. The fires caused **many** to flee.   4. ____
5. Robert felt **tired**.   5. ____
6. She has been **there** before.   6. ____
7. The **primary** goal is to reduce spending.   7. ____
8. The test was hard **but** fair.   8. ____
9. Do you want fries **with** that?   9. ____
10. **That** player is going to be a star.   10. ____
11. **That** is not what I meant.   11. ____
12. You are going **beyond** the rules.   12. ____
13. Is this **your** book?   13. ____
14. The book is **mine**.   14. ____
15. He wants an **education**.   15. ____
16. **Wow**, what a shot that was!   16. ____
17. He agreed to proceed **slowly**.   17. ____
18. They **were sleeping** soundly at noon.   18. ____
19. The candidate selected a **charismatic** running mate.   19. ____
20. She is **unusually** talented.   20. ____
21. **Everyone** joined in the protest.   21. ____
22. Auto workers **are striking** for better pay.   22. ____
23. This is the first major **strike** in several years.   23. ____
24. The workers took a **strike** vote.   24. ____
25. He is the one **whom** I suspect.   25. ____

26. The researcher played a video game **while** waiting for the results.   26. ____
27. What is your **plan**?   27. ____
28. Nancy **is** a feminist.   28. ____
29. No one came **after** ten o'clock.   29. ____
30. Put the book **there**.   30. ____
31. I saw him **once**.   31. ____
32. The **theater** was dark.   32. ____
33. The tribe owns a **factory**.   33. ____
34. Weren't **you** surprised?   34. ____
35. They waited **for** us.   35. ____
36. The vote on the motion was quite **close**.   36. ____
37. Did you pay your **dues**?   37. ____
38. **All** survivors were calm.   38. ____
39. **All** were calm.   39. ____
40. The student read **quickly**.   40. ____
41. She **became** an executive.   41. ____
42. **Well**, what shall we do now?   42. ____
43. He worked **during** the summer.   43. ____
44. **Tomorrow** is her birthday.   44. ____
45. Will she call **tomorrow**?   45. ____
46. **If** I go, will you come?   46. ____
47. The executive stood **behind** her staff.   47. ____
48. He should never **have been advanced** in rank.   48. ____
49. The **wild** party was finally over.   49. ____
50. The corporation reported earnings **falsely**.   50. ____

6. GRAMMAR AND SENTENCES: Parts of Speech

(Study 4–9, The Parts of Speech)

Part 1

In the first blank in each sentence, write a word of your own that **makes sense**.
Then in the blank at the right, tell what **part of speech** your word is (use the abbreviations in parentheses):

noun	adjective (**adj**)	preposition (**prep**)
pronoun (**pro**)	adverb (**adv**)	conjunction (**conj**)
verb		interjection (**inter**)

Example: The singer wore a *gaudy* jacket.   ____adj____

(Collaborative option: Students work in pairs, alternating: one writes the word, the other names the part of speech.)

1. The letter should arrive _____.   1. _____

2. May I _____ you Friday?   2. _____

3. Every night, _____ monsters filled his dreams.   3. _____

4. The weather was gray _____ miserable.   4. _____

5. Is _____ your locker?   5. _____

6. The _____ was deathly quiet.   6. _____

7. _____ vacation proved quite hazardous.   7. _____

8. _____! I dropped my keys down the sewer.   8. _____

9. Many _____ trees are threatened by acid rain.   9. _____

10. This plane goes _____ Cleveland.   10. _____

11. He is a real _____.   11. _____

12. _____ now.   12. _____

13. They put the motion to a vote, but _____ failed.   13. _____

14. Ms. Kostas _____ a registered pharmacist.   14. _____

15. The senator would _____ accept a bribe.   15. _____

16. Approach that pit bull dog _____ carefully.   16. _____

17. Arles is in France, _____ Aachen is in Germany.   17. _____

18. The jury found that she was _____.   18. _____

19. _____ he won the lottery, he was envied.   19. _____

20. _____ of the runners collapsed from the heat.   20. _____

21. Farnsworth sought refuge _____ the storm.   21. _____

22. There in the dirt gleamed a tiny _____.   22. _____

23. The hill folk have always _____ the valley folk.   23. _____

24. Over the mountains and _____ the woods they trekked. 24. _____

25. _____! That's a sweet-looking car. 25. _____

Part 2

In each blank, write the **correct** preposition: **at**, **in**, or **on**.
(In some blanks, either of two prepositions may be correct.)

Example: Franko lives <u>in</u> an apartment <u>on</u> Broadway.

Fran Bradley, a retired banker, lived _____ a pleasant street _____ a small town _____ the Midwest. Most mornings she awakened promptly _____ six, except _____ Sundays, when she slept until eight. Then she would ride to worship _____ her 1987 Ford or _____ her old three-speed bicycle. Often she was the first one _____ her house of worship.

18

7. GRAMMAR AND SENTENCES: Uses of Nouns

(Study 11, The Uses of Nouns)

Part 1

In the blank, tell how the boldfaced word in each sentence is **used** (use the abbreviations in parentheses):

subject (**subj**)	indirect object (**ind obj**)	appositive (**app**)
subjective complement (**subj comp**)	objective complement (**obj comp**)	direct address (**dir add**)
direct object (**dir obj**)	object of preposition (**obj prep**)	

Example: The passenger gave the **driver** a tip.   ___ind obj___

1. The **delegates** gathered for the vote.   1. _____
2. The delegates gathered for the **vote**.   2. _____
3. The **committee** named Kamura treasurer.   3. _____
4. The committee named **Kamura** treasurer.   4. _____
5. The committee named Kamura **treasurer**.   5. _____
6. Russo will be a **member** of the board.   6. _____
7. Fallen leaves covered the **path**.   7. _____
8. **Ladies** and gentlemen, here is the star of tonight's show.   8. _____
9. Ladies and gentlemen, here is the **star** of tonight's show.   9. _____
10. The young star, **Leslie Mahoud**, appeared nervous.   10. _____
11. Antilock brakes give the **driver** more control.   11. _____
12. These brakes have become a **source** of controversy.   12. _____
13. These brakes have become a source of **controversy**.   13. _____
14. Misapplication of these brakes has caused some **accidents**.   14. _____
15. Which **company** will get the contract?   15. _____
16. Which company will get the **contract**?   16. _____
17. Everyone was bored by the speaker's **redundancy**.   17. _____
18. Redundancy, needless **repetition**, can put an audience to sleep.   18. _____
19. Marie, make that **customer** an offer she cannot refuse.   19. _____
20. **Marie**, make that customer an offer she cannot refuse.   20. _____
21. The company named Marie **salesperson** of the month.   21. _____
22. Tiger Woods has already become a legendary **golfer**.   22. _____
23. The other poker players gave **Fred** encouragement to bet high.   23. _____
24. Their cheating made Fred the **loser** in the poker game.   24. _____
25. Fred, a trusting **fellow**, never caught on.   25. _____

In each sentence, fill in the blank with a noun of your own that **makes sense**. Then in the blank at the right, tell how that noun is **used** (use the abbreviations in parentheses):

subject (**subj**)   indirect object (**ind obj**)   appositive (**app**)

subjective complement (**subj comp**)   objective complement (**obj comp**)   direct address (**dir add**)

direct object (**dir obj**)   object of preposition (**obj prep**)

Example: We sang songs far into the _night_.   _obj prep_

(Collaborative option: Students work in pairs, alternating: one writes the word, the other names its use.)

1. First prize was a brand-new _____.   1. _____

2. _____, please make more coffee.   2. _____

3. The new _____ in town should expect a warm welcome.   3. _____

4. Every autumn the region's trees, mostly _____, delight touring leaf-peepers.   4. _____

5. Brad's CD collection contains mostly songs by _____.   5. _____

6. Before the examination Professor Ferrano gave us a(n) _____.   6. _____

7. Paula's attitude made her a(n) _____ to many classmates.   7. _____

8. Warmhearted Pat gave the _____ a hug.   8. _____

8. GRAMMAR AND SENTENCES: Complements

(Study 11B, Complements)

Part 1

In the blank, tell how each boldfaced complement is **used** (use the abbreviations in parentheses). If any complement is an adjective, **circle** it.

subjective complement (**subj comp**)   objective complement (**obj comp**)
direct object (**dir obj**)   indirect object (**ind obj**)

Examples: The ambassador delivered the **ultimatum**.   <u>dir obj</u>
            The queen became (furious.)   <u>subj comp</u>

1. The Lakers have been consistent **winners** in basketball.   1. _____
2. Sarah gave the **bedroom** a new coat of paint.   2. _____
3. The city lost many **jobs** after September 11, 2001.   3. _____
4. Kimiko declared English her **major**.   4. _____
5. The guide gave us **directions** to the Rue de la Paix.   5. _____
6. Friends, Romans, countrymen, lend **me** your ears.   6. _____
7. The soprano completed her **practice**.   7. _____
8. She sounds **happier** every day.   8. _____
9. **Whom** did you meet yesterday?   9. _____
10. Will the company give **John** another offer?   10. _____
11. Politicians will promise **us** anything.   11. _____
12. The group had been studying **anthropology** for three semesters. 12. _____
13. Her former employer gave **her** the idea for the small business. 13. _____
14. I named him my **beneficiary**.   14. _____
15. She is an **instructor** at the community college.   15. _____
16. She became an **administrator**.   16. _____
17. I found the **dictionary** under the bed.   17. _____
18. He considered her **brilliant**.   18. _____
19. Select whatever **medium** you like for your art project.   19. _____
20. The company made her **manager** of the branch office.   20. _____
21. Wasn't Eva's sculpture **stunning**?   21. _____
22. Please bake **me** an apple pie.   22. _____
23. Most women don't understand **menopause**.   23. _____
24. The toddler threw her **boots** against the wall.   24. _____
25. The besieging troops gave the surrounded city an **ultimatum** this morning. 25. _____

In each sentence, fill in the blank with a complement of your own. Then in the blank at the right, **tell what kind** of complement it is.

subjective complement (**subj comp**)   objective complement (**obj comp**)
direct object (**dir obj**)   indirect object (**ind obj**)

Example: The test results were <u>inconclusive</u>.   <u>subj comp</u>

(Collaborative option: Students work in pairs, alternating: one writes the word, the other names the kind of complement.)

1. This prescription drug is _____.   1. _____

2. The columnist received an anonymous _____.   2. _____

3. Santos named Ahmed his _____.   3. _____

4. Last year we did _____ a favor.   4. _____

5. The haunted house attracted curious _____ from far and near.   5. _____

6. This memo has caused _____ much grief.   6. _____

7. To be rid of him, they designated him _____.   7. _____

8. She did not seem particularly _____.   8. _____

9. GRAMMAR AND SENTENCES: Uses of Nouns

(Study 11, The Main Uses of Nouns)

First write a sentence of your own, using the boldfaced verb. Include the parts mentioned in parentheses (in the order given). Then **identify** each of those parts by writing its name under the proper word (use the abbreviations given below):

subject (**subj**)	objective complement (**obj comp**)
subjective complement (**subj comp**)	object of preposition (**obj prep**)
direct object (**dir obj**)	appositive (**app**)
indirect object (**ind obj**)	direct address (**dir add**)

Example: designated (subject, direct object, objective complement) <u>The teacher designated Paul the class librarian.</u>
                      subj   dir obj   obj comp

(Collaborative option: Students work in pairs, alternating: one writes the sentence, the other identifies the uses.)

1. **destroyed** (Subject, direct object) _____

 _____

2. **was** (Subject, subjective complement) _____

 _____

3. **sent** (Subject, indirect object, direct object) _____

 _____

4. **considered** (Subject, direct object, objective complement) _____

 _____

 _____

5. **sat** (Subject, appositive)_____

 _____

6. **get** (Direct address, understood subject, direct object) _____

 _____

7. **has obtained** (Subject, object of preposition, direct object) _____

_____

_____

8. **may show** (Subject, indirect object, direct object) _____

_____

9. **may become** (Subject, object of preposition, subjective complement) _____

_____

_____

10. **made** (Subject, direct object, objective complement) _____

_____

_____

11. **might have been** (Subject, appositive, subjective complement) _____

_____

_____

12. Make up your own verb. (Subject, direct object, appositive, objective complement) _____

_____

_____

_____

10. GRAMMAR AND SENTENCES: Verb Tenses and Forms

(Study 14A, Principal Parts, and 14B, Tense and Form)

Part 1

Identify the tense or other form of the boldfaced verb (use the abbreviations in parentheses):

present (**pres**)   past perfect (**past perf**)
past   future perfect (**fut perf**)
future (**fut**)   conditional (**cond**)
present perfect (**pres perf**)   past conditional (**past cond**)

Example: They **spoke** too fast for us.   ____past____

1. The operation **costs** $15,000.   1._____
2. The plane **will** surely **depart** on time.   2._____
3. Next summer, we **shall have lived** in this house for ten years.   3._____
4. Billingsley Hall, our dormitory, **has acquired** a new coat of paint. 4._____
5. By noon he **will have finished** the whole job.   5._____
6. If the study were flawed, it **would be rejected**.   6._____
7. **Shall** I ever **see** you again?   7._____
8. Patriotic banners **appeared** all over town.   8._____
9. At first they **had** not **believed** the rumors.   9._____
10. If the study had been flawed, it **would have been rejected**.   10._____
11. The company **guaranteed** that the package would arrive in the morning. 11._____
12. Dylan **will begin** cello lessons in the spring.   12._____
13. The children **have created** a snow castle in the front yard.   13._____
14. **Have** you an extra set of car keys?   14._____
15. They **would have passed** if they had studied harder.   15._____
16. I **wrote** a review of the school play.   16._____
17. The family **has planned** a vacation.   17._____
18. The comic **laughed** at his own jokes.   18._____
19. In one week the flu **hit** five staff members.   19._____
20. This Friday **would have been** my grandmother's hundredth birthday. 20._____
21. **Would** you **mind** if we left early?   21._____
22. Congress **will have adjourned** by the time the law expires.   22._____
23. Michael **has applied** for a junior year abroad.   23._____
24. **Is** it fair for you to turn me down?   24._____
25. Firefighters in the forest **braved** high winds and intense heat.   25._____

Part 2

In each blank, write the needed **ending**: **ed** (or **d**), **s** (or **es**), or **ing**. If no ending is needed, leave the blank empty.

Example: Every day the sun rise<u>s</u> later, and I wake_____ up later.

Today the brown cliffs rise_____ directly from the sea; no beach separate_____ the cliffs from the water. The waves have pound_____ the granite base of that cliff for ages but have fail_____ to wear it away. Now, as always, great white gulls are swoop_____ just above the foam; they are seek_____ fish that are destine_____ to become their dinner. Years ago, when my friend Jan and I first gather_____ the courage to approach the cliff's sheer edge and peer over, we imagine_____ what it would be like if we tumble_____ over and fell into that seething surf far below. At that time, the thought fill_____ me with terror.

Today, ten years later, as my friend and I stand_____ atop the cliffs, Jan speak_____ of how she felt then. I can tell that she is try_____ to relive that experience of our youth. We are not feel_____ the same terror now, and we will never feel_____ it again. Still, nothing would make_____ us go closer to the edge. When a man or woman reach_____ age thirty, he or she often attempt_____ to recapture the excitement of youth but rarely succeed_____. In a few moments Jan and I will walk_____ back to where our cars are park_____. We have been pretend_____ to be youngsters again, but now each of us know_____ that we can never repeat the past. Jan look_____ at me with a smile.

11. GRAMMAR AND SENTENCES: Verbs—Kind, Tense, Voice, and Mood

(Study 13, Kinds of Verbs; 14, Using Verbs Correctly; and 15, Avoiding Verb Errors)

Part 1

Write **T** if the verb is **transitive**.
Write **I** if it is **intransitive**.
Write **L** if it is **linking**.

Example: The house **looked** decrepit.   _____L_____

1. Jenny **kissed** me when we met.   1. _____

2. The thunder **sounded** louder each time.   2. _____

3. **Lay** your wet coat by the furnace.   3. _____

4. The roses **opened** early this spring.   4. _____

5. The island **lies** not far off the mainland.   5. _____

6. The last express **has** already **left**.   6. _____

7. **Set** the vase carefully on the table.   7. _____

8. Bosnik **remained** speaker of the Assembly for eight terms.   8. _____

9. Never **interfere** in another person's quarrel.   9. _____

10. The bus **departed** five minutes ago.   10. _____

11. Pollution **accumulated** in the atmosphere.   11. _____

12. The waiters **served** hors d'oeuvres before the main meal.   12. _____

13. My friend **seemed** nervous.   13. _____

14. The new class **ended** abruptly.   14. _____

15. Women **have been** instrumental in maintaining the social structure of the American Protestant
churches.   15. _____

Part 2

Rewrite each boldfaced verb in the tense or form given in parentheses.

Example: Now we **live** in Hamilton Hall. (present perfect)   _For the past year we have lived in Hamilton Hall._

1. Packing material from the box **clutters** the floor. (past)

_____

_____

2. The CIA **appointed** Choi its chief agent in Asia. (present perfect)

_____

_____

3. The CIA **appointed** Choi its chief agent in Asia. (past perfect)

_____

_____

4. The Red Sox **will win** the pennant by next fall. (future perfect)

_____

_____

5. The Everglades **will die** without relief from pollution. (conditional)

_____

_____

6. The Everglades **will die** without relief from pollution. (past conditional)

_____

_____

7. The Piffle Company **seeks** a new vice-president. (present progressive)

_____

_____

8. The Piffle Company **seeks** a new vice-president. (past emphatic)

_____

_____

9. The Piffle Company **seeks** a new vice-president. (present perfect)

_____

_____

10. The Piffle Company **seeks** a new vice-president. (present perfect progressive)

_____

_____

First write **A** if the boldfaced verb is in the **active voice** or **P** if it is in the **passive voice.** Then **rewrite** the sentence in the opposite voice (if it was active, make it passive; if it was passive, make it active). If necessary, supply your own subject.

Examples: A car **struck** the lamppost.   _____A_____
        The lamppost was struck by a car.

        The door **was left** open.   _____P_____
        Someone left the door open.

1. One name **was** inadvertently **omitted** from the list.   1. _____

    _____

    _____

2. The negotiator **carried** a special agreement to the union meeting.   2. _____

    _____

    _____

3. The media **bashed** the incumbent's speech.   3. _____

    _____

    _____

4. The meeting **was called** to order.   4. _____

    _____

    _____

5. The ancient city **was** totally **destroyed** by a volcanic eruption.   5. _____

    _____

    _____

6. An accounting error **was discovered.**   6. _____

    _____

    _____

7. Younger voters **have selected** a presidential candidate.   7. _____

    _____

    _____

8. The prosecutor **subjected** the witness to a vigorous cross-examination.

8. _____

_____

_____

9. By dawn the police **will have barricaded** every road.

9. _____

_____

_____

10. The status report **will be submitted** next week.

10. _____

_____

_____

11. The left fielder **threw out** the runner.

11. _____

_____

_____

12. Environmental activists **have begun** a nationwide antipollution campaign.

12. _____

_____

_____

Part 4

Rewrite each sentence in the **subjunctive** mood.

_____

Example: Today the sky is sunny.
    I wish the sky <u>were</u> sunny today.

1. Benito is on time.

    I wish Benito _____ on time.

2. I am a scuba diver; I search for sunken ships.

    If I _____ a scuba diver, I would search for sunken ships.

3. The customers insisted; their money was returned.

    The customers insisted that their money _____ returned.

12. GRAMMAR AND SENTENCES: Verbals

(Study 14D, Distinguish a Verbal from a Verb)

Part 1

Identify each boldfaced verbal by writing

    inf for infinitive   **pres part** for present participle
    ger for gerund   **past part** for past participle

Example: The **outnumbered** soldiers surrendered.   ___past part___

1. October is the best month **to watch** the falling leaves.   1._____

2. October is the best month to watch the **falling** leaves.   2._____

3. The next item on our agenda is **to select** a new secretary.   3._____

4. The next item on our agenda is **selecting** a new secretary.   4._____

5. I submitted a **typed** application.   5._____

6. **Encouraged** by their initial weight loss, Cecilia and Roy continued their diets.   6._____

7. **To lose** more weight, they had to both diet and exercise.   7._____

8. By **losing** weight, they felt and looked healthier.   8._____

9. **Ignoring** all criticism, Flo defended her friend's actions.   9._____

10. Flo likes **helping** her friends.   10._____

11. The movie *Titanic,* **seen** on a small TV screen, is much less impressive.   11._____

12. **Examining** the report, the consumer decided not to invest.   12._____

13. **Frightened,** he became cautious.   13._____

14. The purpose of the cookbook is **to reduce** the threat of cancer through a healthful diet.   14._____

15. **Reducing** carbon dioxide emissions was a top priority in a recent bill.   15._____

Complete each sentence with a verbal or verbal phrase of your own. Then in the blank at the right, tell how it is **used** (use the abbreviations in parentheses):

subject (**subj**)   object of preposition (**obj prep**)
subjective complement (**subj comp**) adjective (**adj**)
direct object (**dir obj**)   adverb (**adv**)

Examples: <u>Faced with the evidence</u>, the suspect admitted the crime.   <u>adj</u>
The suspect was accused of <u>absconding with company funds</u>.   <u>obj prep</u>

(Collaborative option: Students work in pairs, alternating: one writes the word or phrase, the other identifies its use.)

1. _____ is no way to greet the day.   1. _____

2. Binoy likes _____.   2. _____

3. The _____ crowd rose to its feet.   3. _____

4. The driver got out and opened the hood [for what purpose?]

   _____   4. _____

5. Professor Zullo's obsession is _____

   _____   5. _____

6. I earned an *A* in her course by _____

   _____   6. _____

7. The huge motor home, _____

   _____, lumbered up the mountain road.   7. _____

8. As a last resort the officials tried _____

   _____   8. _____

9. My ambition since childhood has been _____

   _____   9. _____

10. Oddly, _____

    has never been one of my goals.   10. _____

13. GRAMMAR AND SENTENCES: Verbs

(Study 14, Using Verbs Correctly, and 15, Avoiding Verb Errors)

In each blank, write the **correct form** of the verb in parentheses (some answers may require more than one word).

Examples: (prefer) We have always <u>preferred</u> vanilla.
(see) Yesterday all of us <u>saw</u> the rainbow.

1. (use) The young man had never _____ a microwave.

2. (begin) He found the instruction book and _____ to read it.

3. (cross) They were _____ the busy street in the wrong place.

4. (blow) Trees of all sizes were _____ down in the storm.

5. (try) If I had found the courage, I would _____ _____ skydiving.

6. (drink) Drivers who had _____ alcoholic beverages were detained by the police.

7. (fly) By the time I reach Hawaii, I shall _____ _____ for thirteen hours nonstop.

8. (freeze) If they had not brought heavy clothing, they would _____ _____ on the hike.

9. (possess) The Tsar's court felt that Rasputin _____ a strange power over them.

10. (choose) The district has always _____ a Republican for Congress.

11. (bring) Sidney _____ his guitar to school last year.

12. (forbid) Yesterday the resident assistant _____ Sidney to play it after 9 p.m.

13. (lead) Firefighters _____ the children to safety when the smoke became too dense.

14. (lay) When their chores were finished, the weary farmers _____ their pitchforks against the fence and rested.

15. (pay) The company had always _____ its employees well.

16. (ring) The pizza deliverer walked up to the door and _____ the bell.

17. (rise) We are late; the sun has _____ already.

18. (see) Then I _____ him running around the corner.

19. (shine) Before the interview Rod _____ his old shoes.

20. (shine) Rod polished his car until it _____ brightly.

21. (break) The rebels had _____ the peace accord.

22. (shake) The medicine had to be _____ well before being used.

23. (struggle) For hours the fox had _____ _____ [2 words—use progressive form] to escape from the trap.

24. (show) Last week Ford _____ its new models at the automotive exhibition.

25. (mean) Daisy had not _____ to hit Mrs. Wilson.

26. (sink) In 1945 Nimitz's carrier planes _____ much of the enemy's fleet.

27. (drag) When it grew dark the poachers _____ the dead deer to their truck.

28. (speak) If they had known who she was, they never would _____ _____ to her.

29. (swing) In his last at-bat Sosa _____ the bat harder than ever before.

30. (throw) The pitcher had _____ a high fastball.

31. (write) A columnist had _____ that Sosa could not hit a high fastball.

32. (seek) After her divorce, Carla _____ a place of peace and quiet.

33. (admire) For years to come, people _____ _____ _____ [3 words—use progressive form] your paintings.

14. GRAMMAR AND SENTENCES: Using Verbs

(Study 14, Using Verbs Correctly, and 15, Avoiding Verb Errors)

Write **C** if the boldfaced verb is used **correctly**.
Write **X** if it is used **incorrectly**.

Examples: In chapter 1 Greg goes to war, and in chapter 10 he **died.**   _____X_____
                The lake **was frozen** overnight by the sudden winter storm.   _____C_____

1. The longer he stayed, the more he **payed**.   1. _____
2. The pilots **have flown** this route hundreds of times.   2. _____
3. The giant shark **swam** far up the estuary.   3. _____
4. The phone **has rang** five times. Why don't you answer it?   4. _____
5. My hat **was stole** when I left it at the restaurant.   5. _____
6. We **have ridden** the train to Chicago many times.   6. _____
7. The little child **tore** open the present wrapped in bright yellow paper.   7. _____
8. The student **sunk** into his chair to avoid being called on by the professor.   8. _____
9. **Have** you **gone** to see the new *Star Wars* episode?   9. _____
10. We **should have known** that Robert would be late for the meeting.   10. _____
11. The little boy standing by the counter **seen** the man shoplift a watch.   11. _____
12. In the novel *Roscoe*, the main character decides to leave politics but **was drawn** back to it.   12. _____
13. The judge decreed that the abuser **be** sent to prison.   13. _____
14. The medals **shone** brightly on the general's uniform.   14. _____
15. The children **swang** on the swing until their mother called them home for supper.   15. _____
16. When we were small, we **wore** hats and white gloves on special occasions.   16. _____
17. The author **hasn't spoken** to the news media for fifty years.   17. _____
18. Jack **wrote** his essay on the summer spent on his grandfather's farm.   18. _____
19. We **lay** on the couch reading the Sunday newspaper and munching doughnuts.   19. _____
20. Jack **stole** the chocolate candy when his brother left the kitchen.   20. _____
21. When Chillingworth realizes what Dimmesdale has been doing, he **began** to plot revenge.   21. _____
22. Only immigrants who could not afford first- or second-class ship fares **passed** through Ellis Island.   22. _____
23. Begin by taking Route 202 to West Chester; then you **should follow** Route 30 to Lancaster.   23. _____
24. Today long-distance telephone calls cost less than they **costed** forty years ago.   24. _____
25. Stella **has run** five miles along the coastal trail every day this year.   25. _____
26. Within ten minutes after someone broke into our house, the police **were notified** by us.   26. _____
27. The second act complicates the conflict, but the third act **resolves** it.   27. _____
28. Then I **come** up to him and said, "Let her alone!"   28. _____

29. McWilliams **has swum** from the mainland to Catalina Island. 29. _____

30. The pastor suggested that the discussion **be** postponed until the next Parish Council meeting. 30. _____

31. The child **was** finally **rescued** hours after falling into the pit. 31. _____

32. As the curtain rose, Bruce **is** standing alone on stage, peering out a window. 32. _____

33. The scientists' report **awakened** the nation to the dangers of overeating. 33. _____

15. GRAMMAR AND SENTENCES: Adjectives and Adverbs

(Study 16–17, Using Adjectives and Adverbs)

If the boldfaced adjective or adverb is used **correctly,** leave the blank empty.
If the boldfaced adjective or adverb is used **incorrectly,** write the correct word(s) in the blank.

Examples: Her performance was truly **impressive.** _____
The Yankees are playing **good** this year. _____ well _____

1. The rock star sounds **good** on her new CD. 1._____

2. Why let his innocent remarks make you feel **badly**? 2._____

3. She was the **most** talented member of the dance couple. 3._____

4. Her computer is in very **good** condition for such an old machine. 4._____

5. He was very **frank** in his evaluation of his work. 5._____

6. My father spoke very **frankly** with us. 6._____

7. Her mother is the **kindest** of her two parents. 7._____

8. My stomach feels bad, and my back hurts **bad**. 8._____

9. The student looked **cheerful**. 9._____

10. The student looked **wearily** at the computer monitor. 10._____

11. I comb my hair **different** now. 11._____

12. Was the deer hurt **bad**? 12._____

13. He seemed **real** sad. 13._____

14. The learning assistant tried **awful** hard to keep the residence hall quiet during finals week. 14._____

15. Reading Eudora Welty's work is a **real** pleasure. 15._____

16. The teaching assistant glanced **nervously** at the class. 16._____

17. The bus driver seemed **nervous**. 17._____

18. The campus will look **differently** when the new buildings are completed. 18._____

19. Yours is the **clearest** of the two explanations. 19._____

20. The book is in **good** condition. 20._____

21. I did **poor** in organic chemistry this term. 21._____

22. Mario looked **debonair** in his new suit. 22._____

23. Trevor felt **badly** about having to fire the veteran employee. 23._____

24. Daryl's excuse was far **more poorer** than Keith's. 24._____

25. She writes very **well**. 25._____

26. It rained **steady** for the whole month of December in Houston. 26._____

27. The roses smell **sweet**. 27._____

28. He tries **hard** to please everyone. 28._____

29. John is **near** seven feet tall. 29._____

30. He talks **considerable** about his career plans. 30._____

31. She donated a **considerable** sum of money to the project. 31._____

32. The **smartest** of the twins is spoiled. 32._____

33. The **smartest** of the triplets is spoiled. 33._____

34. The coach gazed **uneasily** at her players. 34._____

35. He felt **uneasy** about the score. 35._____

36. Do try to drive more **careful**. 36._____

37. It was Bob's **most unique** idea ever. 37._____

38. The trial was **highly** publicized. 38._____

39. The wood carving on the left is even **more perfect** than the other one. 39._____

40. The house looked **strangely** to us. 40._____

41. She looked **strangely** at me, her brow furrowed. 41._____

42. He was ill, but he is **well** now. 42._____

43. That lobbyist is the most **influential** in Washington. 43._____

44. The orchestra sounded **good** throughout the hall. 44._____

45. Societal violence has **really** reached epidemic proportions in this country. 45._____

46. He seemed very **serious** about changing jobs. 46._____

47. Something in the refrigerator smelled **bad**. 47._____

48. We felt **badly** about missing the farewell party. 48._____

49. The police acted **swiftly** when they received the tip. 49._____

50. However, their response was not **swift** enough. 50._____

16. GRAMMAR AND SENTENCES: Articles and Determiners

(Study 16E, Use Articles and Determiners Correctly)

Part 1

In each blank, write the **correct** article: **a**, **an**, or **the**; or leave the blank empty if **no** article is needed. (In some blanks either of two answers is correct.)

Example: When <u>the</u> moon and <u>a</u> planet come close to each other in <u>the</u> sky, <u>an</u> exciting sight awaits _____ viewers.

_____ exciting play occurred yesterday in _____ big-league baseball game at _____ Dodger Stadium. _____ Dodger player dropped _____ ball in the glare of _____ sun. When _____ ball fell to _____ ground, three Dodger players ran after it; thus _____ nobody was guarding the bases for _____ Dodgers. _____ crowd groaned with _____ disappointment. _____ batter from _____ other team ran around _____ bases with _____ determination. _____ Dodger player retrieved _____ ball and made _____ accurate throw that reached home plate ahead of the runner, who was called "Out!" From the crowd came _____ cheers. The Dodgers won the game and celebrated with _____ champagne from _____ France. This was _____ biggest Dodger victory of the year.

Part 2

For each blank, choose from the list any correct determiner (limiting adjective), and write it in. Try not to use any word on the list more than once.

every	many	other	more	some
each	most	such	(a) little	
either	(a) few	both	much	
another	all	enough	any	

Example: They needed <u>*another*</u> person to help lift the car.

1. _____ country that voted for the United Nations resolution was praised.

2. _____ countries that voted against it were criticized.

3. _____ discussions took place before the vote.

4. _____ of the neutral countries tried to postpone the vote.

5. But _____ pressure was put on these countries to vote.

6. _____ effort to influence the neutral countries' vote was rebuffed.

7. Delegates from _____ countries wanted to get the voting finished.

8-9. _____ delegates had _____ patience.

10. Finally, the Secretary General declared that _____ voting would take place the next day.

17. GRAMMAR AND SENTENCES: Pronoun Kind and Case

(Study 18, The Kinds of Pronouns, and 19, Using the Right Case)

Part 1

Classify each boldfaced pronoun (use the abbreviations in parentheses):

personal pronoun (**pers**)   indefinite pronoun (**indef**)
interrogative pronoun (**inter**)   reflexive pronoun (**ref**)
relative pronoun (**rel**)   intensive pronoun (**intens**)
demonstrative pronoun (**dem**)

Example: **Who** is your partner?   _____inter_____

1. I made him an offer that **he** could not refuse.   1. _____

2. I blame **no one** but myself for the error.   2. _____

3. **This** is all I ask: that you tell me the truth.   3. _____

4. I blame no one but **myself** for the error.   4. _____

5. **Which** of the city newspapers do you read?   5. _____

6. She is the executive **who** makes the key decisions in this company. 6. _____

7. I **myself** have no desire to explore the rough terrain of mountainous regions. 7. _____

8. **Everyone** promised to be on time for the staff meeting.   8. _____

9. They chose three charities and gave a thousand dollars to each of **them.** 9. _____

10. A small motor vehicle **that** can travel on rough woodland trails is called an all-terrain vehicle. 10. _____

11. Be sure that you take care of **yourself** on the expedition.   11. _____

12. These are my biology notes; **those** must be yours.   12. _____

13. **Who** would you say is the most likely culprit?   13. _____

14. If we lose, only we **ourselves** are to blame.   14. _____

15. They requested that **neither** of the parties be informed.   15. _____

Part 2

Write the number of the **correct** pronoun.

Example: The message was for Desmond and (1)**I** (2)**me**.   _____2_____

1. None of (1)**we** (2)**us** plaintiffs felt that we had been adequately compensated. 1. _____

2. Although I tried to be careful around the cat, I still stepped on (1)**its** (2)**it's** tail three times. 2. _____

3. May we—John and (1)**I** (2)**me**—join you for the meeting?   3. _____

4. Between you and (1)**I** (2)**me**, I feel quite uneasy about the outcome of the expedition.

4. _____

5. Were you surprised that the book was written by Jake and (1)**he** (2)**him**?

5. _____

6. It must have been (1)**he** (2)**him** who wrote the article about plant safety for the company newsletter.

6. _____

7. Why not support (1)**we** (2)**us** students in our efforts to have a new student union?

7. _____

8. No one except (1)**she** (2)**her** could figure out the copier machine.

8. _____

9. He is much more talented in dramatics than (1)**she** (2)**her**.

9. _____

10. The attorneys notified (1)**whoever** (2)**whomever** had a listed address.

10. _____

18. GRAMMAR AND SENTENCES: Pronoun Case

(Study 19, Using the Right Case)

In the first blank, write the **number** of the **correct** pronoun.
In the second blank, write the **reason** for your choice (use the abbreviations in parentheses):

subject (**subj**)	indirect object (**ind obj**)
subjective complement (**subj comp**)	object of preposition (**obj prep**)
direct object (**dir obj**)	

Example: The tickets were for Jo and (1)**I** (2)**me**.   _2_ obj prep

1. The queen knew that the traitor could be only (1)**he** (2)**him**.   1. ____ _____

2. Did you and (1)**he** (2)**him** ever find the tickets?   2. ____ _____

3. All of (1)**we** (2)**us** passengers were irked by the long delay at the airport.   3. ____ _____

4. Professor Fini showed (1)**they** (2)**them** the results of the experiment.   4. ____ _____

5. Sam Lewis preferred to be remembered as the person (1)**who** (2)**whom** invented the jalapeño-flavored lollipop.   5. ____ _____

6. I invited (1)**he** (2)**him** to the senior dance.   6. ____ _____

7. Speakers like (1)**she** (2)**her** are both entertaining and informative.   7. ____ _____

8. I was very much surprised when I saw (1)**he** (2)**him** at the art exhibit.   8. ____ _____

9. Have you and (1)**he** (2)**him** completed your research on the origins of American rodeos?   9. ____ _____

10. We asked Joan and (1)**he** (2)**him** about the playground activities.   10. ____ _____

11. The leader of the student group asked, "(1)**Who** (2)**Whom** do you think can afford the 10 percent increase in tuition?"   11. ____ _____

12. All of (1)**we** (2)**us** tourists spent the entire afternoon in a roadside museum.   12. ____ _____

13. It was (1)**he** (2)**him** who made all the arrangements for the dance.   13. ____ _____

14. Television network executives seem to think that ratings go to (1)**whoever** (2)**whomever** broadcasts the sexiest shows.   14. ____ _____

15. My two friends and (1)**I** (2)**me** decided to visit Window Rock, Arizona, headquarters of the Navajo nation.   15. ____ _____

16. This argument is just between Dick and (1)**I** (2)**me**.   16. ____ _____

17. My father always gave (1)**I** (2)**me** money for my tuition.   17. ____ _____

18. The cowboy movie star offered to co-produce a movie with (1)**whoever** (2)**whomever** promised an accurate portrayal of his life.   18. ____ _____

19. If you were (1)**I** (2)**me**, would you consider going on a summer cruise?   19. ____ _____

20. Gandhi, Mother Teresa, and Martin Luther King, Jr., are persons (1)**who** (2)**whom** I think will be remembered as heroes of the twentieth century.   20. ____ _____

21. Everyone was excused from class except Louise, Mary, and (1)**I** (2)**me**.   21. ____ _____

22. The teaching assistant asked (1)**he** (2)**him** about the experiment.   22. ____ _____

23. Nina was as interested as (1)**he** (2)**him** in moving to Florida after their retirement.   23. _____ _____

24. I knew of no one who had encountered more difficulties than (1)**she** (2)**her**.   24. _____ _____

25. Monisha gave (1)**I** (2)**me** the summary for our report.   25. _____ _____

26. The teachers invited (1)**we** (2)**us** parents to a meeting with an educational consultant.   26. _____ _____

27. The dance instructor was actually fifteen years older than (1)**he** (2)**him**.   27. _____ _____

28. (1)**We** (2)**Us** veterans agreed to raise money for a memorial plaque.   28. _____ _____

29. Are you and (1)**she** (2)**her** planning a joint report?   29. _____ _____

30. It is (1)**I** (2)**me** who am in charge of the bake sales for my children's school.   30. _____ _____

31. I am certain that he is as deserving of praise as (1)**she** (2)**her**.   31. _____ _____

32. If you were (1)**I** (2)**me**, where would you spend spring break?   32. _____ _____

33. (1)**Who** (2)**Whom** do you think will be the next mayor?   33. _____ _____

34. Assign the project to (1)**whoever** (2)**whomever** doesn't mind traveling.   34. _____ _____

35. She is a person (1)**who** (2)**whom** is, without question, destined to achieve success.   35. _____ _____

36. He is the author about (1)**who** (2)**whom** we shall be writing a paper.   36. _____ _____

37. Was it (1)**he** (2)**him** who won the contest?   37. _____ _____

38. The only choice left was between (1)**she** (2)**her** and him.   38. _____ _____

39. No one actually read the book except (1)**she** (2)**her**.   39. _____ _____

40. "Were you calling (1)**I** (2)**me**?" Jill asked as she entered the room.   40. _____ _____

41. Both of (1)**we** (2)**us** agreed that the exercise class was scheduled at an inconvenient time.   41. _____ _____

42. Imagine finally meeting (1)**he** (2)**him** after so many years of correspondence!   42. _____ _____

43. The poll takers asked both her friend and (1)**she** (2)**her** some very personal questions.   43. _____ _____

44. Do you suppose that (1)**he** (2)**him** will ever find time to come?   44. _____ _____

45. José sent an invitation to (1)**I** (2)**me**.   45. _____ _____

46. It was the other reviewer who disliked the movie, not (1)**I** (2)**me**.   46. _____ _____

47. The dean's objection to the content of our play caused the directors and (1)**we** (2)**us** much trouble.   47. _____ _____

48. A dispute arose about (1)**who** (2)**whom** would arrange the conference call.   48. _____ _____

49. Number 79 is the player (1)**who** (2)**whom** I believe scored the winning touchdown.   49. _____ _____

50. The flag will be raised by (1)**whoever** (2)**whomever** the mayor selects.   50. _____ _____

19. GRAMMAR AND SENTENCES: Pronoun Reference

(Study 20, Avoiding Faulty Reference)

Part 1

Write **C** if the boldfaced word is used **correctly**.
Write **X** if it is used **incorrectly**.

Example: Gulliver agreed with his master that **he** was a Yahoo.   _____X_____

1. Dawna switched her major to physics; then she broke up with Rashid. None of us expected **this**.   1. _____

2. Ms. Franco met Ms. Kazakovich at the jewelry counter when **she** was buying a necklace for her daughter.   2. _____

3. The President later found his Middle East policy under attack by the opposing party. He resolved to ignore **that** criticism.   3. _____

4. The latest of these attacks was so vicious that the Secretary of State urged the President to respond strongly to **it**.   4. _____

5. The veteran football player practiced with the rookie because **he** wanted to review the new plays.   5. _____

6. In Buffalo, **they** eat chicken wings served with blue cheese dressing and celery.   6. _____

7. I was late filing my report, **which** greatly embarrassed me.   7. _____

8. In the United States, **they** mail approximately 166 billion letters and packages each year.   8. _____

9. She was able to complete college after earning a research assistantship. We greatly admire her for **that**.   9. _____

10. The physician's speech focused on the country's inattention to the AIDS epidemic; the country was greatly surprised by **it**.   10. _____

11. The President's dog was a favorite of the media, and **he** liked all that attention.   11. _____

12. They intended to climb sheer Mount Maguffey; no one had ever accomplished **that** feat before.   12. _____

13. Pat always wanted to be a television newscaster; thus she majored in **it** in college.   13. _____

14. The average American child watches over thirty hours of television each week, **which** is why we are no longer a nation of readers.   14. _____

15. **It** was well past midnight when the phone rang.   15. _____

16. The speaker kept scratching his head, a mannerism **that** proved distracting.   16. _____

17. According to the *Times,* **it** is expected to snow heavily in the upper Midwest this winter.   17. _____

18. When Dan drives down the street in his red sports car, **they** all look on with admiration and, perhaps, just a little envy.   18. _____

19. Aaron told Evan that **he** couldn't play in the soccer game.   19. _____

20. Eric started taking pictures in high school. **This** interest led to a brilliant career in photography.   20. _____

21. **It** is best to be aware of both the caloric and fat content of food in your diet.   21. _____

22. In some vacation spots, **they** add the tip to the bill and give poor service.   22. _____

23. In Russo's novel *Empire Falls*, the main character seems just like **him**. 23._____

24. In some sections of the country **it** seemed as if the drought would never end. 24._____

25. In some sections of the history text, it seems as if **they** ignored women's contributions to the development of this country. 25._____

Part 2

Choose eight items that you marked **X** in part 1. **Rewrite** each correctly in the blanks below. Before each sentence, write its number from part 1.

Example: 26. *Gulliver agreed that he was a Yahoo, as his master said.*

_____

_____

_____

_____

_____

_____

_____

_____

_____

_____

_____

_____

_____

_____

_____

_____

20. GRAMMAR AND SENTENCES: Phrases

(Study 21, Phrases)

Part 1

In the first blank, write the number of the **one** set of words that is a prepositional phrase.
In the second blank, write **adj** if the phrase is used as an adjective, or **adv** if it is used as an adverb.

Example: <u>The starting pitcher</u> <u>for the Dodgers</u> <u>is a left-hander.</u>   <u>2</u> <u>adj</u>
             1   2   3

1. <u>If time permits,</u> <u>the club members</u> <u>will vote today</u> <u>on your proposal.</u>   1. ____ ____
       1   2   3   4

2. <u>The great white pines</u> <u>growing in the northern forests</u> <u>may soon die</u> <u>without more rain.</u>   2. ____ ____
         1   2   3   4

3. <u>The lady</u> <u>wearing the fur stole</u> <u>has been dating</u> <u>an animal activist</u> <u>from Oregon.</u>   3. ____ ____
     1   2   3   4   5

4. <u>What they saw</u> <u>before the door closed</u> <u>shocked them</u> <u>beyond belief.</u>   4. ____ ____
       1   2   3   4

5. <u>The most frequently used word</u> <u>in the English language</u> <u>is the word</u> *the*.   5. ____ ____
           1   2   3

6. <u>The need</u> <u>for adequate child care</u> <u>was not considered</u> <u>when the President addressed the convention.</u> 6. ____ ____
     1   2   3   4

7. <u>The observation</u> <u>that men and women have different courtship rituals</u> <u>seems debatable</u>
     1   2   3

  <u>in a modern postindustrial society.</u>   7. ____ ____
      4

8. <u>At our yard sale,</u> <u>I found out that people will buy almost anything</u> <u>if the price is right.</u>   8. ____ ____
       1   2   3

9. <u>Until the last five minutes</u> <u>our team seemed</u> <u>to have the game won,</u> <u>but we lost.</u>   9. ____ ____
        1   2   3   4

10. <u>After everyone left</u> <u>to attend the meeting,</u> <u>Nugent sneaked back</u> <u>for a long, quiet nap.</u> 10. ____ ____
      1   2   3   4

Some of the boldfaced expressions are verbal phrases; others are parts of verbs (followed by modifiers or complements). In the blank, **identify** each **expression** (use the abbreviations in parentheses):

verbal phrase used as adjective (**adj**)
verbal phrase used as adverb (**adv**)
verbal phrase used as noun (**noun**)
part of verb (with modifiers or complements) (**verb**)

Examples: Singing in the rain can give one a cold.   <u>noun</u>
Gene is **singing in the rain** despite his cold.   <u>verb</u>

1. **Speaking from the Capitol steps** is a favorite act of politicians.   1. _____

2. Senator Claghorn is **speaking from the Capitol steps** today.   2. _____

3. **To prepare his income taxes,** Sam spent several hours sorting through the shoe boxes filled with receipts.   3. _____

4. By age 30, many women begin **sensing a natural maternal need.**   4. _____

5. Both lawyers, **having presented their closing arguments**, nervously awaited the jury's verdict.   5. _____

6. The Clementes were **having the Robertsons to dinner that evening**.   6. _____

7. His idea of a thrill is **driving in stock-car races.**   7. _____

8. **Driving in stock-car races**, he not only gets his thrills but also earns prize money.   8. _____

9. Her congregation is **surviving on a very small income**.   9. _____

10. She earnestly desires **to increase the membership**.   10. _____

21. GRAMMAR AND SENTENCES: Verbal Phrases

(Study 21B, The Verbal Phrase)

Part 1

In each sentence, find a verbal phrase. **Circle** it, and in the small blank at the right, tell how it is used: as adjective (write **adj**), adverb (write **adv**), or **noun**.

Example: A study (conducted by the Yale Medical School) found that smoking is addictive.   ___adj___

1. Holding the flag high, the veterans marched down Main Street.   1. _____

2. Leading the Memorial Day parade was an honored tradition for the veterans.   2. _____

3. Topic selection is the first step in producing a research paper.   3. _____

4. Hopelessly in love, June neglected to go to her science class.   4. _____

5. The Secretary of State, realizing the need for negotiation, arranged a peace conference between the belligerents.   5. _____

6. To provide safe neighborhoods, the police have begun intensified nighttime patrols.   6. _____

7. I can't help liking her even though she isn't interested in my favorite sport, hockey.   7. _____

8. Disappointed with her grades, Sabrina made an appointment with her counselor.   8. _____

9. I appreciate your helping us at the craft fair.   9. _____

10. Our lacrosse team, beaten in the playoffs, congratulated the winners.   10. _____

11. They changed all the locks to feel more safe.   11. _____

12. No matter how often the teacher told them to be quiet, the kindergartners chattered and fidgeted constantly.   12. _____

13. Rock climbing is a sport demanding endurance.   13. _____

14. I passed chemistry by studying past midnight all last week.   14. _____

15. The hunter put down his gun, realizing that the ducks had flown out of range.   15. _____

Part 2

Complete each sentence with a verbal phrase of your own. Then, in the small blank at the right, tell how you used it: **adj**, **adv**, or **noun**.

1. To take the test without _____ was not wise at all.   1. _____

2. Alison organized a group of senior citizens [hint: for what purpose?] _____   2. _____

   _____.

3. Worried about her children, the young mother decided _____   3. _____

   _____.

4. _____ may be

linked to increased risk of rectal and bladder cancer.   4. _____

5. She tried to obtain the information without _____

_____ .   5. _____

6. The student _____ is here to

select a major.   6. _____

7. The best book _____ is one

that helps you escape daily tension.   7. _____

8. _____ , he found an article

that was easy to understand.   8. _____

9. Physicians recommend that patients _____

donate their own blood.   9. _____

10. _____ has been

the cause of too many fires.   10. _____

22. GRAMMAR AND SENTENCES: Review of Phrases

(Study 21, Phrases; also suggested: 25E, Use Reduction)

Part 1

Classify each boldfaced phrase (use the abbreviations in parentheses):

prepositional phrase (**prep**)   gerund phrase (**ger**)
infinitive phrase (**inf**)   absolute phrase (**abs**)
participial phrase (**part**)

Example: The economies **of Asian countries** grew shaky.   ____prep____

1. The reddish bird flitting **among the weeds** is a rufous-sided towhee.   1. _____

2. The reddish bird **flitting among the weeds** is a rufous-sided towhee.   2. _____

3. The first televisions had small round screens encased **in large wooden cabinets.**   3. _____

4. **His insisting that he was right** made him unpopular with his associates.   4. _____

5. The committee voted **to adjourn immediately.**   5. _____

6. **Because of the storm**, the excursion around the lake had to be postponed.   6. _____

7. **To stay awake in Smedley's class** required dedication and plenty of black coffee.   7. _____

8. **During early television programming**, many commercials were five minutes long.   8. _____

9. **Flying a jet at supersonic speeds** has been Sally's dream since childhood.   9. _____

10. We were obliged to abandon our plans, **the boat having been damaged in a recent storm.**   10. _____

11. **Realizing that his back injury would get worse**, the star player retired from professional basketball.   11. _____

12. **To pay for their dream vacation**, Harry and Sue both took on extra jobs.   12. _____

13. The children were successful in **developing their own lawn-mowing company.**   13. _____

14. **The semester completed**, students were packing up to go home.   14. _____

15. The distinguished-looking man **in the blue suit** is the head of the company.   15. _____

16. **Earning a college degree** used to guarantee a well-paying job.   16. _____

17. Approximately one-fourth **of the American work force** has a college degree.   17. _____

18. On May 11, 1939, the first baseball game was telecast **in America.**   18. _____

19. Deciding which car **to buy** is a difficult task.   19. _____

20. Two crates **of oranges** were delivered to the fraternity house.   20. _____

21. **Anticipating an overflow audience**, the custodian put extra chairs in the auditorium.   21. _____

22. A car **filled with students** left early this morning to arrange for the class picnic.   22. _____

23. The agent **wearing an official badge** is the one to see about tickets.   23. _____

24. **Her mind going blank at the last minute**, Madeline could not answer the quizmaster's million-dollar question.   24. _____

25. **Frequently checking one's bank-account balance** can prevent embarrassing "insufficient funds" check returns.　　　　　　　　　　　　　　　　　　25. _____

Combine the following pairs of sentences by reducing one of the sentences to a phrase.

Examples: The new furniture arrived yesterday. It was for the den.
The new furniture for the den arrived yesterday.

Professor Hughes gave us an assignment. We had to find five library references on the Depression.
Professor Hughes gave us an assignment to find five library references on the Depression.

(Collaborative option: Students work in pairs or small groups to suggest ways of combining.)

1. Fred flipped through the channels. He decided that reading the phone book would be more exciting than watching television.

_____

_____

_____

2. Scientists are using artificial life simulation programs. They are doing this for futuristic experimentation.

_____

_____

_____

3. North America, Asia, and Europe must work together. That way they can prevent the holes in the ozone layer from becoming any larger.

_____

_____

_____

4. The international community has changed with the fall of the Soviet Union. It has revamped its undercover operations.

_____

_____

_____

5. In the 1920s Americans often used homemade crystal radio sets. They did this so that they could listen to radio broadcasts.

_____

_____

_____

6. His weekend was ruined. So Alfredo decided to go to bed early.

_____

_____

_____

7. The white potato plant was grown strictly as an ornament in Europe. This was before the 1700s.

_____

_____

_____

8. American widows report that friends and relatives interfere too much. These widows frequently prefer to spend time alone.

_____

_____

_____

23. GRAMMAR AND SENTENCES: Recognizing Clauses

(Study 22, Clauses)

Classify each boldfaced clause (use the abbreviations in parentheses):

independent [main] clause (**ind**)
dependent [subordinate] clause: adjective clause (**adj**)
                                 adverb clause (**adv**)
                                 noun clause (**noun**)

Example: The program will work **when the disk is inserted**.   __adv__

1. The house had a long flight of steps leading to the first floor **because it was built on the side of a hill**.   1. _____

2. Farmers asked the government for disaster relief because of the drought, **which had by then lasted nearly four years**.   2. _____

3. The tourists asked **how they could reach Times Square by subway**.   3. _____

4. The student **who made the top grade in the history quiz** is my roommate.   4. _____

5. **Whether I am able to go to college** depends on whether I can find employment.   5. _____

6. **After Judd had written a paper for his English class**, he watched television.   6. _____

7. Canada celebrates Thanksgiving in October; **the United States celebrates it in November**.   7. _____

8. The career center offers seminars to anyone **who needs help writing a résumé**.   8. _____

9. There is much excitement **whenever election results are announced**.   9. _____

10. The detective listened carefully to the suspect's answers, but **she couldn't find a reason to charge the suspect**.   10. _____

11. Few Americans realize **that their homes are full of minute dust mites**.   11. _____

12. My first impression was **that someone had been in my room quite recently**.   12. _____

13. The actor **who had lost the Oscar** declared through clenched teeth that she was delighted just to have been nominated.   13. _____

14. He dropped a letter in the mailbox; **then he went to the library**.   14. _____

15. The candidate decided to withdraw from the city council race **because she didn't approve of the media's treatment of her mental illness**.   15. _____

16. Why don't you sit here **until the rest of the class arrives**?   16. _____

17. The real estate mogul, **who is not known for his modesty**, has named still another building after himself.   17. _____

18. **Although he is fifty-two years old**, he is very youthful in appearance.   18. _____

19. The Battle of Saratoga is more famous, yet **the Battle of Brandywine involved more soldiers**.   19. _____

20. **Why she never smiles** is a mystery to her colleagues.   20. _____

21. **Why don't you wait** until you have all the facts?   21. _____

22. She is a person **whom everyone respects and admires**.   22. _____

23. The weather is surprisingly warm **even though it is December**.   23. _____

24. My answer was **that I had been unavoidably detained**.   24. _____

25. The cat loved to sleep in the boys' room **because it could stalk their goldfish at night.**   25. _____

26. The trophy will be awarded to **whoever wins the contest.**   26. _____

27. The detective walked up the stairs; **she opened the door of the guest room.**   27. _____

28. Is this the book **that you asked us to order for you?**   28. _____

29. The audience could not believe **that the show would be delayed for an hour.**   29. _____

30. My Aunt Minnie Matilda, **who wrote piano duets for children**, died penniless.   30. _____

31. **Because students are prone to resolving conflicts by fighting with one another**, the principal is working on developing conflict resolution groups.   31. _____

32. The log cabin **where my father was born** is still standing.   32. _____

33. Many Americans realize **that dual-income families are a result of a declining economy rather than gender equality**.   33. _____

24. GRAMMAR AND SENTENCES: Dependent Clauses

(Study 22B, Kinds of Dependent Clauses)

Part 1

Underline the dependent clause in each item. Then, in the blank, **classify** it as an adjective (**adj**), adverb (**adv**), or **noun**.

Example: The textbook explained fully <u>what the instructor had outlined</u>.   ___noun___

1. Although he was an experienced tree-cutter, his miscalculation brought the giant oak down on the garage roof.   1. _____

2. The couple searched the Internet for a vacation spot where cell phones could not reach them.   2. _____

3. The early bicycles weren't comfortable, because they had wooden wheels and wooden seats.   3. _____

4. The student who complained about the food was given another dessert.   4. _____

5. Whether Camille dyes her hair remains a mystery.   5. _____

6. After Jonathan had read the morning paper, he threw up his hands in despair.   6. _____

7. Whoever predicted today's widespread use of computers was truly a prophet.   7. _____

8. Professor George gave extra help to anyone who asked for it.   8. _____

9. There is always much anxiety whenever final exams are held.   9. _____

10. Studies show that calcium intake among American teenagers is often inadequate.   10. _____

Part 2

In each long blank, write a dependent clause of your own. Then, in the small blank, **identify** your clause as adjective (**adj**), adverb (**adv**), or **noun**.

Example: The noted author, <u>who was autographing her books</u>, smiled at us.   ___adj___

(Collaborative option: Students work in pairs, alternating: one writes the clause, the other tells how it is used.)

1. Dr. Jackson,_____,
    declared Burton the winner.   1. _____

2. The award went to the actor _____.   2. _____

3. _____,
    parents are spending less time with their children.   3. _____

4. Luis remarked _____.   4. _____

5. _____,
    Congress voted against the bill.   5. _____

6. It was the only mistake _____.   6. _____

7. During his presentation, Nathan explained _____
    _____.   7. _____

8. Most of the audience had tears in their eyes _____

    _____.   8. _____

9. The United States, _____,

    is still the preferred destination of millions of immigrants.   9. _____

10. The candidate told her followers _____.   10. _____

25. GRAMMAR AND SENTENCES: Noun and Adjective Clauses

(Study 22B, Kinds of Dependent Clauses; also suggested: 25D, Use Subordination)

Combine each of the following pairs of sentences into one sentence. Do this by reducing one of the pair to a noun or adjective clause.

Examples: Something puzzled the police. What did the note mean?
What the note meant puzzled the police.
The X-Files became immensely popular in the late 1990s. It appeared on the Fox TV network.
The X-Files, which appeared on the Fox TV network, became immensely popular in the late 1990s.

(Collaborative option: Students work in pairs or small groups to suggest ways of combining.)

1. One thing remained unresolved. Who was the more accomplished chef? [Hint: try a noun clause.]

_____

_____

2. The programmer retired at age twenty. She had written the new computer game. [Hint: try an adjective clause.]

_____

_____

3. I do not see how anyone could object to that. The senator said it.

_____

_____

4. The laboratory assistant gave the disk to Janine. He had helped Janine learn the word-processing software.

_____

_____

5. They planned something for the scavenger hunt. It seemed really bizarre.

_____

_____

6. We should spend the money on someone. Who needs it most?

_____

_____

7. Large classes and teacher apathy are problems. Most school districts tend to ignore them.

   _____

   _____

8. Every teacher has a worst fear. Her students may hate to read.

   _____

   _____

9. Glenn was a certain kind of person. He seemed to thrive on hard work and tight deadlines.

   _____

   _____

10. Animal rights activists demonstrated in certain states. In these states grizzly bear hunting is allowed.

    _____

    _____

11. Samuel F. B. Morse is famous for pioneering the telegraph. He was also a successful portrait painter.

    _____

    _____

12. Something could no longer be denied. The war was already lost.

    _____

    _____

13. The long black limousine had been waiting in front of the building. It sped away suddenly.

    _____

    _____

14. When Columbus reached America, there were more than three hundred Native American tribes. Together these tribes contained more than a million people.

    _____

    _____

    _____

15. You must decide something now. That thing is critically important.

    _____

    _____

16. My English professor has written a biography of Bret Harte. She is obviously enthralled by this nineteenth-century writer.

_____

_____

17. The jackpot will be won by someone. That person holds the lucky number.

_____

_____

18. Juan has a friendly disposition. It has helped him during tense negotiations at work.

_____

_____

19. The TV news reported things regarding the episode. We were appalled by them.

_____

_____

20. Paleontologists have unearthed a set of bones. They make up the most nearly complete Tyrannosaurus Rex ever found.

_____

_____

_____

26. GRAMMAR AND SENTENCES: Adverb Clauses

(Study 22B, Kinds of Dependent Causes; also suggested: 25D, Use Subordination)

Combine each of the following pairs of sentences into one sentence by reducing one sentence to the kind of adverb clause specified in brackets.

Examples: The sun set. Then the lovers headed home. [time]
When the sun set, the lovers headed home.
Students must score 1400 on their College Boards. Otherwise they will not be admitted. [condition]
Students will not be admitted unless they score 1400 on their College Boards.

(Collaborative option: Students work in pairs or small groups to suggest ways of combining.)

1. [cause] Four hundred thousand Americans each year get skin cancer. Therefore, many parents are teaching their children to avoid overexposure to sunlight.

   _____

   _____

   _____

2. [place] The candidate was willing to speak anywhere. But she had to find an audience there.

   _____

   _____

3. [manner] Carl ran the race. He seemed to think his life depended on it.

   _____

   _____

4. [comparison] Her brother has always been able to read fast. She has always been able to read faster.

   _____

   _____

5. [purpose] This species of tree has poisonous leaves. That way, insects will not destroy it.

   _____

   _____

6. [time] The concert was half over. Most of the audience had already left.

   _____

   _____

7. [comparison] I worked hard on that project. I could not have worked harder.

_____

_____

8. [purpose] He read extensively. His purpose was to be well prepared for the test.

_____

_____

9. [condition] American attitudes must change. Otherwise small family farms will disappear.

_____

_____

10. [concession] Her grades were satisfactory. But she did not qualify for the scholarship.

_____

_____

11. [result] She worried very much. The result was that she could no longer function effectively.

_____

_____

12. [condition] You may accept the position, or you may not. Either way, you should write a thank-you note to the interviewer.

_____

_____

13. [cause] Frosts destroyed Florida's citrus crops this year. So citrus prices will increase significantly.

_____

_____

14. [condition] Do not complete the rest of the form yet. You have to see your advisor first.

_____

_____

15. [concession] Marina was only 5 feet 5 inches tall. But she was determined to be a basketball star.

_____

_____

16. [place] Fahnestock preferred one kind of vacation place. No one stood in line for anything there.

_____

_____

17. [condition] The substance may be an acid. Then the litmus paper will turn red.

_____

_____

18. [cause] Ethnic jokes can be particularly harmful. Such humor subtly reinforces stereotypes.

_____

_____

19. [manner] She smiled in a certain way. Maybe she knew something unsuspected by the rest of us.

_____

_____

20. [comparison] Horgan received good grades. But Schultz usually received better ones.

_____

_____

27. GRAMMAR AND SENTENCES: Kinds of Sentences

(Study 22C, Clauses in Sentences; also suggested: 25B–E, Use Coordination, Compounding, Subordination, and Reduction)

Part 1

Classify each sentence (use the abbreviations in parentheses):

simple (**sim**)	complex (**cx**)
compound (**cd**)	compound-complex (**cdcx**)

Example: Hank opened the throttle, and the boat sped off. _____cd_____

1. The governor maintained that the state's finances were sound. 1._____

2. Ted Williams, whose .406 season batting average has been unequaled for more than sixty years, had amazingly sharp eyes. 2._____

3. Completion of the new library will be delayed unless funds become available. 3._____

4. Consider the matter carefully before you decide; your decision will be final. 4._____

5. This year, either medical companies or discount store chains are a good investment for the small investor. 5._____

6. The play, which was written and produced by a colleague, was well received by the audience. 6._____

7. The storm, which had caused much damage, subsided; we then continued on our hike. 7._____

8. We waited until all the spectators had left the gymnasium. 8._____

9. The argument having been settled, the meeting proceeded more or less amicably. 9._____

10. The prescription was supposed to cure my hives; instead it made my condition worse. 10._____

11. Tired of being a spy, he settled in Vermont and began writing his memoirs. 11._____

12. His chief worry was that he might reveal the secret by talking in his sleep. 12._____

13. The television special accurately portrayed life in the 1950s; critics, therefore, praised the production for its authenticity. 13._____

14. The story appearing in the school paper contained several inaccuracies. 14._____

15. The police officer picked up the package and inspected it carefully. 15._____

16. Because she was eager to get an early start, Sue packed the night before. 16._____

17. By 2080, there will be over one million Americans one hudred years old or older; this significant increase of centenarians will profoundly affect the health care system. 17._____

18. Noticing the late arrivals, the speaker motioned for them to be seated. 18._____

19. A study of people in their eighties revealed that most had a satisfying relationship with a family member or care provider; in other words, these older Americans were not lonely in their old age. 19._____

20. Through the thick fog rolling in from the sea, Courtney could barely discern the tail lights of a car farther along the beach. 20._____

In the long blank, **combine** each set of sentences into one sentence. Then, in the small blank, **classify** your new sentence as simple (**sim**), compound (**cd**), complex (**cx**), or compound-complex (**cdcx**).

Examples: The President flew to Gibraltar. From there she cruised to Malta.
The President flew to Gibraltar and from there cruised to Malta.   _____sim_____

The ballerina's choreography won praise that night. She was not satisfied with it. She spent the next morning reworking it.
Although the ballerina's choreography won praise that night, she was not satisfied with it and spent the next morning reworking it.   _____cx_____

(Collaborative option: Students work in pairs or small groups to suggest ways of combining.)

1. The suspect went to the police station. She turned herself in.   1. _____

_____

_____

_____

2. The little girl won the poetry contest. She plans to be a writer.   2. _____

_____

_____

_____

3. We wanted that house. It was already sold. So we had to look for another one.   3. _____

_____

_____

_____

4. Batik is a distinctive and complex method of dyeing cloth. It was created on the island of Java.   4. _____

_____

_____

_____

5. Scientists convened. They came from all over the world. They wanted to discuss the greenhouse effect. This was a serious problem.   5. _____

_____

_____

_____

6. In the 1920s there were three favorite amusements. They were mahjong, ouija, and crossword puzzles.

6. _____

7. Only one country has a lower personal income tax than the United States. That is Japan. This is among the wealthy nations.

7. _____

8. The American Dream seems inaccessible to many Americans. These Americans have difficulty even making ends meet.

8. _____

9. Some couples marry before age thirty. These couples have a high divorce rate.

9. _____

10. Banks make this promise to their customers. Banking will become more convenient. It will happen through computer technology.

10. _____

11. Erskine smacked his lips. He plowed through another stack of buttermilk pancakes. They were smothered in blueberry syrup.

11. _____

12. The Middle East is the birthplace of three major world religions. One is Judaism. Another is Christianity. The third is Islam.

12. _____

13. The women sat up talking. They did this late one night. They talked about their first dates. Most laughed about their teenage years. These years had been awkward.

13. _____

_____

_____

_____

28. GRAMMAR AND SENTENCES: Subject-Verb Agreement

(Study 23, Subject-Verb Agreement)

Write the number of the **correct** choice.

Example: One of the network's best programs (1)**was** (2)**were** canceled.   ___1___

1. Neither the men's nor the women's room (1)**has** (2)**have** been cleaned today.   1._____

2. Physics (1)**is** (2)**are** among the hardest majors in the curriculum.   2._____

3. Working a second job to pay off my debts (1)**has** (2)**have** become a priority.   3._____

4. Not one of the nominees (1)**has** (2)**have** impressed me.   4._____

5. (1)**Does** (2)**Do** each of the questions count the same number of points?   5._____

6. The number of jobs lost in California's Silicon Valley (1)**has** (2)**have** increased significantly in the past two years.   6._____

7. *Ninety-nine* (1)**is** (2)**are** hyphenated because it is a compound number.   7._____

8. The college president, along with five vice-presidents, (1)**was** (2)**were** ready for the meeting.   8._____

9. Both the secretary and the treasurer (1)**was** (2)**were** asked to submit reports.   9._____

10. Everyone in the audience (1)**was** (2)**were** surprised by the mayor's remarks.   10._____

11. *Women* (1)**is** (2)**are** spelled with an *o* but pronounced with an *i* sound.   11._____

12. Every junior and senior (1)**was** (2)**were** expected to report to the gymnasium.   12._____

13. There (1)**is** (2)**are** a professor, several students, and a teaching assistant meeting to discuss the course reading list.   13._____

14. Ten dollars (1)**is** (2)**are** too much to pay for that book.   14._____

15. (1)**Is** (2)**Are** there any computers available in the lab this morning?   15._____

16. Neither the neighbors nor the police officer (1)**was** (2)**were** surprised by the violent crime.   16._____

17. Each of the crises actually (1)**needs** (2)**need** the President's immediate attention.   17._____

18. (1)**Is** (2)**Are** your father and brother coming to see you graduate tomorrow?   18._____

19. A good book and some chocolate doughnuts (1)**was** (2)**were** all she needed to relax.   19._____

20. There (1)**is** (2)**are** one coat and two hats in the hallway.   20._____

21. (1)**Does** (2)**Do** Coach Jasek and the players know about the special award?   21._____

22. My two weeks' vacation (1)**was** (2)**were** filled with many projects around the house.   22._____

23. The only thing that annoyed me more (1)**was** (2)**were** the children's tracking mud in from the backyard.   23._____

24. (1)**Hasn't** (2)**Haven't** either of the roommates looked for the missing ring?   24._____

25. There (1)**is** (2)**are** a bird and a squirrel fighting over the birdseed in the feeder.   25._____

26. On the table (1)**was** (2)**were** a pen, a pad of paper, and two rulers.   26._____

27. It is remarkable that the entire class (1)**is** (2)**are** taking the field trip.   27._____

28. It (1)**was** (2)**were** a book and a disk that disappeared from the desk.   28._____

29. There (1)**is** (2)**are** many opportunities for part-time employment on campus.

29. _____

30. (1)**Is** (2)**Are** algebra and chemistry required courses?

30. _____

31. One of his three instructors (1)**has** (2)**have** offered to write a letter of recommendation.

31. _____

32. (1)**Does** (2)**Do** either of the books have a section on usage rules?

32. _____

33. Neither my parents' car nor our own old Jeep (1)**is** (2)**are** reliable enough to make the trip.

33. _____

34. Marbles, stones, and string (1)**is** (2)**are** my son's favorite playthings.

34. _____

35. Each of the books (1)**has** (2)**have** an introduction written by the author's mentor.

35. _____

36. The lab report, in addition to several short papers, (1)**was** (2)**were** due immediately after spring break.

36. _____

37. Neither the teacher nor the parents (1)**understands** (2)**understand** why Nathan does so well in math but can barely read first-grade books.

37. _____

38. The old woman who walks the twin Scottish terriers (1)**detests** (2)**detest** small children running on the sidewalk in front of her house.

38. _____

39. At the Boy Scout camp-out, eggs and bacon (1)**was** (2)**were** the first meal the troop attempted to prepare on an open fire.

39. _____

40. (1)**There's** (2)**There are** more butter and mayonnaise in the refrigerator.

40. _____

41. Everyone (1)**was** (2)**were** working hard to finish planting the crops before the rainy season.

41. _____

42. The children, along with their teacher, (1)**is** (2)**are** preparing a one-act play for the spring open house.

42. _____

43. Minnie Olson is one of those people who always (1)**volunteers** (2)**volunteer** to help the homeless.

43. _____

44. Lucy announced that *The Holy Terrors* (1)**is** (2)**are** the title of her next book, which is about raising her three sons.

44. _____

45. The class, along with the teacher, (1)**was** (2)**were** worried about the ailing class pet.

45. _____

46. Five dollars (1)**does** (2)**do** not seem like much to my eight-year-old son.

46. _____

47. Either the choir members or the organist (1)**was** (2)**were** constantly battling with the minister about purchasing fancy new choir robes.

47. _____

48. In the last 200 years, over 50 million people from 140 countries (1)**has** (2)**have** left their homelands to immigrate to the United States.

48. _____

49. Food from different geographic locations and ethnic groups often (1)**helps** (2)**help** distinguish specific cultural events.

49. _____

50. Virtually every painting and every sculpture Picasso did (1)**is** (2)**are** worth over a million dollars.

50. _____

51. There on the table (1)**was** (2)**were** my wallet and my key chain.

51. _____

52. Neither the documentary about beekeeping nor the two shows about Iceland (1)**was** (2)**were** successful in the ratings.

52. _____

53. Each of the new television series (1)**is** (2)**are** about single-parent families.

53. _____

54. Sitting on the sidewalk (1)**was** (2)**were** Amy and her four best friends.

54. _____

55. *Les Atrides* (1)**is** (2)**are** a ten-hour, four-play production of ancient Greek theater.

55. _____

56. A political convention, with its candidates, delegates, and reporters, (1)**seems** (2)**seem** like bedlam.

56. _____

57. In the auditorium (1)**was** (2)**were** assembled the orchestra members who were ready to practice for the upcoming concert.

57. _____

58. Each of the art historians (1)**has** (2)**have** offered a theory for why the Leonardo painting has such a stark background.

58. _____

59. (1)**Was** (2)**Were** either President Smith or Dean Nicholson asked to speak at the awards ceremony?

59. _____

60. Watching local high school basketball games (1)**has** (2)**have** become his favorite weekend activity.

60. _____

61. His baseball and his glove (1)**was** (2)**were** all Jamil was permitted to take to the game.

61. _____

62. Neither my friend nor I (1)**expects** (2)**expect** to go on the overnight trip.

62. _____

63. My coach and mentor (1)**is** (2)**are** Gwen Johnson.

63. _____

64. Cable television, along with VCRs and DVDs, (1)**has** (2)**have** drawn millions of viewers away from traditional network television.

64. _____

65. She is the only one of six candidates who (1)**refuses** (2)**refuse** to speak at the ceremony.

65. _____

66. Neither the systems analyst nor the accountants (1)**was** (2)**were** able to locate the problem in the computer program.

66. _____

29. GRAMMAR AND SENTENCES: Pronoun-Antecedent Agreement

(Study 24, Pronoun-Antecedent Agreement)

Write the number of the **correct** choice.

Example: One of the women fell from (1)**her** (2)**their** horse.   ___1___

1. Agatha Christie is the kind of writer who loves to keep (1)**her** (2)**their** readers guessing until the last page.   1._____

2. Many tourists traveling in the West enjoy stopping at roadside attractions because (1)**you** (2)**they** never know what to expect.   2._____

3. When Lucia found that her young son had caught the measles, she became concerned that (1)**it** (2)**they** would leave marks on his face.   3._____

4. He majored in mathematics because (1)**it** (2)**they** had always been of interest to him.   4._____

5. Lucy edited the news because (1)**it was** (2)**they were** often full of inaccuracies.   5._____

6. He assumed that all of his students had done (1)**his** (2)**their** best to complete the test.   6._____

7. Both Ed and Luis decided to stretch (1)**his** (2)**their** legs when the bus reached Houston.   7._____

8. Ironically, neither woman had considered how to make (1)**her** (2)**their** job easier.   8._____

9. Each of the researchers presented (1)**a** (2)**their** theory about the age of the solar system.   9._____

10. He buys his books at the campus bookstore because (1)**it has** (2)**they have** low prices.   10._____

11. Electronics can be a rewarding field of study, because (1)**it** (2)**they** can lead to good jobs in a number of areas.   11._____

12. Every member of the men's basketball team received (1)**his** (2)**their** individual trophy.   12._____

13. All in the class voted to have (1)**its** (2)**their** term papers due a week earlier.   13._____

14. I like swimming because it develops (1)**one's** (2)**your** muscles without straining the joints.   14._____

15. Neither Aaron nor Marzell has declared (1)**his** (2)**their** major.   15._____

16. Citizens who still do not recycle (1)**your** (2)**their** garbage need to read this news article.   16._____

17. The Zoomation Company has just introduced (1)**its** (2)**their** new 95-gigabyte computer.   17._____

18. Neither the guide nor the hikers seemed aware of (1)**her** (2)**their** danger on the trail.   18._____

19. The faculty has already selected (1)**its** (2)**their** final candidates.   19._____

20. Critics argue that (1)**those kind** (2)**those kinds** of movies may promote violent tendencies in children.   20._____

21. One has to decide early in life what (1)**one wants** (2)**they want** out of life.   21._____

22. Neither the coach nor the players underestimated (1)**her** (2)**their** opponents.   22._____

23. The corporation insists that (1)**its** (2)**their** financial statements have been completely honest.   23._____

24. Students should take accurate and complete notes so that (1)**they** (2)**you** will be prepared for the exam.   24._____

25. Some people prefer trains to planes because trains bring (1)**you** (2)**them** closer to the scenery.   25._____

26. In the next five years, owners of older vehicles polluting the environment can sell (1)**their** (2)**your** cars or trucks for scrap.

26. _____

27. If a stranger tried to talk to her, she would just look at (1)**him** (2)**them** and smile.

27. _____

28. Every one of the trees in the affected area had lost most of (1)**its** (2)**their** leaves.

28. _____

29. Some women can understand (1)**herself** (2)**themself** (3)**themselves** better through reading feminist literature.

29. _____

30. The medical committee was surprised to learn that (1)**its** (2)**their** preliminary findings had been published in the newspaper.

30. _____

31. The campus disciplinary board determined that (1)**its** (2)**their** process for reviewing student complaints was too cumbersome and slow.

31. _____

32. None of the boys should blame (1)**himself** (2)**themself** (3)**themselves** (4)**yourself** for misfortunes that cannot be prevented.

32. _____

33. Rita is a person who cannot control (1)**her** (2)**their** anger when under stress.

33. _____

34. Professor Brown is one of those teachers who really love (1)**his** (2)**their** profession.

34. _____

35. Everyone in the men's locker room grabbed (1)**his** (2)**their** clothes and ran when the cry of "Fire!" came from the hallway.

35. _____

36. Rita is the only one of the singers who writes (1)**her** (2)**their** own music.

36. _____

37. Each of the singers in the newly formed Irish band dreamed of earning (1)**her** (2)**their** first million dollars.

37. _____

38. As part of the Kim family's Vietnamese New Year celebration, each wrote *cau doi*. Memories of home and family are the subject of (1)**this kind of poem** (2)**these kind of poems**.

38. _____

39. During the Christmas season, many Latin American families serve (1)**its** (2)**their** favorite dish—tortillas spread with mashed avocado and roast chicken.

39. _____

40. If everyone in the Women's Club would contribute a small portion of (1)**her** (2)**their** January 1 paycheck, we should be able to purchase the microwave for the staff luncheon room.

40. _____

41. Each cat claimed (1)**its** (2)**their** specific area of the bedroom for long afternoon naps.

41. _____

42. Hearing-impaired people now have a telecommunication device to allow (1)**him** (2)**them** to make phone calls to a hearing person.

42. _____

43. When my professors complain that Americans don't support local school districts, I remind (1)**her** (2)**them** that most families view education as extremely important.

43. _____

44. Drivers of the new Mercolet sedan know that Mercolet has produced the most stylish car that (1)**its** (2)**their** engineers could design.

44. _____

45. If viewers are not happy with public television programming, (1)**you** (2)**they** should write letters to local television stations.

45. _____

46. Researchers have found that the type of relationship couples have can affect (1)**your** (2)**their** overall immune system.

46. _____

47. When climbing a mountain in autumn, amateurs had better take warm clothing in (1)**your** (2)**their** packs to guard against hypothermia.

47. _____

48. Unfortunately, Sid was one of those climbers who neglected to pack (1)**his** (2)**their** winter gear.

48. _____

49. A neighborhood organization of young people is meeting to determine how (1)**it** (2)**they** can help elderly neighbors in the community.

49. _____

50. Either the lead actor or the chorus members missed (1)**his** (2)**their** cue.

50. _____

30. GRAMMAR AND SENTENCES: Agreement Review

(Study 23–24, Agreement)

Write **C** if the sentence is **correct**.
Write **X** if it is **incorrect**.

Example: Nobody in the first two rows were singing.   ____X____

1. The deep blue of the waters seem to reflect the sky.   1. _____

2. All of the fish swim upstream in spring.   2. _____

3. All of the fish tastes good if you grill it properly.   3. _____

4. The strength of these new space-age materials have been demonstrated many times.   4. _____

5. All these experiences, along with the special love and care that my daughter needs, have taught me the value of caring.   5. _____

6. Evan's pants are ripped beyond repair.   6. _____

7. There's a northbound bus and a southbound bus that leave here every hour.   7. _____

8. According to a recent survey, almost every American feels that their self-esteem is important.   8. _____

9. The management now realizes that a bigger budget is needed; they plan to ask for federal assistance.   9. _____

10. When an older student senses that an institution understands nontraditional students, she generally works to her academic potential.   10. _____

11. I found that the thrill of attending college soon leaves when you have to visit the bursar's office.   11. _____

12. Everyone who read the letter stated that they were surprised by the contents.   12. _____

13. You should hire one of those experts who solves problems with computers.   13. _____

14. Two hundred miles was too much for a day trip.   14. _____

15. At school, there are constant noise and confusion at lunch.   15. _____

16. Cleveland or Cincinnati are planning to host the statewide contests.   16. _____

17. Bacon and eggs are no longer considered a healthy breakfast.   17. _____

18. Probably everybody in the computer center, except Colleen and Aaron, know how to run the scanner.   18. _____

19. Neither Chuck nor Arnold are as blessed with talent as Sylvester.   19. _____

20. *Powerpuff Girls* was described as "comic book stuff" by the newspaper's television critic.   20. _____

21. The researcher, as well as her assistants, are developing a study to compare the brain tissue of Alzheimer's disease sufferers and healthy subjects.   21. _____

22. Neither criticism nor frequent failures were enough to retard his progress.   22. _____

23. Where are the end of the recession and the revival of consumer confidence?   23. _____

24. Economics have been the most dismal science I've ever studied.   24. _____

25. Has either of your letters appeared in the newspaper?   25. _____

26. It were the general and the Secretary of State who finally convinced the President that an
    armed conflict might be inevitable.   26. _____

27. Neither Janet nor her parents seem interested in our offer to help.   27. _____

28. He is one of those employees who was always late for work on Monday mornings.   28. _____

29. She is the only one of the experts who solves problems with computers.   29. _____

30. The faculty are squabbling among themselves, disagreeing vehemently about Faculty Senate
    bylaws.   30. _____

31. Such was the hardships of the times that many were forced into begging or stealing to survive. 31. _____

32. The special scissors that was needed for the repair could not be found.   32. _____

33. Billiards were returning to popularity at the time.   33. _____

31. GRAMMAR AND SENTENCES: Effective Sentences

(Study 25, Creating Effective Sentences)

Choose the **most effective** way of expressing the given ideas. Write the letter of your choice (**A**, **B**, or **C**) in the blank.

Example: A. The floods came. They washed away the roadway. They also uprooted trees.

B. The floods came, and they washed away the roadway and uprooted trees.

C. The floods came, washing away the roadway and uprooting trees.   _____C_____

1. A. There was a company in Minneapolis. It shortened its work week from forty hours to thirty-six hours. The company's output increased.

B. A company in Minneapolis shortened its work week from forty hours to thirty-six hours, and this company found out the company's output increased.

C. When a Minneapolis company shortened its work week from forty to thirty-six hours, its output increased.   1. _____

2. A. Broadway has been revived by a new band of actors. These new actors are from Hollywood. They find it refreshing and challenging to perform before a live audience.

B. Broadway has been revived by a new breed of actors—Hollywood stars, who find it refreshing and challenging to perform before a live audience.

C. Broadway has been revived by this new breed of actors, which has seen actors coming from Hollywood; they have found it refreshing and challenging to perform before a live audience.   2. _____

3. A. Recreational tree climbing has become popular. Ecologists hope that a code of tree-climbing ethics will be developed. Such a code may help to prevent damage to the delicate forest ecosystems.

B. Recreational tree climbing has become popular and ecologists hope that a code of tree-climbing ethics will be developed, and such a code may help to prevent damage to the delicate forest ecosystems.

C. Before recreational tree climbing becomes any more popular, ecologists hope that a code of tree-climbing ethics will be developed to prevent permanent damage to delicate forest ecosystems.   3. _____

4. A. Harry Truman, who woke up the next morning to find himself elected President, had gone to bed early on election night.

   B. Harry Truman, who had gone to bed early on election night, woke up the next morning to find himself elected President.

   C. Harry Truman went to bed early on election night, and he woke up the next morning and found himself elected President.

4. _____

5. A. The papers were marked "Top Secret." The term *Top Secret* indicates contents of extraordinary value.

   B. The papers were of extraordinary value, and therefore they were marked "Top Secret."

   C. The papers were marked "Top Secret," indicating their extraordinary value.

5. _____

6. A. The university was noted for its outstanding faculty, its concern for minorities, and the quality of its graduates.

   B. The university was noted for its outstanding faculty, it showed concern for minorities, and how well its graduates did.

   C. The university was known for three things: it had an outstanding faculty, it showed concern for minorities, and the quality of its graduates.

6. _____

7. A. The Broadway theater, which has survived many changes, is changing rapidly again, the change being that wealthy entertainment corporations, which include, for example, Disney's company, are taking over the big theaters as they bring in huge musicals that have vapid content, high prices, and draw audiences away from more challenging plays.

   B. The Broadway theater, having survived many changes, is again changing rapidly as wealthy entertainment corporations such as Disney's take over the big theaters with vapid, high-priced musicals, drawing audiences away from more challenging plays.

   C. The Broadway theater has survived many changes. Once again it is changing rapidly. Wealthy entertainment corporations are taking over the big theaters. One example is Disney. These corporations bring in huge musicals that prove to be vapid as well as high priced. The result is that they draw audiences away from more challenging plays.

7. _____

8. A. Nick moves to Long Island and rents a house, and it is next to Gatsby's, but he does not know Gatsby. One night he sees a shadowy figure on the lawn, and he concludes that it must be Gatsby himself.

   B. Moving to Long Island, Nick rents a house next to Gatsby's. Though he does not know Gatsby, one night he concludes that the shadowy figure he sees on the lawn must be Gatsby himself.

   C. Nick, who moves to Long Island, rents a house which is next to Gatsby's, whom he does not know; one night he concludes that the shadowy figure that he sees on the lawn must be Gatsby himself.

8. _____

9. A. A fungus struck one plant and then another until it had killed nearly all of them, but one of them survived.

   B. A fungus that killed nearly all the plants spread from one to another, yet only one survived.

   C. A fungus spread among the plants, killing all but one.

9. _____

10. A. One family in a heatless building called the welfare office for money to buy an electric heater.

    B. One family lived in a building that had no heat, and so they called the welfare office to get money to buy an electric heater.

    C. One family, calling the welfare office for money to buy an electric heater, lived in a building that was heatless.

10. _____

⚫ **32. GRAMMAR AND SENTENCES:** Effective Sentences

(Study 25B–E, Use Coordination, Compounding, Subordination, Reduction)

Rewrite each of the following sets of sentences in the **most effective** way. Your result may contain one sentence or more. You may add, drop, or change words, but do not omit any information.

Example: The Lions had the ball on the Broncos' ten-yard line, and they attempted four passes, but they could not score, and so they lost the game.

Though the Lions had the ball on the Broncos' ten-yard line, they lost the game because they could not score in four pass attempts.

(Collaborative option: Students work in pairs or small groups to suggest different ways of rewriting.)

1. The Washington Monument was closed to the public. This happened in the spring and fall of 1998. The National Park Service had to repair the structure. That was the reason for the closing.

_____

_____

_____

_____

2. One airline charges an unrestricted fare of $1,734 from Boston to Reykjavik. Reykjavik is in Iceland. The same airline will fly you between the same cities for $298.

_____

_____

_____

_____

3. Many college students have a choice. This is what car-leasing companies report. These college students are the ones who do not have much in savings. One choice is that they can drive an old used car. The other is that they can lease a new car.

_____

_____

_____

_____

4. Computers have become less expensive. They have also become easier to use. And you can get free software. With this you can browse the Internet.

    _____

    _____

    _____

    _____

5. More bodies were pulled from the floodwaters in central Texas. This happened as storms continued their eastward march across the Southwest. The storms were torrential, and the march was deadly. One man was killed. This was because his home was swept away in the floods.

    _____

    _____

    _____

    _____

6. A new report has come out. It states that girls now outnumber boys in secondary schools. This is true in eighteen countries. Most of these countries are in Latin America.

    _____

    _____

    _____

    _____

7. But fifty-one countries still have serious gender gaps in education. In these countries there are 75 million fewer girls than boys in schools. This figure comes from a report by a population research group. The report was released on October 18.

    _____

    _____

    _____

    _____

8. The Surgeon General has announced new plans. The plans unveil the first national strategy for suicide prevention. The Surgeon General says that suicide is a serious public health problem, and it can no longer be ignored.

    _____

    _____

    _____

    _____

9. Gardeners are dealing with an increasingly serious pest. These gardeners are on both sides of the Rocky Mountains. The pest is hungry deer. Some gardeners are spraying odors. The deer do not like these odors. Other gardeners are covering their plants with plastic.

_____

_____

_____

_____

10. Kidnappings have reached record levels around the world. The global economic turmoil is likely to push the figure still higher. A leading institution states this.

_____

_____

_____

_____

33. GRAMMAR AND SENTENCES: Parallel Structure

(Study 25F, Use Parallel Structure)

Part 1

For each sentence: in the first three blanks, **identify** each of the boldfaced elements (use the abbreviations in parentheses):

gerund or gerund phrase (**ger**)   participle or participial phrase (**part**)
prepositional phrase (**prep**)   infinitive or infinitive phrase (**inf**)
clause (**cl**)   adjective (**adj**)
noun [with or without modifiers] (**noun**)
verb [with or without modifiers or complements] (**verb**)

Then, in the last blank, write **P** if the sentence contains **parallel structure**, or **NP** if it does **not**. (If the sentence is parallel, the first three blanks will all have the same answer.)

Examples: Congress rushed **to pass the tax bill, the Medicare bill,**
   and to adjourn.   _inf_ _noun_ _inf_ _NP_
   Shakespeare was **a poet, a playwright, and an actor.**   _noun_ _noun_ _noun_ _P_

1. The job required some knowledge of **word processing, desktop publishing, and to write.**
   1. _____ _____ _____ _____

2. Hector fought with **great skill, epic daring, and superb intelligence.**
   2. _____ _____ _____ _____

3. The mosques of ancient Islamic Spain typically contained **ornate stone screens, long hallways, and the columns looked like spindles.**
   3. _____ _____ _____ _____

4. The castle, **built on a hill, surrounded by farmland, and commanding a magnificent view,** protected the peasants from invasions by hostile forces.
   4. _____ _____ _____ _____

5. A newly discovered primate from the Amazon has **wide-set eyes, a broad nose, and the fur is striped like a zebra.**
   5. _____ _____ _____ _____

6. By nightfall, we were **tired, hungry, and grumpy.**
   6. _____ _____ _____ _____

7. The guerrillas **surrounded the village, set up their mortars, and the shelling began.**
   7. _____ _____ _____ _____

8. Kiesha did not know **where she had come from, why she was there, or the time of her departure.**
   8. _____ _____ _____ _____

9. Her favorite pastimes remain **designing clothes, cooking gourmet meals, and practicing the flute.**
   9. _____ _____ _____ _____

10. Eliot's poetry is **witty, complex, and draws on his vast learning.**
   10. _____ _____ _____ _____

Rewrite each sentence in parallel structure.

Example: The apartment could be rented by the week, the month, or you could pay on a yearly basis.
The apartment could be rented by the week, month, or year.

(Collaborative option: Students work in pairs or small groups to explore possible different parallel options. Each student writes a different version—where possible—in the blanks.)

1. Before 8 a.m., my youngest son had made himself breakfast, a snow fort in the front yard, and tormented his brothers.

---

2. Our new wood-burning stove should keep us warm, save us money, and should afford us much pleasure.

---

3. Christopher Columbus has been remembered as an entrepreneur, an explorer, a sailor, and now perhaps for how he exploited native populations.

---

4. The chief ordered Agent 007 to break into the building, crack the safe, and to steal the plans.

---

5. A good batter knows how to hit to the opposite field and staring down the pitcher.

---

6. When kindergartners were asked how the U.S. President should behave, they said someone who was fair, who shares, and not a hitter.

_____

_____

_____

7. The scouts marched briskly off into the woods, trekked ten miles to Alder Lake, and tents were erected by them.

_____

_____

_____

8. Dean has three main strengths: his ability to listen, he likes people, and his interest in cultural awareness.

_____

_____

_____

9. Global warming may not only increase air and ocean temperatures but also the force of storms.

_____

_____

_____

10. Neither regulating prices nor wages will slow inflation enough.

_____

_____

_____

11. During its early years, Sears, Roebuck and Company sold not only clothes, furniture and hardware, but also customers could buy cars and houses.

_____

_____

_____

12. Charlene practiced shooting from the top of the key as well as how to dribble with either hand.

_____

_____

_____

13. The new ambassador impressed everyone with her wit, charm, her grace, and they liked her intelligence.

_____

_____

_____

14. The experimental group either consisted of white rats or gray ones.

_____

_____

_____

15. But in a larger sense we cannot dedicate this ground, we cannot consecrate it either, nor can it be hallowed by us.

_____

_____

_____

34. GRAMMAR AND SENTENCES: Fragments

(Study 26A, Fragments)

Part 1

Write **S** after each item that is one or more **complete sentences.**
Write **F** after each item that contains a **fragment.**

Example: Luis was offered the job. Having presented the best credentials.   _____F_____

1. When people shut down their computers. It is important for them to wait until the screen flashes "safe to turn off."   1._____

2. Being coached in what is appropriate to do and say in a job interview, so as not to make a disastrous mistake.   2._____

3. The manuscript having been returned, Johanna sat down to revise it.   3._____

4. Harrison desperately wanted the part. Because he believed that this was the film that would make him a star.   4._____

5. The old-fashioned clock stopped ticking. Renée had forgotten to wind it.   5._____

6. He admitted to being a computer nerd. As a matter of fact, he was proud of his computing skills.   6._____

7. More than 50 percent of Americans surveyed felt guilty about their child-care arrangements.   7._____

8. I read all the articles. Then I wrote the first draft of my paper.   8._____

9. Many Americans prefer indirect business levies rather than direct taxation. Where do you stand on this issue?   9._____

10. Maurice kept nodding his head as the coach explained the play. Thinking all the time that it would never work.   10._____

11. Because she was interested in rocks, she majored in geology.   11._____

12. I argued with two of my classmates. First with Edward and then with Harry.   12._____

13. There are many humorous research projects. Such as developing an artificial dog to breed fleas for allergy studies.   13._____

14. Taylor was absolutely positive he would pass. Regardless of having received failing grades on both his essay and the midterm.   14._____

15. Colleen stepped up to the free-throw line; then she made two points to win the game.   15._____

16. Professor Fustie was known for his forgetfulness. For example, leaving his grade book in the men's room.   16._____

17. Because the army could not indefinitely maintain such a long supply line.   17._____

18. She went to the supermarket. After she had made a list of groceries that she needed.   18._____

19. I telephoned Dr. Gross. The man who had been our family physician for many years.   19._____

20. We suspect Harry of the theft. Because he had access to the funds and he has been living far beyond his means.   20._____

Rewrite each item in one or more sentences, eliminating any **fragment(s)**. You may add information, but do not omit any.

Examples: Chief Joseph led his Native Americans on a desperate flight to freedom. A flight doomed to failure.
Chief Joseph led his Native Americans on a desperate flight to freedom, a flight doomed to failure.

Because the patient was near death.
Because the patient was near death, the doctors operated immediately.

(Collaborative option: Students work in pairs or small groups to suggest different ways of rewriting.)

1. Two days before the competition, he felt nervous. However, much more at ease just before the contest.

   _____

   _____

   _____

2. She was a star athlete. A brilliant student besides.

   _____

   _____

   _____

3. If there is no change in the patient's condition within the next twenty-four hours.

   _____

   _____

   _____

4. The Scottish and Irish farmers forced from their land so it could be turned into sheep pastures. More profitable for the landowners.

   _____

   _____

   _____

5. When it becomes too hot to work in the fields. The workers taking a welcomed rest.

   _____

   _____

   _____

35. GRAMMAR AND SENTENCES: Comma Splices and Fused Sentences

(Study 26B, Comma Splices, and 26C, Fused Sentences)

Part 1

Write **S** after any item that is a **complete sentence**.
Write **Spl** after any item that is a **comma splice**.
Write **FS** after any item that is a **fused sentence**.

Example: The mission was a success, everyone was pleased.   _____Spl_____

1. The apartment was up two rickety flights of stairs it was the best they could afford.   1. _____

2. The party broke up at one in the morning, Jack lingered for a few final words with Kathy.   2. _____

3. Battered by insider-trading scandals, the stock market plunged to its yearly low.   3. _____

4. The moon enters the earth's shadow, a lunar eclipse occurs, causing the moon to turn a deep red.   4. _____

5. The ticket agent had sold eighty-one tickets to boarding passengers there were only eleven empty seats on the train.   5. _____

6. Since she was in the mood for a romantic movie, she hired a babysitter and went to see *My Big Fat Greek Wedding.*   6. _____

7. Sheer exhaustion having caught up with me, I had no trouble falling asleep.   7. _____

8. The restaurant check almost made me faint, because I had left my wallet home, I couldn't pay for the meal.   8. _____

9. Those of us who lived in off-campus housing ignored the rule, since we were seniors, we never worried about campus regulations.   9. _____

10. It was a cloudy, sultry afternoon when we sighted our first school of whales, and the cry of "Lower the boats!" rang throughout the ship.   10. _____

11. The war was finally over; however, little could be done to ease the refugees' sense of loss.   11. _____

12. The author described fifty ways to recycle fruitcakes; my favorite is to use slices of fruitcake as drink coasters.   12. _____

13. The three major television networks face stiff competition for ratings, because of cable networks, viewers can decide from among four hundred programs.   13. _____

14. In the film a lawyer reopens an investigation on behalf of a woman imprisoned for murder she convincingly claims that she is innocent.   14. _____

15. Though the teacher believed that it was important for her students to write every day, she did not enjoy grading so many papers.   15. _____

Rewrite each item in one or more **correct sentences,** eliminating any **comma splices** or **fused sentences.** Add or change words as needed. Do not omit any information.

Example: The software game was full of violent scenes, thus it was banned from the school's computer center.
The software game was banned from the school's computer center because it was full of violent scenes.

(Collaborative option: Students work in pairs or small groups to suggest different ways of rewriting.)

1. More Americans are buying the popular sport-utility vehicles, however, they are finding the insurance premiums unexpectedly high.

   _____

   _____

   _____

2. Crime is still a major concern for many Americans so many teenagers are arrested for violent crimes.

   _____

   _____

   _____

3. I waited in line for my turn at the automatic teller machine, I balanced my checkbook.

   _____

   _____

   _____

4. Over 30 percent of children from rural America live in mobile homes, therefore, Congress has established a commission to study mobile home safety and construction standards.

   _____

   _____

   _____

5. A shortage of licensed contractors often exists in the areas hit by natural disasters homeowners quickly learn to wait for a work crew with the proper credentials.

   _____

   _____

   _____

6. The largest Native American reservation is the Navajo it is located mostly in Arizona and covers 16 million acres.

   _____

   _____

   _____

7. According to some researchers, little boys may have different educational experiences from little girls, in other words, even though it may be unintentional, teachers often have subtly different expectations based on the gender of their students.

   _____

   _____

   _____

   _____

8. Tarantulas are large spiders with powerful fangs and a mean bite, they live not only in the tropics but also in the United States.

   _____

   _____

   _____

9. Jogging can reduce fatal heart attacks because it is an aerobic activity, it keeps the arteries from clogging.

   _____

   _____

   _____

10. The software game was full of violent scenes, therefore, it was banned from the school's computer center. [Correct this in a different way from that shown in the example.]

    _____

    _____

    _____

36. GRAMMAR AND SENTENCES: Fragments, Comma Splices, and Fused Sentences

(Study 26, Conquering the "Big Three" Sentence Errors)

Rewrite any item that contains a **fragment**, **comma splice**, or **fused sentence**, so that it contains none of these. You may add words or information as needed, but do not omit any information. If an item is already correct, leave the blank empty.

Examples: When she saw the full moon rising over the hill.
　　　　　　When she saw the full moon rising over the hill, she thought of the night they had met.
　　　　　　When Peary and Henson reached the Pole, they rejoiced.

(Collaborative option: Students work in pairs or small groups to suggest different ways of rewriting.)

1. Because pie, ice cream, and candy bars have practically no nutritional value.

　　_____
　　_____
　　_____

2. When the bindings release, the ski comes off.

　　_____
　　_____
　　_____

3. Which promotes tooth decay when not used properly.

　　_____
　　_____
　　_____

4. Lady Bird Johnson and Barbara Bush, first ladies greatly admired.

　　_____
　　_____
　　_____

5. Whereas older cars run on leaded gas and lack complex pollution controls.

　　_____
　　_____
　　_____

6. Because she was not prepared for the interviewer's questions and felt she would never get the job.

_____

_____

_____

7. By installing smoke detectors, families may someday save family members from perishing in a fire.

_____

_____

_____

8. Watching from the seventh floor during the parade.

_____

_____

_____

9. Which could strengthen your immune system.

_____

_____

_____

10. Stay.

_____

_____

_____

11. We planned the trip carefully, yet we still had a series of disasters.

_____

_____

_____

12. The quarterback signed the largest professional sports contract to date, he will earn $120 million over a six-year period.

_____

_____

_____

13. Native Americans dances and music for every tribal ceremony and social occasion celebrated.

_____

_____

_____

14. Scientists are currently interested in studying polar bears. Because the bears' body chemistry may reveal how pollution has affected the Arctic.

_____

_____

_____

15. Americans, for the moment, may be less concerned about taxes. Polls indicate that Americans would rather balance the federal budget than lower taxes.

_____

_____

_____

16. Until all the workers were able to present their points of view.

_____

_____

_____

17. The community was unaware of the city's plan to tear down a playground, therefore, few citizens attended the city council meetings.

_____

_____

_____

18. Since 1975, over 1.5 million Vietnamese have left their homeland in search of a peaceful life, many have settled in Australia, Canada, Europe, and the United States.

_____

_____

_____

19. Because the founder of a popular fast-food restaurant chain has encouraged corporations to provide financial support for employees adopting children.

_____

_____

_____

20. All my coworkers on diets and won't eat any cookies or cake.

_____

_____

_____

21. What happened to Clyde, Roberta, and Sondra is told in the novel *An American Tragedy* it was written by Theodore Dreiser.

_____

_____

_____

22. He wore a pair of mud-encrusted, flap-soled boots they looked older than he did.

_____

_____

_____

23. For the agency wanted to know how its money was spent.

_____

_____

_____

24. Enrique reread his assignment a dozen times before handing it in. To be absolutely sure his ideas were clear.

_____

_____

_____

25. The executive waited, however, until every worker at the meeting presented a point of view.

_____

_____

_____

26. That she is dead is beyond dispute.

_____

_____

_____

27. "I believe," declared the headmaster. "That you deserve expulsion."

_____

_____

_____

28. The scouts hiked two miles until they reached the falls then they had lunch.

_____

_____

_____

29. The police having been warned to expect trouble, every available officer lined the avenue of the march.

_____

_____

_____

30. In the 1800s, Ireland's vital crop was wiped out by the potato blight, Irish people who owned ten acres of land were disqualified from poor relief.

_____

_____

_____

31. The Irish immigrants did not settle on farms for fear that the potato blight would strike again, but the German immigrants did go into farming they had no fear of this blight.

_____

_____

_____

32. In the first George Bush's administration the Gulf War, and in George W. Bush's the war against terrorism.

_____

_____

_____

33. A victory that is unmatched in the history of amateur sports.

_____

_____

_____

37. GRAMMAR AND SENTENCES: Placement of Sentence Parts

(Study 27A, Needless Separation of Related Sentence Parts)

If the boldfaced words are **in the wrong place**, draw an arrow from these words to the place in the sentence where they should be.

If the boldfaced words are **in the right place**, do nothing.

Examples: Never give a toy to a child **that can be swallowed**.

People who buy cigars **made in Cuba** violate U.S. laws.

1. He ordered a pizza for his friends **covered with pepperoni.**

2. She **only** had enough money to buy two of the three books that she needed.

3. The parents of college students **who have earned scholarships** are indeed fortunate.

4. After asking a few questions, we decided **quickly** to end the conference call because we weren't interested in what the company had to offer.

5. We saw the plane taxi onto the field **that would soon be leaving for Chicago.**

6. Some Americans are spending **almost** a third of their income on rent.

7. The President attempted to prevent the outbreak of war **in the Oval Office.**

8. Unfortunately, the resale shop was full of **wrinkled** little girls' dresses.

9. We hurriedly bought a picnic table from a clerk **with collapsible legs.**

10. We learned that no one could discard anything at the municipal dump **except people living in the community.**

11. The only baseball jacket left was a **green and white** child's starter jacket.

12. The race car driver planned **after the Grand Prix race** to retire before she received another injury.

13. The bride walked down the aisle with her father **wearing her mother's wedding gown.**

103

14. Despite her sincerity and honesty, the candidate failed to **carefully, completely, and with candor** explain why she dropped out of the campaign.

15. The two scientists, **working independently,** achieved the same results.

16. Send, **after you have received all the donations,** the total amount to the organization's headquarters.

17. **Only** one teacher seems able to convince Raymond that he should study.

18. Only a few Olympic athletes can expect lucrative endorsement contracts **with gold medals**.

19. We watched the *Queen Elizabeth II* as she slowly sailed out to sea **from our hotel window**.

20. Indicate **on the enclosed sheet** whether you are going to the class picnic.

21. A **battered** man's hat was hanging on a branch of the tree.

22. Sam, **running out in his robe and slippers to get the morning newspaper on a cold January morning,** slammed the front door shut and then realized that he was locked out.

23. Croaker College, which had **almost** lost all of its football games last year, fired its coach.

24. Albino Ruales realized **when he heard that his grandmother was moving in with his family** that many households are now multigenerational.

25. He replied that they went to Paris **usually** in the spring.

38. GRAMMAR AND SENTENCES: Dangling and Misplaced Modifiers

(Study 27A, Needless Separation of Related Sentence Parts, and 27B, Dangling Modifiers)

Part 1

If the boldfaced words are a dangling or misplaced modifier, **rewrite** the sentence correctly in the blanks below it. If the sentence is correct, do nothing.

Examples: Returning the corrected essays, most students were disappointed by their marks.

When the instructor returned their corrected essays, most students were disappointed by their marks.

Roosevelt and Churchill, **meeting at sea**, drafted the Four Freedoms.

_____

_____

(Collaborative option: Students work in pairs to suggest different ways of rewriting.)

1. **Announcing his first baseball game in 1939,** the late Red Barber began a broadcasting career that would last over fifty years.

_____

_____

_____

2. **Rowing across the lake,** the moon often disappeared behind the clouds.

_____

_____

3. **Having worked on my paper for three hours,** the network went down and my paper was lost in cyberspace.

_____

_____

4. **While on vacation,** the idea for a new play came to him.

_____

_____

5. **Worried about what books their children were borrowing from libraries,** the library finally agreed to develop an on-line rating system for families.

_____

_____

_____

6. **Upon entering college,** he applied for part-time employment in the library.

_____

_____

7. **Practicing every day for five hours,** Dani's expensive music lessons really paid off.

_____

_____

8. **Sleeping in late,** the house seemed incredibly quiet with the boys still in bed.

_____

_____

9. **After sleeping in until noon,** the day seemed to go by too quickly.

_____

_____

10. **When nine years old,** my father took my sister and me on our first camping trip.

_____

_____

11. **At the age of ten,** I was permitted to go, for the first time, to a summer camp.

_____

_____

12. **After putting away my fishing equipment,** the surface of the lake became choppy.

_____

_____

13. **Racing toward the primate section of the zoo,** the chimpanzees' playful laughter drew the children to their cage.

_____

_____

14. **To achieve a goal,** a person must expect to work and to make sacrifices.

_____

_____

15. **Suggesting that the American standard of living has declined,** some American economists predict a gloomy financial status for the next generation.

_____

_____

16. **After hearing of Tom's need for financial aid,** a hundred dollars was put at his disposal.

_____

_____

17. **Pickled in spiced vinegar,** the host thought the peaches would go with the meat.

_____

_____

18. **While running in a local marathon,** the weather was quite uncooperative.

_____

_____

19. **Relieved by her high grade on the first paper,** her next paper seemed less difficult.

_____

_____

20. **To be a happy puppy,** you need to exercise your pet regularly.

_____

_____

21. **As a teenager,** Darlene worked two jobs to help her family financially.

_____

_____

22. **After eating too much chocolate,** my scale revealed that I had gained ten pounds.

_____

_____

23. Finally, **after working for days,** the garden was free of weeds.

_____

_____

24. **To proofread my paper,** I reread it several times and used the grammar- and spell-checking functions of my word-processing software.

_____

_____

25. **After finishing my assignment,** the dog ate it.

_____

_____

26. **To get ready for summer vacation,** camp registrations had to be completed this week.

_____

_____

27. **Realizing that the unemployment rate was still over 10 percent,** most workers were not changing jobs.

_____

_____

28. **To get a passing grade in this course,** the professor's little quirks must be considered.

_____

_____

Part 2

In the first set of blanks, write a sentence of your own containing a humorous dangling or misplaced modifier. In the second set of blanks, **rewrite** the sentence correctly.

Example: _Hanging stiffly from the clothesline, Mother saw that the wash had frozen overnight._
Mother saw that the wash, hanging stiffly from the clothesline, had frozen overnight.

(Collaborative option: Students work in pairs or small groups to invent and correct sentences.)

1. _____

_____

_____

_____

_____

_____

2. _____

3. _____

4. _____

5. _____

39. GRAMMAR AND SENTENCES: Effective Sentences Review

(Study 25–27, Effective Sentences)

If an item is **incorrect** or **ineffective** in any of the ways you learned in sections 25–27, **rewrite** it correctly or more effectively in the blanks below it.

If an item is **correct** and **effective** as is, do nothing.

Examples: The lakes were empty of fish. Acid rain had caused this.
Acid rain had left the lakes empty of fish.
Working in pairs, the students edited each other's writing.

(Collaborative option: Students work in pairs to suggest ways of rewriting incorrect or ineffective sentences.)

1. If one drives a car without thinking, you are more than likely to have an accident.

    _____

    _____

2. The entire class was so pleased at learning that Dr. Turner has rescheduled the quiz.

    _____

    _____

3. The author decided to forthrightly, absolutely, unequivocally, and immediately deny the allegations of plagiarism.

    _____

    _____

4. A study revealed that vigorous exercise may add only one or two years to a person's life. This study used Harvard graduates.

    _____

    _____

    _____

5. The film director, thinking only about how he could get the shot of the erupting volcano, endangered everyone.

    _____

    _____

6. With her new auditory implant, Audrey heard so much better.

_____

_____

7. Watching the star hitter blast a home run over the fence, the ball smashed a windshield of an expensive sports car.

_____

_____

8. The owner of the team seems to insult her players and fans and mismanaging the finances.

_____

_____

9. The witness walked into the courtroom, and then she wishes she could avoid testifying.

_____

_____

10. An increase in energy taxes causes most people to consider carpooling and improving energy-conservation practices in their homes.

_____

_____

_____

11. According to historians, settlers traveling westward used prairie schooners, not Conestoga wagons, and they used oxen and mules instead of horses to pull the wagons, and they did not pull their wagons into a circle when under an attack.

_____

_____

_____

_____

12. He told me that he was going to write a letter and not to disturb him.

_____

_____

13. Ajay Smith is a senior, and he just won national recognition for his poetry.

_____

_____

14. In the 1400s many English villages held football competitions, an inflated animal bladder was kicked or shoved between two distant points by opposing teams.

_____

_____

_____

15. If a student knows how to study, you should achieve success.

_____

_____

16. He went to his office. He sat down. He opened his briefcase. He read some papers.

_____

_____

17. Summer is a time for parties, friendships, for sports, and in which we can relax.

_____

_____

18. I met the new dorm counselor in my oldest pajamas.

_____

_____

19. Being a ski jumper requires nerves of steel, you have to concentrate to the utmost, and being perfectly coordinated.

_____

_____

_____

20. The plane neither had enough fuel nor proper radar equipment.

_____

_____

21. The instructor wondered when did the students begin sneaking out of class.

_____

_____

22. Because they would not worship the Roman gods meant that Christians might be thrown to the lions.

_____

_____

23. In baseball, a sacrifice is when the batter allows himself to be put out in order to advance a base runner.

_____

_____

24. Saddened by the collapse of his marriage, Fassbinder's mansion now seemed an ugly, echoing cavern.

_____

_____

25. By the coach putting Robinson at quarterback would have given the team a chance at the title.

_____

_____

40. SENTENCES AND GRAMMAR: Review

(Study 1–27, Sentences and Grammar)

Part 1

Write **T** for each statement that is **true**.
Write **F** for each statement that is **false**.

Example: A **present participle** ends in *-ing* and is used as an adjective.   _____T_____

1. Both a **gerund** and a **present participle** end in *-ing*.   1. _____
2. The greatest number of words ever used in a **verb** is four.   2. _____
3. **Parallel structure** is used to designate ideas that are unequal in importance.   3. _____
4. A **dangling participle** may be corrected by being changed into a dependent clause.   4. _____
5. *It's* is a contraction of *it is; its* is the **possessive** form of the pronoun *it*.   5. _____
6. The **verb precedes the subject** in a sentence beginning with the expletive *there*.   6. _____
7. A **preposition** may contain two or more words; *because of* is an example.   7. _____
8. The **principal parts of a verb** are the *present tense*, the *future tense*, and the *past participle*.   8. _____
9. A **collective noun** may be followed by either a singular or plural verb.   9. _____
10. A **prepositional phrase** may be used only as an adjective modifier.   10. _____
11. A **compound sentence** is one that contains two or more independent clauses.   11. _____
12. Not all **adverbs** end in *-ly*.   12. _____
13. The verb **be** is like an equal sign in mathematics.   13. _____
14. A **noun clause** may be introduced by the subordinating conjunction *although*.   14. _____
15. An **adjective clause** may begin with *when* or *where*.   15. _____
16. Both **verbals** and **verbs** may have modifiers and complements.   16. _____
17. The terminal punctuation of a declarative sentence is the **exclamation point.**   17. _____
18. *Without* is a **subordinating conjunction**.   18. _____
19. A sentence may begin with the word *because*.   19. _____
20. The **verb** of a sentence can consist of a past participle alone.   20. _____
21. A **subjective complement** may be a noun, a pronoun, or an adverb.   21. _____
22. A **direct object** may be a noun or a pronoun.   22. _____
23. When there is an **indirect object**, it must precede the direct object.   23. _____
24. When there is an **objective complement**, it must precede the direct object.   24. _____
25. Pronouns used as appositives are called **intensive pronouns.**   25. _____
26. The word *scissors* takes a **singular verb**.   26. _____
27. An **antecedent** is the noun for which a pronoun stands.   27. _____

115

28. A **simple sentence** contains two or more independent clauses.　　　28. _____

29. A pronoun following the verb *be* needs the **objective case**.　　　29. _____

30. A **complex sentence** contains at least one independent clause and one dependent clause.　　　30. _____

31. A **sentence fragment** is not considered a legitimate unit of expression; a **nonsentence** is.　　　31. _____

32. **Adjectives** never stand next to the words they modify.　　　32. _____

33. Not all words ending in -*ly* are **adverbs**.　　　33. _____

34. An **indefinite pronoun** designates no particular person.　　　34. _____

35. The words *have* and *has* identify the **present perfect tense** of a verb.　　　35. _____

36. A statement with a subject and a verb can be a fragment if it follows a **subordinating conjunction.**　　　36. _____

37. An **adverb** may modify a noun, an adjective, or another adverb.　　　37. _____

38. **Verbs** are words that assert an action or a state of being.　　　38. _____

39. The **indicative mood** of a verb is used to express a command or a request.　　　39. _____

40. The function of a **subordinating conjunction** is to join a dependent clause to a main clause.　　　40. _____

41. The **subjunctive mood** expresses doubt, uncertainty, a wish, or a supposition.　　　41. _____

42. An **adjective** may modify a noun, a pronoun, or an adverb.　　　42. _____

43. A **gerund** is a verb form ending in -*ing* and used as a noun.　　　43. _____

44. A **clause** differs from a **phrase** in that a clause always has a subject and a predicate.　　　44. _____

45. **Adjectives** tell *what kind, how many,* or *which one;* **adverbs** tell *when, where, how,* or *to what degree.*　　　45. _____

46. A **comma splice** is a grammatical error caused by joining two independent clauses with a comma.　　　46. _____

47. **Coordinating conjunctions** *(and, but, or, nor, for, yet, so)* join words, phrases, and clauses of equal importance.　　　47. _____

48. **Pronouns in the objective case** *(him, me, . . .)* should be used as direct objects of verbs and verbals.　　　48. _____

49. **Mixed construction** occurs when two sections of a sentence that should be grammatically compatible are not.　　　49. _____

50. A **simple short sentence** can be a forceful expression in a passage.　　　50. _____

Part 2

Write **C** if the item is **correct**.
Write **X** if it is **incorrect**.

Example: Was that letter sent to Paul or **I**?　　　___X___

1. **Having been notified to come at once,** there was no opportunity to call you.　　　1. _____

2. I suspected that his remarks were directed to Larry and **me**.　　　2. _____

3. He, **thinking that he might find his friends on the second floor of the library,** hurried.　　　3. _____

4. If a student attends the review session, **they** will do well on the first exam.　　　4. _____

5. In the cabin of the boat **was** a radio, a set of flares, and a map of the area.　　　5. _____

6. The Queen, standing beside her husband, children, and grandchildren, **were** waving regally at the crowd.

6. _____

7. She is a person **who** I think is certain to succeed as a social worker.

7. _____

8. **Is** there any other questions you wish to ask regarding the assignment?

8. _____

9. The driver had neglected to fasten his seat belt, **an omission that cost him a month in the hospital.**

9. _____

10. He particularly enjoys **playing softball** and **to run** a mile every morning.

10. _____

11. Forward the complaint to **whoever** you think is in charge.

11. _____

12. Every girl and boy **was** to have an opportunity to try out for the soccer team.

12. _____

13. Neither the bus driver nor the passengers **were** aware of their danger.

13. _____

14. Within the next five years, personal computers will be **not only** smaller **but also** more affordable.

14. _____

15. Not everyone feels that **their** life is better since the 1960s civil rights movement.

15. _____

16. Homemade bread tastes **differently** from bakery bread.

16. _____

17. Not **having had** the chance to consult his lawyer, Larry refused to answer the officer's questions.

17. _____

18. **Is** either of your friends interested in going to Florida over spring break?

18. _____

19. He enrolled in economics because **it** had always been of interest to him.

19. _____

20. The snow fell **steady** for two days.

20. _____

21. Burt paced nervously up and down the corridor. **Because he was concerned about the weather.**

21. _____

22. **A heavy rain began without warning,** the crew struggled with the tarpaulin.

22. _____

23. **To have better control over spending,** the checkbook is balanced each week.

23. _____

24. Casey **asked for time, stepped out of the batter's box, and his finger was pointed** toward the bleachers.

24. _____

25. **By investing in real estate at this time** can earn you a substantial profit in several years.

25. _____

41. SENTENCES AND GRAMMAR: Review

(Study 1–27, Sentences and Grammar)

On your own paper, **rewrite** each of the following paragraphs so that it is **free of errors** and more **effective**. You may change or reduce wording, combine sentences, and make any other necessary changes, but do not omit any information.

(Collaborative option: Students work in pairs or small groups to suggest ways of improving the paragraphs. They evaluate or edit each other's work.)

1. Neither the strength nor the wisdom of Clyde Griffiths' parents were sufficient to bring up their family properly. He grew ashamed of his parents, his clothes, and he had to live in ugly surroundings. Clyde grew older, he dreamed of a life of wealth and elegance. Spending most of his money on clothes and luxuries for himself, his parents were neglected by him. One night when Clyde's uncle invited him to dinner. He met beautiful, wealthy Sondra Finchley. Determined to have her, she was too far above his social position. So Clyde starts going with a factory worker, her name was Roberta, and she became pregnant by him, but it was decided by Clyde that just because of Roberta was no reason he had to give up his pursuit of Sondra.

2. The novel *Slaughterhouse-Five* tells of a man named Billy Pilgrim, who is a prisoner in World War II and later traveled to the planet Tralfamadore. In one particularly amusing episode, the Tralfamadoreans throw Billy into a cage in one of their zoos, along with a sexy Earthling actress named Montana Wildhack. The Tralfamadoreans crowd around the cage to watch the lovemaking between he and her. The less interesting sections of the novel depict the middle-class civilian life of Billy. Who grows wealthy despite having little awareness of what is going on. Billy acquires his wealth by becoming an optometrist, he marries his employer's daughter, and giving lectures on his space travels. I like most of the book because its the most unique novel I have ever read and because it makes you realize the horrors of war and the hollowness of much of American life. However, after reading the entire book, Kurt Vonnegut, Jr., the author, disappointed me because I, enjoying science fiction, wish they had put more about space travel into it.

3. In reading, critical comprehension differs from interpretive comprehension. Critical comprehension adds a new element. That element was judgment. On the interpretive level a student may understand that the author of a poem intends a flower to represent youth, on the critical level they evaluate the author's use of this symbol. The student evaluate the quality of the poem too. For example. On the interpretive level a student would perceive that the theme of a story is "If at first you don't succeed, try, try again"; on the critical level the student judges whether the saying is valid. Critical comprehension includes not only forming opinions about characters in stories but also judgments about them. By learning to comprehend critically, the student's overall reading ability will increase markedly.

4. Studying the woodland ground with my magnifying glass, I grew astonished. First I saw a column of tiny leaves marching along a two-inch-wide road. Peering through the glass, each leaf was being carried like an umbrella in the jaws of an ant far more smaller than the leaf itself. I began to notice other ant trails, all leading to tiny mounds of earth, they looked like miniature volcanoes. Up the mounds and into the craters trod endless parades of ants, each holding aloft its own parasol, which made my spine tingle with excitement. When I heard a faint buzzing made me look around. Above the ant-roads swarmed squadrons of tiny flies. As if on signal they dived straight down to attack the ants. If a person saw this, they would not have believed it. The ants, their jaws clamped upon the giant leaves, had no means of defense. Yet, as if answering air-raid sirens, you could see an army of smaller ants racing toward the leaf-carriers, who they strove to protect.

5. Because the leaf-carrying ants now had some protection did not mean that the attack was over by the flies. As the first attacking fly dived upon a leaf-carrier, the tiny protector ants reached and snapped at the aerial raider with their formidable jaws and they drove it away, but then all along the leaf-carrying column other flies joined the attack. Now I could see that atop each moving leaf a tiny protector ant was "riding shotgun" through my magnifying glass. Whenever a fly dive-bombed a leaf-carrier was when the shotgun ant on the leaf reached out and bit the fly. One shotgun ant grasped a fly's leg in its jaws and sends the winged enemy spinning to the ground. The ant's comrades swarmed all over the helpless fly, and it was soon reduced to a lifeless shell by them. Similar scenes were taken place all over the miniature battlefield. Finally the squadrons of flies, unable to penetrate the ants' defenses, rised, seemingly in formation, and droned back to their base. Would they mourn their casualties, I wondered. Will their leader have to report the failed attack to an angry insect general?

42. SENTENCES AND GRAMMAR: Review for Non-Native English Speakers

**(Study 7, Words that Connect; 14, Using Verbs Correctly; and
16E, Use Articles and Determiners Correctly)**

Part 1

In each ⎣box⎦ , write the **correct** preposition: **at**, **in**, or **on**.

On each blank line, write the **correct** verb ending: **ed** (or **d**), **s** (or **es**), or **ing**. If no ending is needed, leave the line empty.

In each set of brackets [   ], write the correct **article**: **a**, **an**, or **the**. If no article is needed, leave the brackets empty.

Example: [The] newest building ⎣in⎦ our city is [an] apartment house. It was construct<u>ed</u> for senior citizens.

1. Living ⎣   ⎦ [   ] large city requires strong nerves and [   ] outstanding sense of humor. This is especially true ⎣   ⎦ Mondays. When I wait ⎣   ⎦ my corner for [   ] bus that take____ me to work, I hear [   ] screams of ambulances and fire engine____ as they speed by. When I am finally ⎣   ⎦ my office building, I am push____ into [   ] elevator by [   ] crowd. I manage to get off ⎣   ⎦ [   ] twelfth floor. But when I give [   ] cheery "Good morning!" to [   ] first coworker I meet, I am often answer____ with [   ] grouchy remark. The people at my former job, ⎣   ⎦ 1999, treat____ me much better. I stay____ there only a year, but it was [   ] best job I have had since be____ [   ] America.

2. In [   ] depth of winter ⎣   ⎦ 1925, ⎣   ⎦ [   ] small Alaskan town called Nome, [   ] epidemic of [   ] deadly disease diphtheria start____. The people were shock____ to hear that there was no medicine available to stop [   ] disease from spread____. The ice-locked town was completely block____ off from the outside world: no boat or plane could reach____ it, and no roads or rail lines had yet been construct____ there. Only [   ] dogsleds might possibly rush____ the medicine to Nome in time. But [   ] nearest supply of medicine was ⎣   ⎦ the city of Anchorage, a thousand miles away. ⎣   ⎦ Nome's tiny telegraph office, the town's doctor transmitt____ [   ] desperate message: "Nome need____ diphtheria medicine at once!"

3. Officials in Anchorage round____ up all the available medicine and had it shipped ☐ [   ] train to the end of the line ☐ Nenana, still 674 miles from Nome. From there relays of dogsled teams took over. The first team's drivers trudge____ through the white wilderness to [   ] tiny hamlets of Tolovana and Bluff. ☐ Bluff, Gunnar Kaasen's team, headed by the dog Balto, began [   ] next leg of [   ] journey. Through raging blizzards, thirty-below-zero cold, and missed relay stations, [   ] Balto led Kaasen's team all the way to Nome. ☐ just 5½ days the dog teams had cover____ what was normally [   ] month's journey. Nome had been save____.

Part 2

In each blank, write any correct determiner (limiting adjective) from the list. Try not to use any word on the list more than once.

every	many	other	more	some	several
each	most	such	(a) little	another	all
(n) either	(a) few	both	much	enough	any

Example: They needed _another_ person to help lift the car.

_____ day last week there were _____ alarming stories in the newspapers. _____ of them made _____ sense. One story said that soon there would not be _____ fish left in the oceans or lakes. _____ story warned that global warming would soon drown or boil us all. _____ of these stories gave me nightmares.

43. PUNCTUATION: The Comma

(Study 30–32, The Comma)

Part 1

If **no comma** is needed in the bracketed space(s), leave the blank empty. If **one or more commas** are needed, write in the **reason** from the list below (only one reason per blank; use the abbreviations in parentheses).

independent clauses joined by
    conjunction (**ind**)
introductory adverb clause (**intro**)
series (**ser**)
parenthetical expression (**par**)
nonrestrictive clause (**nr**)

appositive (**app**)
absolute phrase (**abs**)
direct address (**add**)
mild interjection (**inter**)
direct quotation (**quot**)

Examples: The New England states include Vermont[ ] Maine[ ] and New Hampshire.   _____ser_____
        The Secretary of State[ ] held a press conference.   _____

1. *The Gardens of Kyoto*[ ] a novel by Kate Walbert[ ] recounts a woman's coming of age in America, Paris, and Japan during the 1950s.   1._____

2. Professors[   ] who assign too many long papers[   ] may have small classes.   2._____

3. Well[ ] it looks as if I'll have to go to plan B.   3._____

4. If the new security system fails to work[ ] we could be in deep trouble.   4._____

5. In truth[ ] Dr. Faust[ ] your future does not look good.   5._____

6. Phillip's father[   ] who is a religious man[   ] disapproves of many teenage antics.   6._____

7. Dan and Marilyn[   ] however[   ] are hopeful for a 2008 victory.   7._____

8. John Fitzgerald Kennedy[   ] the thirty-fifth President of the United States[   ] was assassinated on November 22, 1963.   8._____

9. The Chinese are trained to write with their right hands[   ] for it is difficult to do Chinese calligraphy with the left hand.   9._____

10. Before you meet clients for the first time[   ] learn all that you can about their company, their style, and their risk-taking ability.   10._____

11. He sat down at his desk last evening[   ] and made a preliminary draft of his speech.   11._____

12. Julie went into the library[   ] but she hurried out a few minutes later.   12._____

13. Lincoln spoke eloquently about government of the people[   ] by the people[   ] and for the people.   13._____

14. After she had listened to her favorite album[   ] she settled down to study.   14._____

15. The candidate gave a number of speeches in Illinois[   ] where she hoped to win support.   15._____

16. She had always wanted to visit the small village[   ] where her father lived, but she knew neither its name nor its location.   16._____

17. My instructor[   ] Dr. Ursula Tyler[   ] outlined the work for the current semester.   17._____

18. What you need[   ] David[   ] is a professional organizer to straighten out your office.   18._____

independent clauses joined by
    conjunction (**ind**)
introductory adverb clause (**intro**)
series (**ser**)
parenthetical expression (**par**)
nonrestrictive clause (**nr**)

appositive (**app**)
absolute phrase (**abs**)
direct address (**add**)
mild interjection (**inter**)
direct quotation (**quot**)

---

19. "Is this[   ]" she asked[   ] "the only excuse that you have to offer?"

    19. _____

20. Castles were cold and filthy[   ] according to historians[   ] because castles were built more for protection than convenience.

    20. _____

21. His hands swollen from five fire-ant bites[   ] John swore that he would rid his yard of all ant hills.

    21. _____

22. Both potato and corn crops had a major impact on the life expectancy of Europeans[   ] living in the eighteenth century.

    22. _____

23. Ford's first Model T sold for $850 in 1908[   ] but the price dropped to $440 in 1915 because of mass production.

    23. _____

24. We were asked to read *The Grapes of Wrath*[   ] which John Steinbeck wrote in the 1930s.

    24. _____

25. Lorraine Hansberry[   ] the author of *A Raisin in the Sun*[   ] died at age thirty-five.

    25. _____

Part 2

If **no comma** is needed in the bracketed space(s), leave the blank empty. If **one or more commas** are needed, write in the **reason** from the list below (only one reason per blank; use the abbreviations in parentheses).

parenthetical expression (**par**)
after yes and no (**y/n**)
examples introduced by *such as*,
    especially, or *particularly* (**examp**)
contrast (**cont**)
nonrestrictive clause (**nr**)

omission (**om**)
confirmatory question (**ques**)
direct address (**add**)
date (**date**)
state or country (**s/c**)

Examples: He came from New York; she[   ] from Maine.

    _____om_____

    The Secretary of State[   ] held a press conference.

    _____

1. On New York's number 7 subway line[   ] which runs through Queens[   ] one can hear more languages spoken than almost anywhere in the world.

    1. _____

2. Senator[   ] would you comment on reports that you will not run again?

    2. _____

3. Menlo Park[   ] New Jersey[   ] was Edison's home.

    3. _____

4. Seashells are an exquisite natural sculpture[   ] aren't they?

    4. _____

5. The decision to have the surgery[   ] of course[   ] should be based on several doctors' opinions.

    5. _____

6. Clarissa Denton[   ] who wrote that note to you[   ] needs a lesson in manners!

    6. _____

7. The person[   ] who wrote that note to you[   ] needs a lesson in manners!

    7. _____

8. For this production, John played Robert; Judith[   ] Harriet.

    8. _____

9. Is it true[   ] sir[   ] that you are unwilling to be interviewed by the press?

    9. _____

10. Marisa Martinez came all the way from San Antonio[   ] Texas[   ] to attend college in Cleveland.

    10. _____

parenthetical expression (**par**)   omission (**om**)
after yes and no (**y/n**)   confirmatory question (**ques**)
examples introduced by *such as*,   direct address (**add**)
   especially, or *particularly* (**examp**)   date (**date**)
contrast (**cont**)   state or country (**s/c**)
nonrestrictive clause (**nr**)

---

11. Frank graduated from the University of Michigan; Esther[   ] from Columbia University.   11. _____

12. Students[   ] who work their way through college[   ] learn to value their college training.   12. _____

13. She said, "No[   ] I absolutely refuse to answer your question."   13. _____

14. Latin America has many types of terrain[   ] such as lowlands, rain forests, vast plains, high
plateaus, and fertile valleys.   14. _____

15. On September 11[ ] 2001[ ] people throughout the world were horrified by what they saw on
television.   15. _____

16. The film had been advertised as a children's picture[   ] not a production full of violence.   16. _____

17. We were fortunate[   ] nevertheless[   ] to have recovered all of our luggage.   17. _____

18. The average person in the Middle Ages never owned a book[   ] or even saw one.   18. _____

19. You will join us at the art museum[   ] won't you?   19. _____

20. I've already told you[   ] little boy[   ] that I'm not giving back your ball.   20. _____

21. Ralph Ellison[   ] who wrote *Invisible Man*[   ] is also well known for his essays, interviews,
and speeches.   21. _____

22. Not everyone[   ] who objected to the new ruling[   ] signed the petition.   22. _____

23. It was[   ] on the other hand[   ] an opportunity that he could not turn down.   23. _____

24. William Clinton[   ] who was our forty-second President[   ] was only the third to face
impeachment hearings.   24. _____

25. She enjoys several hobbies[   ] especially collecting coins and writing verse.   25. _____

44. PUNCTUATION: The Comma

(Study 30–32, The Comma)

Part 1

If **no comma** is needed in the bracketed space(s), leave the blank empty. If **one or more commas** are needed, write in the **reason** from the list below (only one reason per blank; use the abbreviations in parentheses).

independent clauses joined by conjunction (**ind**)	appositive (**app**)
introductory clause or phrase(s) (**intro**)	absolute phrase (**abs**)
series (**ser**)	coordinate adjectives (**adj**)
contrast (**cont**)	mild interjection (**inter**)
	direct quotation (**quot**)

Examples: The New England states include Vermont[ ] Maine[ ] and New Hampshire.   _____ser_____

The Secretary of State[ ] held a press conference.   _____

1. In India one can be treated for mental illness in a hospital[ ] or at a healing temple.   1._____

2. Oh[ ] don't worry about the traffic at this time of day.   2._____

3. Ernest Hemingway[ ] a distinctive stylist[ ] endured countless parodies of his writing.   3._____

4. Confused by the jumble of direction signs at the intersection[ ] Lomanto pulled into a gas station to ask for help.   4._____

5. The concert having ended[ ] the fans rushed toward the stage.   5._____

6. He hoped to write short stories[ ] publish his poems[ ] and plan a novel.   6._____

7. If the fog continues[ ] we'll have to postpone our trip.   7._____

8. Many people had tried to reach the top of the mountain[ ] yet only a few had succeeded.   8._____

9. Equipped with only an inexpensive camera[ ] she succeeded in taking a prize-winning picture.   9._____

10. During times of emotional distress and heightened tensions[ ] Lee remains calm.   10._____

11. To prepare for her finals[ ] Cathy studied in the library all week.   11._____

12. Recognizing that his position was hopeless[ ] James resigned.   12._____

13. Airbags in cars have saved many lives during crashes[ ] but they can be dangerous for children under twelve.   13._____

14. Mr. Novak found himself surrounded by noisy[ ] exuberant students.   14._____

15. "We are[ ]" she said[ ] "prepared to serve meals to a group of considerable size."   15._____

16. The study found that the experimental medication did not significantly reduce blood pressure[ ] nor did it lower patients' heart rates.   16._____

17. To improve a child's diet[ ] add more beans and green vegetables to the meal.   17._____

18. Although Derek was an excellent driver[ ] he still had difficulty finding a sponsor for the race.   18._____

19. "You must be more quiet[ ] or the landlord will make us move," she said.   19._____

20. Dave Smithers[   ] the sophomore class president[   ] campaigned for an increase in campus activities.

21. I could not decide whether to attend college[   ] or to travel to Nigeria with my aunt.

22. Built on a high cliff[   ] the house afforded a panoramic view of the valley below.

23. Our phone constantly ringing[   ] we decided to rely on the answering machine to avoid interruptions during supper.

24. The professor raised his voice to a low roar[   ] the class having apparently dozed off.

25. Her courses included Russian[   ] organic chemistry[   ] and marine biology.

20. _____

21. _____

22. _____

23. _____

24. _____

25. _____

Part 2

Write **C** if the punctuation in brackets is **correct**.
Write **X** if it is **incorrect**.
(Use only one letter for each answer.)

Example: The New England states include Vermont[,] Maine[,] and New Hampshire.   _____C_____

1. All the art classes visited the Museum of Modern Art[,] when it held its long-awaited Matisse retrospective.

2. We traveled to Idaho[,] and went down the Snake River.

3. "The records show," the clerk declared[,] "that there is a balance due of $38.76."

4. As they trudged deeper into the woods[,] they recalled legends of the Great Brown Bear and began to look around warily.

5. You expect to graduate in June[,] don't you?

6. O'Connor started the second half at linebacker[,] Bryant having torn his knee ligaments.

7. O'Connor started the second half at linebacker[,] Bryant had torn his knee ligaments.

8. Trying to concentrate[,] Susan closed the door and turned off the television set.

9. "My fellow Americans[,] I look forward to the opportunity to serve this country," he said.

10. The newly elected President, on the eve of his inauguration, declared[,] "Saving Social Security and Medicare will receive priority in my administration."

11. Helen, who especially enjoys baseball, sat in the front row[,] and watched the game closely.

12. "Are you going to a fire?"[,] the police officer asked the speeding motorist.

13. Two of the students left the office[,] the third waited to see the dean.

14. Angela and two of her friends[,] recently performed at the student talent show.

15. "I won't wait any longer," she said[,] picking up her books from the bench.

16. His tough[,] angry attitude was only a way to prevent others from knowing how scared he was about failing.

17. The relatively short drought[,] nonetheless[,] had still caused much damage to the crops.

18. The apartment they rented[,] had no screens or storm windows.

19. According to the polls, the candidate was losing[,] he blamed the media for the results.

20. The challenger[,] said the incumbent[,] was a tax evader. [The incumbent was making a statement about the challenger.]

1. _____

2. _____

3. _____

4. _____

5. _____

6. _____

7. _____

8. _____

9. _____

10. _____

11. _____

12. _____

13. _____

14. _____

15. _____

16. _____

17. _____

18. _____

19. _____

20. _____

21. The challenger[,] said the incumbent[,] was a tax evader. [The challenger was making a statement about the incumbent.]   21. _____

22. "Did you know," the financial aid officer replied[,] "that each year thousands of scholarships go unclaimed?"   22. _____

23. Her English professor[,] who was having difficulty getting to class on time[,] requested that the class move to a different building.   23. _____

24. F. Scott Fitzgerald[,] the author of *The Great Gatsby*[,] grew up in Minnesota.   24. _____

25. Next summer she hopes to fulfill a lifelong wish[,] to travel to Alaska by ship.   25. _____

45. PUNCTUATION: The Comma

(Study 30–32, The Comma)

Part 1

Write **C** if the punctuation in brackets is **correct**.
Write **X** if it is **incorrect**.
(Use only one letter for each blank.)

Example: Since they had no further business there[,] they left.   ___C___

1. The campaign hit a new low when the candidates began accusing each other of embezzlement[,] tax fraud[,] and even marital infidelity.   1. _____

2. Piloting a barge towboat on the Mississippi[,] requires not just skill but a thorough knowledge of the river.   2. _____

3. In the haste of the evacuation[,] civilian personnel had to leave most of their valuables behind.   3. _____

4. Having turned on her word processor[,] Colleen began her great American novel.   4. _____

5. Haven't you any idea[,] of the responsibility involved in running a household?   5. _____

6. First-graders now engage in writing journals[,] in problem-solving activities[,] and in brief science experiments.   6. _____

7. Shaking hands with his patient, the physician asked[,] "Now what kind of surgery are we doing today?"   7. _____

8. Peter's goal was to make a short film in graduate school[,] and not worry about a future career.   8. _____

9. Erron and Nakita determined to find a less painful[,] but effective diet.   9. _____

10. The American cowboys' hats actually had many purposes besides shielding their faces from the sun and rain[,] for many cowboys used their hats as pillows and drinking cups.   10. _____

11. During conversations about controversial topics[,] our faces often communicate our thoughts, especially our emotional responses.   11. _____

12. Harry Rosen[,] a skilled, polished speaker[,] effectively used humor during his speeches.   12. _____

13. To understand how living arrangements affect student relationships[,] the psychology department completed several informal observational studies on campus.   13. _____

14. Many music lovers insist[,] that the now-obsolete vinyl LP record produces better sound quality than the currently popular CD.   14. _____

15. The states with the greatest numbers of dairy cows are Wisconsin[,] and California.   15. _____

16. Young Soo's mother was preparing *kimchi*[,] a pickled cabbage dish that is commonly eaten with Korean meals.   16. _____

17. Having friends must be an important aspect of our culture[,] for many popular television series focus on how a group of characters care for their friendships with one another.   17. _____

18. People beginning an intimate relationship use a significant number of affectionate expressions[,] but the frequency of these expressions drops as the relationship matures.   18. _____

19. Working hard to pay the mortgage, to educate their children, and to save money for retirement[,] many of America's middle class now call themselves the "new poor." 19. _____

20. The children could take martial arts classes near home[,] or they could decide to save their money for summer camp. 20. _____

21. Now only 68 percent of American children live with both biological parents[,] 20 percent of children live in single-parent families[,] and 9 percent live with one biological parent and a stepparent. 21. _____

22. Jeff was hungry for a gooey[,] chocolate brownie smothered in whipped cream and chocolate sauce. 22. _____

23. His thoughts dominated by grief[,] Jack decided to postpone his vacation for another month. 23. _____

24. "Oh[,] I forgot to bring my report home to finish it tonight," sighed Mary. 24. _____

25. People exercise because it makes them feel good[,] they may even become addicted to exercise. 25. _____

Part 2

In each sentence the brackets show where a comma may or may not be needed. In the blank, write the **number of commas** needed. If none, write **0**.

Example: Lucy ordered a hamburger[ ] a salad[ ] and a soda[ ] with plenty of ice. __2__

1. In addition to your college application form[ ] you need[ ] an official high school transcript[ ] three letters of recommendation[ ] and a check for the fee. 1. _____

2. Although it is not required[ ] by state law[ ] the presence of a lifeguard would have prevented[ ] the nearly fatal accident. 2. _____

3. According to Robert Darnton's research[ ] the story "Little Red Riding Hood[ ]" may reveal[ ] some information[ ] about the anxieties and issues of eighteenth-century French peasants. 3. _____

4. My fellow Americans[ ] can we allow other nations[ ] who pose as friends[ ] to threaten our security? 4. _____

5. I wanted[ ] to go[ ] to Harvard; Terry[ ] to Yale. 5. _____

6. I[ ] didn't realize[ ] that four Latin American writers[ ] have won the Nobel Prize for Literature. 6. _____

7. Unlike the Maya[ ] and Aztecs[ ] the Incas had no written language[ ] but instead[ ] they kept records on knotted strings called *quipus*. 7. _____

8. The chairman[ ] who had already served two terms in Congress[ ] and one in the State Assembly[ ] declared his candidacy again. 8. _____

9. Jack was born on December 1[ ] 1990[ ] in Fargo[ ] North Dakota[ ] during a blizzard. 9. _____

10. I consider him[ ] to be[ ] a hard-working student, but[ ] I may be wrong. 10. _____

11. Audrey Starke[ ] a woman[ ] whom I met last summer[ ] is here[ ] to see me. 11. _____

12. Having an interest[ ] in anthropology[ ] she frequently audited[ ] Dr. Irwin's class[ ] that met on Saturdays. 12. _____

13. Native Americans were the first to grow corn[ ] potatoes[ ] squash[ ] pumpkins[ ] and avocados. 13. _____

14. Well[ ] I dislike her intensely[ ] but[ ] she is quite clever[ ] to be sure. 14. _____

15. To solve[ ] her legal problems[ ] she consulted an attorney[ ] that she knew[ ] from college.

15. _____

16. "To what[ ]" he asked[ ] "do you attribute[ ] your great popularity[ ] with the students?"

16. _____

17. From Native Americans[ ] the world learned about cinnamon[ ] and chocolate[ ] and chicle[ ] the main ingredient in chewing gum.

17. _____

18. "Blowin' in the Wind[ ]" a folk song written by Bob Dylan[ ] in 1962[ ] promises[ ] that life will get better through time.

18. _____

19. Many filmmakers are creating[ ] serious movies[ ] about their cultural heritage; however, there are[ ] few commercially successful movies about Asian American cultures.

19. _____

20. "You haven't seen my glasses[ ] have you?" Granny asked[ ] the twins[ ] thinking they had hidden them[ ] somewhere in the living room.

20. _____

21. The car having broken down[ ] because of a dirty carburetor[ ] we missed the first act[ ] in which[ ] Hamlet confronts his father's ghost.

21. _____

22. After she had paid her tuition[ ] she checked in at the residence hall[ ] that she had selected[ ] and soon began[ ] unloading her suitcases and boxes.

22. _____

23. The space launch went so punctually[ ] and smoothly[ ] that the astronauts began their voyage[ ] relaxed[ ] and confident.

23. _____

24. Chinese porcelain[ ] which is prized for its beauty[ ] and its translucence[ ] was copied[ ] by seventeenth-century Dutch potters.

24. _____

25. The road to Brattleboro[ ] being coated with ice[ ] we proceeded[ ] slowly[ ] and cautiously.

25. _____

46. PUNCTUATION: The Comma

(Study 30–32, The Comma)

Part 1

Either **insert** or **cross out a comma** to make the sentence correct. In the blank, write the word that comes just **before** the inserted or crossed-out comma.

Examples: When the soldiers looked around, the stranger had vanished. ___around___
The cloud-hidden sunX gave us no clue as to which way was south. ___sun___

1. They could have followed the state highway to the right but they chose the local road to the left. 1._____

2. The 1990s will be remembered by most Americans, as a decade of rising prosperity. 2._____

3. Having examined and reexamined the ancient manuscript the committee of scholars declared it genuine. 3._____

4. If the weather is pleasant and dry, we will march in the St. Patrick's Day parade, and then dance at a parish party. 4._____

5. Amanda has decided to write a cookbook, remodel her kitchen and travel through California. 5._____

6. Many Americans now prefer news sources, that offer human interest stories. 6._____

7. The country, that receives the most media attention often is the recipient of the most aid from the United Nations. 7._____

8. Coaching soccer, and teaching part-time at a local college keep me quite busy. 8._____

9. George and Robert thoroughly and painstakingly considered, what had to be done to defuse the bomb. 9._____

10. If ever there was the law on one side, and simple justice on the other, here is such a situation. 10._____

11. Ann Tyler, who won a Pulitzer Prize in 1988 has recently written a novel about a woman in her forties who runs away from her family. 11._____

12. Hillary Clinton, will be remembered as the first First Lady to gain an elective office of her own. 12._____

13. Claiming that he was just offering good advice Ace frequently would tell me which card to play. 13._____

14. What gave Helen the inspiration for her short story, was her mother's account of growing up on a farm. 14._____

15. Owen's baseball cards included such famous examples as Willie Mays's running catch in the 1954 World Series, and Hank Aaron's record-breaking home run. 15._____

16. The volume that was the most valuable in the library's rare book collection, was a First Folio edition of Shakespeare's plays. 16._____

17. *Gone with the Wind*, a film enjoyed by millions of people throughout the world was first thought unlikely to be a commercial success. 17._____

18. Because the material was difficult to understand Monica decided to hire a tutor. 18._____

19. The study asserted that parents in sending their infant children to day care, may be slowing their youngsters' mental development.

19. _____

20. Although everyone was ready for the test no one complained when Professor Smith canceled it.

20. _____

Part 2

Write an original sentence that contains an example of the comma used as stated in the brackets. **Circle** the comma(s) so used.

_____

Example: [Setting off a parenthetical expression] <u>This course, it seems to me, requires too much work.</u>

(Collaborative option: Students work in pairs or small groups to suggest and comment on different examples.)

1. [between two independent clauses] _____

_____

_____

2. [with an introductory adverb clause] _____

_____

_____

3. [with coordinate adjectives] _____

_____

_____

4. [with a long introductory prepositional phrase or a series of introductory prepositional phrases or an introductory verbal phrase] _____

_____

_____

5. [with an absolute phrase] _____

_____

_____

6. [with a parenthetical expression] _____

_____

_____

7. [with an expression of contrast] _____

    _____

    _____

8. [with a date or address] _____

    _____

    _____

9. [with a nonrestrictive clause or phrase] _____

    _____

    _____

10. [with a direct quotation] _____

    _____

    _____

11. [to prevent misreading] _____

    _____

    _____

12. [in direct address] _____

    _____

    _____

13. [Write a sentence with a *restrictive* clause—one that does **not** use commas.] _____

    _____

    _____

47. PUNCTUATION: The Period, Question Mark, and Exclamation Point

(Study 33–34, The Period; 35–36, The Question Mark; and 37–38, The Exclamation Point)

Write **C** if the punctuation in brackets is **correct**.
Write **X** if it is **incorrect**.

Example: Is there any word from the Awards Committee yet[?]   _____C_____

1. The judge would never hold me in contempt, would she[?]   1. _____
2. "Move back; the fire's advancing!" shouted the forest ranger[!]   2. _____
3. The police officer calmly inquired whether I had the slightest notion of just how fast I was backing up[?]   3. _____
4. Mr. Hall and Miss[.] James will chair the committee.   4. _____
5. The chem[.] test promises to be challenging.   5. _____
6. Where is the office? Down the hall on the left[.]   6. _____
7. Good afternoon, ma'am[.] May I present you with a free scrub brush?   7. _____
8. "How much did the owners spend on players' salaries?" the reporter asked[?]   8. _____
9. His next question—wouldn't you know[?]—was, "What do you need, ma'am?"   9. _____
10. "Wow! Does your computer have a video camera too[!]"   10. _____
11. "What a magnificent view you have of the mountains[!]" said he.   11. _____
12. Who said, "If at first you don't succeed, try, try again" [?]   12. _____
13. Would you please check my computer for viruses[?]   13. _____
14. HELP WANTED: Editor[.] for our new brochure.   14. _____
15. Pat, please type this memo[.] to the purchasing department.   15. _____
16. What? You lent that scoundrel Snively $10,000[?!]   16. _____
17. I asked her why, of all the men on campus, she had chosen him[?]   17. _____
18. Why did I do it? Because I respected her[.] Jackie worked hard to finish her degree.   18. _____
19. "Footloose and Fancy Free[.]" [title of an essay]   19. _____
20. Would you please send me your reply by e-mail[.]   20. _____
21. "The Lakers win[!!]" the announcer screamed as O'Neal's jump shot slipped through the net at the buzzer.   21. _____
22. Charlie was an inspiring [(?)] date. He burped all through dinner.   22. _____
23. My supervisor asked how much equipment I would need to update the computer center[.]   23. _____
24. The essay was "Computers: Can We Live Without Them[?]"   24. _____
25. I heard the news on station W[.]I[.]N[.]K.   25. _____
26. The postmark on the package read "Springfield, MA[.] 01102."   26. _____
27. The monarch who followed King George VI[.] was Queen Elizabeth II.   27. _____

28. According to Ramsey, "The election drew a light turnout[.] . . . Predictably, the Socialist Party won."   28. _____

29. You lost your wallet again[?] I don't believe it.   29. _____

30. The duke was born in 1576[(?)] and died in 1642.   30. _____

31. What[!?] What did you just call me?   31. _____

32. Do you know when I may expect my refund[?]   32. _____

33. Could I have committed the crime? Never[.] I was on a business trip to St. Louis at the time.   33. _____

48. PUNCTUATION: The Semicolon

(Study 39, The Semicolon)

Part 1

Write the **reason** for the semicolon in each sentence (use the abbreviations in parentheses). Use only one reason for each sentence.

between clauses lacking a coordinating conjunction (**no conj**)
between clauses joined by a conjunctive adverb (**conj adv**)
between clauses having commas within them (**cl w com**)
in a series having commas within the items (**ser w com**)

Example: It was a glorious day for the North; it was a sad one for the South.   <u>no conj</u>

1. Pressure-treated wood has been popular for decks because it resists the elements for years; however, its arsenic content has made it environmentally undesirable.   1._____

2. The farmers are using an improved fertilizer; thus their crop yields have increased.   2._____

3. Still to come were Perry, a trained squirrel; Arnold, an acrobat; and Mavis, a magician.   3._____

4. "Negotiations," he said, "have collapsed; we will strike at noon."   4._____

5. Tacoma's Museum of Glass is one of only two museums in the country dedicated to exhibiting glass; the other is the Corning Museum of Glass in upstate New York.   5._____

6. The average Internet user spends about six hours a week online; the majority of these users reach the Internet from work.   6._____

7. Pam, who lives in the suburbs, drives her car to work each day; yet Ruben, her next-door neighbor, takes the bus.   7._____

8. Changing your time-management habits requires determination; therefore, begin by writing down your goals.   8._____

9. The play was performed in Altoona, Pennsylvania; Buckhannon, West Virginia; and The Woodlands, Texas.   9._____

10. Flight 330 stops at Little Rock, Dallas, and Albuquerque; but Flight 440, the all-coach special, is an express to Phoenix.   10._____

Part 2

If a **semicolon is needed** in the brackets, write the **reason** in the blank, as you did in part 1 (**no conj, conj adv, cl w com, ser w com**). If **no semicolon** is needed, leave the blank empty.

Examples: He would not help her get the job[ ] moreover, he could not.   <u>conj adv</u>
After the rap concert[ ] we drove to Salty's.   _____

1. The Puritans banned the Christmas holiday when they settled in North America[ ] the holiday was not revived until the 1880s.   1._____

2. Shall I telephone to find out the time[ ] when the box office opens?   2._____

between clauses lacking a coordinating conjunction (**no conj**)
between clauses joined by a conjunctive adverb (**conj adv**)
between clauses having commas within them (**cl w com**)
in a series having commas within the items (**ser w com**)

3. A recent study indicates that saccharin does not cause cancer in humans[   ] the only consumers who should worry are laboratory rats.

3. _____

4. The lake suffered from a buildup of stream-borne silt[ ] until it became so shallow that it had to be dredged.

4. _____

5. The lake suffered from a buildup of stream-borne silt[ ] it became so shallow that it had to be dredged.

5. _____

6. The surprises in the team's starting lineup were Garcia, the second baseman[   ] Hudler, the shortstop[   ] and Fitzgerald, the catcher.

6. _____

7. The national public education system needs to redefine its expectations[   ] because most schools do not expect all of their students to succeed.

7. _____

8. Hollywood has always portrayed the Union soldiers as dressed in blue and the Confederate troops in gray[   ] however, for the first year of the Civil War, most soldiers wore their state militia uniforms, which came in many colors.

8. _____

9. Soft drinks are a traditional beverage in the United States[   ] flavored soda water first appeared in 1825 in Philadelphia.

9. _____

10. Orville and Wilbur Wright ran a bicycle shop in Dayton, had no scientific training, and never finished high school[   ] yet, by inventing the airplane, they revolutionized transportation worldwide.

10. _____

11. She is very bright[ ] at twenty, she is the owner of a successful small business.

11. _____

12. John uses a video conferencing network to conduct business[   ] instead of spending time flying all over the world for meetings.

12. _____

13. Exercising is quite beneficial[   ] because it helps to reduce physical and psychological stress.

13. _____

14. Our representatives included Will Leeds, a member of the Rotary Club[   ] Augusta Allcott, a banker[   ] and Bill Rogers, president of the Chamber of Commerce.

14. _____

15. Peter lives in Minnesota[   ] Howard, in New York.

15. _____

49. PUNCTUATION: The Semicolon and the Comma

(Study 30–32, The Comma, and 39, The Semicolon)

Write **com** if you would insert a **comma** (or commas) in the brackets.
Write **semi** if you would insert a **semicolon** (or semicolons).
If you would insert nothing, leave the blank empty.
Write only one answer for each blank.

Example: The milk had all gone sour[ ] we could not have our cappucino. ____semi____

1. The flood waters rose steadily throughout the night[ ] by dawn our kitchen was flooded to the countertops. 1._____

2. Many Americans have financial plans for retirement[ ] but stock-market turmoil has made them rethink those plans. 2._____

3. Dr. Jones[ ] who teaches geology[ ] graduated from MIT. 3._____

4. The Dr. Jones[ ] who teaches geology[ ] graduated from MIT. 4._____

5. I met the woman[ ] who is to be president of the new junior college. 5._____

6. She likes working in Washington, D.C.[ ] she hopes to remain there permanently. 6._____

7. To the east we could see the White Mountains[ ] to the west, the Green. 7._____

8. Read the article carefully[ ] then write an essay on the author's handling of the subject. 8._____

9. One of my grandmother's most prized possessions is an antique glass bowl[ ] that was made in Murano, Italy. 9._____

10. The game being beyond our reach[ ] the coach told me to start warming up. 10._____

11. We're going on a cruise around the bay on Sunday[ ] and we'd like you to come with us. 11._____

12. If Amy decides to become a lawyer[ ] you can be sure she'll be a good one. 12._____

13. Customer satisfaction is important[ ] the owners, therefore, hired a consulting firm to conduct a customer survey. 13._____

14. Li-Young registered for an advanced biology course[ ] otherwise, she might not have been admitted to medical school. 14._____

15. The newest computers[ ] moreover[ ] are cheaper than last year's less powerful models. 15._____

16. Cell phones are rapidly gaining in popularity[ ] but these phones do not work from some remote areas. 16._____

17. He began his speech again[ ] fire engines having drowned out his opening remarks. 17._____

18. The best day of the vacation occurred[ ] when we took the children sledding. 18._____

19. Let me introduce the new officers: Phillip Whitaker, president[ ] Elaine Donatelli, secretary[ ] and Pierre Northrup, treasurer. 19._____

20. We thought of every possible detail when planning the dinner party[ ] yet we didn't anticipate our cat's jumping into the cake. 20._____

21. We have known the Floyd Archers[ ] ever since they moved here from New Jersey. 21._____

22. The actor Hal Holbrook has successfully portrayed Mark Twain[ ] everywhere in the country for more than forty years. 22._____

23. The drama coach was a serene person[ ] not one to be worried about nervous amateurs. 23._____

24. To turn them into professional performers was[ ] needless to say[ ] an impossible task. 24. _____

25. "Yes, I will attend the review session," Jack said[ ] "if you can guarantee that the time spent will be worthwhile." 25. _____

26. Call the security office[ ] if there seems to be any problem with the locks. 26. _____

27. Couples with severe disabilities may have difficulty raising a family[ ] there are few programs to help disabled parents with their children. 27. _____

28. Britain was the first Common Market country to react[ ] others quickly followed suit. 28. _____

29. The American troops stormed ashore at Omaha and Utah beaches[ ] the British, at Sword, Gold, and Juno. 29. _____

30. Perhaps because the weather was finally warm again[ ] I didn't want to stay inside. 30. _____

31. American couples are examining their lifestyles[ ] many are cutting back in their work schedules to spend more time with their children. 31. _____

32. The World Series hadn't yet begun[ ] however, he had equipped himself with a new radio. 32. _____

33. I could not remember ever having seen her as radiantly happy[ ] as she now was. 33. _____

34. No, I cannot go to the game[ ] I have a term paper to finish. 34. _____

35. Kristi Yamaguchi[ ] in fact, is a fourth-generation Japanese American. 35. _____

36. Victor, on the other hand[ ] played the best game of his career. 36. _____

37. Home-grown products are common in rural farming communities[ ] on the other hand, such products can command high prices in urban areas. 37. _____

38. "There will be no rain today[ ]" she insisted. "The weather forecaster says so." 38. _____

39. Swimming is an excellent form of exercise[ ] swimming for twenty-six minutes consumes 100 calories. 39. _____

40. Though the American flag had only forty-eight stars in 1944[ ] the war movie mistakenly showed a fifty-star flag. 40. _____

41. The short story[ ] that impressed me the most[ ] was written by a thirty-five-year-old police officer. 41. _____

42. Mary constantly counts calories and fat content in the food she eats[ ] yet she never loses more than a pound. 42. _____

43. Many cultures follow different calendars[ ] for example, the Jewish New Year is celebrated in the fall, the Vietnamese and Chinese New Year at the beginning of the year, and the Cambodian New Year in April. 43. _____

44. "My fraternity[ ]" stated Travis, "completes numerous community service projects throughout the school year." 44. _____

45. All the students were present for the final, but[ ] most were suffering from the flu. 45. _____

46. Muslim students on campus asked the administration for a larger international student center[ ] and a quiet place for their daily prayers. 46. _____

47. Whenever Sam is feeling sad and discouraged about his job[ ] he puts on a Tony Bennett record and dances with the dog. 47. _____

48. Barry and I were planning a large farewell party for Eugene within the next month[ ] but certainly not next week. 48. _____

49. To read only mysteries and novels[ ] was my plan for the holiday break. 49. _____

50. Most Amish reside in Pennsylvania[ ] however, there are settlements also in Ohio and upstate New York. 50. _____

50. PUNCTUATION: The Semicolon and the Comma

(Study 30–32, The Comma, and 39, The Semicolon)

Write an **original sentence** illustrating the use of the semicolon or comma stated in brackets.

Example: [two independent clauses with no coordinating conjunction between them]
  Five students scored A on the exam; four scored D.

(Collaborative option: Students work in pairs or small groups to suggest and comment on different examples.)

1. [two independent clauses with *furthermore* between them] _____

   _____

   _____

2. [two independent clauses joined by *and*, with commas within the clauses]_____

   _____

   _____

3. [two independent clauses joined by *yet*] _____

   _____

   _____

4. [three items in a series, with commas within each of the items] _____

   _____

   _____

5. [two independent clauses with *then* between them] _____

   _____

   _____

6. [two independent clauses with no word between them] _____

   _____

   _____

7. [an introductory adverb clause] _____

   _____

   _____

8. [a nonrestrictive clause] _____

    _____

    _____

9. [two independent clauses with *in fact* or *also* between them]_____

    _____

    _____

10. [two independent clauses with *however* inside the second clause (not between the clauses)] _____

    _____

    _____

51. PUNCTUATION: The Apostrophe

(Study 40–42, The Apostrophe)

In the first blank, write the number of the **correct** choice (**1** or **2**). In the second blank, write the **reason** for your choice (use the abbreviations in parentheses; if your choice for the first blank has no apostrophe, leave the second blank empty).

singular possessive (**sing pos**)   contraction (**cont**)
plural possessive (**pl pos**)   plural of letter or symbol used as a word (**let/sym**)

_____

Examples: The fault was (1) **Jacob's** (2) **Jacobs'**.   <u> 1 </u> sing pos
           The fault was (1) **your's** (2) **yours**.   <u> 2 </u> _____

1. It (1)**wasn't** (2)**was'nt** the weather that caused the delay; it was an electrical failure.   1. ____ _____

2. The (1)**Smith's** (2)**Smiths** have planned a murder-mystery party.   2. ____ _____

3. The (1)**James'** (2)**Jameses** are moving to Seattle.   3. ____ _____

4. My (1)**brother-in-law's** (2)**brother's-in-law** medical practice is flourishing.   4. ____ _____

5. The (1)**Novotny's** (2)**Novotnys'** new home is spacious.   5. ____ _____

6. (1)**Its** (2)**It's** important to exercise several times a week.   6. ____ _____

7. (1)**Who's** (2)**Whose** responsible for the increased production of family-oriented movies?   7. ____ _____

8. The two (1)**girl's** (2)**girls'** talent was quite evident to everyone.   8. ____ _____

9. Some economists fear that the Social Security system may be bankrupt by the (1)**2020s** (2)**2020's**.   9. ____ _____

10. It will be a two-(1)**day's** (2)**days'** drive to Galveston.   10. ____ _____

11. The dispute over the last clause caused a (1)**weeks** (2)**week's** delay in the contract signing.   11. ____ _____

12. Mary accidentally spilled tea on her (1)**bosses** (2)**boss's** report.   12. ____ _____

13. After the long absence, they fell into (1)**each others'** (2)**each other's** arms.   13. ____ _____

14. Each woman claimed that the diamond ring was (1)**her's** (2)**hers**.   14. ____ _____

15. Geraldine uses too many (1)*ands* (2)*and's* in most of her presentations.   15. ____ _____

16. Bumstead never dots his (1)**I's** (2)**Is**.   16. ____ _____

17. (1)**Wer'ent** (2)**Weren't** you surprised by the success of her book?   17. ____ _____

18. Which is safer, your van or (1)**ours** (2)**our's**?   18. ____ _____

19. Georgiana insisted, "I (1)**have'nt** (2)**haven't** seen Sandy for weeks."   19. ____ _____

20. He bought fifty (1)**cents** (2)**cents'** worth of bubblegum.   20. ____ _____

21. The back alley was known to be a (1)**thieve's** (2)**thieves'** hangout.   21. ____ _____

22. (1)**Paul's and David's** (2)**Paul and David's** senior project was praised by their advisor.   22. ____ _____

23. The (1)**children's** (2)**childrens'** kitten ate our goldfish.   23. ____ _____

24. "The (1)**evenings** (2)**evening's** been delightful," Lily said. "Thank you."   24. ____ _____

25. The local (1)**coal miner's** (2)**coal miners'** union was the subject of Bill's documentary.   25. ____ _____

52. PUNCTUATION: The Apostrophe

(Study 40–42, The Apostrophe)

For each bracketed apostrophe, write **C** if it is **correct**; write **X** if it is **incorrect**. Use the first column for the first apostrophe, the second column for the second apostrophe.

<u>C</u>   <u>X</u>

Example: Who[']s on first? Where is todays['] lineup?

1. Everyone else[']s opinion carries less weight with me than your[']s.   1. ____ ____

2. Mrs. Jackson[']s invitation to the Williams[']es must have gone astray.   2. ____ ____

3. He would[']nt know that information after only two day[']s employment.   3. ____ ____

4. Were[']nt they fortunate that the stolen car wasn't their[']s?   4. ____ ____

5. It[']s a pity that the one bad cabin would be our[']s.   5. ____ ____

6. We[']re expecting the Wagner[']s to meet us in Colorado for a ski trip.   6. ____ ____

7. Home-baked pizza[']s need an oven temperature in the upper 400[']s.   7. ____ ____

8. Does[']nt the governor see that most voters won[']t support her cuts in farm aid?   8. ____ ____

9. The two sisters had agreed that they[']d stop wearing each others['] shoes.   9. ____ ____

10. She[']s not going to accept anybody[']s advice, no matter how sound it might be.   10. ____ ____

11. The three students['] complaints about the professor[']s attitude in class were finally addressed by the administration.   11. ____ ____

12. He[']s hoping for ten hours['] work a week in the library.   12. ____ ____

13. The idea of a cultural greeting card business was not our[']s; it was Lois[']s.   13. ____ ____

14. There are three *i*[']s in the word *optimistic;* there are two *r*[']s in the word *embarrass.*   14. ____ ____

15. The computer printout consisted of a series of 1[']s and 0[']s.   15. ____ ____

16. Their advisor sent two dozen yellow rose[']s to the Women Student Association[']s meeting.   16. ____ ____

17. I really did[']nt expect to see all of the drivers['] finish the race.   17. ____ ____

18. Hav[']ent you heard about the theft at the Jone[']s house?   18. ____ ____

19. The popular mens['] store, established in 1923, was[']nt able to compete with the large discount stores in the nearby mall.   19. ____ ____

20. I'm sure that, if he[']s physically able, he[']ll be at the volunteer program.   20. ____ ____

21. The responsibility for notifying club members is her[']s, not our[']s.   21. ____ ____

22. Can[']t I persuade you that you[']re now ready to move out of the house?   22. ____ ____

23. Both lawyers['] used hard-hitting tactic[']s to explain why their company should not be required to pay damages.   23. ____ ____

24. Everyones['] agreeing that in the 1990[']s too many stocks were overvalued.   24. ____ ____

25. Marie Stockton sought her sister-in-law[']s advice when she considered opening a women[']s fitness salon.   25. ____ ____

53. PUNCTUATION: The Apostrophe

(Study 40–42, The Apostrophe)

In the paragraphs below, most words ending in **s** are followed by a small number. At the right are blanks with corresponding numbers. In each blank, write the **correct ending** for the word with that number: **'s** or **s'** or **s**.

Example: We collected our days$_{51}$ pay after cleaning the tables$_{52}$.
　　　　　　　　　　　　　　　　　　　　　　　　　　　　　51. _'s_　52. _s_

All young performers$_1$ dream of gaining recognition from their audiences$_2$ and of seeing
their names in lights$_3$ on Broadway. These were Annie Smiths$_4$ dreams when she left her
parents$_5$ home and ran off to New York City. At age eighteen, however, Annie was not prepared
for the difficulties$_6$ of living alone and working in a large city. Her wages$_7$ as a waitress barely
covered a months$_8$ rent. And she still needed to buy groceries$_9$ and pay her utilities$_{10}$. It took
Annie several months$_{11}$ time to find two suitable roommates, who would share the rent and
other bills. However, the roommates$_{12}$ also helped in other important ways, for when Annie
felt that she couldn't go for another audition, her roommates$_{13}$ encouragement to continue
helped bolster Annies$_{14}$ determination. Annie realized that for anyones$_{15}$ dream to happen, a
great deal of hard work had to come first.

One evening in the late 1990s$_{16}$, as she was clearing away the last two customers$_{17}$
dishes at Carusos$_{18}$ Restaurant, she heard a distinguished-looking woman asking the head
waiter, "Whos$_{19}$ that young lady? She moves$_{20}$ with such grace, and shes$_{21}$ got the poise
and features$_{22}$ of a movie star; I'm a film director, and I'd like to speak to her."

This storys$_{23}$ ending is a happy one, for in a years$_{24}$ time Annie became a star.
All that she had dreamed of was now hers$_{25}$.

1.___　2.___
3.___　4.___
5.___
6.___　7.___
8.___　9.___　10.___
11.___
12.___
13.___
14.___　15.___

16.___　17.___
18.___
19.___　20.___　21.___
22.___
23.___　24.___
25.___

54. PUNCTUATION: Italics

(Study 43, Italics [Underlining])

Part 1

Write the **reason** for each use of italics (use the abbreviations in parentheses):

title of printed, performed, or electronic work (**title**)
name of ship, train, plane, or spacecraft (**craft**)
title of painting or sculpture (**art**)
foreign word not Anglicized (**for**)
word, letter, symbol, or figure referred to as such (**wlsf**)
emphasis (**emph**)

Example: Does this library subscribe to *Smithsonian*?   **title**

1. The *Andrea Doria* sank after a collision with the *Stockholm*.   1. ____

2. Channel 8 seems to show nothing but reruns of *Frasier* and *Seinfeld*.   2. ____

3. For many years, the *Manchester Guardian* has been a leading newspaper in England.   3. ____

4. Norman Rockwell painted *The Four Freedoms* during World War II.   4. ____

5. The directions on the test indicated that all questions were to be answered with *1*s or *2*s.   5. ____

6. Dozens of English words connected with dining come from the French—*cuisine, à la mode,* and *hors d'oeuvres,* to name just a few.   6. ____

7. Susan learned to spell the word *villain* by thinking of a "villa in" Italy.   7. ____

8. "Are you sure you won't *ever* cheat on me?" she asked.   8. ____

9. The statue *The Women of Belfast* is on loan from the Ulster Museum.   9. ____

10. An article had been written recently about the submarine *Nautilus*.   10. ____

11. N. Scott Momaday's *1969: The Way to Rainy Mountain* recounts the Kiowa Indians' migration to the American plains.   11. ____

12. How many *s*'s and *i*'s are there in your last name?   12. ____

13. Though *Oklahoma!* was first performed more than sixty years ago, it is still a favorite of local theater groups.   13. ____

14. The American pronunciation of *vase* is *vayss* or *vaze*; the British pronunciation is *vahz*.   14. ____

15. Richard Rodriguez's autobiography, *Hunger of Memory*, helped me understand some of the issues surrounding bilingual education.   15. ____

16. American women are learning to say a strong *no* to many professional demands so that they have time for family and friends.   16. ____

17. Aboard the *Enterprise*, the captain made plans to return to the planet Zircon to rescue Mr. Spock.   17. ____

18. Michelangelo's *David* was originally mounted outdoors but was moved into a museum to protect the stone from erosion.   18. ____

19. Now that I have *Ace Anti-virus Protector* software installed on my hard drive, I have no worries about computer viruses.   19. ____

20. The first American to orbit the earth was John Glenn in *Friendship 7*.   20. ____

21. Her printed *R*'s and *B*'s closely resemble each other.   21. ____

22. Although he never held office, Lopez was the *de facto* ruler of his country.   22. ____

23. Some people spell and pronounce the words *athlete* and *athletics* as if there were an *e* after *th* in each word.   23. ____

24. The movie *A Family Thing* addresses racial issues in the United States in a thought-provoking and sensitive manner.   24. ____

25. The original meaning of the word *mad* was "disordered in mind" or "insane."   25. ____

In each sentence, **underline** the word(s) that should be in italics.

Example: The cover of <u>Newsweek</u> depicted African refugees.

1. Dave Matthews and his group performed items from their new CD, Busted Stuff.

2. Deciding to come home by ship, we made reservations on the Queen Elizabeth II.

3. Geraldine went downtown to buy copies of Esquire and Field and Stream.

4. "It's time for a change!" shouted the candidate during the debate.

5. Proof, David Auburn's drama about love, fear, genius, and madness, won both the Pulitzer Prize and a Tony Award.

6. The New York Times must have weighed ten pounds last Sunday.

7. The Mystery! series on public television promises amateur sleuths a weekly escape into murder and intrigue.

8. Browsing through recent fiction at the library, Ms. Kovalchik came across The Red Tent, by Anita Diamant.

9. Among the magazines scattered in the room was a copy of Popular Mechanics.

10. Maya Angelou's first published work, I Know Why the Caged Bird Sings, is an autobiography describing her first sixteen years.

11. When I try to pronounce the word statistics, I always stumble over it.

12. I still have difficulty remembering the difference between continual and continuous.

13. "I'll never stop fighting for my rights," Megan Morton thundered. "And I mean never."

14. Picasso's Guernica depicts the horrors of war.

15. The Thinker is a statue that many people admire.

16. Spike Lee's film Malcolm X inspired me.

17. You'll enjoy reading "The Man of the House" in the book Fifty Great Short Stories.

18. The British spelling of the word humor is h-u-m-o-u-r.

19. "How to Heckle Your Prof" was an essay in John James's How to Get Thrown Out of College.

20. Michelangelo's Last Judgment shows "the omnipotence of his artistic ability."

21. The source of the above quotation is the Encyclopaedia Britannica.

22. They were able to download the entire program Master Chess from the Internet.

23. He had been a noted braumeister in Germany.

24. Perry won the spelling bee's award for creative expression with his rendition of antidisestablishmentarianism.

25. The instructor said that Sam's 7s and his 4s look very much alike.

55. PUNCTUATION: Quotation Marks

(Study 44–48, Quotation Marks)

Insert quotation marks at the proper places in each sentence.

Example: She wrote "Best Surfing Beaches" for *Outdoor* magazine.

1. Readers were mesmerized by the article Halle Berry: A True Survivor Story in *Good Housekeeping*.

2. The young couple read the *Better Homes and Gardens* article No Need to Cook.

3. W. C. Fields's dying words were, I'd rather be in Philadelphia.

4. The poem The Swing was written by Robert Louis Stevenson.

5. Be prepared, warned the weather forecaster, for a particularly harsh winter this year.

6. Childhood Memories is a chapter in the reader *Growing Up in the South*.

7. In Kingdom of the Skies, in the magazine *Arizona Highways*, Joyce Muench described the unusual cloud formations that enhance Arizona's scenery.

8. The word *cavalier* was originally defined as a man on a horse.

9. One of the most famous American essays is Emerson's Self-Reliance.

10. One of my favorite short stories is Eudora Welty's A Worn Path.

11. The song The Wind Beneath My Wings was sung to inspire mentors to stay with the literary program.

12. The World Is Too Much with Us is a poem by William Wordsworth.

13. The New Order is an article that appeared in *Time* magazine.

14. An article that appeared in the *Washington Post* is Can We Abolish Poverty?

15. Cousins' essay The Right to Die poses the question of whether suicide is ever an acceptable response to life circumstances.

16. The Love Song of J. Alfred Prufrock is a poem by T. S. Eliot.

17. The dictionary of slang defines *loopy* as slightly crazy.

18. The concluding song of the evening was Auld Lang Syne.

19. We read a poem by Alice Walker entitled Women.

20. Today's local newspaper ran an editorial titled Save the Salmon.

21. What we have here, the burly man said, is a failure to communicate.

22. Never in the field of human combat, said Winston Churchill, has so much been owed by so many to so few.

23. She read Julio Cortazar's short story The Health of the Sick.

24. *Discography* means a comprehensive list of recordings made by a particular performer or of a particular composer's work.

25. How rude of him to say, I don't care to see you!

56. PUNCTUATION: Quotation Marks

(Study 44–48, Quotation Marks)

Write **C** if the punctuation in brackets is **correct**.
Write **X** if it is **not**.

Example: "What time is it["?] wondered Katelyn.   ___X___

1. The stadium announcer intoned[, "]Ladies and gentlemen, please rise for our national anthem."   1. _____

2. The television interviewer shoved a microphone in the mother's face and demanded, "How did you feel when you heard that your little girl had been kidnapped[?"]   2. _____

3. In the first semester we read Gabriel García Márquez's short story "Big Mama's Funeral[".]   3. _____

4. "Where are you presently employed?[",] the interviewer asked.   4. _____

5. "When you finish your rough draft," said Professor Grill[, "]send it to my e-mail address."   5. _____

6. Who was it who mused, "Where are the snows of yesteryear["?]   6. _____

7. Dr. Nelson, our anthropology teacher, asked, "How many of you have read *The Autobiography of Malcolm X* [?"]   7. _____

8. "We need more study rooms in the library[,"] declared one presidential candidate in the student government debate.   8. _____

9. "Write when you can[,"] Mother said as I left for the airport.   9. _____

10. To *dissuade* means "to persuade someone not to do something[."]   10. _____

11. "Ask not what your country can do for you[;"] ask what you can do for your country."   11. _____

12. He said, "Our language creates problems when we talk about race in America.[" "]We don't have enough terms to explain the complexities of cultural diversity."   12. _____

13. "Do you remember Father's saying, 'Never give up['?"] she asked.   13. _____

14. She began reciting the opening line of one of Elizabeth Barrett Browning's sonnets: "How do I love thee? Let me count the ways[."]   14. _____

15. Gwendolyn Brooks's poem ["]The Bean Eaters["] is one of her best.   15. _____

16. ["]*The Fantasticks*,["] which ran for more than forty years, is the longest-running musical play in American theater.   16. _____

17. "Want to play ball, Scarecrow[?"] the Wicked Witch asked, a ball of fire in her hand.   17. _____

18. "Shall I read aloud Whitman's poem 'Out of the Cradle Endlessly Rocking['?"] she asked.   18. _____

19. Have you read Adrienne Rich's poem "Necessities of Life[?"]   19. _____

20. When Susan saw the show about America's homeless, she exclaimed, "I have to find a way to help[!"]   20. _____

21. The noun *neurotic* is defined as "an emotionally unstable individual[".]   21. _____

22. "I'm going to the newsstand," he said[, "]for a copy of *Sports Illustrated*."   22. _____

23. "Do you believe in fairies[?"] Peter Pan asks the children.   23. _____

24. How maddening of her to reply calmly, "You're so right["!]   24. _____

25. "I need you in my office right away," the comptroller barked over the phone[. "]The FBI has subpoenaed our books."　　　　　　　　　　　　　　　　　　　　25. _____

26. The city's Department of Investigation used hotel rooms specially ["]salted["] with money and jewelry to bait their traps for the criminals.　　　　　　　26. _____

27. "The Lottery[,"] a short story by Shirley Jackson, was discussed in Janet's English class.　　　　　　　　　　　　　　　　27. _____

28. The reporter said[, "]Thank you for the lead on the story," and ran off to track down the source.　　　　　　　　　　28. _____

29. "Was the treaty signed in 1815[?"] the professor asked, "or in 1814?"　　　　　29. _____

30. The mayor said, "I guarantee that urban renewal will move forward rapidly[;"] however, I don't believe him.　　　　　　　　　　　　　　　　　　30. _____

31. Richard Rodriguez writes: "Only when I was able to think of myself as an American, no longer an alien in *gringo* society, could I seek the rights and opportunities necessary for full public individuality[".]　　　　　　　　　　　　　　　　　　　　31. _____

32. "Have you seen the rough draft of the article?" asked Jackie[?]　　　32. _____

33. "You blockhead[,"] screamed Lucy[!]　　　　　　　　　　33. _____

57. PUNCTUATION: Italics and Quotation Marks

(Study 43, Italics, and 44–48, Quotation Marks)

Write the number of the **correct** choice.

Example: A revival of Lerner and Lowe's show (1)*My Fair Lady*
(2)"My Fair Lady" is playing at Proctor's Theater.   _____1_____

1. The Broadway hit play (1)*Rent* (2)"Rent" was based on a Puccini opera.   1._____

2. That opera was (1)*La Boheme* (2)"La Boheme."   2._____

3. Keats's poem (1)"Ode on a Grecian Urn" (2) *Ode on a Grecian Urn* is required reading.   3._____

4. Paul Kennedy's book (1)*The Rise and Fall of the Great Powers* (2)"The Rise and Fall of the Great Powers" discusses how nations become politically and militarily dominant.   4._____

5. I just remembered the title of that article in *Prevention* magazine. It is (1)*The New Science of Eating to Get Smart* (2)"The New Science of Eating to Get Smart."   5._____

6. The closing song of the concert was (1)"R-e-s-p-e-c-t" (2)*R-e-s-p-e-c-t.*   6._____

7. (1)*A Haunted House* (2)"A Haunted House" is a short story by Virginia Woolf.   7._____

8. The brevity of Carl Sandburg's poem (1)*Fog* (2)"Fog" appealed to her.   8._____

9. Jack received (1)*A's* (2) "A's" in three of his classes this fall.   9._____

10. She used too many (1)*ands* (2)"ands" in her introductory speech.   10._____

11. (1)*Science and Religion* (2)"Science and Religion" is an essay by Albert Einstein.   11._____

12. He has purchased tickets for the opera (1)"Faust" (2)*Faust.*   12._____

13. Sharon didn't use a spell-check program and, therefore, unfortunately misspelled (1)*psychology* (2)"psychology" throughout her paper.   13._____

14. Dr. Baylor spent two classes on Wallace Stevens's poem (1)"The Idea of Order at Key West" (2)*The Idea of Order at Key West.*   14._____

15. His favorite newspaper has always been the (1)*Times* (2)"Times."   15._____

16. (1)"Our Town" (2)*Our Town* is a play by Thornton Wilder.   16._____

17. The word *altogether* means (1)"wholly" or "thoroughly." (2)*wholly* or *thoroughly.*   17._____

18. (1)*What Women Want* (2)"What Women Want" is an essay by Margaret Mead.   18._____

19. James Thurber's short story (1)*The Secret Life of Walter Mitty* (2)"The Secret Life of Walter Mitty" amused her.   19._____

20. The Players' Guild will produce Marlowe's (1)*Dr. Faustus* (2)"Dr. Faustus" next month.   20._____

21. Who do you think will ever publish your article (1)*The Joy of Fried Earthworms* (2)"The Joy of Fried Earthworms"?   21._____

22. (1)*Biology: Science of Life* (2)"Biology: Science of Life" is our very expensive textbook for biochemistry class.   22._____

23. One of the first assignments for our African American history classes was James Baldwin's book (1)*Notes of a Native Son* (2)"Notes of a Native Son."   23._____

24. Our film class saw Truffaut's (1)*Shoot the Piano Player* (2)"Shoot the Piano Player" last week.   24. _____

25. She read (1)*Dover Beach,* (2)"Dover Beach," a poem by Matthew Arnold.   25. _____

26. (1)*Pygmalion* (2)"Pygmalion" is a play by George Bernard Shaw.   26. _____

27. You fail to distinguish between the words (1)*range* and *vary.* (2)"range" and "vary."   27. _____

28. I read a poem by Yeats titled (1)"The Cat and the Moon." (2)*The Cat and the Moon.*   28. _____

29. Madeline decided to treat herself by ordering a subscription to (1)*Time* (2)"Time."   29. _____

30. (1)*Fragmented* (2)"Fragmented" is a play by my colleague Prester Pickett.   30. _____

31. I used (1)"Do Lie Detectors Lie?" (2)*Do Lie Detectors Lie?* from *Science* magazine to write my report on famous murder trials.   31. _____

32. Through Kevin Coyne's book (1)"A Day in the Night of America," (2)*A Day in the Night of America*, readers have a chance to see how 7.3 million Americans spend their time working a night shift.   32. _____

33. The last section of the textbook is titled (1)*Paragraphs and Papers.* (2)"Paragraphs and Papers."   33. _____

58. PUNCTUATION: Colon, Dash, Parentheses, and Brackets

(Study 49–50, The Colon; 51, The Dash; 52–53, Parentheses; and 54, Brackets)

Part 1

Write **C** if the colon is used **correctly**.
Write **X** if it is used **incorrectly**.

Example: This bus runs via: Swan Street, Central Avenue, and North Main. ___X___

1. The President of the International Olympic Committee stepped to the podium and declared: "Let the games begin." 1._____

2. The coach signaled the strategy: we would try a double steal on the next pitch. 2._____

3. Dear Sir:
   My five years' experience as a high school English teacher qualifies me to be the editor of your newsletter. 3._____

4. Dearest Rodney:
   My heart yearns for you so greatly that I can hardly bear the days until we're in each other's arms again. 4._____

5. The following soldiers will fall out for guard duty: Pierce, Romano, Foster, and Sanchez. 5._____

6. The carpenter's tools included: saw, hammer, square, measuring tape, and nails. 6._____

7. College students generally complain about things such as: their professors, the cafeteria food, and their roommates. 7._____

8. She began her letter to Tom with these words: "I'll love you forever!" 8._____

9. Her train reservations were for Tuesday at 3:30 p.m. 9._____

10. The dean demanded that: the coaches, the players, and the training staff meet with him immediately. 10._____

11. Tonight's winning numbers are: 169, 534, and 086. 11._____

12. She was warned that the project would require two qualities: creativity and perseverance. 12._____

13. The project has been delayed: the chairperson has been hospitalized for emergency surgery. 13._____

14. If Smith's book is titled *The World Below the Window: Poems 1937–1997*, must I include both the title and subtitle in my Works Cited list? 14._____

15. I packed my backpack with: bubble bath, a pair of novels, and some comfortable clothes. 15._____

Set off the boldfaced words by inserting the correct punctuation: **dash(es)**, **parentheses**, or **brackets**.

Example: Senator Aikin **(Dem., Maine)** voted for the proposal.

1. In my research paper I quoted Wilson as observing, "His **Fitzgerald's** last years became a remarkable mix of creative growth and physical decline." [Punctuate to show that the boldfaced expression is inserted editorially by the writer of the research paper.]

2. Holmes had deduced **who knew how?** that the man had been born on a moving train during the rainy season. [Punctuate to indicate a sharp interruption.]

3. He will be considered for **this is between you and me, of course** one of the three vice-presidencies in the firm. [Punctuate to indicate merely incidental comment.]

4. I simply told her **and I'm glad I did!** that I would never set foot in her house again. [Punctuate to indicate merely incidental comment.]

5. Campbell's work on *Juvenal* **see reference** is an excellent place to start.

6. At Yosemite National Park we watched the feeding of the bears **from a safe distance, you can be sure.** [Punctuate to achieve a dramatic effect.]

7. Her essay was entitled "The American Medical System and It's **sic** Problems."

8. The rules for using parentheses **see page 7** are not easy to understand.

9. We traveled on foot, in horse-drawn wagons, and occasionally **if we had some spare cash to offer, if the farmers felt sorry for us, or if we could render some service in exchange** atop a motorized tractor. [Punctuate to indicate that this is *not* merely incidental comment.]

10. The statement read: "Enclosed you will find one hundred dollars **$100** to cover damages."

11. David liked one kind of dessert **apple pie.**

12. **Eat, drink, and be merry** gosh, I can hardly wait for senior week.

13. The essay begins: "For more than a hundred years **from 1337 until 1453** the British and French fought a pointless war." [Punctuate to show that the boldfaced expression is inserted editorially.]

14. The concert begins at **by the way, when does the concert begin**?

15. Getting to work at eight o'clock every morning **I don't have to remind you how much I dislike getting up early** seemed almost more than I cared to undertake. [Punctuate to indicate merely incidental comment.]

16. She said, "Two of my friends **one has really serious emotional problems** need psychiatric help." [Punctuate to achieve a dramatic effect.]

17. Within the last year, I have received three **or was it four?** letters from her. [Punctuate to indicate merely incidental comment.]

18. Julius was born in 1900 **?** and came west as a young boy.

59. PUNCTUATION: The Hyphen and the Slash

(Study 55, The Hyphen, and 56, The Slash)

Write **C** if the use or omission of a hyphen or slash is **correct**.
Write **X** if it is **incorrect**.

Example: Seventy six trombones led the big parade.   ___X___

1. Emily Dickinson wrote, "Because I could not stop for **Death, / He** kindly stopped for me."   1. _____

2. "I've **n-n-never** been so **c-c-cold**," stammered Neil, stumbling to shore after the white-water raft overturned.   2. _____

3. He's a true **show-must-go-on** kind of actor.   3. _____

4. One refers to the monarch of Britain as "**His/Her** Majesty."   4. _____

5. The speaker was **well known** to everyone connected with administration.   5. _____

6. The **well-known** author was autographing his latest novel in the bookstore today.   6. _____

7. The team averaged over **fifty-thousand** spectators a game.   7. _____

8. The contractor expects to build many **five-** and **six-room** houses this year.   8. _____

9. The senator composed a **carefully-worded** statement for a press conference.   9. _____

10. I sent in my subscription to a new **bi-monthly** magazine.   10. _____

11. Sam's **brother-in-law** delighted in teasing his sister by belching at family dinners.   11. _____

12. We'll have a chance to see two top teams in action at tonight's **Bulls/Pistons** game.   12. _____

13. He made every effort to **recover** the missing gems.   13. _____

14. After the children spilled blueberry syrup on her white sofa, Letitia had to **recover** it.   14. _____

15. At **eighty-four**, Hartley still rides his motorcycle in the mountains on sunny days.   15. _____

16. Charles will run in the **hundred yard** dash next Saturday.   16. _____

17. "The children are not to have any more **c-a-n-d-y**," said Mother.   17. _____

18. After he graduated from college, he became a manager of the **student-owned** bookstore.   18. _____

19. The idea of a **thirty hour** week appealed to the workers.   19. _____

20. Baird played **semi-professional** baseball before going into the major leagues.   20. _____

21. Customers began avoiding the **hot-tempered** clerk in the shoe department.   21. _____

22. Al's main problem is that he lacks **self-confidence**.   22. _____

23. The **brand-new** vacuum cleaner made a loud squealing noise every time we turned it on.   23. _____

24. The word processing software was **brand new**.   24. _____

25. Mr. Pollard's major research interest was **seventeenth-century** French history.   25. _____

60. PUNCTUATION: Review

(Study 30–56, Punctuation)

Write **T** for each statement that is **true**.
Write **F** for each that is **false**.

Example: A **period** is used at the end of a declarative sentence.		_____T_____
1. **Three spaced periods** are used to indicate an omission (ellipsis) in quoted material.		1._____
2. **Possessive personal pronouns** contain an apostrophe.		2._____
3. The **question mark** is always placed inside closing quotation marks.		3._____
4. The sentence "Dellene searched for her friend, Mitch," means that Dellene has only one friend.		4._____
5. A **dash** is used before the author's name on the line below a direct quotation.		5._____
6. **Parentheses** are used to enclose editorial remarks in a direct quotation.		6._____
7. A **restrictive clause** is not set off within commas.		7._____
8. A **semicolon** is used to set off an absolute phrase from the rest of the sentence.		8._____
9. The use of **brackets** around the word *sic* indicates an error occurring in quoted material.		9._____
10. Mild interjections should be followed by an **exclamation point**; strong ones, by a **comma**.		10._____
11. An indirect question is followed by a **period**.		11._____
12. A **semicolon** is used after the expression *Dear Sir*.		12._____
13. The title of a magazine article should be underlined to designate the use of **italics**.		13._____
14. *Ms.* may take a **period** but *Miss* does not.		14._____
15. **Single quotation marks** are used around a quotation that is within another quotation.		15._____
16. Both *Mr. Jones'* and *Mr. Jones's* are acceptable **possessive forms** of *Mr. Jones*.		16._____
17. The title at the head of a composition should be enclosed in **double quotation marks**.		17._____
18. **No apostrophe** is needed in the following greeting: "Merry Christmas from the Palmers."		18._____
19. The **possessive** of *somebody else* is *somebody's else*.		19._____
20. The **possessive** of *mother-in-law* is *mother's-in-law*.		20._____
21. A **semicolon** is normally used between two independent clauses joined by *and* if one or both clauses contain internal commas.		21._____
22. A quotation consisting of several sentences takes **double quotation marks** at the beginning of the first sentence and at the end of the last sentence.		22._____
23. A quotation consisting of several paragraphs takes **double quotation marks** at the beginning and end of each paragraph.		23._____
24. Generally, a **foreign word** is not italicized if it can be found in a reputable American dictionary.		24._____
25. The word *the* is **italicized** in the name of a newspaper or a magazine.		25._____
26. A polite request in the form of a question is followed by a **period**.		26._____

27. **Single quotation marks** may be substituted for double quotation marks around any quoted passage.   27. _____

28. The **comma** is always placed outside quotation marks.   28. _____

29. The **colon** and **semicolon** are always placed outside quotation marks.   29. _____

30. A **comma** is always used to separate the two parts of a compound predicate.   30. _____

31. The expression *such as* is normally followed by a **comma**.   31. _____

32. The **nonsentence** is a legitimate unit of expression and may be followed by a **period**.   32. _____

33. An **exclamation point** and a **question mark** are never used together.   33. _____

34. **Parentheses** are used around words that are to be deleted from a manuscript.   34. _____

35. A **comma** is used between two independent clauses not joined by a coordinating conjunction.   35. _____

36. A **semicolon** is used after the salutation of a friendly letter.   36. _____

37. The subject of a sentence should be separated from the predicate by a **comma**.   37. _____

38. An overuse of **underlining** (italics) for emphasis should be avoided.   38. _____

39. The **contraction** of the words *have not* is written thus: *hav'ent*.   39. _____

40. Nonrestrictive clauses are always set off with **commas**.   40. _____

41. **Double quotation marks** are used around the name of a ship.   41. _____

42. A **comma** is used before the word *then* when it introduces a second independent clause.   42. _____

43. The prefix *semi* always requires a **hyphen**.   43. _____

44. **No comma** is required in the following sentence: "Where do you wish to go?" he asked.   44. _____

45. A **dash** is a legitimate substitute for all other marks of punctuation.   45. _____

46. A **slash** is used to separate two lines of poetry quoted in a running text.   46. _____

47. A **dash** is placed between words used as alternatives.   47. _____

48. Every introductory prepositional phrase is set off by a **comma**.   48. _____

49. An introductory adverbial clause is usually set off with a **comma**.   49. _____

50. A **colon** may be used instead of a **semicolon** between two independent clauses when the second clause is an explanation of the first.   50. _____

Name _____ Class _____ Date _____ Score (R____ x 1.33) _____

61. PUNCTUATION: Review

(Study 30–56, Punctuation)

Part 1

Write **C** if the punctuation in brackets is **correct**.
Write **X** if it is **incorrect**.

Example: The last question on the test [,] counted 30 points. _____X_____

1. Abner Fenwick found, to his chagrin, that Physics 101 was quite difficult[;] but, because he put in maximum effort, he earned a *B*. 1._____

2. Pritchett left the casino in despair[,] his last hundred dollars lost on a wrong call in blackjack. 2._____

3. I wondered why we couldn't get rid of the computer virus[?] 3._____

4. Dear Dr. Stanley[;] Thank you for your letter of May 10. 4._____

5. Rafael enjoyed inviting his friends[,] and preparing elaborate meals for them; however, most of his attempts were disasters. 5._____

6. When the benefits officer described the new medical insurance package, everyone asked, "How much will this new policy cost us["?] 6._____

7. I remembered the job counselor's remark: "If you send out three hundred inquiry letters in your hometown without even one response, relocate[."] 7._____

8. "Despite the recession," explained the placement counselor[,] "health care, construction, and teaching still promise an increase in employment opportunities." 8._____

9. A novella by Conrad, a short story by Lawrence, and some poems of Yeats[,] were all assigned for the last week of the semester. 9._____

10. Despite the population loss in many Great Plains towns, Fargo, North Dakota[,] is thriving. 10._____

11. Why is it that other children seem to behave better than our[']s? 11._____

12. The relief workers specifically requested food, blankets, and children['s] clothing. 12._____

13. Approximately seven million Americans visit their doctor each year[;] seeking an answer for why they feel so tired. 13._____

14. Whenever he speaks, he's inclined to use too many *and-uh*[']s between sentences. 14._____

15. The auditor requested to review[:] the medical receipts, our child-care expenses, and any deductions for home improvement. 15._____

16. The last employee to leave the office is responsible for the following[,] turning off the machines, extinguishing all lights, and locking all executives' office doors. 16._____

17. Everywhere there were crowds shouting anti[-]American slogans. 17._____

18. Private colleges and universities are concerned about dwindling enrollment[;] because their tuition costs continue to climb while requests for substantial financial aid are also increasing. 18._____

19. During the whole wretched ordeal of his doctoral exams[;] Charles remained outwardly calm. 19._____

20. More than twenty minutes were cut from the original version of the film[,] the producers told neither the director nor the writer. 20._____

167

21. The mock-epic poem "Casey at the Bat" was first published June 3, 1888[,] in the *Examiner*.   21. _____

22. We were married on June 5, 2003[,] in Lubbock, Texas.   22. _____

23. The temperature sinking fast as dusk approached[;] we decided to seek shelter for the night.   23. _____

24. By the year 2000, only about half of Americans entering the workforce were native born and of European stock[;] thus this country is truly becoming a multiracial society.   24. _____

25. My only cousin[,] who is in the U.S. Air Force[,] is stationed in the Arctic.   25. _____

26. Any U.S. Air Force officer[,] who is stationed in the Arctic[,] receives extra pay.   26. _____

27. Hey! Did you find a biology book in this classroom[?!]   27. _____

28. Charles Goodyear, the man who gave the world vulcanized rubber, personified the qualities of the classic American inventor[:] he spent nine years experimenting to find a waterproof rubber that would be resistant to extreme temperatures.   28. _____

29. To reach the Museum of Natural History you take the D train to Columbus Circle[;] then you transfer to the C train.   29. _____

30. The first well-known grocery store group was[,] the Atlantic and Pacific Tea Company, founded in 1859.   30. _____

31. Fernando jumped and squealed with delight[,] because he found a new pair of roller blades under his bed as a present from his family's Three Kings celebration.   31. _____

32. The movies[,] that I prefer to see[,] always have happy endings.   32. _____

33. At the powwow Anna and her friends entered the Fancy Shawl Dance competition[;] for they wanted to dance in their new dresses and moccasins.   33. _____

Part 2

In the following paragraphs, **insert** the correct punctuation mark(s) in each set of brackets. If no punctuation is needed, leave the brackets empty.

Example: Shawaun said[, "]Do it yourself[,"] and stormed out.

1. The published writings of F. Scott Fitzgerald[ ] range from youthful short stories[ ] such as [ ]Bernice Bobs Her Hair[ ] to novels of well[ ]to[ ]do Long Islanders[ ] American expatriates[ ] and Hollywood movie moguls. His most famous novel[ ] [The Great Gatsby][ ] set on Long Island[ ] depicts the rise and fall[ ] of a [nouveau-riche] young man[ ] who was born James Gatz[ ] in the Midwest. Gatsby[ ]s failed attempt to recapture his past has captivated Americans[ ]it also became a play and a film[ ]for nearly four generations.

2. Sojourner Truth [ ]c. 1797–1883[ ] was born a slave[ ] and was never taught to read or write[ ] but she became a noted abolitionist and campaigner for women[ ]s rights. How could such a transformation occur[ ] This remarkable woman[ ] who was born in upstate New York[ ] was originally named Isabella. After she was freed[ ] she worked as a servant in New York City[ ] during which time she heard voices from heaven. Responding to the voices[ ] she changed her name to Sojourner Truth [ ][sojourner] means [ ]one who stops briefly at a place[ ][ ] and began traveling through the North. She soon gained fame as a fiery[ ] articulate[ ] and moving public speaker. In 1878 a book was written about her[ ] [Narrative of Sojourner Truth][ ] by Olive Gilbert. What an inspiration Sojourner Truth has been to all[ ] who believe in equality[ ]

62. MECHANICS: Capitals

(Study 60–62, Capitalization)

Write **C** if the boldfaced word(s) are **correct** in use or omission of capital letters.
Write **X** if the word(s) are **incorrect**.

Example: Cajuns speak a dialect of **french**.   __X__

1. Hundreds attended the **Muslim** prayer service.   1. _____

2. The recession changed many seniors' plans to attend an expensive **College**.   2. _____

3. The **turkish** bath is closed.   3. _____

4. She wanted to become a Methodist **Minister**.   4. _____

5. When will **Congress** convene?   5. _____

6. She is a **Junior** at the University of Houston.   6. _____

7. Gregory looked forward eagerly to visiting his **Mother-in-Law.**   7. _____

8. He always disliked **Calculus**.   8. _____

9. Joe constantly reads about the **Civil War.**   9. _____

10. I made an appointment with **Professor** Allen.   10. _____

11. She met three **Professors** today.   11. _____

12. "Did you save your paper on the disk?" **she** asked.   12. _____

13. Each **Spring** I try a new sport.   13. _____

14. The deaths were reported in the *Times*.   14. _____

15. I worked in the **Southwest**.   15. _____

16. Her **Aunt Miriam** has returned.   16. _____

17. He's late for his **anthropology** class.   17. _____

18. John was **Secretary** of his class.   18. _____

19. Woods was promoted to **Major**.   19. _____

20. The bookstore has a special sale on Hewlett-Packard **Computers**.   20. _____

21. I enrolled in **english** and physics.   21. _____

22. He began his letter with "My **Dear** Mrs. Johnson."   22. _____

23. He ended it with "Yours **Truly**."   23. _____

24. We once lived in the **Northwest**.   24. _____

25. I passed German but failed **Biology**.   25. _____

26. He plans to attend **Medical School** next year.   26. _____

27. Harold believes there is life on **venus**.   27. _____

28. I asked **Mother** for some legal advice.   28. _____

29. He goes to **Roosevelt High School**.   29. _____

30. Has the **senate** in Washington elected a majority leader yet?   30. _____

31. The year that actually began the **twenty-first century** was 2001, not 2000.   31. _____

32. I listen to **wfbg** every morning.   32. _____

33. We are planning a picnic on Memorial **day**.   33. _____

34. I spent the fall break with my **Aunt**.   34. _____

35. Her favorite subject is **German**.   35. _____

36. The tourists visited **Niagara Falls**.   36. _____

37. The **President's** veto of the most recent bill has angered Congress.   37. _____

38. He enrolled in **Physics 215**.   38. _____

39. This is a Lutheran **Church**.   39. _____

40. I am writing a book; **My** editor wants the first chapter soon.   40. _____

41. This is **NOT** my idea of fun.   41. _____

42. I think **mother nature** was particularly cruel this winter.   42. _____

43. She earned a **Ph.D.** degree.   43. _____

44. The **Championship Fight** was a disappointment.   44. _____

45. It is a **Jewish** custom for men to wear skull caps during worship.   45. _____

46. His father fought in the Korean **war**.   46. _____

47. The chairperson of the **Department of History** is Dr. Mo.   47. ____

48. He said simply, **"my** name is Bond."   48. ____

49. **"Sexual Harassment: The Price of Silence"** is a chapter from my composition reader.   49. ____

50. She spent her **Thanksgiving** vacation in Iowa with her family.   50. ____

63. MECHANICS: Capitals

(Study 60–62, Capitalization)

In the first blank write the number of the **first** correct choice (**1** or **2**).
In the second blank write the number of the **second** correct choice (**3** or **4**).

Example: Wandering (1)**West** (2)**west**, Max met (3)**Milly** (4)**milly**. <u>2</u> <u>3</u>

1. Investors lost millions in accounting scandals at the Megabux (1)**Company** (2)**company**, which produces Zoomfast (3)**Cars** (4)**cars**. 1. ____ ____

2. Her (1)**Father** (2)**father** went (3)**South** (4)**south** on business. 2. ____ ____

3. The new (1)**College** (2)**college** is seeking a (3)**Dean** (4)**dean**. 3. ____ ____

4. Children are taught to begin letters with "My (1)**Dear** (2)**dear** (3)**Sir** (4)**sir**." 4. ____ ____

5. Business letters often end with "Very (1)**Truly** (2)**truly** (3)**Yours** (4)**yours**." 5. ____ ____

6. After (1)**Church** (2)**church**, we walked across the Brooklyn (3)**Bridge** (4)**bridge**. 6. ____ ____

7. The (1)**Politician** (2)**politician** declared that the protester was (3)**Un-American** (4)**un-American**. 7. ____ ____

8. The young (1)**Lieutenant** (2)**lieutenant** prayed to the (3)**Lord** (4)**lord** for courage in the battle. 8. ____ ____

9. My (1)**Cousin** (2)**cousin** now lives in the (3)**East** (4)**east**. 9. ____ ____

10. The (1)**President** (2)**president** addresses (3)**Congress** (4)**congress** tomorrow. 10. ____ ____

11. Joan Bailey, (1)**M.D.**, (2)**m.d.**, once taught (3)**Biology** (4)**biology**. 11. ____ ____

12. Dr. Mikasa, (1)**Professor** (2)**professor** of (3)**English** (4)**english**, is writing a murder mystery. 12. ____ ____

13. The (1)**Comet** (2)**comet** can be seen just below (3)**The Big Dipper** (4)**the Big Dipper**. 13. ____ ____

14. "I'm also a graduate of North Harris (1)**College** (2)**college**," (3)**She** (4)**she** added. 14. ____ ____

15. The (1)**Rabbi** (2)**rabbi** of (3)**Temple** (4)**temple** Beth Emeth is a leader in interfaith cooperation. 15. ____ ____

16. Vera disagreed with the review of "(1)**The** (2)**the** War Chronicles" in (3)*The* (4)*the New York Times*. 16. ____ ____

17. The club (1)**Treasurer** (2)**treasurer** said that the financial report was "(3)**Almost** (4)**almost** complete." 17. ____ ____

18. The (1)**Girl Scout** (2)**girl scout** leader pointed out the (3)**Milky Way** (4)**milky way** to her troop. 18. ____ ____

19. Students use the textbook *Writing (1)For (2)for Audience (3)And (4)and Purpose*. 19. ____ ____

20. Educational Support Services is in (1)**Room** (2)**room** 110 of Yost (3)**Hall** (4)**hall**. 20. ____ ____

21. At the (1)**Battle** (2)**battle** of Gettysburg, Confederate troops actually approached from (3)**North** (4)**north** of the town. 21. ____ ____

22. I think it's never (1)**O.K.** (2)**o.k.** to ignore a summons from the (3)**Police** (4)**police**. 22. ____ ____

23. The correspondent described the (1)**Pope** (2)**pope** as looking "(3)**Frail** (4)**frail** and unsteady." 23. ____ ____

24. "Maria, look up at the (1)**Moon** (2)**moon**," Guido said softly, "(3)**And** (4)**and** drink in its beauty." 24. ____ ____

25. "Maria, look up at (1)**Venus** (2)**venus**," Guido said softly. "(3)**Drink** (4)**drink** in its beauty." 25. ____ ____

64. MECHANICS: Numbers and Abbreviations

(Study 65–67, Numbers, and 68–69, Abbreviations)

Write the number of the **correct** choice.

Example: The book was (1)3 (2)**three** days overdue.   ____2____

1. (1)**135** (2)**One hundred thirty-five** votes was the official margin of victory.   1._____

2. The odometer showed that it was (1)**5½** (2)**five and one-half** miles from the campus to the beach.   2._____

3. (1)**Prof.** (2)**Professor** Hilton teaches Asian philosophy.   3._____

4. Their parents went to (1)**Fla.** (2)**Florida** for the winter.   4._____

5. Builders are still reluctant to have a (1)**thirteenth** (2)**13th** floor in any new buildings.   5._____

6. The exam will be held at noon on (1)**Fri.** (2)**Friday**.   6._____

7. The (1)**P.O.** (2)**post office** on campus always has a long line of international students mailing letters and packages to their friends and families.   7._____

8. Judd has an interview with the Sherwin Williams (1)**Co.** (2)**Company**.   8._____

9. Nicole will study in Germany, (1)**Eng.** (2)**England**, and Sweden next year.   9._____

10. Evan Booster, (1)**M.D.**, (2)**medical doctor,** is my physician.   10._____

11. Frank jumped 22 feet, (1)**3** (2)**three** inches at the Saturday meet.   11._____

12. For the laboratory, the department purchased permanent markers, legal pads, pencils, (1)**etc.** (2)**and other office supplies.**   12._____

13. For (1)**Xmas** (2)**Christmas**, the Fords planned a quiet family gathering rather than their usual ski holiday.   13._____

14. Travis needed to leave for work at exactly 8:00 (1)**a.m.** (2)**o'clock**.   14._____

15. John's stipend was (1)**$2,145** (2)**two thousand one hundred forty-five dollars**.   15._____

16. She will graduate from medical school June (1)**2**, (2)**2nd**, 2006.   16._____

17. He and his family moved to Vermont last (1)**Feb.** (2)**February**, didn't they?   17._____

18. Over (1)**900** (2)**nine hundred** students attend Roosevelt Junior High School.   18._____

19. Brad loved all of his (1)**phys. ed.** (2)**physical education** electives.   19._____

20. Next year, the convention will be held on April (1)**19**, (2)**19th**, (3)**nineteenth**, in Burlington.   20._____

21. The service included an inspiring homily by the (1)**Rev.** (2)**Reverend** Spooner.   21._____

22. The lottery prize has reached an astonishing (1)**twenty-four million dollars** (2)**$24 million**.   22._____

23. The family next door adopted a (1)**two-month-old** (2)**2-month-old** baby girl from China.   23._____

24. We had an opportunity to meet (1)**Sen.** (2)**Senator** Lester at the convention.   24._____

25. The diagram was on (1)**pg.** (2)**page** 44.   25._____

26. One of my friends will do her student teaching in (1)**TX.** (2)**Texas** this spring.   26._____

27. When we offered tickets to a baseball game for our raffle, we had (1)**one-third** (2)**1/3rd** of the employees purchase tickets.   27. _____

28. Jack's dissertation was (1)**two hundred fifty** (2)**250** pages.   28. _____

29. The plane expected from (1)**LA early this a.m.** (2)**Los Angeles early this morning** is late.   29. _____

30. The bus arrives at 10:55 a.m. and leaves at (1)**11:00** (2)**eleven** a.m.   30. _____

31. The Elks Club raised $265, the Moose $126, and the Beavers Lodge (1)**ninety dollars** (2)**$90**.   31. _____

32. Rachel's name was (1)**twenty-sixth** (2)**26th** on the list of high school graduates.   32. _____

33. Private Bailey wanted a (1)**3-day** (2)**three-day** pass to see Lorena.   33. _____

65. MECHANICS: Capitals, Numbers, and Abbreviations

(Study 60–62, Capitalization; 64–66, Numbers; and 67–68, Abbreviations)

In the first blank write the number of the **first** correct choice (**1** or **2**).
In the second blank write the number of the **second** correct choice (**3** or **4**).

Example: I have only (1)**three and one-half** (2)**3½** years until (3)**Graduation** (4)**graduation**. <u> 2 </u> <u> 4 </u>

1. After much squabbling (1)**Congress** (2)**congress** finally passed the Tax Reduction (3)**Act** (4)**act**. 1. ____ ____

2. Many of those who died when (1)*The Titanic* (2)the *Titanic* went down are buried in a (3)**Cemetery** (4)**cemetery** in Halifax, Nova Scotia. 2. ____ ____

3. My (1)**Supervisor** (2)**supervisor** said our presentation was (3)**"Insightful!"** (4)**"insightful."** 3. ____ ____

4. "I expect," he said, (1)**"To** (2)**"to** get an *A* in my (3)**Chem.** (4)**chemistry** class." 4. ____ ____

5. On June (1)**6,** (2)**6th,** 2001, she spoke at St. Paul's (3)**High School** (4)**high school.** 5. ____ ____

6. The new college (1)**President** (2)**president** greeted the (3)**Alumni** (4)**alumni** during the graduation ceremonies. 6. ____ ____

7. An (1)**American Flag** (2)**American flag** from the World Trade (3)**Center** (4)**center** was flown at the memorial service. 7. ____ ____

8. The (1)**treasurer** (2)**Treasurer** of the (3)**Junior Accountants' Club** (4)**junior accountants' club** has absconded with our dues. 8. ____ ____

9. (1)**308** (2)**Three hundred eight** students passed the test out of (3)**427** (4)**four hundred twenty-seven** who took it. 9. ____ ____

10. She likes her (1)**english** (2)**English** and (3)**science** (4)**Science** classes. 10. ____ ____

11. I soon realized that (1)**spring** (2)**Spring** means rain, rain, and more rain in northeastern (3)**Ohio** (4)**ohio.** 11. ____ ____

12. Industry in the (1)**South** (2)**south** is described in this month's (3)*Fortune* (4)*fortune* magazine. 12. ____ ____

13. Victor is going to take an (1)**english** (2)**English** course this semester instead of one in (3)**History** (4)**history.** 13. ____ ____

14. She was ecstatic; (1)**Her** (2)**her** boyfriend had just bought her a 1996 General Motors (3)**Pickup Truck** (4)**pickup truck.** 14. ____ ____

15. The new (1)**doctor** (2)**Doctor** has opened an office on Main (3)**Street** (4)**street.** 15. ____ ____

16. The (1)**korean** (2)**Korean** students have planned their (3)**3rd** (4)**third** annual International Dinner. 16. ____ ____

17. I spent (1)**New Year's Day** (2)**new year's day** with (3)**mother** (4)**Mother.** 17. ____ ____

18. Her (1)**Japanese** (2)**japanese** instructor is touring the American (3)**Midwest** (4)**midwest** over the summer. 18. ____ ____

19. I need a (1)**Psychology** (2)**psychology** book from the (3)**Library** (4)**library.** 19. ____ ____

20. The (1)**class** (2)**Class** of '75 honored the (3)**Dean of Men** (4)**dean of men.** 20. ____ ____

21. Leslie enrolled in (1)**Doctor** (2)**Dr.** Newell's history course; she is majoring in (3)**social science** (4)**Social Science.** 21. ____ ____

22. Jim moved to eastern Arizona; (1)**He** (2)**he** bought over (3)**400** (4)**four hundred** acres of land.　　22. ____ ____

23. She knows (1)**four** (2)**4** students who are going to (3)**College** (4)**college** this fall.　　23. ____ ____

24. Many (1)**hispanic** (2)**Hispanic** students have immigrated to this (3)**country** (4)**Country** because of political turmoil in their homelands.　　24. ____ ____

25. In (1)**Chapter four** (2)**chapter 4**, (3)**Chief Inspector Morse** (4)**chief inspector Morse** discovers the professor's body.　　25. ____ ____

66. SPELLING

(Study 69–73, Spelling)

Write the number of the **correctly spelled** word.

Example: A knowledge of (1)**grammar** (2)**grammer** is helpful.   ____1____

1. A large (1)**quantity** (2)**quanity** of illegal drugs was seized by customs inspectors at the border.   1._____

2. The company's lawyers said that it was (1)**alright** (2)**all right** to sign the contract.   2._____

3. No one thought that a romance would (1)**develope** (2)**develop** between those two.   3._____

4. Mrs. Smith will not (1)**acknowlege** (2)**acknowledge** whether she received the check.   4._____

5. I love to (1)**surprise** (2)**suprise** the children with small presents.   5._____

6. After three well-played quarters, the Bruins had a (1)**disasterous** (2)**disastrous** fourth quarter.   6._____

7. One of the volunteers will be (1)**ninety** (2)**ninty** (3)**nintey** years old next week.   7._____

8. The salary will depend on how (1)**competant** (2)**competent** the employee is.   8._____

9. I loved listening to Grandpa's tales about his childhood because he always (1)**exagerated** (2)**exaggerated** the details.   9._____

10. It's important to accept valid (1)**criticism** (2)**critcism** without taking the comments personally.   10._____

11. It was (1)**ridiculous** (2)**rediculous** to expect Fudgley to arrive on time.   11._____

12. (1)**Approximately** (2)**Approximatly** fifty families attended the adoption support group meeting.   12._____

13. The murder was a (1)**tradegy** (2)**tragedy** (3)**tradgedy** felt by the entire community.   13._____

14. The Statue of Liberty is a (1)**symbel** (2)**symbol** of the United States.   14._____

15. Everyone could hear the (1)**argument** (2)**arguement** between the two young lovers.   15._____

16. Tim asked several questions because he wasn't sure what the professor (1)**ment** (2)**meant** by a "term paper of reasonable length."   16._____

17. The professor was offended by the (1)**ommission** (2)**omission** of his research data.   17._____

18. Carrying a portable telephone seems a (1)**necessary** (2)**neccessary** precaution.   18._____

19. Every time I visit Aunt Nan, she likes to (1)**reminisce** (2)**reminice** about her youth.   19._____

20. Meeting with a tutor for an hour before the examination was a (1)**desperate** (2)**desparate** attempt by Tom to pass his math class.   20._____

21. Susan was excited about her (1)**nineth**- (2)**ninth**-grade graduation ceremony.   21._____

22. Sally needed a lot of (1)**repetition** (2)**repitition** in order to memorize the formulas for her next chemistry test.   22._____

23. How (1)**definite** (2)**defenite** is their decision to return to Texas?   23._____

24. The weight loss program offered a (1)**guarantee** (2)**garantee** that I would lose at least ten pounds.   24._____

25. Jake hoped his temporary job would become a (1)**permenent** (2)**permanent** position.   25._____

26. I always bring back a (1)**souvenir** (2)**suvinir** for my children when I travel on business.   26._____

27. We were glad that the (1)**auxilary** (2)**auxiliary** lights came on during the severe thunderstorm.   27. _____

28. Rodney, unfortunately, had not (1)**fulfilled** (2)**fullfilled** the requirements for graduation.   28. _____

29. In our state, students in the (1)**twelth** (2)**twelfth** grade must pass a basic skills test.   29. _____

30. This year, our five-year-old son began to question the (1)**existance** (2)**existence** of the tooth fairy.   30. _____

31. When Loretta turned (1)**forty** (2)**fourty**, her office mates filled her office with balloons and threw her a surprise party.   31. _____

32. Alex said that one of the worst aspects of life in Russia was the government's (1)**suppression** (2)**suppresion** of religious activity.   32. _____

33. Jack (1)**use to** (2)**used to** run a mile five times a week.   33. _____

34. Unfortunately, I find chocolate—any chocolate—(1)**irresistable** (2)**irresistible**.   34. _____

35. All three of my children are heading towards (1)**adolescence** (2)**adolesence**.   35. _____

36. The (1)**phychologist** (2)**psychologist** arranged a group program for procrastinators.   36. _____

37. My mother's suggestion actually seemed quite (1)**sensible** (2)**sensable**.   37. _____

38. The (1)**Sophomore** (2)**Sophmore** Class voted to sponsor a dance next month.   38. _____

39. They chose the restaurant that had a (1)**late-nite** (2)**late-night** special.   39. _____

40. The high school's star athlete was a (1)**conscientous** (2)**conscientious** student.   40. _____

41. The (1)**rythm** (2)**rhythm** of the song was perfect for our skating routine.   41. _____

42. My friend decided to (1)**persue** (2)**pursue** a degree in sociology.   42. _____

43. I don't have time for (1)**questionaires** (2)**questionnaires**.   43. _____

44. Robert's (1)**perseverance** (2)**perserverence** led to his ultimate success in the theater.   44. _____

45. She has a (1)**tendancy** (2)**tendency** to do her best work early in the day.   45. _____

46. Her services had become (1)**indispensible** (2)**indispensable** to the firm.   46. _____

47. A reception was held for students having an (1)**excellent** (2)**excellant** scholastic record.   47. _____

48. Glen hopes to add (1)**playright** (2)**playwright** (3)**playwrite** to his list of professional credits.   48. _____

49. You will find no (1)**prejudice** (2)**predjudice** in our organization.   49. _____

50. Caldwell is (1)**suppose to** (2)**supposed to** deliver the lumber sometime today.   50. _____

67. SPELLING

(Study 69–73, Spelling)

If the word is spelled **incorrectly**, write the **correct spelling** in the blank.
If the word is spelled **correctly**, leave the blank empty.

Examples: hindrance   _____
              vaccum   ___vacuum___

1. unusualy	1._____	26. sincereley	26._____
2. oppinion	2._____	27. saftey	27._____
3. criticize	3._____	28. synonim	28._____
4. familar	4._____	29. catagory	29._____
5. proceedure	5._____	30. imaginery	30._____
6. thru	6._____	31. managment	31._____
7. pursue	7._____	32. amateur	32._____
8. accross	8._____	33. reguler	33._____
9. confident	9._____	34. hygiene	34._____
10. maneuver	10._____	35. cemetery	35._____
11. relieve	11._____	36. heros	36._____
12. absense	12._____	37. bookkeeper	37._____
13. sacrefice	13._____	38. monkeys	38._____
14. mischievious	14._____	39. persistant	39._____
15. prevalent	15._____	40. curiosity	40._____
16. parallel	16._____	41. stimulent	41._____
17. noticeable	17._____	42. villian	42._____
18. disasterous	18._____	43. knowledge	43._____
19. indepindent	19._____	44. optimism	44._____
20. bussiness	20._____	45. embarass	45._____
21. acquire	21._____	46. eighth	46._____
22. truly	22._____	47. maintenence	47._____
23. government	23._____	48. father-in-laws	48._____
24. appologize	24._____	49. happyness	49._____
25. controlling	25._____	50. crisises	50._____

68. SPELLING

(Study 69–73, Spelling)

Part 1

In the blank, write the **missing letter(s)** (if any) in the word.
If no letter is missing, leave the blank empty.

Examples: gramm_*a*_r
ath____lete

1. suppr____ssion
2. piano____s
3. kni____s [sharp instruments]
4. bus____ly
5. defin____te
6. permiss____ble
7. perm____nent
8. guid____nce
9. d____scription
10. fascinat____ing
11. gu____rantee
12. abs____nce
13. appar____nt
14. hindr____nce
15. crit____cism
16. benefit____ed
17. confer____ed
18. am____teur
19. argu____ment
20. me____nt

21. math____matics
22. pre____judice
23. par____llel
24. erron____ous
25. prev____lent
26. rest____urant
27. rep____tition
28. nec____ssary
29. sacr____fice
30. compet____nt
31. com____ing [arriving]
32. tru____ly
33. chimn____s
34. excell____nt
35. sch____dule
36. independ____nt
37. immediat____ly
38. consc____entious
39. op____ortunity
40. dis____atisfied

In the blank, write the **missing letters** in each word: **ie** or **ei**.

Example: bel_ie_ve

1. h____r
2. ach____ve
3. rec____ve
4. c____ling
5. w____rd

6. v____n
7. ch____f
8. l____sure
9. hyg____ne
10. w____gh

69. MECHANICS AND SPELLING: Review

(Study 60–73, Mechanics and Spelling)

In each of the following paragraphs, correct all errors in **capitalization**, **number form**, **abbreviations**, **syllabication**, and **spelling**. Cross out the incorrect form and write the correct form above it.

(Collaborative option: Students work in pairs or small groups to find and correct errors.)

1. Martha's Vineyard is an Island off cape Cod, Mass., that covers aproximately two hundred sixty sq. kilometers. The first europeans to settle there were the English, in 1642. In the 18th Century Fishing and Whaling came into existance as its cheif sources of employment. By the 18 ninteys its developement as a Summer resort was under way. Wealthy people from N.Y. and Boston vacationed on its beaches and sailed around its harbors. John D. Rockefeller, jr., and other socialites visited there, usually in Aug., the most populer vacation month. It was a favorite spot of the Kennedy Family. Today the year-round population is about 6 thousand. Its communitys include Oak Bluffs, Tisbury, and W. Tisbury. Martha's Vineyard also contains a State Forest.

2. A hurricane is a cyclone that arises in the Tropics, with winds exceeding seventy-five mph, or 121 kilometers per hour. The term *Hurricane* is usually applied to cyclones in the N. Atlantic ocean, whereas those in the western Pacific are called typhoons. Some hurricanes, however, arise in the eastern Pacific, off the West coast of Mexico, and move Northeast. In an average yr. three point five hurricanes will form off the east coast of North America, maturing in the Caribbean sea or the gulf of Mexico. Such hurricanes are most prevelent in Sept. One of the most destructive of these storms slammed into the United States in 1938, causing 100s of deaths in the Northeast. In the nineteen-nineties Hurricane Andrew devastated southern Fla., including Everglades national park. Homes, Churches, schools, and wharfs were ripped apart. Hurricanes can last from 1 to thirty days, weakening as they pass over land. Over the warm Ocean, however, their fury intensifies, and they often generate enormous waves that engulf Coastal areas. To learn more about hurricanes, read *Hurricanes, Their Nature And History.*

3. Turkey is a unique Country. Though partly in Europe, it is ninety seven % in Asia; thus it combines elements of European, middle eastern, and Asiatic cultures. Though the country's Capital is Ankara, its most-famous city is Istanbul, which was for 100s of yrs. called Constantinople and before that Byzantium. To the west of Turkey lies the Aegean sea; to the s.e. lie Iran, Iraq, & Syria. The vast majority of Turks are Muslim, but there are also small

numbers of christians and Spanish Speaking Jews. Modern Turkey came into being after the downfall of the Ottoman empire in world war I; its present boundaries were established by the treaty of Lausanne in nineteen twenty-three. 17 years later the nation switched from the arabic to the roman Alphabet. In Government Turkey has a two house Legislature and a head of State.

70. WORD CHOICE: Conciseness, Clarity, and Originality

(Study 80, Be Concise, Clear, and Original)

Rewrite each sentence in the space below it, **replacing** or **eliminating** all redundant, overblown, vague, or clichéd expressions. You may use a dictionary, and you may invent specifics if necessary.

Examples: We find our general consensus of opinion to be that the governor should resign.
Our consensus is that the governor should resign.

She looked really nice.
She wore jet-black jeans and a trim white blouse, and her broad smile would melt an iceberg.

(Collaborative option: Students work in pairs or small groups to examine sentences and suggest improvements.)

1. The director she believes that within a few months that she can increase profits by 25 percent.

   _____

   _____

2. As a small child of three years of age, I was allowed outside to play only during the hours from eight to eleven a.m. in the morning and from three to five p.m. in the afternoon.

   _____

   _____

3. Lady Macbeth returned back to the deadly murder scene to leave the daggers beside the grooms.

   _____

   _____

4. In the Bible it says that we should not make a judgment about others.

   _____

   _____

5. Except for the fact that my grandmother is on Medicaid, she would not be able to afford living in her very unique senior citizens' residential facility.

   _____

   _____

6. The deplorable condition of business is due to the nature of the current conditions relevant to the economic situation.

   _____

   _____

7. The thing in question at this point in time is whether the initial phase of the operation is proceeding with a sufficient degree of efficiency.

   _____

   _____

8. She jumped off of the wall and continued on down the lane so that she could meet up with me outside of my domicile.

   _____

   _____

9. The house was blue in color and octagonal in shape.

   _____

   _____

10. She couldn't hardly lose her way, due to the fact that the road was intensely illuminated.

    _____

    _____

11. On the basis of this report, it leads me to come to the conclusion that the recruitment process at this office is in need of amelioration.

    _____

    _____

12. The next thing our speaker will speak about is the problem of the transportation situation.

    _____

    _____

13. We are voting to elect Barnett because of the fact that she has a great attitude and so many nice qualities.

    _____

    _____

14. In this day and age things can happen out of a clear blue sky, quick as a wink, to upset one's apple cart.

    _____

    _____

15. The patient fell on his gluteus maximus when we IV'd him in pre-op.

    _____

    _____

16. We have reached the conclusion that the men and women who fly our planes need further training in finding their way from one location to another.

_____

_____

17. I saw my father stumble out of the drinking establishment and walk in an unsteady way down the alley.

_____

_____

18. It is a known fact that people who have undergone the training process in emergency rescue procedures necessarily have to know how to take over in a crisis situation.

_____

_____

19. In the event that inclement weather becomes a factor, the game may be postponed until a later date.

_____

_____

20. We have lost our way, but however, we may connect up with our friends if we utilize our heads to find the right road.

_____

_____

71. WORD CHOICE: Standard, Appropriate American Usage

(Study 81, Use Standard, Appropriate American English)

Part 1

Write the number of the **correct** choice (use standard, formal American English).

Example: Lincoln had no doubt (1)**but that** (2)**that** the South would secede.   _____2_____

1. The dictator determined to attack across the border (1)**irregardless** (2)**regardless** of the consequences.   1. _____

2. That year Einstein conceived his most (1)**revolutionary** (2)**terrific** theory, that of general relativity.   2. _____

3. (1)**Hopefully,** (2)**We hope that** the instructor will post our grades before we leave for the holidays.   3. _____

4. Juan used (1)**these kind of tools** (2)**these kinds of tools** to repair the roof.   4. _____

5. We were disappointed (1)**somewhat** (2)**some** at the poor quality of the color printer.   5. _____

6. We heard the same report (1)**everywhere** (2)**everywheres** we traveled.   6. _____

7. Eileen and Bob (1)**got married** (2)**were married** on a tropical beach at sunrise.   7. _____

8. Do (1)**try to** (2)**try and** spend the night with us when you are in town.   8. _____

9. The diplomat was (1)**most** (2)**almost** at the end of her patience.   9. _____

10. I (1)**had ought** (2)**ought** to have let her know the time of my arrival.   10. _____

11. Will you be sure to (1)**contact** (2)**telephone** me tomorrow?   11. _____

12. He (1)**seldom ever** (2)**hardly ever** writes to his sister.   12. _____

13. The (1)**children** (2)**kids** in my class are interested in the field trip.   13. _____

14. The lawyer was not (1)**enthused** (2)**enthusiastic** about her new case.   14. _____

15. The supervisor (1)**should of** (2)**should have** rewritten the memo.   15. _____

16. The van needed a new battery (1)**besides** (2) **plus** an oil change.   16. _____

Write **C** if the boldfaced expression is **correct**.
Write **X** if it is **incorrect**.

Example: Lincoln had no doubt **but that** the
South would secede.   __X__

1. You **hadn't ought** to sneak into the show.   1. ____
2. We were **plenty** surprised by the outcome of our survey.   2. ____
3. He studied **a lot** for the biology lab exam.   3. ____
4. Susan is **awfully** depressed.   4. ____
5. I **sure** am sore from my exercise class.   5. ____
6. He **better** get here before noon.   6. ____
7. She is a **real** hard worker.   7. ____
8. I admire **that kind** of initiative.   8. ____
9. He has **plenty** of opportunities for earning money.   9. ____
10. The damage was **nowhere near** as severe as it was originally estimated to be.   10. ____
11. His finances are in bad **shape.**   11. ____
12. **Due to** the pollution levels, the city banned incinerators.   12. ____
13. The horrors of war drove him **mad.**   13. ____
14. She was **terribly** pleased at winning the contest.   14. ____
15. Be sure **and** review your class notes before the examination.   15. ____
16. I am a neat person, **aren't I**?   16. ____
17. He wrote essays, short stories, **etc.**   17. ____
18. There was a **bunch** of people in the waiting room.   18. ____
19. Sue's balloon had **bursted.**   19. ____

20. I am sure that he will be **O.K.**   20. ____
21. The students created a mock exam **theirselves.**   21. ____
22. They plan to visit Munich **and/or** Salzburg.   22. ____
23. **Being as how** the bank was closed, Sonya could not withdraw her money.   23. ____
24. She needed the money so **badly** that she cried.   24. ____
25. This has been an auspicious day for you and **me.**   25. ____
26. He **couldn't help but** wonder at her attitude.   26. ____
27. When the **cops** came, everyone was relieved.   27. ____
28. Her **funny** way of speaking made them wonder where she had grown up.   28. ____
29. January employment figures had an **impact** on the stock market.   29. ____
30. His proposal made her **so** happy.   30. ____
31. Do you plan to go to **that there** party?   31. ____
32. We tried to find Illyria, but there was **no such place.**   32. ____
33. The vote was 97–31, **so** the treaty was approved.   33. ____
34. They **have got** a solution to the puzzle.   34. ____

72. WORD CHOICE: Standard, Appropriate American Usage

(Study 81, Use Standard, Appropriate American English)

Most of the following sentences contain one or more lapses from standard, formal English. In the blanks below, **rewrite** the sentence in standard, formal American English. If a sentence needs no change, leave the blanks empty.

Example: You better not bring drugs to campus, seeing as how this is a drug-free school.
You had better not bring drugs to campus, because this is a drug-free school.

(Collaborative option: Students work in pairs to discuss ways sentences could be rewritten.)

1. It was funny how Clem couldn't scarcely outrun the cops this time.

   _____

   _____

2. Just between you and I, he better get into shape before the marathon.

   _____

   _____

3. If and when they would have had kids, they would have been a lot happier.

   _____

   _____

4. They considered it okay for him to drive home, being that he had not drunk anything.

   _____

   _____

5. Irregardless of what the critics think, the new CD by the Mossy Stones will sell a half a million copies.

   _____

   _____

6. Clara was sort of hungry after them guests had eaten all her food.

   _____

   _____

7. Hopefully, this new tax will not impact on the poor an awful lot.

_____

_____

8. If he had of known that the authorities had contacted a bunch of his friends, he would of left town without waiting on a bus.

_____

_____

9. It being clear that everyone outside of John knew the truth, his friends planned on telling him.

_____

_____

10. They had seldom ever seen the manager so awful mad at anyone anywheres.

_____

_____

11. If Farley's appendix busts, there will be no doubt but that the family better rush him to a hospital.

_____

_____

12. The diplomats agreed that if they signed the treaty, you could be sure they'd avoid a confrontation in the Balkans.

_____

_____

13. "Aren't I lucky?" the woman exclaimed. She looked as if she couldn't help but bust with joy, being as how she had just won the lottery.

_____

_____

14. They had got an inkling that Raspley would try and foreclose the mortgage if and when the lovers married.

_____

_____

15. The generals read in the intelligence reports where the enemy forces had spread themselves every which way across the battlefront; plus, their troops must have been some fatigued after a couple days of forced marches.

_____

_____

_____

_____

73. WORD CHOICE: Nondiscriminatory Terms

(Study 82, Use Nondiscriminatory Terms)

Each sentence contains a sexist or other discriminatory term. **Circle** that term. Then, in the blank, write a nondiscriminatory replacement. (If the circled term should be deleted without a replacement, leave the line empty.)

Example: (Every citizen must use his) right to vote.

All citizens must use their

1. How I admire those brave pioneer men who brought civilization to the West!

   _____

2. Every student must bring his textbook to class.

   _____

3. All policemen are expected to be in full uniform while on duty.

   _____

4. Man's need to survive produces some surprising effects.

   _____

5. The speaker asserted that every gal in his audience should make her husband assume more household responsibilities.

   _____

6. The stewardess assured us that we would land in time for our connecting flight.

   _____

7. The female truck driver stopped and asked us for directions.

   _____

8. The innkeeper, his wife, and his children greeted us when we arrived at the inn.

   _____

9. The repairman's estimate was much lower than we had expected.

   _____

10. Everyone hoped that his or her proposal would be accepted.

_____

11. The spinster who lives upstairs never attends the block parties.

_____

12. The victim was shot by an unknown gunman.

_____

13. The new lady mathematics professor has published several textbooks.

_____

14. The college has a large ratio of Oriental students.

_____

15. The ecumenical worship service was open to all faiths, Christian and non-Christian.

_____

16. All kinds of persons with disabilities were there, including the mentally deficient.

_____

17. Our South Side neighborhood was home to many Italians and colored people.

_____

18. Why would you want to blacken your reputation by doing something like that?

_____

19. In our country people may attend whatever church they choose.

_____

20. Early in the fall the senior men began inviting girls to the graduation dance.

_____

74. WORD CHOICE: Words Often Confused

(Study 83, Distinguish Between Similar Words)

Write the number of the **correct** choice.

Example: He sought his lawyer's (1)**advise** (2)**advice**.   2

1. Take my (1)**advice** (2)**advise**, Julius; stay home today.   1. _____

2. If you (1)**break** (2)**brake** the car gently, you won't feel a jolt.   2. _____

3. Camping trailers with (1)**canvas** (2)**canvass** tops are cooler than hardtop trailers.   3. _____

4. The diamond tiara stolen from the museum exhibit weighed more than three (1)**carets** (2)**carats**.   4. _____

5. The Dean of Student Affairs doubted whether the young man was a (1)**credible** (2)**creditable** witness to the fight in the dining hall.   5. _____

6. Over the (1)**course** (2)**coarse** of the next month, the committee will review the sexual harassment policy.   6. _____

7. Helping Allie with history was quite a (1)**descent** (2)**decent** gesture, don't you agree?   7. _____

8. This little (1)**device** (2)**devise** will revolutionize the personal computer industry.   8. _____

9. The professor made an (1)**illusion** (2)**allusion** to a recent disaster in Tokyo when describing crowd behavior.   9. _____

10. She was one of the most (1)**eminent** (2)**imminent** educators of the decade.   10. _____

11. We knew that enemy troops would try to (1)**envelop** (2)**envelope** us.   11. _____

12. Go (1)**fourth** (2)**forth**, graduates, and be happy as well as successful.   12. _____

13. Despite their obvious differences, the five students in Suite 401 had developed real friendship (1)**among** (2)**between** themselves.   13. _____

14. The software game created by Frank really was (1)**ingenious** (2)**ingenuous**.   14. _____

15. She tried vainly to (1)**lesson** (2)**lessen** the tension in the house.   15. _____

16. The style of furniture is actually a matter of (1)**personal** (2)**personnel** taste.   16. _____

17. Even though Gary studied hard and attended every class, he discovered that he was (1)**disinterested** (2)**uninterested** in majoring in chemistry.   17. _____

18. The judge (1)**respectfully** (2)**respectively** called for the bailiff to read the jury's questions.   18. _____

19. When the grand marshal gave the signal, the parade (1)**preceded** (2)**proceeded.**   19. _____

20. Middle-aged professionals are forsaking their high-powered lifestyles for a (1)**quiet** (2)**quite** existence in the country.   20. _____

21. (1)**Weather** (2)**Whether** to pay off all her creditors was a big question to be resolved.   21. _____

22. We were so overweight that we bought a (1)**stationary** (2)**stationery** bicycle for our fifth anniversary.   22. _____

23. The laser printer produces a much sharper image (1)**than** (2)**then** the older dot-matrix printer.   23. _____

24. The computer operator read (1)**thorough** (2)**through** most of the manual before finding a possible solution.   24. _____

25. The ability to pass doctoral qualifying exams is essentially a (1)**rite** (2)**right** of passage. 25. _____

26. She is the first (1)**woman** (2)**women** to umpire in this league. 26. _____

27. (1)**Your** (2)**You're** aware, aren't you, that the play is sold out? 27. _____

28. This scanner will (1)**complement** (2)**compliment** your computer. 28. _____

29. The student was (1)**anxious** (2)**eager** to receive his award at the banquet. 29. _____

30. It will take me (1)**awhile** (2)**a while** to finish these calculations. 30. _____

31. The best advice is to take a long walk if you (1)**lose** (2)**loose** your temper. 31. _____

32. Some of the television programs needed to be (1)**censored** (2)**censured** by parents because they were showing extreme violence before 9 p.m. 32. _____

33. Nobody (1)**accept** (2)**except** Gloria would stoop so low. 33. _____

34. Sam unplugged his phone, locked his door, and worked (1)**continuously** (2)**continually** on his research paper. 34. _____

35. Her approach for preparing for the history final was (1)**different from** (2)**different than** my strategy. 35. _____

36. (1)**Everyone** (2)**Every one** of the computers was destroyed by the flood. 36. _____

37. If John (1)**passed** (2)**past** the physics final, it must have been easy. 37. _____

38. The library copy of the magazine had lost (1)**its** (2)**it's** cover. 38. _____

39. For (1)**instance**, (2)**instants**, this computer doesn't have enough memory to run that particular word-processing package. 39. _____

40. The firm is (1)**already** (2)**all ready** for any negative publicity from the outcome of the lawsuit. 40. _____

41. Can you name the (1)**capitals** (2)**capitols** of all fifty states? 41. _____

42. His physical condition showed the (1)**affects** (2)**effects** of inadequate rest and diet. 42. _____

43. My gregarious little niece was (1)**eager** (2)**anxious** to go to the party. 43. _____

44. Shall we dress (1)**formally** (2)**formerly** for the Senior Ball? 44. _____

45. To be an effective teacher had become her (1)**principal** (2)**principle** concern. 45. _____

46. More than a million people (1)**emigrated** (2)**immigrated** from Ireland during the nineteenth-century potato famine. 46. _____

47. The Farkle family were (1)**altogether** (2)**all together** in the living room when the grandmother announced that she was willing her money to a nearby cat sanctuary. 47. _____

48. The student (1)**inferred** (2)**implied** from the professor's expression that the final exam would be challenging. 48. _____

49. "I, (1)**to** (2)**too** (3)**two**, have a statement to make," she said. 49. _____

50. Homelessness—(1)**its** (2)**it's** no longer just an American problem. 50. _____

75. WORD CHOICE: Words Often Confused

(Study 83, Distinguish Between Similar Words)

Write the number of the **correct** choice.

Example: He sought his lawyer's (1)**advise** (2)**advice**. _____2_____

1. Chris feels (1)**good** (2)**well** about the results of the faculty survey. 1._____

2. He said, "(1)**Their** (2)**There** (3)**They're** is no reason for you to wait." 2._____

3. "(1)**Whose** (2)**Who's** there?" she whispered. 3._____

4. The cat ran behind my car, and I accidentally ran over (1)**its** (2)**it's** tail. 4._____

5. The consultant will (1)**ensure** (2)**insure** that the audit is completed on time. 5._____

6. The twins (1)**formally** (2)**formerly** attended a private college in California. 6._____

7. The mere (1)**cite** (2)**site** (3)**sight** of Julia made his heart soar. 7._____

8. Will people be standing in the (1)**aisles** (2)**isles** at the dedication ceremony? 8._____

9. Dr. Smith is (1)**famous** (2)**notorious** for her educational research. 9._____

10. "Sad movies always (1)**affect** (2)**effect** me that way," lamented Kay. 10._____

11. The (1)**thorough** (2)**through** commission report indicated that approximately forty percent
    of American schools do not have enough textbooks in their classrooms. 11._____

12. Jonathon had the (1)**presence** (2)**presents** of mind to make a sharp right turn and to step on
    the accelerator. 12._____

13. The principal expected the students' behavior to (1)**correspond to** (2)**correspond with** the
    school district's expectations. 13._____

14. If you rehearse enough, you're (1)**likely** (2)**liable** to get the lead role in the play. 14._____

15. The family has (1)**born** (2)**borne** the noise and dust of the nearby highway construction for
    several months. 15._____

16. (1)**Their** (2)**They're** (3)**There** leasing a truck because they can't afford the down payment to
    purchase a new one. 16._____

17. The time capsule (1)**may be** (2)**maybe** the best way for the general public to understand how
    people lived one hundred years ago. 17._____

18. Some (1)**individual** (2)**person** dropped off a package at the mailroom. 18._____

19. The track coach told me that he wanted to (1)**discuss** (2)**discus** my performance at the last meet. 19._____

20. The voters are (1)**apt** (2)**likely** to vote for a candidate who promises to reduce unemployment. 20._____

21. (1)**Who's** (2)**Whose** theory do you believe regarding the geographical origin of humankind? 21._____

22. The (1)**council** (2)**counsel** (3)**consul** met to decide the fate of the student who cheated on the
    psychology final. 22._____

23. A tall tree has fallen and is (1)**laying** (2)**lying** across the highway. 23._____

24. A significant (1)**percent** (2)**percentage** of Americans still smoke. 24._____

25. In *The Oxbow Incident*, the wrong man is (1)**hung** (2)**hanged**. 25._____

26. Did you ask if he will (1)let (2)leave you open a charge account?   26. _____

27. The new dance had (1)to (2)too (3)two many steps to remember.   27. _____

28. Sarah promised to (1)learn (2)teach me some gardening techniques.   28. _____

29. Shooting innocent bystanders is one of the most (1)amoral (2)immoral street crimes committed.   29. _____

30. The alfalfa milkshake may taste unpleasant, but it is (1)healthy (2)healthful.   30. _____

31. The tennis player always (1)lays (2)lies down before an important match.   31. _____

32. When (1)your (2)you're in love, the whole world seems beautiful.   32. _____

33. On high school basketball courts Sam was often (1)compared to (2)compared with the young Michael Jordan.   33. _____

34. The (1)amount (2)number of trees needed to produce a single book should humble any author.   34. _____

35. This medication will (1)lessen (2)lesson the pain until we reach the emergency room.   35. _____

36. The newspaper was soggy because it had (1)laid (2)lain in a rain puddle all morning.   36. _____

37. After spending $1,000 on repairs, we hope that the van finally works (1)like (2)as it should.   37. _____

38. The couple (1)adapted (2)adopted a baby girl from Bulgaria.   38. _____

39. Cindy was (1)besides (2)beside herself with anger.   39. _____

40. The agreement was (1)among (2)between Harry and me.   40. _____

41. Do not (1)set (2)sit the floppy disk on top of the computer monitor.   41. _____

42. The play was from (1)classical (2)classic Rome.   42. _____

43. The curtain was about to (1)raise (2)rise on the last act of the senior play.   43. _____

44. The camp is just a few miles (1)farther (2)further along the trail.   44. _____

45. The news report (1)convinced (2)persuaded me to join a volunteer organization that renovates homes in low-income neighborhoods.   45. _____

46. The author of that particular book was (1)censored (2)censured for his views by a national parenting group.   46. _____

47. You may borrow (1)any one (2)anyone of my books if you promise to return it.   47. _____

48. Compared (1)to (2)with the Steelers, the Raiders have a weaker defense but a stronger offense.   48. _____

49. The linebacking unit was (1)composed (2)comprised of Taylor, Marshall, and Burt.   49. _____

50. The three children tried to outrun (1)each other (2)one another.   50. _____

76. WORD CHOICE: Review

(Study 80–83, Word Choice)

Each sentence may contain an inappropriate or incorrect expression. **Circle** that expression, and in the blank write an appropriate or correct replacement. Use standard, formal American English. If the sentence is correct as is, leave the blank empty.

Examples: He sought his lawyer's (advise.) ___advice___

The director reported that the company (was fine and dandy.) ___had doubled its profits.___

Whose idea was it? _____

1. Those sort of books are expensive. _____

2. The cabin was just like I remembered it from childhood vacations. _____

3. Some children look like their parents. _____

4. I was surprised that the banquet was attended by lots of people. _____

5. Its time for class. _____

6. You too can afford such a car. _____

7. I can't hardly hear the speaker. _____

8. Randy promised me that he is over with being angry with me. _____

9. Irregardless of the result, you did your best. _____

10. Will he raise your salary? _____

11. Try to keep him off of the pier. _____

12. I usually always stop at this corner meat market when I am having dinner guests. _____

13. You are selling vanilla, chocolate, and black cherry? I will take the latter. _____

14. His efforts at improving communication among all fifty staff members will determine his own success. _____

15. Her success was due to hard work and persistence. _____

16. I'm invited, aren't I? _____

17. Their house is now for sale. _____

18. Henry and myself decided to start a small business together. _____

19. The club lost its president. _____

20. Did he lay awake last night? _____

21. The professor's opinion differed with the teaching assistant's perspective. _____

22. Bob laid the carpet in the hallway. _____

23. The cat has been laying on top of the refrigerator all morning.
_____

24. He has plenty of opportunities for earning money. _____

25. San Francisco offers many things for tourists to do. _____

26. Most all her friends are married. _____

27. He always did good in English courses. _____

28. The low price of the printer plus the modem prompted me to buy both.
_____

29. Because her supervisor seemed unreasonable, Sue finally decided to resign.
_____

30. Max has less enemies than Sam. _____

31. The speaker inferred that time management depended more on attitude than skill.
_____

32. Glenn has a long way to travel each week. _____

33. Did he loose his wallet and credit cards? _____

34. She looked like she was afraid. _____

35. Walking to school was a rite of passage in our home.
_____

36. Who were the principals in the company? _____

37. Have you written in regards to an appointment? _____

38. Elaine adopted her novel for television. _____

39. Damp weather affects her sinuses. _____

40. A lion hunting its prey is immoral. _____

41. The troop was already to leave for camp. _____

42. The men and girls on the team played well. _____

43. We split the bill between the three of us. _____

44. Be sure to wear causal clothes to tonight's party.
_____

45. The hum of the air conditioner was continual. _____

46. The informer was hanged. _____

47. The air conditioner runs good now. _____

48. The child is too young to understand. _____

49. Regardless of his shortcomings, she loves him. _____

50. Where is the party at? _____

51. Please bring these plans to the engineering department. _____ _____

52. The salesman was looking forward to the sale. _____

53. The sun will hopefully shine today. _____

54. Their political strategy failed in the end. _____

55. The workmen complained that the work site was unsafe.

    _____

56. His chances for a promotion looked good. _____

57. The twins frequently wear one another's clothing. _____

58. A twisted branch was laying across our path. _____

59. She was disinterested in the boring play. _____

60. Send a cover letter to the chair of the department. _____

61. This line is for shoppers with ten items or fewer. _____

62. The hostile countries finally effected a compromise.

    _____

63. The professor was somewhat annoyed at the girls in his class.

    _____

64. Sam differed from Gina about the issue of increasing social services.

    _____

65. I meant to lay down for just an hour. _____

66. Let us think further about it. _____

67. He is the most credible person I have ever met. He will believe anything.

    _____

68. He enjoys the healthy food we serve. _____

69. Paul's conversation was sprinkled with literary illusions.

    _____

70. The husband and wife were both pursuing law degrees.

    _____

71. Her position in the company was most unique. _____

72. He has already departed. _____

73. I will have to rite a letter to that company. _____

74. There's was an informal agreement. _____

75. Durnell is a student which always puts his studies first.

    _____

76. Their is always another game. _____

77. The gold locket had lain on the floor of the attic for ten years.

    _____

78. Foyt lead the race from start to finish. _____

79. By the tone of her writing, the news reporter implied that the politician was guilty of fraud.

    _____

80. To find the missing watch, we ventured further into the crowd.

    _____

77. WORD CHOICE: Review

(Study 80–83, Word Choice)

On your own paper, **rewrite** each paragraph below so that it displays all the word-choice skills you have learned, but none of the word-choice faults you have been cautioned against. Use standard, formal American English.

(Collaborative option: Students work in pairs or small groups to discuss each paragraph, suggest new wording, and edit one another's work.)

1. It has been brought to our attention that company personnel have been engaging in the taking of unauthorized absences from their daily stations. The affect of this action is to leave these stations laying unattended for durations of time extending up to a quarter of an hour. In this day and age such activity is inexcusable. Therefore the management has reached the conclusion that tried and true disciplinary measures must necessarily be put into effect. Thus, commencing August 5, workmen who render theirselves absent from their work station will have a certain amount of dollars deducted from the wages they are paid.

2.   Needless Required College Courses [title of essay]

   The topic of which I shall write about in this paper is needless required college courses. I will show in the following paragraphs that many mandatory required courses are really unnecessary. They have no purpose due to the fact that they are not really needed or wanted but exist just to provide jobs for professors which cannot attract students by themselfs on there own. It is this that makes them meaningless.

3. In my opinion, I think that the general consensus of opinion is usually always that the reason why lots of people fail to engage in the voting procedure is because they would rather set around home then get off of they're tails and get down to the nearest voting facility. In regards to this matter some things ought to be done to get an O.K. percent of the American people to vote.

4. Each and every day we learn, verbally or from newspapers, about business executives having heart attacks and every so often ending up dead. The stress of high management-type positions is said to be the principle casual factor in such attacks. But a search threw available data shows that this is a unfounded belief. For awhile it was universally expected that persons in high-level jobs experienced the most stress. But yet this is such a misconception. It is in the low-echelon jobs that more strain and consequently more heart attacks occur.

5. The Bible's Book of Exodus relates the flight of the ancient Jews from Egypt to Israel. The narrative says where God sent ten plagues upon the land, the reason why being to punish the rulers for not letting the Jewish people go. Moses then lead his people across the Red Sea, who's waves parted to leave them go through. There trek thorough the desert lasted weaks, months, and than years. The people's moral began to sag. However, Moses brought them the Ten Commandments from Mt. Sinai, and they emigrated safely into the Promised Land. Moses, though, died before he could enter this very fine country. Some question the historic accuracy of the narrative, but others find it entirely credulous. If you except it fully or not, its one of the world's most engrossing stories.

78. PARAGRAPHS AND PAPERS: Topic Sentences and Paragraph Unity

(Study 92A, The Topic Sentence, and 92D, Unity and Emphasis)

First, **circle** the topic sentence of each paragraph. Then find one or more sentences that violate the **unity** of the paragraph (that do not relate directly to the topic). Write the number(s) of the sentence(s) in the blank at the end of the paragraph.

1. (1)From a pebble on the shore to a boulder on a mountainside, any rock you see began as something else and was made a rock by the earth itself. (2)Igneous rock began as lava that over hundreds of years hardened far beneath the earth's surface. (3)Granite is an igneous rock that is very hard and used for buildings and monuments. (4)Sedimentary rock was once sand, mud, or clay that settled to the bottom of a body of water and was packed down in layers under the ocean floor. (5)All rocks are made up of one or more minerals. (6)Metamorphic rock began as either igneous rock or sedimentary rock whose properties were changed by millions of years of exposure to the heat, pressure, and movement below the earth's crust.

   1. _____

2. (1)Although we normally associate suits of armor with the knights of medieval Europe, the idea of such protective coverings is much older and more pervasive than that. (2)Some knights even outfitted their horses with metal armor. (3)As long as 3,500 years ago, Assyrian and Babylonian warriors sewed pieces of metal to their leather tunics the better to repel enemy arrows. (4)A thousand years later, the Greeks wore metal helmets, in addition to large metal sheets over their chests and backs. (5)Native Americans of the Northwest wore both carved wooden helmets and chest armor made from wood and leather. (6)Nature protects the turtle and the armadillo with permanent armor. (7)Even with body armor largely absent from the modern soldier's uniform, the helmet still remains as a reminder of the vulnerability of the human body.

   2. _____

3. (1)Mention the name of George Washington and most Americans envision a larger-than-life hero, who, even as a little boy, could not tell a lie. (2)However, it turns out that Washington was more human than his biographers would have us believe. (3)His contemporaries described Washington as moody and remote. (4)He was also a bit vain, for he insisted that his fellow officers address him as "Your Excellency." (5)He refused to allow himself to be touched by strangers. (6)Washington was also known to weep in public, especially when the Patriots' war effort was sagging. (7)Washington was even plagued with traitors, who gave the British advice on how to beat the Americans. (8)He was not even a gifted military officer. (9)Rather than being a hero of the French and Indian War, Washington may have provoked the French to go to war by leading an unnecessary and irrational attack against a group of Frenchmen. (10)While Washington was certainly a brave man, dedicated to freeing the colonists from British tyranny, he was not the perfect man that early biographers described.

   3. _____

4. (1)In the mid-1800s, an apple or a pear was considered too dangerous to eat. (2)In fact, any fresh vegetable or fruit was considered too risky because one bite might lead to cholera, dysentery, or typhoid. (3)During cholera epidemics, city councils often banned the sale of fruits and vegetables. (4)The only safe vegetable was a boiled potato. (5)A typical breakfast might include black tea, scrambled eggs, fresh spring shad, wild pigeons, pig's feet, and oysters. (6)Milk was also considered a perilous beverage because many people died from drinking spoiled milk. (7)Milk really was a threat to people's health, because it was processed and delivered to home with little regard for hygiene. (8)Children and those who were ill were often malnourished because the foods with the most nutrients were also the most deadly. (9)Until the invention of the icebox in the 1840s, rich and poor people alike risked their health and even their lives every time they ate a meal.

4. _____

5. (1)Infant sacrifice must be clearly differentiated from infanticide. (2)The latter practice, growing out of economic want, was not uncommon among primitive peoples whose food supply was inadequate. (3)Even in most of the Greek city-states, in Rome, and among the Norsemen before they accepted Christianity, it was the father's right to determine whether his newborn child should be accepted and nurtured or instead be abandoned—simply left to perish from exposure. (4)In the eighth and ninth centuries the Norse invaded Britain and left elements of their linguistic and cultural heritage. (5)But in infant sacrifice a father offered this most precious gift to the gods. (6)Thus Abraham was told: "Take now thy son, thine only son Isaac, whom thou lovest, and get thee into the land of Moriah; and offer him there for a burnt offering upon one of the mountains, which I will tell thee of."

—Constance Irwin (adapted)

5. _____

79. PARAGRAPHS AND PAPERS: Paragraph Development

(Study 92B, Adequate Development)

Each paragraph below is inadequately developed. Choose **one**, and, on the back of this page or on your own paper, **rewrite** it to develop the topic sentence (boldfaced) adequately. Use six to nine sentences, adding your own facts and ideas as needed. (You may change the topic sentence to express a different viewpoint.)

(Collaborative option: Students work in pairs or small groups to pool information, discuss how to develop paragraphs, and review or edit one another's work.)

1. Young people today see how their parents act and how they feel about the world today. Since they feel their parents are wrong, they rebel because they do not want to become a carbon copy of their elders. Young people want to be treated as persons, not just kids who do not know what they are talking about and who should not express their own ideas because they are too young to understand. **Young people today want to do and think as they please.**

2. **Traffic is choking our community.** Every morning and evening at rush hour the streets are clogged with commuters' cars. There are just too many of them. I see this every weekday. And on weekends the roads to nearby resorts are almost as crowded. This problem is ruining our community, and something must be done about it.

3. **I like the old movies shown on TV better than the recent releases shown in theaters.** The old films contain more-dramatic plots and more-famous actors. They are exciting and fast paced. The actors are widely known for their acting ability. Today's films are boring or mindless and have less-famous actors.

80. PARAGRAPHS AND PAPERS: Paragraph Coherence

(Study 92C, Coherence)

In each blank, write the transitional expression from the list below that fits most logically. For some blanks there is more than one correct answer. Try not to use any expression more than once.

afterward	meanwhile	more important	however
consequently	nevertheless	therefore	likewise
even so	on the other hand	thus	in particular
formerly	finally	as a result	that is

Example: Thousands of workers were heading home by car, bus, and train. <u>Meanwhile,</u> at home, their spouses were readying supper.

1. The rescue workers searched for victims in the debris for more than twelve hours until relieved; _____, as they sat and wearily sipped coffee, they vented their frustration at finding no survivors.

2. Arthur, in Jacksonville, was trying frantically to call Rachel about the news; _____, in Sacramento, Rachel was desperately trying to reach him.

3. Today we take a common United States currency for granted. Money, _____, was not standardized in the U.S. until the Civil War, when the federal government produced its first paper money.

4. In postwar America people were enjoying a strong economy, which provided plenty of jobs and high wages; _____, life seemed secure and promising.

5. The term *teenager* entered the language only as recently as 1941; _____, teenagers were not really a recognized presence in American society.

6. When we speak to family members, we use an informal and intimate language. When we are speaking to a large group, _____, we are more likely to choose different words and a different tone of voice.

7. If you toss a coin repeatedly and it comes up heads each time, common sense tells you to expect tails to turn up soon. _____, the chances of heads coming up remain the same for each toss of the coin.

8. The first real movie—_____, one that actually had a story line—was the film entitled *The Great Train Robbery.*

9. American children spend about a quarter of their waking time watching television; _____, it is important to monitor what young children are watching.

10. A seven-hundred pound microwave oven, called the Radarange, was first produced by Tappan in 1955. _____, Americans were not interested in purchasing a microwave oven until the late 1960s, when the appliance was much smaller and more reliable.

11. The young singing group tried again and again to produce a hit recording, without success; _____, they struck gold with "Gotta Have Your Love."

12. Whelan's stocks soared 350 points in a day; _____, she felt that she could buy an expensive new car.

13. Some baseball records have been thought unbreakable; _____, few people expected Lou Gehrig's 2,130-consecutive-game streak ever to fall.

14. Nineteen forty-one marked the beginning of Franklin Roosevelt's third term; _____, it was the year that the U.S. was plunged into World War II.

81. PARAGRAPHS AND PAPERS: Paragraph Review, Netiquette

(Study 92C, Coherence, and 93, Netiquette)

Part 1

Go back to the paragraph you wrote in exercise 79.

On your own paper or in the space below, **rewrite** it, being sure that it has a controlling structure, appropriate transitions, and repeated key words or phrases as needed to give it coherence. **Circle** your transitions and repeated key words or phrases. At the end, skip a line and **write a sentence** briefly stating what your controlling structure is.

(Collaborative option: Students work in pairs or small groups to assist one another in revising.)

A student sent the following e-mail in response to an online job offer. On the lines below, **rewrite** the e-mail in appropriate English, changing, deleting or adding text as necessary. (This is a preliminary contact, not a full application letter.)

FROM: Pat Benson

TO: Online Services Company

SUBJECT: Your Website

DATE: October 18, 2005

Hi there!

I got real enthused when I saw your job offer on your Website August 5. In this day and age you won't find many as UNIQUE as me. Being as how I have taken a lot of computer courses and plan on taking more, I can, IMHO, handle ANY THING I could meet up with in your service department. If you can utilize my services, contact me. I'll be waiting anxiously. Gotta go now. Bye! ☺

Pat

FROM: _____

TO: _____

SUBJECT: _____

DATE:_____

_____

_____

_____

_____

_____

_____

_____

_____

_____

_____

_____

_____

_____

_____

_____

_____

_____

_____

82. PARAGRAPHS AND PAPERS: The Thesis Sentence

(Study 94B, Forming a Thesis)

First, from the list below, **identify** the main weakness in each thesis sentence, and write the letter of that weakness in the short blank. Then, in the long blanks, **rewrite** the thesis sentence so that it is usable for an essay. (For the purposes of this exercise, you may invent facts or ideas as needed.)

A–no assertion   C–too broad, too vague, or unsupportable
B–split focus   D–stale, uninteresting to U.S. collegians

Examples: Our nation's social problems need solving now.   _____C_____
Our college has a moral obligation to use part of its endowment to relieve local poverty.

The Antarctic is one of the coldest places on earth.   _____A_____
Despite its inhospitable climate, new scientific advances hold promise for making the
_   Antarctic a desirable place for people to live._

(Collaborative option: Students work in pairs or small groups to evaluate thesis sentences and compose new ones.)

1. Canada is actually larger than the United States.   1._____

       _____

       _____

2. Radio can be more entertaining than television, and radio commercials are more profitable for sponsors than those on TV.   2._____

       _____

       _____

3. Spring is a lovely time of year.   3._____

       _____

       _____

4. Studying the types of sand in Mongolia's Gobi Desert can be rewarding.   4._____

       _____

       _____

5. A drug experience really affects a person.   5._____

       _____

       _____

6. The state's welfare system is inhumane, and its housing rehabilitation program is in shambles.   6. _____

_____

_____

7. I experienced the birth of twins.   7. _____

_____

_____

8. I have strong feelings about euthanasia.   8. _____

_____

_____

9. Air pollution will kill off the human race.   9. _____

_____

_____

10. The scourge of acne must be eliminated.   10. _____

_____

_____

83. PARAGRAPHS AND PAPERS: Planning the Essay

(Study 94A–D, Before Starting to Write)

Follow the directions below.

(Collaborative option: Students work in pairs or small groups to share knowledge and ideas and to offer suggestions.)

A. Assume that you have been assigned to write an essay in one of these broad subject areas: popular music, man-woman relationships, improving this college. Choose one. On your own paper, **brainstorm, freewrite,** or **cluster** whatever ideas you can generate on this subject. From those ideas, produce **three** limited topics suitable for a 2- to 4-page essay. List those topics here:

1._____

2._____

3._____

B. From each of these topics, develop the best tentative thesis sentence you can for this 2- to 4-page essay:

Topic 1:_____

_____

Topic 2:_____

_____

Topic 3:_____

_____

From these three, choose the one that, considering your knowledge, ideas, and interests, you can best develop into an essay. Refine that thesis sentence and write it here:

Topic #____: _____

_____

C. Consider which approach seems most workable for this topic and thesis: narration, description, explanation, persuasion, problem-solution, effect-cause (or vice versa), comparison/contrast. State your most likely approach:

_____

D. List below the major divisions (subtopics) you see for your essay (three is the most common number, but others may work better for your topic). If you chose the persuasive approach, for example, each division would probably be a separate reason.

Divisions:

_____

_____

_____

_____

84. PARAGRAPHS AND PAPERS: The Essay Outline

(Study 94E, Outlining)

In the space below, write a detailed **outline** for the essay you began preparing in exercise 83. Continue on the back or on your own paper if necessary. Use any of the methods mentioned in section 94E. Make it detailed enough so that you can write an essay from it.

(Collaborative option: Students work in pairs or small groups to construct the outline or evaluate one another's outlines.)

85. PARAGRAPHS AND PAPERS: The Essay Introduction and Conclusion

(Study 95, Writing and Revising the Essay; Review 91–92, Paragraphs)

Part 1

Each sentence is the opening of an essay. In the blank, write **Y** (for **yes**) if the sentence is an **effective interest-arouser**. Write **N** (for **no**) if it is **not**.

Example: The United States faces many problems today.   _____N_____

(Collaborative option: Students in pairs or small groups discuss the effectiveness of each sentence.)

1. Throughout my life I have encountered many interesting situations and experiences.   1. _____

2. It is 7 a.m. this bitter cold December day, and the line outside the employment agency has grown bigger since I arrived at 5.   2. _____

3. There are many things that bring a person joy in life, and many things that bring sadness.   3. _____

4. One of the biggest-selling items in this city's public housing project supermarkets is dog food—yet no dogs are allowed in the projects.   4. _____

5. In the past few years Americans generally have lost interest in threats from global warming and the ozone hole, yet both are still on the increase and no one is sounding the alarm.   5. _____

6. The United States is a very different place from what it was a hundred years ago, or even fifty, all because of different circumstances that have surrounded people in each era.   6. _____

7. What would you do if your life savings were suddenly wiped out and you and all your family lost their jobs, as happened to our great-grandparents in the Depression?   7. _____

8. Our hike that day started just like many others we had taken through the Appalachians, until we stepped around a fallen tree and recoiled in horror.   8. _____

9. The figures are awesome: a world population today of more than 6 billion, with more than a million newcomers every week.   9. _____

10. In my short but checkered career I have worked at a wide variety of jobs, some indoors and some out, some easy and some hard.   10. _____

Part 2

Each item contains the closing sentence(s) of an essay. In the blank, write **Y** (for **yes**) if the sentence is an **effective closing**. Write **N** (for **no**) if it is **not**.

Example: And so we must do something about this problem.   _____N_____

(Collaborative option: Students in pairs or small groups discuss the effectiveness of each sentence.)

1. On the whole, as I said before, my experience was one of the many things I remember as significant in my life.   1. _____

2. I can still remember how I felt as I stood there with all those people looking at me while I had to pay my restaurant bill with borrowed nickels and dimes.   2. _____

3. The dining hall in which we must all eat, and to whose employees we must entrust our health, is, as I have shown, a pigsty. The time has come to start doing something about it.   3. _____

4. This experience changed my whole attitude toward money, making me ruthlessly determined never again to be embarrassed by lack of funds. We all seem to suffer more from such humiliations than from any other kind of defeat, even a physical beating, a job lost, or a romance gone sour. Along with hunger and sex, the fear of looking bad in the eyes of others is one of the most basic of human motives.   4. _____

5. Therefore, every citizen should go to the polls this election day and vote for Maryann Rivera for governor. Another reason is that her opponent is old and may die in office.   5. _____

6. The federal government, then, must cut the money supply before it is too late; otherwise, as in Germany in the 1920s, our money may literally not be worth the paper on which it is printed.   6. _____

7. Every spring since that day my parents first took me to Wrigley Field, my heart pounds with anticipation when I hear the cry "Play ball!" Keep your football, basketball, and hockey; baseball will always be America's game.   7. _____

8. Thus, since there are more people in the world today than can be fed, and a million more arriving weekly, it is up to the United Nations to take the bull by the horns and find a solution to the problem.   8. _____

9. Since poverty will never disappear, it is up to leaders of the prosperous nations, particularly the United States, to shift their focus from assuring middle-class comfort to making laws and programs that will create a vast new Marshall Plan to feed the hungry at home and worldwide. Perhaps then those in our housing projects will not have to subsist on dog food.   9. _____

10. [For this item, supply your own effective conclusion.]
    I, then, am one of those who have grown up as so-called victims of society. _____

    _____

    _____

    _____

Part 3

On your own paper: First, write an effective **introductory paragraph** for the essay you outlined in exercise 84. Then write an effective **concluding paragraph** for the same essay. (These may be considered drafts until you complete exercise 86. Or your instructor may have you defer writing the conclusion until you have completed exercise 86.)

(Collaborative option: Students in pairs or small groups critique one another's paragraphs.)

86. PARAGRAPHS AND PAPERS: The Essay Body

(Study 95, Writing and Revising the Essay; Review 91–92, Paragraphs)

Part 1

In the blanks, write an effective **topic sentence** for each body paragraph of the essay you have been planning and writing in exercises 83, 84, and 85. (You do not have to use all five sets of blanks; use as many body paragraphs as the structure of your essay demands.) Include a transitional expression or sentence that links each paragraph to the preceding one.

Example: _At first, in the late 1970s and early '80s, personal computers were incredibly primitive by today's standards._

(Collaborative option: Students in pairs or small groups discuss the effectiveness of each sentence.)

1. _____

   _____

   _____

2. _____

   _____

   _____

3. _____

   _____

   _____

4. _____

   _____

   _____

5. _____

   _____

   _____

Part 2

On your own paper, complete the **body paragraphs** of your essay. Remember what section 91B stated about supporting evidence.

(Collaborative option: Students in pairs or small groups critique one another's work.)

Part 3

Bring your introductory, body, and concluding paragraphs together. On your own paper, **revise**, **edit**, and **proofread** your essay until you are satisfied with its quality. Submit a clean copy.

(Collaborative option: Students in pairs or small groups critique, edit, and proofread one another's work.)

87. PARAGRAPHS AND PAPERS: Research Paper Topics and Theses

(Study 96A, Choosing and Limiting a Topic, and 96B, Forming a Thesis)

Part 1

If the topic is **suitable** for a research paper, write **Y** (for **yes**) in the blank. If the topic is **not suitable**, write the **letter of the reason** in the blank. (For some items there is more than one possible correct answer.)

A—too broad, vague, or speculative
B—not researchable or completable with available resources
C—unable to be treated objectively

Example: Poland's transition from communism to capitalism, 1985–1995 _____Y_____

1. Muslim-Christian conflicts since the year 1000   1. _____

2. Is there intelligent life in our solar system?   2. _____

3. Attitudes toward wealth in F. Scott Fitzgerald's story "The Rich Boy"   3. _____

4. The coming triumph of feminism   4. _____

5. The torch is passed: the effects of the 1960 presidential election on the Cold War   5. _____

6. Symbols in the writings of Willa Cather   6. _____

7. New treatments of athletic knee injuries in girls and women   7. _____

8. Devil worship: the one true religion   8. _____

9. Voting patterns of current freshmen at this college in public elections 9. _____

10. The collapse of communism as seen through the eyes of selected average Russians   10. _____

11. The ruination of America's moral standards through unrestricted use of the Internet   11. _____

12. The decline in foreign trade as a cause of the Great Depression   12. _____

13. The Great Depression: it could happen again   13. _____

14. The interstate highway system in our state, 1955–1975: boon or boondoggle?   14. _____

Part 2

In the short blank, write **Y** (for **yes**) if the thesis sentence is a **workable** one for a research paper. Write **N** (for **no**) if it is **not**. Then, after each sentence for which you wrote **N**, write in the long blank a workable thesis sentence on the same topic.

Example: Roosevelt's China policy before World War II was bad. _____N_____
 Roosevelt's China policy helped draw the U.S. into World War II.

(Collaborative option: Students in pairs or small groups suggest workable thesis sentences.)

1. F. Scott Fitzgerald, in his story "The Rich Boy," conveys ambivalent attitudes toward wealth.   1. _____

_____

_____

_____

2. The U.S. election of 1960 radically changed the history of the world.   2. _____

_____

_____

_____

3. Throughout *Huckleberry Finn*, the Mississippi River can be seen as Mark Twain's symbol for life and freedom.   3. _____

_____

_____

_____

4. The lost continent of Atlantis lies just off the sea in the Bermuda Triangle, waiting to be discovered.   4. _____

_____

_____

_____

5. The Second Vatican Council played only a secondary role in the widespread defection of U.S. Catholics in the 1960s and '70s.   5. _____

_____

_____

_____

6. Despite being maligned by critics, current afternoon TV talk shows tend to raise the cultural level of all their viewers.   6. _____

_____

_____

_____

7. The present family court system in this county often harms the very people it is intended to help.   7. _____

_____

_____

_____

8. The U.S. Electoral College should be replaced by direct popular vote because the College system distorts vote totals.

8. _____

_____

_____

_____

9. Shakespeare's works really had to have been written by Francis Bacon because Bacon was educated and Shakespeare was not.

9. _____

_____

_____

_____

10. Devil worship has grown in the U.S. in recent years primarily because of its appeal to alienated youth.

10. _____

_____

_____

_____

11. Massive urban renewal projects in the 1960s and 1970s failed throughout the nation; I know because I grew up in one of them.

11. _____

_____

_____

_____

88. PARAGRAPHS AND PAPERS: Locating and Evaluating Sources

(Study 96C and E, Locating and Evaluating Sources, and 98, Works Cited/References)

Part 1

Choose one of the subjects below. In the library, **locate** five useful printed (not electronic) sources on your subject. In the space on this page, **list** the needed bibliographical information for each, in either **MLA** or **APA** form (or the **COS** alternative for either), as your instructor directs. (For books, also make a note of the call number.)

U.S. responses to the al-Qaeda
   threat before September 11, 2001
Honky-tonk music
The relation of dinosaurs to birds

U.S. national parks since 1950
U.S. automobile safety since 1950
F. Scott Fitzgerald's early short stories
   (up to 1925)

(Collaborative option: Students in pairs or small groups share research tasks and findings.)

Follow the directions for part 1, using the same topic you chose there. This time use electronic instead of printed sources. Be sure to include Internet addresses or other needed information for retrieval, where necessary.

Evaluate the usability of each of the following sources for a research paper titled "The Choosing of U.S. Vice-presidential Candidates, 1980–2000." In the blank, write the number of your evaluation:

1—source is unquestionably usable

2—source is usable but needs to be balanced with a source giving a different viewpoint

3—source is not usable

Example: A 1982 pamphlet urging people to boycott the election   ___3___

1. The *Congressional Record*, published regularly by the U.S. Congress, including speeches of several future Vice-presidents   1. _____

2. Your parent's high school American history textbook   2. _____

3. A Pulitzer Prize-winning book by a noted historian, on the American Vice-presidency, published 2001   3. _____

4. An article, "A Day with the Vice-president," in the current issue of a magazine sold at your supermarket checkout   4. _____

5. An article on the 1988 election in the *Journal of Political Science*   5. _____

6. A Website sponsored by the Republican National Committee   6. _____

7. A book, *The American Vice-presidency*, published by Harvard University Press in 1993   7. _____

8. An e-mail from a friend who is a Washington intern in the Vice-president's secretary's office   8. _____

9. A 1996 article in the *Nation* magazine, titled "A Liberal's-eye View of Recent Vice-presidents"   9. _____

10. A Website named "Traitorous Capitalist Pigs in the White House"   10. _____

89. PARAGRAPHS AND PAPERS: Taking Notes, Citing, Avoiding Plagiarism

(Study 96F, Taking Notes; 97A, Citing Within the Paper; and 97B, Avoiding Plagiarism)*

Part 1

Each item below contains an original passage from a source, followed by a student's note card on the passage. The student has done one or more of the following:

(1) incorrectly or inadequately keyed the note card to its corresponding bibliography card,

(2) misrepresented or distorted what the source said,

(3) come too close to the source's wording (plagiarized).

In the blank note card, **write** a correctly keyed, accurate, unplagiarized note on the passage.

Example: [From page 197 of a book by Antonia Fraser called *The Wives of Henry VIII*. (Catherine is the former Queen; Henry's new Queen is Anne.)]

ORIGINAL: Meanwhile at the court these days, there were indications to encourage Catherine's supporters that all was not well between the King and the new Queen.

STUDENT:

```
Wives of Hen. 8th—197

At this time in the palace, indications
encouraged Catherine's supporters that
things were not well between the King and
his new Queen.
```

YOUR CORRECTED NOTE CARD:

```
Fraser 197

Those who favored the former Queen noticed
and were heartened by signs of friction
between Henry and Anne.
```

*APA-based material in exercises 89 and 90 copyright © 2001 by the American Psychological Association. Adapted with permission.

1. [From page 334 of a book by Walter Cronkite titled *A Reporter's Life*]

   ORIGINAL: He [Daniel Ellsberg] had worked on the Pentagon's detailed history of our involvement in the Vietnam War. He became so incensed over what he considered the dirty secrets therein that he made off with hundreds of pages of papers and offered them to the news media.

   STUDENT:

   > Cronkite, chap. 7
   >
   > Daniel Ellsberg, who had worked on the Pentagon's involvement in Vietnam, became incensed over the dirty secrets he found and stole hundreds of papers, which he offered to the news media.

2. [From page 2 of an anonymous online article from the Sierra Club, "Endangered Species and Their Habitats"; last update, 4 Nov. 1995, retrieved 8 May 2000]

ORIGINAL: In fact, it is massive overcutting—along with automation and the industry's practice of exporting logs for processing with cheap, non-U.S. labor—that has wiped out over 90 percent of America's ancient forests.

STUDENT:

Sierra Club 2

"Massive overcutting" has caused the destruction of nine-tenths of our old-growth forests.

3. [From page 1319 of an article in the October 29, 1998, issue of the *New England Journal of Medicine*. The article is "Therapeutic Strategies for HIV Infection—Time to Think Hard" by David A. Cooper and Sean Emery. You may need a dictionary to rephrase medical jargon.]

ORIGINAL: We must think hard about the implications and practicalities of a medical strategy based on aggressive early intervention with lifelong, complex regimens of antiretroviral therapy to preserve immunocompetence after the suppression of a cytopathic virus.

STUDENT:

Emery, p. 1319

We must think hard about the implications and practical effects of a medical approach based on early aggressive intervening with "lifelong, complex regimens of antiretroviral therapy to preserve immuno-competence" after a cytopathic virus is suppressed.

4. [From a newspaper article by Jane Brody on page 18 of section A of the late edition of the *New York Times* of 6 September 2002. The article is "U.S. Panel Urges Hour of Exercise Each Day."]

ORIGINAL: As for carbohydrates, there is again a wide range of recommended intakes—45 to 65 percent of calories—to allow for dietary flexibility. This range recognizes that both the high-carbohydrate, low-fat diet of many Asian peoples and the higher-fat diet of Mediterranean peoples . . . are . . . associated with good health.

STUDENT:

Good health

When you eat carbohydrates, a wide range of intakes is recommended, from half to 3/4 of your calories. This gives your diet flexibility. This range takes into account that high fat Asian diets and low fat Mediterranean diets both lead to good health.

5. [From page 13 of a book by Philip Zelikow and Condoleezza Rice titled *Germany Unified and Europe Transformed: A Study in Statecraft*, published in 1995. (Gorbachev was the leader of the Soviet Union.)]

ORIGINAL: By far the most important man in Gorbachev's entourage was, like him, an outsider, with no foreign policy expertise. Eduard Shevardnadze, the foreign minister, had been too young to serve in World War II. . . .

STUDENT:

Rice and Zelikov 1995

It is my belief that both Gorbachev and his foreign minister, Eduard Shevardnadze, had very little experience in foreign policy. Shevardnadze had been even too young for World War II military service.

Write the material from each of your note cards in part 1, as you would write it in an actual research paper. Use MLA or APA style as indicated, or as your instructor directs. Be accurate, cite correctly, and do not plagiarize.

Example (MLA): *According to Fraser, those who favored the former Queen noticed and were heartened by signs of friction between Henry and Anne (197).*

1. (MLA) _____

_____

_____

_____

_____

_____

2. (MLA) _____

_____

_____

_____

_____

_____

3. (APA) _____

_____

_____

_____

_____

_____

4. (APA) _____

_____

_____

_____

_____

5. (MLA) _____

_____

_____

_____

_____

_____

90. PARAGRAPHS AND PAPERS: The Works Cited/Reference List

(Study 98, The Works Cited/Reference List)

(Open book) Write a **correct** bibliographical entry for each item. Use **MLA** or **APA** style, or both, as your instructor directs. (If both, make a copy of this page before starting, or use your own paper, for the APA entries.) Your instructor may specify the COS alternative. In some items, more information may be given than is needed.

Example (MLA): Book: The Essential Heart Book for Women, by Morris Notelovitz and Diana Tonnessen, published in New York in 1996 by St. Martin's Griffin.

Notelovitz, Morris, and Diana Tonnessen. <u>The Essential Heart Book for Women</u>. New York: St. Martin's Griffin, 1996.

1. Book: Hateship, Friendship, Courtship, Loveship, Marriage. Author: Alice Munro. Published: 2001 by Alfred A. Knopf in New York.

   _____

   _____

   _____

   _____

2. Journal article: Therapeutic Strategies for HIV Infection—Time to Think Hard. Authors: David A. Cooper and Sean Emery. Journal: New England Journal of Medicine. Published: October 29, 1998, on pages 1319–1321 (pages consecutive throughout volume) of volume 339, number 18.

   _____

   _____

   _____

   _____

3. Online article by Linda Greenhouse in the New York Times Online. It is entitled Citizens' Rights: Justices Ban Two-Tiered Welfare. It was in the issue of May 18, 1999, and was retrieved (accessed) on June 3, 2000. There are no page or edition numbers. The Web address is http://www.nytimes.com/library/politics/scotus/articles/051899welfare-benefits.html

   _____

   _____

   _____

   _____

4. Encyclopedia article: Title: Honoré de Balzac [a writer]. No author of article given. Published on pages 851–852 of vol. 1 of The New Encyclopaedia Britannica: Micropaedia, 15th edition. The publisher is Encyclopaedia Britannica, Inc., of Chicago, and the date is 1998.

_____

_____

_____

_____

5. Magazine article: Going Her Own Way. Author: Candace Ord Manroe. Magazine: Better Homes and Gardens. Published: September 2002 on pages 49–58 of volume 80, number 9.

_____

_____

_____

_____

6. Online article from personal Web site: Article title: The Mythic Role of Space Fiction. Name of site: Welcome from Sylvia Engdahl. Author: Sylvia Engdahl. Written August 17, 1998. No pages. No publishing organization given. Retrieved September 30, 1999. Web address: http://www.teleport.com/~sengdahl/spacemyth.htm

_____

_____

_____

_____

7. Newspaper article: Title: Talks on Global Warming Treaty Resuming Today. Author: John H. Cushman, Jr. Published on page 6 of section A of the late edition of the New York Times on November 2, 1998.

_____

_____

_____

_____

8. Compact disk: Title: Overtures. Composer: Ludwig van Beethoven. Performed by the Philharmonia Orchestra conducted by Otto Klemperer. Recorded by EMI Records of Hayes Middlesex, England, in 1990.

_____

_____

_____

_____

9. Story in collection: Story title: Girls Like You. Author: Jennifer Moses. Collection title: New Stories from the South: The Year's Best, 1998. Editor of collection: Shannon Ravenel. Story on pages 143–152 of collection. Collection published by Algonquin Books of Chapel Hill, Chapel Hill, North Carolina, in 1998.

———————————————————————————————————————

———————————————————————————————————————

———————————————————————————————————————

———————————————————————————————————————

10. Book: Title: Germany Unified and Europe Transformed: A Study in Statecraft. Authors: Philip Zelikow and Condoleezza Rice. Published in 1995 by the Harvard University Press in Cambridge, Massachusetts.

———————————————————————————————————————

———————————————————————————————————————

———————————————————————————————————————

———————————————————————————————————————

91. REVIEW: Proofreading

(Study 30–73, Punctuation; Mechanics and Spelling)

Proofread the following paragraphs for **typographical errors, omitted** or **doubled words,** and errors in **punctuation, capitalization, number form, abbreviations,** and **spelling.** (You may use a dictionary.) Make all corrections neatly above the line.

(Collaborative option: Students work in pairs to detect and correct errors.)

1. It all began when I joined the U.S. Army. My atittude toward life changed completely. I had just truned twenty one and in the prime of my life. I had every thing, that I had always wanted when I was a Teenager. My parents had given me: no responsibiity, no realistic outlook on life and no understanding of what it meant to go form a small country town into a huge army Camp.

2. I came to know what predjudice really means when I was 11 yrs. old. I when with my family to a motel in Bleakville called the Welcome House motel. The clerk at the registration desk said to my Father "you're wife and children can stay here, but you cant. My sisters and I could'nt figure out why my father was being refused a room, untill my mother told us it was because he was a little darker then we were. I remember that the clock said eight-o-two p.m. There was no place else in town to stay, but we all picked up our bags and marched out of the Welcome House Motel.

3. The word grammer strikes fear or loathing into some student's hearts, but, such need not be the case. Its concepts can be simplifyed. For example take the 8 parts of speech; If you except Interjections which are grammatically unconnected to the rest of the sentence there are only four kinds of words; naming words (nouns), doing-being words (verbs), modifiers (adjs. and advs.) and connectors [prepositions and conjunctions.]

4. Ebbets field, home of the famed Brooklyn Dodgers baseball team is today hallowed in the memorys of of many as the perfect old ballpark—the ideal place to watch a game—. Though this small Stadium fell victim to the wreckers ball more than forty year's ago, after the Summer of nineteen fifty-seven it can still generate spasms of nostalgia in true baseball buffs. Actually, it was a dumpy old place with miserable parking, uncomfortable seats, poles that blocked ones view, inadequate lighting, and facilities were by today's standards decidedly primitive.

5.   Darlene gazed lovingley at Michael. "Oh, Michael I can't bear the though that you have to leave", she whis-

pered. "must you go back to So. Hadleyville so soon."

   "It's a 2 & one half hour trip," Michael replied. "and a snowstorm is blowing in from the North."

   "If you go there is my ring!" she cried, pulling the gold band from her finger, and hurling it to the floor.

Michael wondered whether she was serious?

92. REVIEW: Editing & Proofreading

(Study 1–83: Grammar and Sentences; Punctuation; Mechanics and Spelling; Word Choice)

Edit and proofread the paragraphs below. Look not only for **mechanical errors** as in exercise 91, but also for **weak or faulty sentence structure**, and for **grammatical** and **word-choice errors**. Use standard, formal American English. Rewrite the paragraphs on your own paper.

(Collaborative option: Students work in pairs to detect and correct errors and weaknesses; students critique each other's rewrites.)

1. The birth of my son changed my whole life I found myself for the first time responsable for another human being. Because of complications in my pregnancy, I had to have a Cesarean, but everything work out without to much difficulty. Since then Roger has made me forget that pain. Being a healthy boy of five today, I find him a joy, even thorough he can occasionally be annoying. Everyone of his friends have a delightful time when playing with him. Because he has such a sunny disposition. Whenever a new child moved into the neighborhood, Roger is the first to run over and offer them his toys to play with.

2. The high pay earned by many athletes are ruining professional sports. These jerks are being paid outrageous amounts of money just to run around a field or a court for a few months. For example, Kevin Brown, the Dodger's pitcher, was given a $100 million contract, and Michael Jordan has earned more than he can ever count. Most players' salarys exceed a million dollars. In baseball, ever since the players became free agents, they have recieved exorbitant paychecks. I believe such sums are being award to the players without regard to us fans. How many of us can afford to pay a price that amounts to a total of $35 for a seat at a basketball game. Moreover, the players are rarely exerting maximum effort to justify their seven-figure incomes. I, for one, will not return back to Municipal Stadium or Central Coliseum unless they have scheduled a good college game there. It is up to we fans to reverse this thing by boycotting professional games.

3. Most Americans are woefully ignorant of the World's geography. Where a country is located, what kind of resources does it have, and how far away it is are questions that bring a puzzled frown to many? Angola, for example. Where is it? Is it an island? What do they produce there? What are it's people like, do they think and act like we do? Just because a land is distant from us doesn't mean that it's not importent. Being so far from our shores, Americans should not ignore other countries.

4. Commuting to a city College from your home may seem less attractive then to live on a rolling green campus in the hills. Yet they have many advantages. Lower cost is an obvious one, home cooked meals is another. It gives a student the freedom to go wherever they want after classes: to movies or shows, museums, or even get a part-time job. Citys are full of exciting places to find excitement in, therefore they form a welcome antidote for boring classes.

5. There are a right way and a wrong way to walk when hiking. First of all, stay relaxed and no slouching. Swing you're arms, this will help relaxation and momentum it will also make you feel good. You should maintain straightness in your shoulders & hips to. When carrying a pack, leaning a bit forward will help a person center their weight over their feet. Carry plenty of water, and stop to rest after a half an hour.

93. REVIEW: Revising, Editing, and Proofreading

(Study 1–92: Grammar and Sentences; Punctuation; Mechanics and Spelling; Word Choice; Paragraphs)

On your own paper, **revise**, **edit**, and **proofread** the paragraphs below. Correct **all errors** and make any necessary **improvements**, including strengthening **paragraph structure**. Use all the skills and knowledge you have learned in *English Simplified*.

(Collaborative option: Students work in pairs to detect and correct errors and weaknesses; students critique each other's rewrites.)

1. The government must take strong action against polluters, they are slowly killing us all. By poisoning our air and our water. The big business lobbies control Congress, and so it passes few antipollution laws. Its frightening, for example to find that Midwest smokestacks fill the air with acid. Rain becomes filled with this acid. It is killing fish in Adirondack Mountain lakes. These lakes are hundreds of miles to the east. Elsewhere, pesticides are being used. They are being sprayed on potato fields. The chemicals seep underground, and then the local well water becomes contaminated. In Texas, polluted water almost killed ten thousand cattle. The corpses of these cattle had to be burned to prevent the spread of the poison. The government is responding far too weak to this pollution crisis. Congress must resist the lobbyists, and strong antipollution laws must be passed by it.

2. "Warning: The surgion general has determined that cigarette smoking is dangerous to your health," or a similar message printed on all packs of cigaretes but people still smoke. Did you ever wonder why? Every smoker know that cigarettes are harmful to their lungs, but this doesn't stop them. I believe that people just don't care any more about their selves. Years ago people they tried to live longer and not to do any thing that would damage their health now things have changed. I feel that people feel that their is nothing to live for and if your going to die it has to be from something even if its not cancer. Also people relizing that life is not to easy. Price's are going up jobs are hard to find and who wants to live in a world where many things are difficult to get. Being a non smoker myself, cigarettes should be banned from society. If they were baned from society, smokers will live longer. In spite of themself.

3. Enlisting in the army was an experience that changed my life. June 10, 2001 was the day it began. The plane left Boston for Georgia. The plane arrived three hours later. A group of us recruits waited for four more hours for the bus to come to take us to Ft. Gordon. The bus came and took us to camp. We were processed in and given sleeping quarters. They were cold and drafty. The first week was spent in a reception center. It was nice there, too nice. The next week we were transfered to our duty stations to start the armys new basic training program. The sergents appeared to look nice but that was just a dream, For five weeks we did the same grueling thing everyday. At four o'clock we woke up, did lots of exercises, etc, and did the things that had to be done that day. The training was easy but the sergent's made it hard. Some of the other guys were there made it worst. They resisted the discipline. The last week had come and I was a new person. I have learned that, I will never join anything with out thinking twice about it.

4. Cross-country skiing is not as popular as downhill skiing. But it has been slowly but steadily gaining in popularity in the United States, and this is a good thing because it is a much more aerobic sport that is, it gives the skier a better cardiovascular workout and is therefore better for your all-around health, which, in turn, can lead to a person having a longer life. Cross-country skiing will burn up to 9 hundred calories an hour, moreover the upper body muscles are developed as well as the lower body, which is the only part that running or cycling develop. Beside this it developes coordination. And they have less risk of injury then downhill skiers.

5. One of the biggest dangers in writing a paragraph is straying from the topic. Our history professor does this often when lecturing. Another is fused sentences they sneak up on you, so do comma splices. By dangling a modifier, your composition can sound silly. Do not shift voice or mood, for your essay can be made confusing by it. Lack of agreement between subjects and verbs reveal a careless writer at work. Worst of all is to have no topic sentence or cohesion.

94. ACHIEVEMENT TEST: Grammar, Sentences, and Paragraphs

Part 1: Sentences

Write **S** if the boldfaced expression is one complete, correct **sentence**.
Write **F** if it is a **fragment** (incorrect: less than a complete sentence).
Write **Spl** or **FS** if it is a **comma splice** or **fused sentence** (incorrect: two or more sentences written as one).

Example: The climbers suffered from hypothermia. **Having neglected to bring warm clothing.** _____F_____

1. Senator Wanhope won a stunning victory. **After trailing badly in nearly every pre-election poll.** 1._____

2. The nation's airlines are instituting stricter security measures. **Such as arming pilots, securing cockpit doors, and placing plainclothes federal marshals aboard most flights.** 2._____

3. **He will attend college his high school grades are good enough.** 3._____

4. The rancher sold most of his livestock. **Then he turned his property into a profitable dude ranch.** 4._____

5. **Elated at news of the victory.** Borrelli broke out a bottle of her finest champagne. 5._____

6. **When does abstract art become just scribbles?** 6._____

7. **Our guests having arrived, we sat down to dinner.** 7._____

8. Nine families joined the pollution study. **They will wear carbon-filter badges, this device will monitor the air that they are breathing.** 8._____

9. **The storm having washed out the bridge.** We had to spend the night in town. 9._____

10. Sir Thisby invited me to play cricket. **A game I had never even watched.** 10._____

11. **The high humidity forced us to move the picnic inside it was just too hot to eat outside.** 11._____

12. **The student who tape-records the physics lecture.** 12._____

13. **I continued to watch the baseball game on television even though I had not started my calculus homework that was due the next day.** 13._____

14. Allen used a week's vacation. **To sand and refinish the hardwood floors in his home.** 14._____

15. **Children in the experimental group improved their reading scores by nearly a full grade, however, six-month follow-up studies showed that the gains did not last.** 15._____

16. **Though armed and considered dangerous, he surrendered without a struggle.** 16._____

Part 2: Grammar

In the blank, write **C** if the boldfaced expression is used **correctly**.
Write **X** if it is used **incorrectly**.

Example: There **was** dozens of dinosaur bones on the site. _____X_____

1. It was a personal matter between Terry and **myself**. 1._____

2. During the summer, she trained horses, **which** assisted her financially. 2._____

3. At the building supply store there **were** insulation, Sheetrock, and flashing. 3. _____

4. Cousin Max, along with his twin daughters and their cats, **were** waiting at my front door. 4. _____

5. The interviewer asked each of the politicians to explain **their** position on taxation. 5. _____

6. Copies of *Esquire* and *Rolling Stone* **was** in Dr. Moore's waiting room. 6. _____

7. The taxi driver gave Tony and **I** a scornful glance. 7. _____

8. The ticket agent gave Ed and **me** seats that were behind home plate. 8. _____

9. Every committee member **was** given a copy of the report. 9. _____

10. **Refusing to pay high interest**, consumers are cutting up their credit cards. 10. _____

11. Parking restrictions apply **not only** to students **but also** to visitors. 11. _____

12. His mother wanted him to become a corporate lawyer. **This** kept Leonard in college. 12. _____

13. Students should meet their professors, so that if **you** have questions about class, **you** will feel comfortable approaching a professor during office hours. 13. _____

14. **Having an hour to kill,** there was time to stroll through the village. 14. _____

15. His plans included landing a well-paying internship and **to spend** as much time as possible with his girlfriend. 15. _____

16. **Being nervous about the speech**, the microphone amplified my quavering voice. 16. _____

17. We wondered why the list of courses **was** not posted yet. 17. _____

18. There are few people who write in a personal journal as much as **her**. 18. _____

19. I like swimming in icy lakes as well as **to relax in the warm sunshine**. 19. _____

20. Aggression **is when** one nation attacks another without provocation. 20. _____

21. **Is** either of the two bands ready to go on? 21. _____

22. Was it you **who** wrote the editorial titled "Ban Smoking on Campus"? 22. _____

23. Everyone who plays the lottery hopes that **their** ticket will win the million-dollar jackpot. 23. _____

24. It is up to **us** freshmen to demand better food in the dining hall. 24. _____

25. All of **we** residents living in the Sussex area were upset when a fast-food restaurant was built nearby. 25. _____

26. Financial aid will be made available to **whoever** shows a need for it. 26. _____

27. Every student should understand that it is up to **you** to find the strategies to do well academically. 27. _____

28. He coached Little League and joined two service clubs. **It** was expected of him by his associates. 28. _____

29. Harry and **myself** solved the crime of the missing coffee pot in the lounge. 29. _____

30. **Who** do you think will apply for the position of dean of students? 30. _____

31. There **was** a Buick, a Toyota, and a Mercedes parked in the lot. 31. _____

32. Between you and **I**, Martin has only a slim chance of promotion this year. 32. _____

33. **Is** there any objections to your opening a nightclub on campus? 33. _____

34. My advisor suggested that I take Russian. **That** was fine with me. 34. _____

35. Each of the players **has** two passes for all home games. 35. _____

36. Neither Joan nor her two attendants **was** asked to appear on television. 36. _____

37. He is one of the engineering students who **are** interning this summer. 37. _____

38. You will never find anyone more responsible than **her**. 38. _____

39. Jenette is the **friendliest** of the two resident assistants in my building. 39. _____

40. When the Buffalo Bills and the Pittsburgh Steelers play, I know **they** will win. 40. _____

41. Why not give the keys to **whomever** you think will be in charge? 41. _____

42. Did the committee approve of **his** assuming the chair position? 42. _____

43. The study **not only** disproved Blunt's theory **but also** McDavid's. 43. _____

44. In his backpack **were** a notebook computer, an umbrella, and his lunch; he was prepared for a day on campus. 44. _____

45. The coach, as well as the manager and players, **was** sure of winning. 45. _____

46. **Knowing of his parents' disapproval**, it seemed wise for him to reconsider his plan to drop out of school to become a skydiving instructor. 46. _____

47. Daniel decided to **only** purchase three new fish for his aquarium. 47. _____

48. If he **were** more tactful, he would have fewer enemies. 48. _____

49. **Because Lee lost at Gettysburg** did not mean that the South would cease fighting. 49. _____

50. Neither the camp director nor the hikers **was** aware of their danger. 50. _____

Part 3: Paragraphs (not included in scoring)

In the space below or on the back, write a **paragraph** of six to eight sentences on **one** of the following topics (you may use scrap paper also):

_____

The thrill of _____ (something you have done)

A friend I will never forget

If I could go on television for five minutes

The best (or worst) book I have read in the past year

A sorely needed law

95. ACHIEVEMENT TEST: Punctuation

Write **C** if the punctuation in brackets is **correct**.
Write **X** if it is **incorrect**.
(Use only one number in each blank.)

Example: Regular exercise[,] and sound nutrition are essential for good health. _____X_____

1. Santee, South Carolina[,] is the site of a huge outlet mall.   1. _____

2. Residents[,] who own barking dogs[,] refuse to do anything about the noise.   2. _____

3. He thought the idea was mine, but it was their[']s.   3. _____

4. The strike having been averted[,] the workers returned to their jobs.   4. _____

5. He asked me where I had bought my surfboard[?]   5. _____

6. When I open it[']s favorite cat food, the cat races into the kitchen.   6. _____

7. Haven't you often heard it said, "Haste makes waste["?]   7. _____

8. Wouldn't you like to go to the rally with us?"[,] asked the girl across the hall.   8. _____

9. He said, "Let's walk across the campus.[" "]It's such a warm evening."   9. _____

10. Enrollment is up to three[-]thousand students this quarter.   10. _____

11. Twenty[-]six students have volunteered to serve on various committees.   11. _____

12. Dear Sir[;] I have enclosed my application and résumé.   12. _____

13. After you have finished your sociology assignment[,] let's go to a movie.   13. _____

14. Billy Budd struck Claggart[;] because he could not express himself any other way.   14. _____

15. You did agree to give the presentation[,] didn't you?   15. _____

16. We were early[;] as a matter of fact, we were first among the guests to arrive.   16. _____

17. "If you really look closely," the art critic commented[,] "you'll see a purple turtle in the middle of the painting."   17. _____

18. Dr. Johnson had little praise for the current health care system[;] calling it an elitist structure.   18. _____

19. The band recorded its first album in the spring[,] and followed it with a summer concert tour.   19. _____

20. She had hoped to arrange a two month[']s tour of Korea and Japan.   20. _____

21. We hope[,] Ms. Foster[,] that your office will be satisfactory.   21. _____

22. The next stockholders' meeting is scheduled for August 9, 2007[,] but it will be open only to major investors.   22. _____

23. My youngest sister[,] who is fourteen[,] is already shopping for a college.   23. _____

24. Because she played cards until midnight[;] she overslept.   24. _____

25. Jane Cox[,] a biochemistry major[,] won the top scholarship.   25. _____

26. Professor Thomas was asked to create a course for the Women[']s Studies Department.   26. _____

27. The little boy in the center of the old photograph[,] would later write five novels.   27. _____

28. "As for who has written the winning essay[—]well, I haven't as yet heard from the judges," said Mr. Hawkins.   28. _____

29. What he described about the massive oil spill[,] filled us with horror.   29. _____

30. I asked Elizabeth what we should do about our vacation plans[?]   30. _____

31. The newly elected officers are Denzell Jones, president[;] Ruby Pillsbury, vice president[;] and Maria Garcia, secretary.   31. _____

32. Before the radical group surrendered[;] they attempted to negotiate their freedom.   32. _____

33. We followed the trail over several ridges[,] and along the edge of two mountain lakes.   33. _____

34. Before touring Europe, I had many matters to attend to[;] such as making reservations, buying clothes, and getting a passport.   34. _____

35. Having a good sense of humor helps one put problems into perspective[;] certainly it's better than brooding.   35. _____

36. The ticket agent inquired ["]if we were planning to stop in Paris.["]   36. _____

37. Once retired, Ensel painted portraits of family pets[,] and played bingo every Thursday and Saturday.   37. _____

38. Marcia learned that all foods[,] which are high in fat[,] should be eaten in moderation.   38. _____

39. We were told to read ["]Ode to a Nightingale,["] a poem by Keats.   39. _____

40. The alumni magazine had a column cleverly entitled ["]Grad-Tidings.["]   40. _____

41. A civilian conservation corps could provide[:] education, training, and work for thousands of unemployed teenagers.   41. _____

42. Some people wish to have ["]America, the Beautiful["] become our national anthem.   42. _____

43. She hurried towards us[,] her books clasped under her arm[,] to tell us the good news.   43. _____

44. The audience wanted him to sing one more song[;] however, he refused.   44. _____

45. They must be the only ones who have visited New York in recent years and not seen the show ["]The Lion King[."]   45. _____

46. She found a note in her mailbox: "Sorry to have missed you. The Lawson[']s."   46. _____

47. His mother wanted him to major in chemistry[;] he wanted to major in music.   47. _____

48. Chris decided that he wanted a quiet vacation[,] not one full of schedules and guided tours.   48. _____

49. He had gone to the library[. B]ecause he wanted to borrow some videos.   49. _____

50. Her program included courses in English[,] social science[,] and chemistry.   50. _____

51. Every child knows "Twinkle, twinkle, little star[/]How I wonder what you are." 51. _____

52. To prepare for the baseball tryouts[,] Sam practiced every night.   52. _____

53. Ms. Whitney, who is a physical education instructor, came to the rally[;] with Mr. Martin, who is the football coach.   53. _____

54. When the tornado hit eastern Ohio[,] it caused millions of dollars of damage. 54. _____

55. "Some of the seniors wer[']ent able to pay their dues," she said.   55. _____

56. Frank Anderson[,] who is on the tennis team[,] is an excellent athlete.   56. _____

57. "All motorists[,] who fail to stop at the crosswalk[,] should be put in jail!" declared an angry parent.   57. _____

58. Looking at me sweetly, Mark replied, "No[,] I will not lend you a thousand dollars." 58. _____

59. George enrolled in a course in home economics; Elsa[,] in woodworking.   59. _____

60. "Haven't I met you somewhere before?"[,] he asked.   60. _____

61. "It's most unlikely["!] she said, turning away.   61. _____

62. A student[,] whom I met at the banquet[,] would like to work in our department next semester.   62. _____

63. It is a monumental task to build a highway[,] where 10,000-foot mountains block the way.   63. _____

64. He moved to Denver[,] where he worked as a freelance photographer.   64. _____

65. We were[,] on the other hand[,] not surprised by his decision.   65. _____

66. I bought a special type of paintbrush to reach those hard[-]to[-]reach spots near the rain gutters.   66. _____

67. Listen to the arguments of both speakers[,] then decide which side you favor.   67. _____

68. Four generations of Spencer[']s will attend the family reunion.   68. _____

69. Susannah is familiar with many customs of Sweden [(]her father's homeland[)] and can prepare many Swedish dishes.   69. _____

70. Our ex[-]mayor pleaded guilty to a speeding violation.   70. _____

71. The conference sponsored by our fraternity was successful[,] especially the sessions concerning community-service projects.   71. _____

72. The Liberty scored ten points in the last minute but[,] the Sparks held on for the win.   72. _____

73. Jack displayed a unique [(?)] talent when he created a collage of spaghetti sauce, pickles, and pancakes.   73. _____

74. The children[,] on the other hand[,] were content to wear last year's coats and boots.   74. _____

75. The teenager used the word [*like*] throughout her conversation.   75. _____

96. ACHIEVEMENT TEST: Mechanics, Spelling, and Word Choice

Part 1: Capitalization

In each blank, write **C** if the boldfaced word(s) **follow** the rules of capitalization.
Write **X** if the word(s) **do not follow** the rules.

Example: The Mormons settled in what is now **Salt Lake City**.   _____C_____

1. Please meet Charles Ebbings, **Professor** of psychology at Yale.   1. _____
2. It was relaxing to spend a few days away from **college**.   2. _____
3. She is **President** of her class.   3. _____
4. I belong to a **Science Club.**   4. _____
5. He plays for Ohio **State**.   5. _____
6. We saluted the **american** flag.   6. _____
7. We flew over the **french Alps**.   7. _____
8. I asked **Grandmother** to lend me the family album.   8. _____
9. He enjoys living in the **Southwest**.   9. _____
10. The **East** side of the house needs to be repainted.   10. _____
11. I visited an **indian** village while on vacation.   11. _____
12. He naps in his **history** class.   12. _____
13. We heard that **Aunt Harriet** had eloped with the butcher.   13. _____
14. The note began, "My **Dear** John."   14. _____
15. I am going to be a **Medical Anthropologist**.   15. _____
16. The magazine recommended buying a Dell **Computer**.   16. _____
17. I lost points because the answer key contained an error; **nevertheless**, Professor Pruyn refused to change my grade.   17. _____
18. History 303 studies the Russian **Revolution**.   18. _____
19. We toured the lakes and woodlands where the novel ***The Last Of The Mohicans*** took place.   19. _____
20. Walter prayed that **god** would let him win the lottery.   20. _____

Part 2: Abbreviations and Numbers

Write **C** if the boldfaced abbreviation or number is used **correctly**.
Write **X** if it is used **incorrectly**.

Example: They drove through **Tenn.**   ___X___

1. **Fifty million** voters stayed away from the polls.   1. _____

2. Thank God it's **Fri**.   2. _____

3. My cat named Holiday is **5**.   3. _____

4. We live on Sutherland **Rd.**   4. _____

5. She was born on July **6th**, 1980.   5. _____

6. Please meet me at **10** o'clock.   6. _____

7. He released **two hundred** pigeons at the picnic.   7. _____

8. After a brief investigation, we discovered that **13** students were involved in the prank.   8. _____

9. The train leaves at **8** p.m.   9. _____

10. Dinner was served at **six** o'clock.   10. _____

11. Joan Allen, **Ph.D.,** spoke first.   11. _____

12. Lunch cost **12** dollars!   12. _____

13. **Ms.** Martin, please chair the meeting today.   13. _____

14. Lloyd's monthly salary is now **$3,200.50.**   14. _____

15. They lived at **six seventy-four** Ninth Avenue.   15. _____

Part 3: Spelling

In each sentence, one boldfaced word is **misspelled**. Write its number in the blank.

Example: (1)**Its** (2)**too** late (3)**to** go.   ___1___

1. Horace's (1)**peculiar** expression of boredom was his way of making a (2)**statment** about the quality of the (3)**equipment**.   1. _____

2. The open (1)**cemetary** gates permitted an (2)**excellent** (3)**opportunity** for Karloff's laboratory assistant.   2. _____

3. My (1)**psychology** (2)**proffessor** assigns a weekly (3)**written** report.   3. _____

4. He needed (1)**permission** from the (2)**commitee** to participate in the (3)**competition**.   4. _____

5. The (1)**bookkeeper** learned that a (2)**knowledge** of (3)**grammer** is helpful.   5. _____

6. A (1)**fourth** such disaster threatens the very (2)**existance** of the Alaskan (3)**environment**.   6. _____

7. Arthur (1)**definately** considered it a (2)**privilege** to help write the (3)**article**.   7. _____

8. The (1)**principal** (2)**complimented** her for her (3)**excellant** performance.   8. _____

9. It was (1)**apparent** that she was (2)**desparate** because she was listening to his (3)**advice**.   9. _____

10. We (1)**imediately** became (2)**familiar** with the requirements for a (3)**license**.   10. _____

11. Is it (1)**permissable** to ask him to (2)**recommend** me for a (3)**government** position?   11. _____

260

12. (1)**Personaly**, I didn't believe his (2)**analysis** of the result of the (3)**questionnaire**.   12. _____

13. The test pilot felt enormous (1)**optimism** after her third (2)**repitition** of the dangerous (3)**maneuver**.   13. _____

14. It's (1)**ridiculus** that Sue became so angry about the (2)**criticism** of her friend, the (3)**playwright**.   14. _____

15. She was not (1)**conscious** of being (2)**unnecessarily** (3)**persistant** about the matter.   15. _____

Part 4: Word Choice

To be correct, the boldfaced expression must be standard, formal American English and must not be sexist or otherwise discriminatory.
Write **C** if the boldfaced word is used **correctly**.
Write **X** if it is used **incorrectly**.

Examples: The counsel's **advice** was misinterpreted.   ___C___
They **could of** made the plane except for the traffic.   ___X___

1. I was not **altogether** amused.   1. _____

2. Aaron looked **sort of** tired after the test.   2. _____

3. They are all old; for **instants**, Grayson is eighty-six.   3. _____

4. Billy cried when his balloon **burst**.   4. _____

5. Next time, plan to invite **fewer** guests.   5. _____

6. He earned no interest on his **principal**.   6. _____

7. **Can** I add your name as a contributor to the scholarship fund?   7. _____

8. The judge would hear no **farther** arguments.   8. _____

9. I am in real trouble, **aren't I**?   9. _____

10. The team was **plenty** angry.   10. _____

11. The parent **persuaded** her child to take out the garbage.   11. _____

12. He notified **most** of his creditors.   12. _____

13. She knows **less** people than I.   13. _____

14. Saul made an **illusion** to *Hamlet*.   14. _____

15. Was the murderer **hanged**?   15. _____

16. The team **would of** done much better with a different quarterback.   16. _____

17. The **kids** were excited.   17. _____

18. I had **already** signed the check.   18. _____

19. John sounds **like** he needs a vacation.   19. _____

20. I can't stand **those kind** of jokes.   20. _____

21. He is **real** happy about winning the contest.   21. _____

22. The cat is **lying** by the fire.   22. _____

23. She **generally always** works hard.   23. _____

24. He does **good** in math courses.

24. _____

25. His speech **implied** that he would raise taxes.

25. _____

26. We took the tour because of its awesome **things to see**.

26. _____

27. On the bus were city people, suburbanites, and **hillbillies**.

27. _____

28. The chair was about **thirty inches in width**.

28. _____

29. **Due to the fact of his escape**, the police have set up roadblocks.

29. _____

30. Our vacation was even **more perfect** than you can imagine.

30. _____

The following chart gives brief definitions and examples of the grammatical terms you will read about most often in these exercises. Refer to *English Simplified* for more information.

Term	What It Is or Does	Examples
Adjective	Describes a noun	a **fast** runner (describes the noun **runner**)
Appositive	A noun that renames another	Tom Wolfe, **the writer**, lives in New York. (The appositive follows the man's name.)
Adverb	Describes a verb, adjective, or another adverb	He runs **fast** (describes the verb **run**). He runs **very** fast (describes the adverb **fast**). He is an **extremely** fast runner (describes the adjective **fast**).
Clause	A group of words with a subject and a predicate. An independent clause can stand by itself and make complete sense; a dependent clause must be attached to an independent clause.	**He is a fast runner.** (An independent clause) **if he is a fast runner** (A dependent clause that must be attached to some independent clause, such as **He will win.**)
Complement	Completes the meaning of the verb	Direct Object: He threw the **ball**. (Says what was thrown.) Indirect Object: He threw **me** the ball. (Says to whom the ball was thrown.) Subjective Complement: He is a **pitcher**. (Renames the subject **He** after the linking verb **is**.) Objective Complement: The team named Rodgers **coach**. (Follows the direct object **Rodgers** and renames it.)
Conjunction	A word that joins	Coordinating Conjunction: Joins things of equal importance: Men **and** women. Poor **but** honest. Subordinating Conjunction: Joins a dependent clause to a main clause: I left **when** she arrived.
Fragment	A group of words that cannot stand by itself as a sentence	**when I saw them** (a dependent clause); **from Maine to California** (a prepositional phrase)
Noun	Names a person, place, animal, thing, or idea	**Tom, Denver, cat, book, love, truth**
Phrase	A group of related words without a subject and a verb	**from California** (a prepositional phrase); **to see the king** (an infinitive phrase); **built of bricks** (a participial phrase); **building houses** (a gerund phrase)
(Complete) Predicate	The part of the sentence that speaks about the subject	The man **threw the ball.** (says what the subject did)
Pronoun	A word that replaces a noun (or replaces a word group acting as a noun)	**He** will be here soon. (**He** takes the place of the man's name.)
Subject	The person or thing about whom the sentence speaks	**Polly** writes children's books.
Verb	Says what the subject either does or is	She **buys** seashells. She **is** a doctor.